Petroleum Exploration, Drilling and Production

Petroleum Exploration, Drilling and Production

Edited by Michael Dedini

☐ SYRAWOOD
PUBLISHING HOUSE

New York

Published by Syrawood Publishing House,
750 Third Avenue, 9th Floor,
New York, NY 10017, USA
www.syrawoodpublishinghouse.com

Petroleum Exploration, Drilling and Production
Edited by Michael Dedini

International Standard Book Number: 978-1-68286-687-0 (Hardback)

Cataloging-in-Publication Data

Petroleum exploration, drilling and production / edited by Michael Dedini.
    p. cm.
Includes bibliographical references and index.
ISBN 978-1-68286-687-0
1. Petroleum--Prospecting. 2. Boring. 3. Oil fields--Production methods.
4. Petroleum engineering. I. Dedini, Michael.
TN271.P4 P48 2019
622.182 8--dc23

TABLE OF CONTENTS

PREFACE

Over the recent decade, advancements and applications have progressed exponentially. This has led to the increased interest in this field and projects are being conducted to enhance knowledge. The main objective of this book is to present some of the critical challenges and provide insights into possible solutions. This book will answer the varied questions that arise in the field and also provide an increased scope for furthering studies.

Petroleum is a naturally occurring mixture of hydrocarbons found in geological formations that is refined to produce fuels like petrol, diesel and paraffin. The demand for petroleum exploration and drilling has consistently grown over the years, as petroleum is a major energy source. Petroleum exploration involves substantial geological evaluation through seismic data and subsurface analyses. It is being fused with advanced geomechanical simulations for efficient detection and evaluation of petroleum reserves. Drilling and production processes, are being modernized for maximum resource utilization, better affordability and low ecological impact. This book covers varied aspects of petroleum exploration, drilling and production in a comprehensive way. The various studies that are constantly contributing towards advancing technologies and evolution of this field are examined in detail. Petrochemical engineers, geologists, students and researchers will benefit alike from this book.

I hope that this book, with its visionary approach, will be a valuable addition and will promote interest among readers. Each of the authors has provided their extraordinary competence in their specific fields by providing different perspectives as they come from diverse nations and regions. I thank them for their contributions.

Editor

A new method for evaluating the injection effect of chemical flooding

Jian Hou[1,2] · Yan-Hui Zhang[3] · Nu Lu[2] · Chuan-Jin Yao[2] · Guang-Lun Lei[2]

Abstract Hall plot analysis, as a widespread injection evaluation method, however, often fails to achieve the desired result because of the inconspicuous change of the curve shape. Based on the cumulative injection volume, injection rate, and the injection pressure, this paper establishes a new method using the ratio of the pressure to the injection rate (RPI) and the rate of change of the RPI to evaluate the injection efficiency of chemical flooding. The relationship between the RPI and the apparent resistance factor (apparent residual resistance factor) is obtained, similarly to the relationship between the rate of change of the RPI and the resistance factor. In order to estimate a thief zone in a reservoir, the influence of chemical crossflow on the rate of change of the RPI is analyzed. The new method has been applied successfully in the western part of the Gudong 7th reservoir. Compared with the Hall plot analysis, it is more accurate in real-time injection data interpretation and crossflow estimation. Specially, the rate of change of the RPI could be particularly suitably applied for new wells or converted wells lacking early water flooding history.

Keywords Ratio of the pressure to the injection rate · Rate of change of the RPI · Injection efficiency · Chemical crossflow

1 Introduction

Chemical flooding, a rapidly developed tertiary oil recovery technique, is applied successfully and widely both in the Daqing Oilfield and the Shengli Oilfield (Zhang et al. 2010; Hou et al. 2011, 2013; Shaker Shiran and Skauge 2013; Dag and Ingun 2014). Polymer flooding and surfactant-polymer flooding (SP flooding) are considered two of the most mature chemical methods (Chang et al. 2006; Vargo et al. 2000; Li et al. 2012; Delamaide et al. 2014; Sheng et al. 2015). Polymer flooding uses high-molecular weight polymers to increase the viscosity of the injection fluid, and to decrease the oil–water mobility ratio, while the SP flooding further improves the oil recovery by adding a surfactant to the injection fluid to reduce the interfacial tension and then increase the oil displacement efficiency (Shen et al. 2009; Urbissinova et al. 2010). However, due to the high prices of chemicals, in most cases, the implementation of chemical flooding is usually costly. Thus, a timely and accurate evaluation of the displacement efficiency of chemical floods, which can verify the validity of the chemical injection in advance, is urgently needed (Kaminsky et al. 2007; Dong et al. 2009; Seright et al. 2009; AlSofi and Blunt 2014; Ma et al. 2007).

Usually, the injection evaluation of chemical flooding is conducted by the variation of dynamic injection data such as the rise of the injection pressure and the drop of the injection rate. The increase in the flow resistance to the injection fluid indicates the increment of the chemical efficiency. After the chemical solution is injected into a

✉ Jian Hou
houjian@upc.edu.cn

✉ Chuan-Jin Yao
ycj860714@163.com

[1] State Key Laboratory of Heavy Oil Processing, China University of Petroleum, Qingdao 266580, Shandong, China

[2] School of Petroleum Engineering, China University of Petroleum, Qingdao 266580, Shandong, China

[3] CNOOC LTD., Tianjin Bohai Oilfield Institute, Tianjin 300452, China

Edited by Yan-Hua Sun

reservoir, due to the high-molecular weight polymer dissolved in the injection fluid, the flow resistance increases, causing an increase in the injection pressure and a decline in the injection rate (Cheng et al. 2002; Li 2004). As a result, the variation of the injection pressure and the injection rate can be used to evaluate whether the chemical injection is effective.

The Hall plot describes the relationship between the time integral of the injection pressure difference and the cumulative injection volume. The Hall plot analysis was firstly used to evaluate the performance of waterflood wells (Hall 1963) and gradually was used as a simple, effective method for diagnosing the injection efficiency (DeMarco 1969). Since then, researchers have improved the Hall plot analysis. Based on the Hall plot analysis, the slope analysis method which produces an estimation of the pressure at the average water-bank radius in Hall's formula was put forward by Silin et al. (2005). The slope analysis is proved to be more accurate for all data needed is available from oilfields. After that, Izgec and Kabir (2009) presented a complete reformulation of the Hall plot analysis by updating the pressure at the average influence radius after each computing time step and studied the difference between the Hall plot and the derivative of the Hall slope, under the condition of both transient and pseudo-steady states. Compared with the Hall plot, the derivative of the Hall slope could overcome the smooth effect which is caused by integral involved in the Hall plot analysis.

With the development of chemical flooding, Hall plot analysis was used to evaluate the flow behavior of non-Newtonian fluids in the field of polymer flooding (Moffitt and Menzie 1978). Later, the analytical expressions of the Hall slope analysis for polymer floods, resistance factor, and the residual resistance factor were derived by Buell et al. (1990). Considering that the reservoir permeability is an important factor affecting the resistance factor, Kim and Lee (2014) used the effective permeability changing with the water saturation instead of regarding permeability as a constant, improving the accuracy of the Hall plot analysis. Li et al. (2011) investigated the application of the Hall plot analysis in gel flooding. Gradually, the resistance factor and the residual resistance factor have become quantitative indexes to evaluate the injection effect of chemical flooding (Honarpour and Tomutsa 1990; Sugai and Nishikiori 2006; Ghosh et al. 2012). The resistance factor is defined as the ratio of the Hall plot slope for the chemical flood to that for the water flood. As the change of the slope in the Hall plot is usually not significant in the early period of chemical flooding, it is hard to use the Hall plot analysis to obtain the resistance factor. In fact, the Hall plot analysis only represents the average injection effect over a period, rather than to reflect the real-time characteristics.

In view of this, a new method, derived from injection performance data, is proposed for the real-time characterization of the chemical flood. The assumption of the new method is the same as the Hall slope analysis [a two-phase, radial flow of Newtonian liquids (Buell et al. 1990)]. Compared to the Hall plot analysis, it is simple and accurate in parameter calculation and high permeability zone estimation. The first pilot test of SP flooding in China was implemented in the western part of the Gudong 7th reservoir and has achieved great success. In this paper, the dynamic data of this pilot test are used to verify the new method.

2 New approach for injection effect evaluation

2.1 Ratio of pressure to injection rate (RPI)

2.1.1 Theoretical basis

During the period of water flooding, the water injectivity index represents the real-time characteristics of injection wells. The water injectivity index is defined as the ratio of the injection rate to the injection pressure difference. In field applications, it is difficult to measure the injection pressure difference, so the wellhead pressure is often used to replace the injection pressure difference to determine the injectivity index. Similarly for chemical floods, after the chemical solutions are injected into the well, both the injection pressure and the injection rate will change due to an increase in flow resistance in reservoirs, and thus, the dynamic injection data in the chemical flooding can be characterized by the variation of the wellhead pressure and the injection rate. The higher the injection pressure is, the greater the flow resistance will be, and the better the displacement efficiency the chemical flood will achieve. Based on this, the RPI refers to the ratio of the wellhead pressure to the injection rate, describing as follows:

$$\beta = \frac{P_{wh}}{q}, \tag{1}$$

where β is the ratio of the wellhead pressure to the injection rate (RPI) in MPa d/m^3; P_{wh} is the wellhead pressure in MPa; and q is the injection rate in m^3/d.

Figure 1 shows the relationship between the RPI and the cumulative volume of fluids injected into two injection wells (well I34-3166 and well I30-146) in the western part of the Gudong 7th reservoir. As is shown, injection well I34-3166 has experienced three stages of displacement, including water flooding, chemical flooding, and subsequent water flooding; while injection well I30-146 is a converted well (an injection well converted from a production well) and only experienced chemical flooding stage

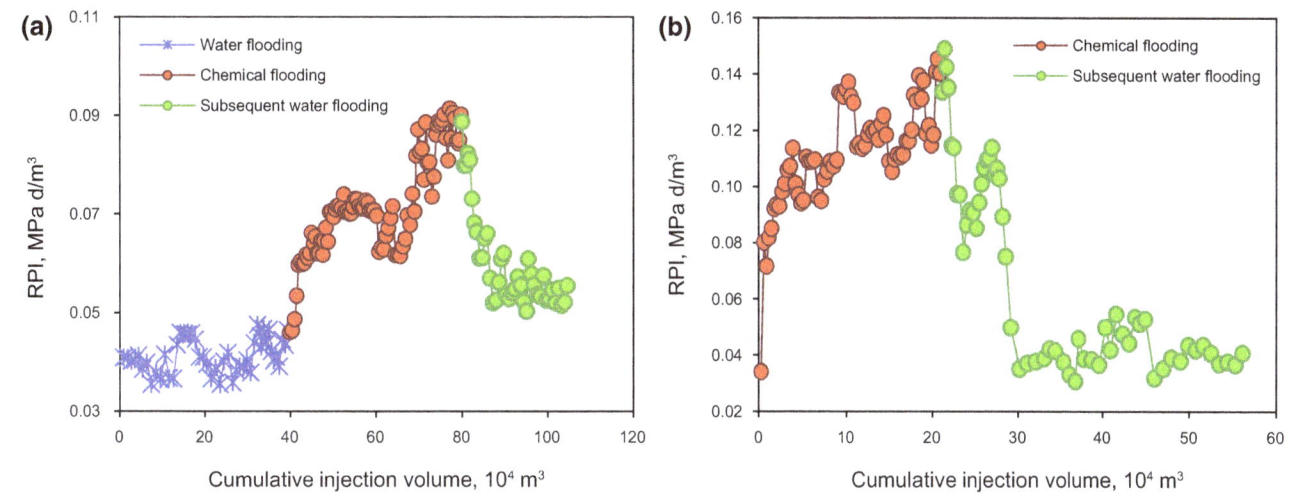

Fig. 1 Examples of the RPI under two different conditions. **a** Injection well I34-3166 with water flooding history. **b** Converted well I30-146 without water flooding history

and subsequent water flooding stage. It can be observed that at the stage of water flooding, the RPI fluctuates around a constant value; then, the curve rises rapidly and tends to vary gently with the increasing injection volume of the chemical solution; at the last stage of subsequent water flooding, the curve begins to fall fast, and eventually fluctuates steadily around a certain constant.

The variation of the RPI can be interpreted by the variation of resistance to flow in the reservoir. Due to the high viscosity of the polymer solution, the flow resistance will increase after the injection of the chemical solution. While at the stage of the subsequent water flooding, the flow resistance will decline gradually due to a decrease in the viscosity of the injection fluid. As a result, the RPI may be a reflection of the flow resistance in the reservoir at different displacement processes.

2.1.2 Relationship with apparent (residual) resistance factor

The RPI is defined as the ratio of the wellhead pressure to the injection rate in chemical flooding as well as in water flooding. It aims to calculate the apparent resistance factor and apparent residual resistance factor by the variation of the RPI, which is called the RPI method. The apparent resistance factor and the apparent residual resistance factor are used to characterize the difference of injectivity between water flooding and chemical flooding.

When the injection rate equals or nearly equals the production rate and the static reservoir pressure is constant or changes only slightly during the evaluation time, the bottom pressure can be approximated to the difference of the static water pressure and the friction pressure loss. The RPI can be presented as follows:

$$\beta \approx \frac{(P_{wh} + \rho g H - \Delta P_f) - P_e}{q} = \frac{P_{wf} - P_e}{q}, \qquad (2)$$

where ρ is the average fluid density in kg/m³; g is the gravity acceleration in m/s², $g = 9.81$ m/s²; H is the height of the wellhole liquid column in meters; ΔP_f is the friction pressure loss in MPa; P_{wf} is the bottom-hole pressure in MPa; and P_e is the static reservoir pressure in MPa.

Darcy's law for single-phase, steady-state, Newtonian flow was used to analyze the performance of the water injection wells by the Hall method. So, the RPI can be represented as follows:

$$\beta_w \approx \frac{P_{wf} - P_e}{q} = \frac{1.867 B_w \mu_w}{K_e h}\left(\ln \frac{R_e}{R_w} + S\right), \qquad (3)$$

where β_w is the RPI of water flooding in MPa d/m³; K_e is the effective permeability in 10^{-3} μm²; R_e is the distance between the injector and the producer in meters; R_w is the wellhole radius in meters; S is the skin factor; B_w is the volume factor of water; μ_w is the water viscosity in mPa s; and h is the effective thickness in meters.

After a chemical flood is injected into a well, there will be several fluid banks between the injector and the producer. A simplified method to solve such a case is applying Darcy's law in a serial form and the resistance factor is defined to quantitatively characterize the variations of displacing fluid viscosity and reservoir permeability (Buell et al. 1990). Because of extensive water flooding in this reservoir, an oil bank does not form, and two banks, a water bank and polymer bank, can be taken into consideration. Thus, the RPI after the injection of the chemical fluid can be represented as

$$\beta_p \approx \frac{1.867 B_w \mu_w}{K_e h}\left\{R_f\left(\ln \frac{R_{b1}}{R_w} + S\right) + \ln \frac{R_e}{R_{b1}}\right\}, \qquad (4)$$

where β_p is the RPI of chemical flooding in MPa d/m^3; R_f is the resistance factor; and R_{b1} is the radius of the chemical displacement front in meters.

Based on Eqs. (3) and (4), the ratio of the RPI of the chemical flooding to the RPI of the water flooding is described by Eq. (5):

$$\frac{\beta_p}{\beta_w} = \frac{R_f\left(\ln\frac{R_{b1}}{R_w} + S\right) + \ln\frac{R_e}{R_{b1}}}{\ln\frac{R_e}{R_w} + S}. \tag{5}$$

As shown in Eq. (5), under the condition that the radius of the chemical slug R_{b1} is equal to the distance between the injector and the producer R_e, the resistance factor equals the ratio of the RPI at the stage of the chemical flooding to the RPI at the stage of the water flooding. However, the size of the chemical slug is often less than the injector–producer distance, so the resistance factor does not equal the ratio of the RPI between the chemical flooding stage and the water flooding stage. Therefore, the apparent resistance factor is defined as

$$R'_f = \frac{\beta_p}{\beta_w}, \tag{6}$$

where R'_f is the apparent resistance factor. When the size of the chemical slug equals the injector–producer distance, the apparent resistance factor does equal the resistance factor.

The flow resistance in the subsequent water flooding process can also be calculated in series form. The enhanced oil recovery at this stage is due to the reservoir permeability reduction induced by the residual polymer, and therefore, the residual resistance factor is proposed to quantitatively characterize this mechanism. The residual resistance factor refers to the ratio of the reservoir permeability before chemical flooding to the permeability after the chemical injection. Thus, the RPI at the stage of subsequent water flooding can be represented as

$$\beta_{ww} \approx \frac{1.867B_w\mu_w}{K_e h}\left\{R_{rf}\left(\ln\frac{R_{b2}}{R_w} + S\right) + R_f\ln\frac{R_{b1}}{R_{b2}} + \ln\frac{R_e}{R_{b1}}\right\}, \tag{7}$$

where β_{ww} is the RPI of subsequent water flooding in MPa d/m^3; R_{rf} is the residual resistance factor which is defined as the ratio of the absolute permeability before SP flooding to the absolute permeability after SP flooding (Buell et al. 1990); and R_{b2} is the radius of the displacement front of subsequent water flooding in meters.

The ratio of the RPI at the stage of subsequent water flooding to that at the stage of initial water flooding can be defined as

$$\frac{\beta_{ww}}{\beta_w} = \frac{R_{rf}\left(\ln\frac{R_{b2}}{R_w} + S\right) + R_f\ln\frac{R_{b1}}{R_{b2}} + \ln\frac{R_e}{R_{b1}}}{\ln\frac{R_e}{R_w} + S}. \tag{8}$$

As shown in Eq. (8), if the displacement radius R_{b2} of subsequent water flooding is equal to the injector–producer distance R_e, the residual resistance factor equals the ratio of the RPI at the stage of subsequent water flooding to that in the process of initial water flooding. The apparent residual resistance factor can be defined as

$$R'_{rf} = \frac{\beta_{ww}}{\beta_w}, \tag{9}$$

where R'_{rf} is the apparent residual resistance factor.

2.1.3 Advantages and disadvantages

The Hall plot analysis can only describe the average injection effect over a period of time, while the RPI is a reflection of the instantaneous value at every displacement moment. Thus, the RPI is more sensitive to the chemical injection and has more advantages over the Hall plot analysis, especially when the variation of Hall plot slope is not obvious, such as at the early stage of chemical flooding.

The disadvantage of the RPI is that it does not have a smoothing effect on the data when the injection pressure and rate have big and frequent fluctuations. An effective way to solve this problem is to apply some mathematical curve smoothing methods such as the data average method and the linear iterative method.

2.2 The rate of change of RPI

2.2.1 Theoretical basis

The RPI describes the variation range of the ratio of the injection pressure to the injection rate. However, the injection efficiency is related not only to the variation range of injection pressure and injection rate but also to the variation rate of injection pressure and injection rate. Under the condition of the same chemical injection rate, the faster the pressure increases, the more effective the chemical injection will be.

The rate of change of RPI refers to the variation of the RPI per unit injection volume. It can also be rearranged to the derivative value of the RPI to the cumulative injection volume:

$$\gamma = \frac{d\beta}{dW}, \tag{10}$$

where γ is the rate of change of the RPI in MPa d/m^6; and W is the cumulative injection volume in m^3.

If the injection rate between 2 months keeps constant or changes only slightly, the rate of change of the RPI can be obtained as follows:

$$\gamma = \frac{\beta_2 - \beta_1}{W_2 - W_1} = \frac{p_{wh2} - p_{wh1}}{q^2(t_2 - t_1)}, \tag{11}$$

where t_1 and t_1 are the injection time in days.

Based on Eq. (11), the rate of change of the RPI refers to the ratio of the wellhead pressure increment per unit time to the square of the injection rate.

The relationship between the RPI and the cumulative injection volume is illustrated in Fig. 2. It is observed that at the polymer pre-protection slug injection stage, the RPI increases along with an increase in the cumulative injection, while the rate of change of RPI behaves in a reverse manner. After curve fitting, there is a logarithmic relationship between the RPI and the cumulative injection volume, while the rate of change of the RPI has a linear relation with the reciprocal of the cumulative injection volume.

2.2.2 Relationship with resistance factor

Based on the definition Eq. (10) of the rate of change of the RPI and the expression Eq. (4) of the RPI during chemical flooding, the derivative of the RPI can be expressed as

$$\gamma = \frac{d\beta}{dW} = \frac{d\left(\frac{1.867\mu_w B_w}{K_e h}\left(R_f(\ln(r/R_w) + S) + \ln(R_e/r))\right)\right)}{dW}, \tag{12}$$

where r is the radius of the chemical displacement front in meters.

A small circular unit in a circular formation is selected as shown in Fig. 3. The circular unit formation can be

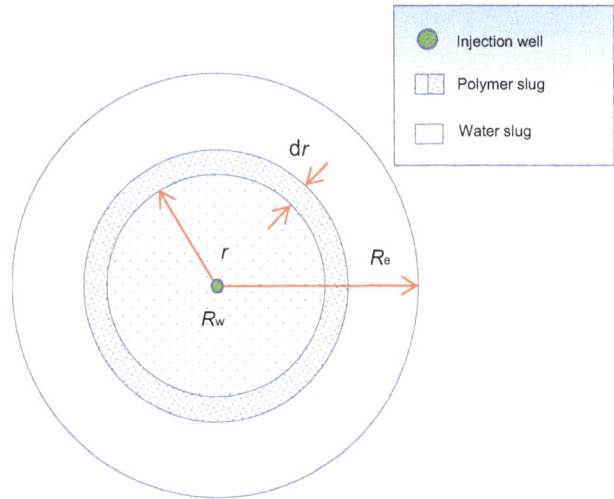

Fig. 3 Schematic diagram of the polymer injection model

regarded as a homogeneous formation with a constant thickness and constant effective permeability. The differential formula of the cumulative injection volume can be expressed as

$$dW = 2\pi r dr \cdot h \cdot \phi \cdot \phi_D, \tag{13}$$

where ϕ is the formation porosity; and ϕ_D is the fraction of the accessible pore volume of the polymer solution.

Substituting Eq. (13) into Eq. (12) gives the simplified derivation:

$$\gamma = \frac{0.9335\mu_w B_w}{\pi K_e h^2 \phi \cdot \phi_D} \cdot \frac{(R_f - 1)}{r^2}. \tag{14}$$

The accumulative volume of the chemical solution injected into the reservoir is

$$W = \pi \cdot r^2 \cdot h \cdot \phi \cdot \phi_D. \tag{15}$$

(a)

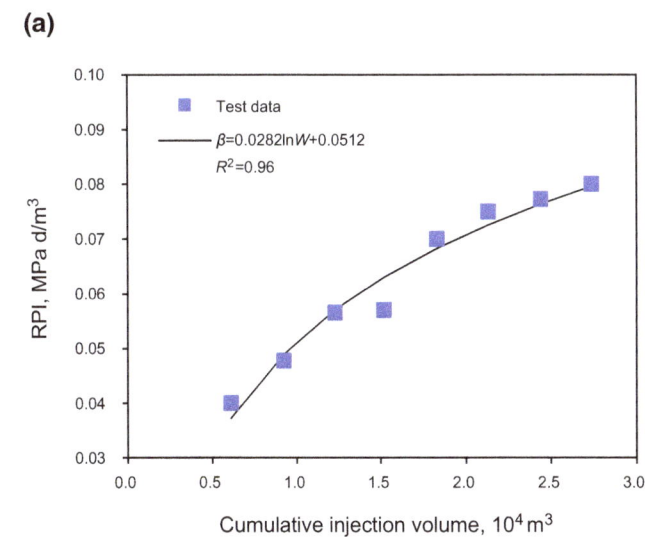

(b)

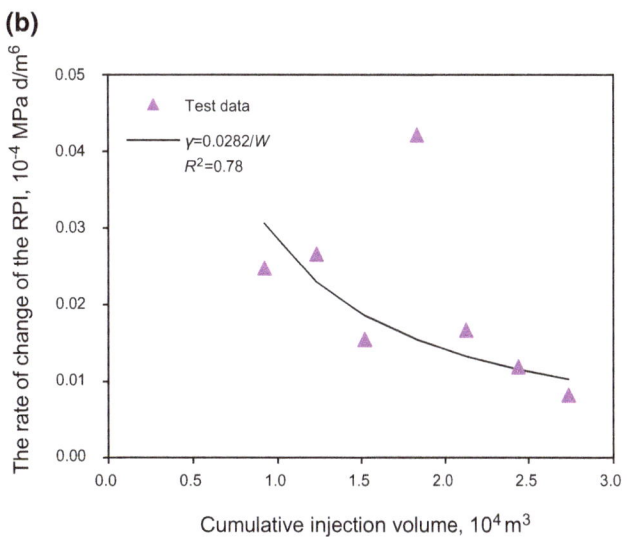

Fig. 2 Statistical relationship of new evaluation indexes and the cumulative injection volume. a RPI. b The rate of change of the RPI

The relationship between the rate of change of the RPI and the cumulative injection volume can be obtained by substituting Eq. (15) into Eq. (14):

$$\gamma = \frac{0.9335\mu_w B_w (R_f - 1)}{K_e h} \cdot \frac{1}{W}. \tag{16}$$

Equation (16) indicates that the rate of change of the RPI varies inversely with the cumulative injection volume, which also proves the validity of the regression expression in Fig. 2a. The expression of the resistance factor can be derived from Eq. (16):

$$R_f = 1 + \frac{K_e h}{0.9335\mu_w B_w} \cdot W \cdot \gamma. \tag{17}$$

Under the condition of the same cumulative injection volume, the greater the rate of change of the RPI is, the better the efficiency of the chemical injection.

2.2.3 Advantages

Dynamic data such as the injection pressure and the injection rate are indispensable to evaluate the resistance factor by both the Hall plot method and the RPI method during water flooding. However, for new wells or converted wells without water-flooding history, the rate of change of the RPI is the only effective way for injection effect evaluation.

3 Field applications

3.1 Field trial of SP flooding

The first pilot application of SP flooding in China is located in the western part of Gudong 7th reservoir, covering an area of 0.94 km², with nearly 277×10^4 tons oil in place. The target zone consists of three layers (Ng5⁴, Ng5⁵, and Ng6¹), with a depth of 1261–1294 m. The average porosity is 34 %, the permeability is 1320×10^{-3} μm², and the permeability variation coefficient is 0.58. Prior to production, the viscosity of the crude oil is 45 mPa s, and the initial oil saturation is 0.72. The initial reservoir pressure is 12.4 MPa, and the reservoir temperature is 68 °C, and the salinity of the formation water is 3152 mg/L.

A pilot water flood was initiated in this target zone in July 1986. Until August 2003, there were 21 production wells (20 of them were open) and 9 injection wells (8 of them were open), with an average daily fluid production rate of 123.2 t/d and an average daily injection rate of 205 m³/d. The average daily oil production rate was 2.95 t/d and the average water cut was 97.6 %, while the average daily injection pressure was 11.6 MPa and the

injection to production ratio was 0.88 with a cumulative injection to production ratio of 1.04. The oil recovery of reservoir was up to 34.5 %.

A pilot SP flood was started in September 2003 and ended in January 2010, including 27 wells, 17 production wells and 10 injection wells. As is shown in Fig. 4, a line drive pattern was adopted during oil production, while the row space is 300 m and the well space is 150 m.

The injection of the SP flood included four slugs: polymer slug (pre-slug), main SP slug I, main SP slug II, and polymer slug (post-slug). The pre-slug started in September 2003, while the main SP slug I started in June 2004, main SP slug II in June 2007, and post-slug in April 2009. After the SP flooding, the subsequent water flooding stage went into operation in January 2010. The total cumulative injection of chemicals was up to 0.635 PV, including polymer 5496 t, surfactant 8727 t, and auxiliary 3024 t. The detailed information is shown in Table 1.

3.2 Calculation of the apparent resistance factor by the RPI

The apparent resistance factor can be defined as the ratio of the RPI at the chemical flooding stage to that at the water flooding stage. Therefore, it is necessary to calculate the RPI at the water flooding stage. Since the RPI at the water flooding stage will not keep constant, as shown in Fig. 1, the average value over a period of time (in general, 1 month is enough) is used in calculation. Using the above method, the apparent resistance factor and the apparent residual resistance factor of injection well I34-3166 are shown in Fig. 5 and Table 2.

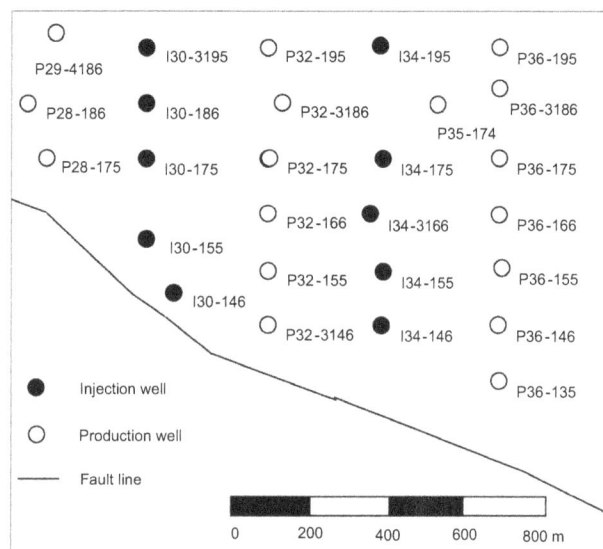

Fig. 4 Location of studied wells

Table 1 Injection data of SP flooding

Injection slug	Starting time	End time	Injection concentration			Slug size, PV
			Polymer, mg/L	Surfactant, mg/L	Auxiliary, mg/L	
Polymer slug (pre-slug)	Sep. 2003	May 2004	1934	0	0	0.078
Main SP slug I	Jun. 2004	May 2007	1856	4618	1610	0.302
Main SP slug II	Jun. 2007	Apr. 2009	1713	3056	1043	0.188
Polymer slug (post-slug)	Apr. 2009	Jan. 2010	1500	0	0	0.067

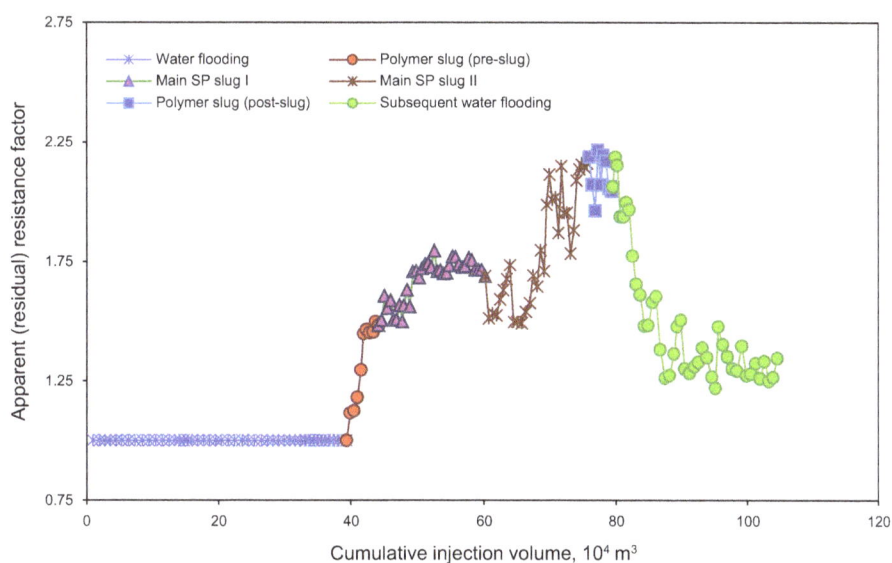

Fig. 5 Apparent (residual) resistance factor of injection well I34-3166 derived by the RPI method

Table 2 Apparent (residual) resistance factor of injection well I34-3166

Injection stage	Apparent (residual) resistance factor			
	RPI method			Hall plot analysis method
	Minimum	Maximum	Mean	
SP flooding				
Polymer slug (pre-slug)	1.1	1.5	1.35	1.65
Main SP slug I	1.5	1.8	1.67	
Main SP slug II	1.5	2.2	1.82	
Polymer slug (post-slug)	2.0	2.2	2.09	
Subsequent water flooding	1.2	2.2	1.48	1.40

Figure 5 indicates that the apparent resistance factor increases from 1.0 to 1.5 rapidly during the injection of the polymer slug (pre-slug), with an average value of 1.35. During the injection of the main SP slug I, the apparent resistance factor rises gradually from 1.5 to 1.8, and the average value is 1.67. However, when the main SP slug II is injected into the reservoir, the apparent resistance factor decreases a little with the cumulative injection volume first and then increases gradually to the maximum value of 2.2, with an average value of 1.82. During the injection of the

polymer slug (post-slug), the apparent resistance factor changes relatively little, changing from 2.0 to 2.2, with an average value of 2.09. At last, during the subsequent water flooding, the residual apparent resistance factor decreases slowly until reaching a steady level near 1.3.

Meanwhile, the apparent resistance factors for the chemical flooding and the subsequent water flooding were also calculated from the Hall plot analysis, as shown in Fig. 6 and Table 2. There is a small difference in the apparent resistance factors calculated from these two

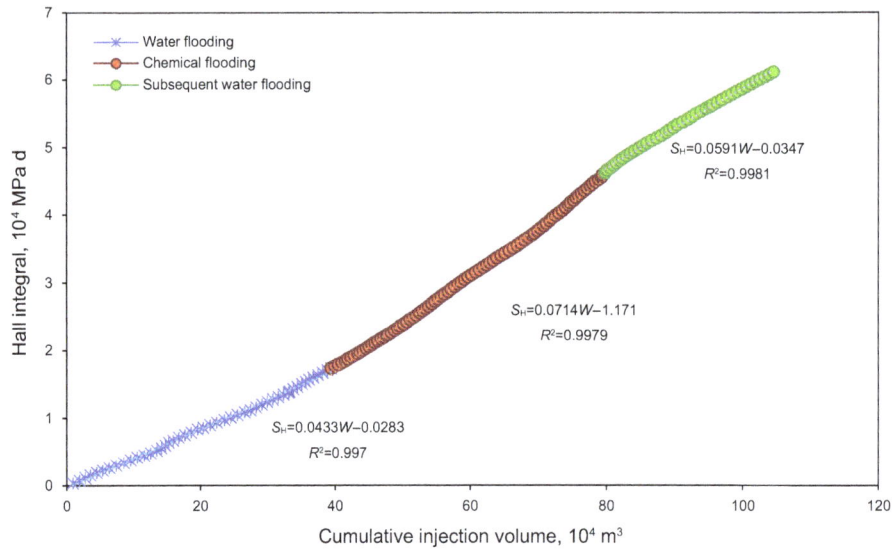

Fig. 6 Hall plot of injection well I34-3166

methods. In fact, both methods are based on the assumption of steady flow, but the RPI can reflect the instantaneous value at every displacement moment and be more sensitive than the Hall plot method. Thus, this method is more helpful for dynamic analysis, especially when the slope difference of the Hall plot is not obvious, i.e., at the early stage of chemical flooding.

3.3 Prediction of crossflow of the chemical floods by the RPI

If the concentration of chemicals fluctuates slightly in the process of the chemical injection, the RPI would increase gradually or remain unchanged with an increase in the injection volume. However, a sharp drop in the RPI indicates a decrease in the resistance to flow in the reservoir, and the possibility of chemical crossflow.

SP flooding pilot tests were conducted in the Shengli Oilfield, and Fig. 7 shows the RPI of injection well I34-175 and the concentration of polymer in the produced fluids from the surrounding production wells. When the cumulative volume of the SP solution injected into the reservoir reached 8×10^4 m^3, there was a sharp drop in the RPI, and 2 months after the drop, polymer was detected in the produced fluids from production well P32-3186 with a polymer concentration of 181 mg/L. A second drop in the RPI occurred when the cumulative injection volume came to 16×10^4 m^3, and 2 months later, polymer of high concentration was produced from wells P32-175 and P32-166. In addition, when the cumulative injection volume amounted to 30×10^4 m^3, a third drop in the RPI appeared and a significantly increase appeared in the concentration of polymers in the produced fluids from wells P32-175, P32-166, P35-174, P36-175, and P36-166.

Based on the analysis above, it is observed that after the RPI drops for some time, polymer will appear in production fluids. For a production well, both the early production of chemical fluids and the rapid growth of the polymer concentration might be caused by polymer crossflow. The existence of thief zones is the most common case causing chemical crossflow. To verify the prediction method by the RPI theoretically, a heterogeneous reservoir model with a production well and an injection well is shown in Fig. 8. The model is composed of two stratified homogeneous layers. The first layer has a low permeability k_1 and a thickness of h_1, while the second layer has a high permeability k_2 and a thickness of h_2. Figure 8 describes the flow without or with crossflow.

According to Darcy's law, the flow resistance can be defined as the ratio of the injection pressure difference to the injection rate. To compare the flow resistance in two cases, the reservoir is divided into four parts, and the length of each part is L_1, L_2, L_3, and L_4. Under the condition that no chemical crossflow exists, the flow resistance of these four part are R_1, R_2, R_3, and R_4, while with chemical crossflow, the flow resistance of these four part are R_1', R_2', R_3', and R_4'. Since under both conditions, a chemical solution is only in the first part with a length of L_1, R_1 is equal to R_1'. Similarly, for the fourth part with a length of L_4 in which only water flows, so R_4 is also equal to R_4'. When the cumulative injection volume of polymer is equal in two cases, the relationship between parameters can be described by Eq. (18):

$$L_2(h_1 + h_2)B\phi = (L_2 + L_3)h_2B\phi, \tag{18}$$

where B is the formation width in meters.

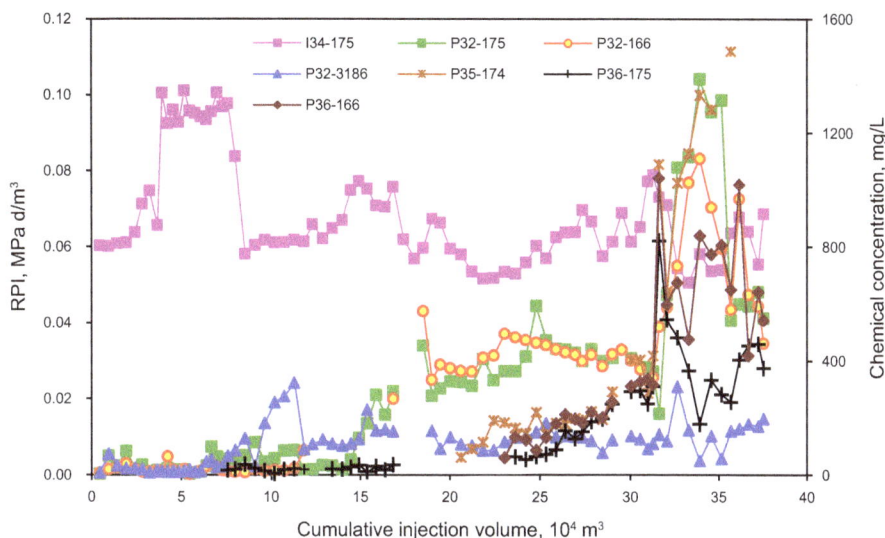

Fig. 7 RPI of injector I34-175 and chemical concentrations of surrounding producers

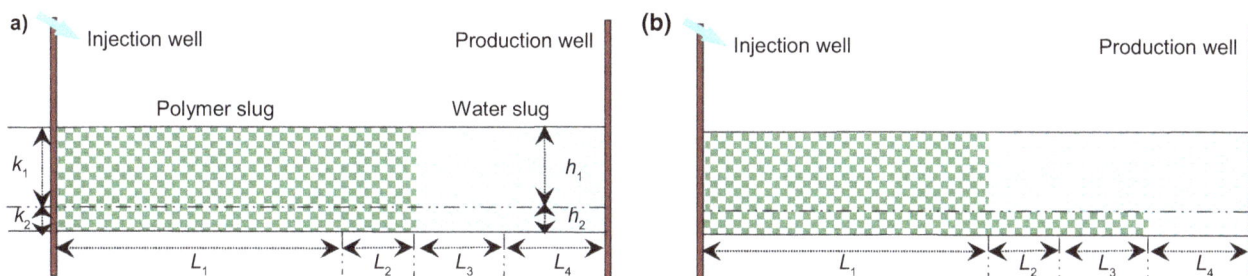

Fig. 8 Schematic diagram of the chemical flood in a reservoir. **a** No chemical crossflow. **b** Chemical crossflow

The difference of the flow resistance in two cases is expressed as

$$
\begin{aligned}
R - R' \\
= (R_1 + R_2 + R_3 + R_4) - (R'_1 + R'_2 + R'_3 + R'_4) \\
= R_2 + R_3 - R'_2 - R'_3 \\
= \frac{1}{\frac{k_1 B_w h_1}{\mu_p L_2} + \frac{k_2 B_w h_2}{\mu_p L_2}} + \frac{1}{\frac{k_1 B_w h_1}{\mu_w L_3} + \frac{k_2 B_w h_2}{\mu_w L_3}} \\
- \frac{1}{\frac{k_1 B_w h_1}{\mu_w L_2} + \frac{k_2 B_w h_2}{\mu_p L_2}} - \frac{1}{\frac{k_1 B_w h_1}{\mu_w L_3} + \frac{k_2 B_w h_2}{\mu_p L_3}} \\
= \frac{L_2 h_1 (\mu_p - \mu_w)(k_1 \mu_p - k_2 \mu_w)}{B_w (k_1 h_1 + k_2 h_2)(k_1 h_1 \mu_p + k_2 h_2 \mu_w)}
\end{aligned}
\qquad (19)
$$

where R is the flow resistance without chemical crossflow; R' is the flow resistances with chemical crossflow; and μ_p is the viscosity of the chemical fluid in mPa s.

In actual cases, because the viscosity ratio of the chemical fluid to water is larger than the permeability ratio of the two layers, R is greater than the R'. It can also be set out that when the chemical crossflow exists, the RPI

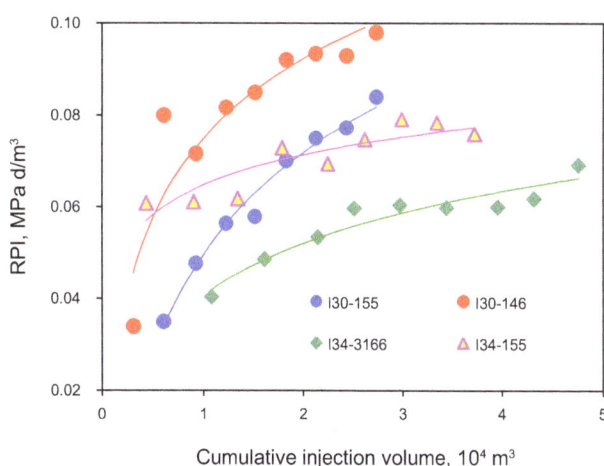

Fig. 9 The relationship between the RPI and the cumulative injection volume

decreases as a result of the drop in the flow resistance. Based on this theory, the chemical crossflow can be estimated by the variation of the RPI.

Table 3 Input parameters and the (apparent) resistance factor

Injection well	Input parameters				(Apparent) resistance factor	
	Thickness, m	Water viscosity, mPa s	Permeability, μm^2	Regression coefficient	RPI change rate method	RPI method
I30-146	11.2	0.45	0.99	0.025	1.68	Conversion well (no water flooding history)
I30-155	11.4	0.45	1.40	0.033	2.23	
I34-155	18.0	0.45	0.78	0.009	1.30	1.41
I34-3166	17.5	0.45	0.51	0.016	1.34	1.35

3.4 Calculation of the resistance factor by the rate of change of RPI

The relationship between the resistance factor and the rate of change of the RPI is given by Eq. (17). By calculating the resistance factor of the polymer slug (pre-slug), the curve in Fig. 9 shows the RPI versus the cumulative injection volume of wells I30-155, I30-146, I34-3166, and I34-155. It is observed that the RPI of the four injection wells have a good logarithmic relationship with the cumulative injection volume, which further verifies Eq. (16).

The calculation of the apparent (residual) resistance factor is as follows:

Firstly, the regression expression of the RPI versus the cumulative injection volume is calculated:

Injection well I30-155

$$\beta = 0.033 \ln W + 0.050, \quad R^2 = 0.94 \tag{20}$$

Injection well I30-146

$$\beta = 0.025 \ln W + 0.075, \quad R^2 = 0.84 \tag{21}$$

Injection well I34-155

$$\beta = 0.009 \ln W + 0.064, \quad R^2 = 0.81 \tag{22}$$

Injection well I34-3166

$$\beta = 0.016 \ln W + 0.040, \quad R^2 = 0.91, \tag{23}$$

where R^2 is the coefficient of determination.

Secondly, the rate of change of the RPI versus the cumulative injection volume is calculated. For injection well I30-155, the equation is

$$\gamma = \frac{d\beta}{dW} = \frac{A}{W} = \frac{0.033}{W} \tag{24}$$

where A is the slope of the regression curve.

Lastly, the resistance factor is obtained based on the relational expression of the rate of change of the RPI and the cumulative injection volume (Table 3). For injection well I30-155, the expression of the resistance factor is

$$R_f = 1 + \frac{K_e h}{0.9335 \mu_w B_w} \cdot A = 1 + 0.033 \frac{K_e h}{0.9335 \mu_w B_w}$$
$$= 2.23. \tag{25}$$

Similarly, the resistance factor of the remaining three injection wells is shown in Table 3. Meanwhile, the RPI is also used to calculate the apparent resistance factor. Since both injection wells I30-146 and I30-155 in Table 3 are converted wells, the traditional Hall plot analysis is not applicable for the resistance factor calculation. The resistance factor of the remaining two wells shows no obvious difference between the RPI method and RPI change rate method. However, for new wells without water flooding history or converted wells, the RPI change rate method is an effective way for injectivity evaluation.

4 Conclusion

(1) A new method (RPI and RPI change rate) is developed for evaluating the efficiency of chemical flooding. It is proved to be a more sensitive and rapid method for calculating parameters than the Hall plot method. Also, it is proved to be an effective way in estimating the existence of chemical crossflow.

(2) The RPI characterizes the real-time injection performance of chemical flooding in a simple form. It is verified that the apparent resistance factor and the apparent residual resistance factor can be obtained by the RPI at different displacement stages, and this method is most attractive at the early stage of chemical injection or when no obvious effect is observed.

(3) The RPI will fall sharply when the resistance to flow declines (caused by chemical crossflow). According to this theory, the RPI can be used to indicate the existence of chemical crossflow.

(4) The rate of change of the RPI can quantitatively describe the response rate of chemical flooding. The greater the rate of change of the RPI is, the faster the

chemical flooding takes effect. Since only injection data of the chemical flooding stage is needed for evaluation, the rate of change of the RPI is most suitable for wells without water flooding history.

Acknowledgments The authors greatly appreciate the financial support from the National Natural Science Foundation of China (Grant No. 51574269), the Important National Science and Technology Specific Projects of China (Grant No. 2016ZX05011-003), the Fundamental Research Funds for the Central Universities (Grant No. 15CX08004A, 13CX05007A), and the Program for Changjiang Scholars and Innovative Research Team in University (Grant No. IRT1294).

References

AlSofi AM, Blunt MJ. Polymer flooding design and optimization under economic uncertainty. J Pet Sci Eng. 2014;124:46–59. doi:10.1016/j.petrol.2014.10.014.

Buell RS, Kazemi H, Poettmann FH. Analyzing injectivity of polymer solutions with the Hall plot. SPE Reserv Eng. 1990;5(1):41–6. doi:10.2118/16963-PA.

Chang HL, Zhang ZQ, Wang QM, et al. Advances in polymer flooding and alkaline/surfactant/polymer processes as developed and applied in the People's Republic of China. J Pet Technol. 2006;58(2):84–9. doi:10.2118/89175-JPT.

Cheng J, Wang D, Li Q. Field test performance of alkaline surfactant polymer flooding in the Daqing oil field. Acta Pet Sin. 2002;23(6):37–40 (in Chinese).

Dag CS, Ingun S. Literature review of implemented polymer field projects. J Pet Sci Eng. 2014;122:761–75. doi:10.1016/j.petrol.2014.08.024.

Delamaide E, Zaitoun A, Renard G, et al. Pelican Lake field: first successful application of polymer flooding in a heavy-oil reservoir. SPE Reserv Eval Eng. 2014;17(03):340–54. doi:10.2118/165234-PA.

DeMarco M. Simplified method pinpoints injection well problems. World Oil. 1969;168(5):97.

Dong M, Ma S, Liu Q. Enhanced heavy oil recovery through interfacial instability: a study of chemical flooding for Brintnell heavy oil. Fuel. 2009;88(6):1049–56. doi:10.1016/j.fuel.2008.11.014.

Ghosh B, Bemani AS, Wahaibi YM, et al. Development of a novel chemical water shut-off method for fractured reservoirs: laboratory development and verification through core flow experiments. J Pet Sci Eng. 2012;96:176–84. doi:10.1016/j.petrol.2011.08.020.

Hall HN. How to analyze waterflood injection well performance. World Oil. 1963;157(5):128–33.

Honarpour MM, Tomutsa L. Injection/production monitoring: an effective method for reservoir characterization. In: SPE/DOE enhanced oil recovery symposium, 22–25 April, Tulsa, Oklahoma; 1990. doi:10.2118/20262-MS.

Hou J, Du Q, Lu T, et al. The effect of interbeds on distribution of incremental oil displaced by a polymer flood. Pet Sci. 2011;8(2):200–6. doi:10.1007/s12182-011-0135-z.

Hou J, Pan G, Lu X, et al. The distribution characteristics of additional extracted oil displaced by surfactant–polymer flooding and its genetic mechanisms. J Pet Sci Eng. 2013;112:322–34. doi:10.1016/j.petrol.2013.11.021.

Izgec B, Kabir CS. Real-time performance analysis of water-injection wells. SPE Reserv Eval Eng. 2009;12(1):116–23. doi:10.2118/109876-PA.

Kaminsky RD, Wattenbarger RC, Szafranski RC, et al. Guidelines for polymer flooding evaluation and development. In: International petroleum technology conference, 4–6 Dec, Dubai, UAE; 2007. doi:10.2523/11200-MS.

Kim Y, Lee J. Evaluation of real-time injection performance for polymer flooding in heterogeneous reservoir. In: SPE the twenty-fourth international ocean and polar engineering conference, 15–20 June, Busan, Korea; 2014.

Li Z. Industrial test of polymer flooding in super high water cut stage of central No. 1 Block Gudao Oilfield. Pet Exp Dev. 2004;31(2):119–21 (in Chinese).

Li Y, Su Y, Ma K, et al. Study of gel flooding pilot test and evaluation method for conventional heavy oil reservoir in Bohai bay. In: SPE annual technical conference and exhibition, 30 Oct–2 Nov, Denver, Colorado, USA; 2011. doi:10.2118/146617-MS.

Li Z, Zhang A, Cui X, et al. A successful pilot of dilute surfactant-polymer flooding in Shengli oilfield. In: SPE improved oil recovery symposium, 14–18 April, Tulsa, Oklahoma, USA; 2012. doi:10.2118/154034-MS.

Ma S, Dong M, Li Z, et al. Evaluation of the effectiveness of chemical flooding using heterogeneous sandpack flood test. J Pet Sci Eng. 2007;55:294–300. doi:10.1016/j.petrol.2006.05.002.

Moffitt PD, Menzie DE. Well injection tests of non-newtonian fluids. In: SPE rocky mountain regional meeting, 17–19 May, Cody, Wyoming, USA; 1978.

Seright RS, Seheult JM, Talashek T. Injectivity characteristics of EOR polymers. In: SPE annual technical conference and exhibition, 21–24 Sept, Denver, Colorado, USA; 2009. doi:10.2118/115142-MS.

Shaker Shiran B, Skauge A. Enhanced oil recovery (EOR) by combined low salinity water/polymer flooding. Energy Fuels. 2013;27(3):1223–35. doi:10.1021/ef301538e.

Shen P, Wang J, Yuan S, et al. Study of enhanced-oil-recovery mechanism of alkali/surfactant/polymer flooding in porous media from experiments. SPE J. 2009;14(02):237–44. doi:10.2118/126128-PA.

Sheng JJ, Leonhardt B, Azri N. Status of polymer-flooding technology. J Can Pet Technol. 2015;54(2):116–26. doi:10.2118/174541-PA.

Silin DB, Holtzman R, Patzek TW. Monitoring waterflood operations: Hall method revisited. In: SPE western regional meeting, 30 Mar–1 April, Irvine, California, USA; 2005. doi:10.2118/93879-MS.

Sugai K, Nishikiori N. An integrated approach to reservoir performance monitoring and analysis. In: SPE Asia Pacific oil & gas conference and exhibition, 11–13 Sept, Adelaide, Australia; 2006. doi:10.2118/100995-MS.

Urbissinova TS, Trivedi JJ, Kuru E. Effect of elasticity during viscoelastic polymer flooding: a possible mechanism of increasing the sweep efficiency. J Can Pet Technol. 2010;49(12):49. doi:10.2118/133471-PA.

Vargo J, Turner J, Bob V, et al. Alkaline-surfactant-polymer flooding of the Cambridge Minnelusa field. SPE Reserv Eval Eng. 2000;3(06):552–8. doi:10.2118/68285-PA.

Zhang H, Dong M, Zhao S. Which one is more important in chemical flooding for enhanced court heavy oil recovery, lowering interfacial tension or reducing water mobility? Energy Fuels. 2010;24(3):1829–36. doi:10.1021/ef901310v.

2

The relationship between international crude oil prices and China's refined oil prices based on a structural VAR model

Song Han[1] · Bao-Sheng Zhang[1] · Xu Tang[1] · Ke-Qiang Guo[1]

Abstract With the frequent fluctuations of international crude oil prices and China's increasing dependence on foreign oil in recent years, the volatility of international oil prices has significantly influenced China domestic refined oil price. This paper aims to investigate the transmission and feedback mechanism between international crude oil prices and China's refined oil prices for the time span from January 2011 to November 2015 by using the Granger causality test, vector autoregression model, impulse response function and variance decomposition methods. It is demonstrated that variation of international crude oil prices can cause China domestic refined oil price to change with a weak feedback effect. Moreover, international crude oil prices and China domestic refined oil prices are affected by their lag terms in positive and negative directions in different degrees. Besides, an international crude oil price shock has a significant positive impact on domestic refined oil prices while the impulse response of the international crude oil price variable to the domestic refined oil price shock is negatively insignificant. Furthermore, international crude oil prices and domestic refined oil prices have strong historical inheritance. According to the variance decomposition analysis, the international crude oil price is significantly affected by its own disturbance influence, and a domestic refined oil price shock has a slight impact on international crude oil price changes. The domestic refined oil price variance is mainly caused by international crude oil price disturbance, while the domestic refined oil price is slightly affected by its own disturbance. Generally, domestic refined oil prices do not immediately respond to an international crude oil price change, that is, there is a time lag.

Keywords International crude oil prices · China's refined oil prices · VAR model · Granger causality · Impulse response · Variance decomposition

1 Introduction

As an important raw material and energy source in the global industrial economy, petroleum has been playing a key role in promoting the development of the global economy and society (Timilsina 2015). The international crude oil price has a significant impact on China's national economy and the refined oil price level. Figure 1 depicts the variation of international crude oil and China's refined oil prices with time from 2011 to 2015. The relationship between them deserves to be further explored.

Considering the frequent fluctuation of international oil prices and China's rising dependence on oil imports, more attention has been paid to oil security since it significantly influences the China national economy and public life. However, the China domestic refined oil pricing mechanism is not fully market-oriented and it is subject to direct government control (Liao et al. 2016). Consequently, there is an urgent need for qualitative and quantitative analysis of the relationship between international crude oil prices and domestic oil prices as well as the feedback mechanism (Zhang and Xie 2016; Broadstock et al. 2012; Du et al. 2010). As the China domestic refined oil price has been strongly influenced by the fluctuating international oil price, the current pricing mechanism fails to fully reflect the real supply and demand situation and petrochemical

✉ Bao-Sheng Zhang
  bshshysh@cup.edu.cn

[1] School of Business Administration, China University of Petroleum, Beijing 102249, China

Edited by Xiu-Qin Zhu

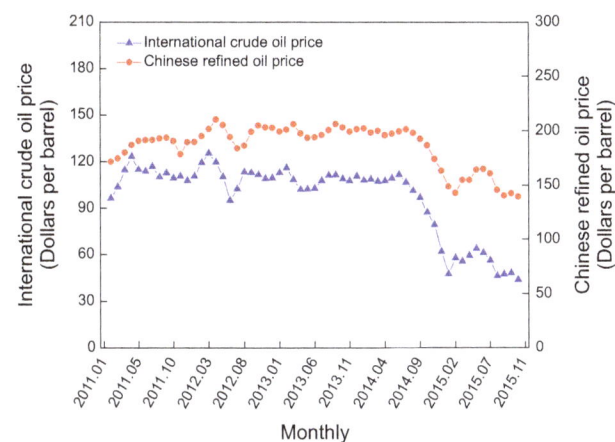

Fig. 1 Variation of international crude oil prices and China's refined oil prices

enterprise costs. Therefore, more factors that influence oil prices should be taken into consideration comprehensively to improve the oil pricing mechanism, so that oil price can reflect the actual supply and demand structure as well as the enterprise costs. What is more, reformation of the refined oil pricing mechanism can lay a solid foundation for future domestic oil market pricing.

Extant research has focused on detailed analysis of relationships between international oil prices and China domestic oil prices.

It is demonstrated that international oil prices not only affect domestic crude oil prices, but also influence the economy in China (Jiang and Jiang 2005). There is a long-term stable relationship between China domestic and international crude oil prices in the quantitative analysis of the price variation during the period of January 1997 to December 2004 (Jiao et al. 2004). Conclusions have been reached that the transmission and feedback mechanism between prices of crude oil and retail refined oil is currently unavailable through an empirical study based on the vector autoregression (VAR) model (Jiang 2013). Statistical test analysis on international crude oil prices and China domestic refined oil prices was conducted to study their interactions, which indicates that there is a structural lag in the international oil price response under the refined oil pricing mechanism in China, and the refined oil pricing mechanism seldom influences the short-term price information transmission (Wang 2014). The anti-symmetry of gasoline/diesel to crude oil price fluctuation is investigated with the anti-symmetry VECM (vector error correction model), and the result demonstrates that the response of gasoline/diesel price to an increase in crude oil price is quick and transient, while the response of gasoline/diesel price to a decrease in crude oil price is delayed and hysteretic (Jiao et al. 2006). It is proposed that the crude oil price is weakly exogenous, its spread is constant in some

but not all relationships, and the link between crude oil prices and several refined product prices implies market integration (Asche et al. 2003). Cross-correlations between crude oil and refined product prices based on the well-known detrended cross-correlation analysis are shown to be significant and strong. Furthermore, multifractality is also revealed in cross-correlations (Liu and Ma 2014).

This paper aims to investigate the transmission and feedback mechanisms between international crude oil prices and China's refined oil from January 2011 to November 2015 by using a Granger causality test, the vector autoregression model, an impulse response function and variance decomposition methods. Understanding of volatility transmission mechanisms is significant for the government to make political decisions. Therefore, the paper provides reasonable suggestions for the policy makers.

The paper is structured as follows. It begins with a data description and a brief introduction of the VAR model in Sect. 2, followed by the empirical analysis in Sect. 3. In Sect. 4, we discuss the results in the context of the Chinese situation and draw some conclusions.

2 Data and econometric methodology

2.1 Data descriptions

This study sets two variables, the international crude oil prices (IC) and China's refined oil prices (RD). We use monthly spot price data of the Brent crude oil price to represent the international crude oil price and average China national monthly retail price of −10# diesel to represent the refined oil price in China. Settings of variables and sources of data are shown in Table 1. Since variable units are not uniform, we have to first normalize the units into US dollars/barrel. According to *Volume and Weight of Crude Oil Conversion Table* from International Oil Network (http://oil.in-en.com/html/oil-67979.shtml), China's refined oil conversion coefficient is 7.31 barrels/tonne (Li and Guo 2013). Then, according to the monthly exchange rate data published by the International Monetary Fund (http://www.imf.org/external/np/fin/data/param_rms_mth.aspx), we convert the unit of yuan/tonne into the unit of US dollars/barrel.

On March 26, 2009, China's National Development and Reform Commission issued a new refined oil pricing mechanism. However, due to incomplete implementation that caused the information on China's refined oil price to be absent during the period from March 2009 to December 2010, the sample data in this work cover the period from January 2011 to November 2015. In order to eliminate the influence of seasonal factors in the index sequence, we introduce the X-13 Census method to adjust the two

Table 1 Variable settings and data sources

Variable	Definition	Unit	Data sources	Variable symbol
Crude oil price	Brent crude oil price	Dollars/ barrel	US Energy Information Administration (http://www.eia.gov/dnav/ pet/hist/LeafHandler.ashx?n=PET&s=RBRTE&f=D)	IC
Refined oil price	Average national monthly retail price of -10# diesel	Yuan/ tonne	China Petroleum and Chemical Industry Economic Data	RD

variables to make the sequence more realistic and objective. The quantitative analysis software, Eviews8.0, is used in this study to carry out the analysis.

2.2 Vector autoregressive model

Without imposing theoretical restrictions on endogeneity among variables, a vector autoregression procedure is appropriate to establish the dynamics between crude oil prices and exchange rates. These two variables are jointly treated as endogenous and are assumed to have no restrictions on the structural relationships in the present analysis.

The VAR is commonly used in systems forecasting interrelated time series and analyzing the dynamic impact of random disturbances on a system of variables. The VAR procedure avoids the need for structural modeling by treating every endogenous variable in the system as a function of lagged values of all of the endogenous variables in the system (Brahmasrene et al. 2014).

The mathematical representation of a VAR model is:

$$y_t = c_t + \sum_{i=1}^{k} \alpha_i y_{t-i} + \sum_{i=1}^{k} \beta_i x_{t-i} + u_t \tag{1}$$

where y_t is an endogenous variable; y_{t-i} $(i = 1, 2,..., p)$ is a lagged exogenous variable; x_{t-i} is an endogenous variable; c_t $(c_1, c_2, c_3,..., c_t)^T$ is a constant term; α_i and β_i are matrices of coefficients to be estimated; u_t is a random disturbance term.

3 Empirical analysis

3.1 Stationary tests of international crude price (IC) and local refined crude price (RD)

The vast majority of econometric models require that economic time series should be stationary. Since most economic variables are non-stationary sequences (Gallagher et al. 2015), it is necessary to implement a stationary test before establishing the model. The standard method of checking sequence stationary is the unit root test (Xu and Lin 2016; Blanco et al. 2013). Table 2 provides the results

of unit root tests based on the augmented Dickey–Fuller (ADF) test methods (Dickey and Fuller 1979).

T-statistics values of IC and RD are greater than the critical value of the 10% test level. Therefore, the test results cannot reject the original hypothesis, which means it is a non-stationary sequence. Then the stationary test of all the variables in the first-order difference is conducted. The augmented Dickey–Fuller (ADF) tests indicate that the null hypothesis of a unit root in the first-order difference can be rejected for all the variables at the 1% significance level (Table 2). Thus, it can be concluded that all the variables are stationary in the first-order difference and can be conducted with a cointegration test.

3.2 Cointegration tests of IC and RD

3.2.1 The optimal lag order analysis

A cointegration test is to examine whether there is a long-term stable relationship between variables (Engle and Granger 1987; Bondia et al. 2016; Ouyang and Lin 2015). First of all, we need to establish the VAR model of the two variables and conduct the optimal lag order analysis. We choose a lag of 2 as dictated, and there are five evaluation criteria, namely likelihood ratio test (LR), final prediction error (FPE), Schwartz information criterion (SC), Akaike information criterion (AIC) and Hannan–Quinn information criterion (HQ). Results are given in Table 3.

3.2.2 Johansen cointegration test

In this paper, the Johansen test method (Johansen 1988; Johansen and Juselius 1990; Moore and Copeland 1995) is used to carry out the cointegration test, and results are shown in Table 4. Under the assumption that the trace statistic is equal to 6.7905, the critical value is equal to 15.4947 at the 5% significance level. The trace statistic value is smaller than the 0.05 critical value, so it cannot reject the original hypothesis that there is no cointegration relationship. Thus, "0" in the first column denotes that the null hypothesis of a unit root is rejected at the 5% significance level. In other words, there is no long-term

Table 2 Results of ADF unit root tests

Series	Test form (c, t, n)	T-statistics	Test critical values	Prob.*	Conclusions
IC	(c,t,0)	−1.6733	−3.1731***	0.7505	Unstable
$\triangle$IC	(c,t,0)	−6.0253	−4.1273*	0.0000	Stable
RD	(c,t,0)	−1.7702	−3.1739***	0.7061	Unstable
$\triangle$RD	(c,t,0)	−5.9441	−4.1273*	0.0000	Stable

$\triangle$: First difference of variables

c Constant, t Trend, n Lag order

* The null hypothesis of a unit root is rejected at the 1% significance level

*** The null hypothesis of a unit root is rejected at the 10% significance level

Table 3 Results of lag selection criteria

Lag	Log likelihood	LR	FPE	AIC	SC	HQ
0	−426.4085	NA	26668.4200	15.8670	15.9407	15.8954
1	−326.6574	188.4188	769.0211	12.3206	12.5416	12.4059
2	−305.6141	38.18969*	409.4221*	11.6894*	12.0577*	11.8315*
3	−302.9547	4.629229	431.0586	11.7391	12.2547	11.9380
4	−297.7721	8.639171	413.9538	11.6952	12.3582	11.9509
5	−295.0053	4.405051	435.6073	11.7409	12.5513	12.0535

* The lag order selected by the criterion

Table 4 Results of Johansen cointegration test

Hypothesized: number of cointegration vectors	Eigenvalue	Trace statistic	0.05 Critical value	Prob.*
0	0.1126	6.7905	15.4947	0.6021
At most 1	0.0018	0.0991	3.8415	0.7529

Trace test indicates no cointegration at the 0.05 level

* Mackinnon-Haug-Michelis (1999) *P* values

stable equilibrium relationship between international crude oil prices and refined oil prices in China.

3.3 Granger causality tests of IC and RD

In order to know the price transmission mechanism between the two variables, the Granger causality test method (Granger 1969, 1980) is used to test the causal relationship between variables. Results are shown in Table 5. According to the data, there is significant unidirectional Granger influence of international crude oil prices on China's refined oil prices in the short run, while it is 2nd-order lag and the significance level is 10%.

3.4 Vector autoregressive models of IC and RD

The vector autoregressive (VAR) model (Christopher 1980) generalizes the univariate autoregressive model to the multivariate case. This provides beneficial features such as estimation of the dynamic interrelation between variables and indifference about the choice of dependent

Table 5 Results of Granger causality tests

Null hypothesis	F-statistic	Prob.
RD does not Granger Cause IC	0.3342	0.7174
IC does not Granger Cause RD	56.5717	9.E−14*

variables (Brahmasrene et al. 2014). By applying the optimal model order, Table 6 presents the results of VAR estimates and model diagnostic tests. The results are as follows: *R*-squared *IC* = 0.9399, while *R*-squared *RD* = 0.9717; Adjusted *R*-squared *IC* = 0.9353, while Adjusted *R*-squared *RD* = 0.9695. The equations are fitted well, and the overall effect of the model is good.

$$IC_t = 1.2300IC_{t-1} - 0.1575IC_{t-2} - 0.0535RD_{t-1} - 0.0496RD_{t-2} + 11.2583 \tag{2}$$

$$RD_t = 0.8433IC_{t-1} - 0.6977IC_{t-2} + 0.8318RD_{t-1} - 0.0037RD_{t-2} + 17.7596 \tag{3}$$

From Eqs. (2) and (3), it can be seen that two variables are not only affected by their own lags, but also influence

Table 6 Results of VAR estimates and model diagnostic tests

Parameter estimates for each equation of the model			Test results for each equation of the model		
	IC	RD	R^2	0.9399	0.9717
IC(−1)	1.2300	0.8433	Adj. R^2	0.9353	0.9695
	(0.1425)	(0.0814)	Sum sq. residuals	1916.4000	624.4234
	[8.6297]	[10.3660]	SE equation	6.0707	3.4653
IC(−2)	−0.1575	−0.6977	F-statistic	203.2994	445.6015
	(0.1813)	(0.1035)	Log likelihood	−181.0613	−149.1022
	[−0.8685]	[−6.7410]	Akaike AIC	6.5285	5.4071
RD(−1)	−0.0535	0.8318	Schwarz SC	6.7077	5.5863
	(0.1782)	(0.1017)	Mean dependent	97.3623	186.1442
	[−0.3003]	[8.1787]	SD dependent	23.8619	19.8332
RD(−2)	−0.0496	-0.0037	Test of the overall effect of the model		
	(0.1480)	(0.0845)	Determinant residual covariance (dof adj.)		415.0074
	[−0.3351]	[−0.0435]	Determinant residual covariance		345.3924
C	11.2583	17.7596	Log likelihood		−328.3324
	(14.4974)	(8.2754)	Akaike information criterion		11.8713
	[0.7766]	[2.1461]	Schwarz criterion		12.2297

Standard errors in (); T-statistics in []

another variable's lag terms. Influence directions and degrees during different periods are not the same. Among them, according to Eq. (2) for the international crude oil price vector autoregression that the international crude oil price is positively affected by its own 1 period lag term and the influence degree is 1.2300; the international crude oil price is negatively affected by its own 2 period lag term, and the influence degree is 0.1575. The 1 and 2 period lag terms of domestic refined oil variables have weak and negative impact on the international crude oil price, with respective effects of 0.0535 and 0.0496. Similarly, analytical results of Eq. (3) are obtained. Therefore, the international crude oil price is the primary factor that affects the price of China's refined oil, and the international crude oil price is negatively affected by China's refined oil price. The international crude oil price has a strong historical inheritance. On the other hand, China's refined oil price has weaker historical inheritance. In quantity, if the international crude oil price in the lag last period increased by 1%, the current crude oil price will increase by 1.23% and China's refined oil price will increase by 0.84%. It is suggested that a rise in the international crude oil price will increase the domestic refined oil price level, and the impact of international oil prices on their own is also very significant.

Overall, the impact of the international crude oil price on China's refined oil price is greater than the impact of the latter on the former. The main reason is that the crude oil is a type of product located in the upstream of the refined oil industry chain (Zhang et al. 2015; Liu and Ma 2014), and

the direction of price transmission is mainly transmitted from the international crude oil price to the refined oil price in China. At the same time, the pricing mechanism of China's refined oil is not fully market-oriented and subject to the regulation by a government department. Therefore, there are time lags in terms of price changes.

3.5 Impulse response functions of IC and RD

The impulse response function is used to investigate the dynamic effects on the system when a variable is subjected to a certain impact (Xu and Lin 2016). It can be used to analyze the time profile of effects of shocks on the future behavior of international crude oil prices and China's refined oil prices. Figure 2 presents the impulse response for international crude oil prices and China's refined oil prices from one-standard deviation.

According to the above analysis, international crude oil prices and domestic refined oil prices interact with each other in domestic and international crude oil market environments and in the process regulated by the National Development and Reform Commission. A change of the international crude oil price will affect China's refined oil price, and vice versa. In other words, the change of China's refined oil price may also give feedback to the international crude oil price. Therefore, after the establishment of the VAR model, the first step is to analyze the impulse response of China's refined oil price when the international crude oil price is subjected to a shock, results of which are shown in Fig. 2a. The second step is to analyze the impulse

(a)

Response of RD to nonfactorized
one S.D. IC innovation

(b)

Response of IC to nonfactorized
one S.D. RD innovation

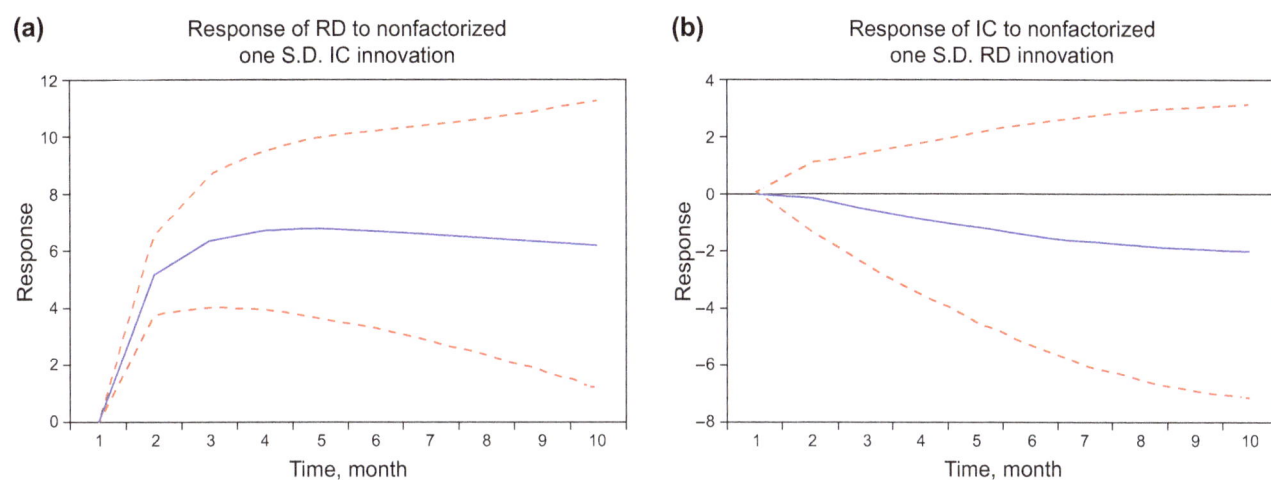

Fig. 2 Impulses response the shock analysis. **a** Response of RD to IC (The *vertical axis* represents the dependent variable: RD). **b** Response of IC to RD (the *vertical axis* represents the dependent variable: IC). The *solid lines* indicate mean responses to a one-standard deviation shock, while the *dotted lines* represent ±2 standard deviations of the responses. The *horizontal axis* represents the number of lag periods of the shock (unit:month).The *vertical axis* represents the response of relevant dependent variables to independent variable impact

response of international crude oil price when China's refined oil price is subject to a shock, results of which are shown in Fig. 2b.

Figure 2a shows the response of China's refined oil price to an international crude oil price shock. An international crude oil price shock has a significantly positive impact on China's refined oil price. The graph shows that the response of China's refined oil price to a shock in the international crude oil price starts to increase in the first month, and it reaches the maximum value of the impulse response in the fourth month, 7.11%, and then decreases gradually. Figure 2b displays the impulse response of the international crude oil price variable to China's refined oil price shock. The shock has a negative impact on China's refined oil price from the second month and then increases gradually in the negative direction.

It can be concluded through comparison of Fig. 2a and b that the impact of international crude oil price volatility on China's refined oil price is greater than the impact of China's refined oil price volatility on the international crude oil price.

3.6 Variance decomposition of IC and RD

Variance decomposition refers to the decomposition of a mean square error of a quantitative shock into the contribution of each variable in the system. Then their contributions are, respectively, calculated to further compute the proportion of their contributions in the total contribution to reflect the relative importance of each variable (Sørensen and Yosha 1998; Hoang and Hoxha 2015).

Table 7 presents results of variance decomposition. A period of ten months is selected to analyze the variance decomposition. Under Columns (3) and (4) in the first month, 100% of the variability in international crude oil price is explained by its own disruptions. In the subsequent prediction, the variance of the international crude oil price is gradually reduced by its own disruptions, and the part of the domestic refined oil price disruptions gradually increases, but the range is small. This shows that the forecast variance change of international crude oil prices is mainly caused by its own disruptions and the effect is lasting. During the ten months, the influence is generally great, with an average of 98.6%. However, the domestic refined oil price disruption also has a certain impact and the average value is 1.37%. Under Columns (7) and (8) in the first month, 93.8% of the variability in China's refined oil prices can be explained by its own disruptions and the part of the international crude oil price disruptions is 6.22%.

The variance decomposition of China's refined oil prices has a great change in the second month, of which 64.6% is attributed to international crude oil price changes, and 35.3% is attributed to its own disruptions. In the subsequent prediction, the variance of China's refined oil prices is greatly decreased by its own disruptions, while the part of the domestic refined oil price disruptions is gradually enhanced. The results illustrate that the price of refined oil in China is influenced greatly by the change of international crude oil prices. Results of the first and second months show that China's refined oil prices do not change with fluctuations of international crude oil prices, which indicates that the impact of international crude oil price on China's refined oil price is not timely. In this period, the international crude oil price has greater influence with an average of 77.8%, while the domestic refined oil price has a certain influence with the average of 22.2%.

Table 7 Results of variance decomposition

Decompositions of IC				Decompositions of RD			
(1) Month	(2) SE	(3) IC	(4) RD	(5) Month	(6) SE	(7) IC	(8) RD
1	6.0707	100.0000	0.0000	1	6.0707	6.2228	93.7772
2	9.5892	99.9650	0.0350	2	9.5892	64.6462	35.3538
3	12.3826	99.7912	0.2088	3	12.3826	77.5959	22.4041
4	14.6546	99.4858	0.5142	4	14.6546	83.7041	16.2959
5	16.5629	99.0815	0.9185	5	16.5629	87.2647	12.7353
6	18.2085	98.6140	1.3860	6	18.2085	89.5454	10.4546
7	19.6571	98.1129	1.8871	7	19.6571	91.0556	8.9444
8	20.9520	97.6006	2.3994	8	20.9520	92.0556	7.9444
9	22.1230	97.0926	2.9074	9	22.1230	92.7013	7.2987
10	23.1913	96.5996	3.4004	10	23.1913	93.0957	6.9043

4 Conclusions and policy implications

This paper aims to investigate the transmission and feedback mechanism between international crude oil prices and China's refined oil prices for the time span from January 2011 to November 2015 by using the Granger causality test, a vector autoregression model, an impulse response function and variance decomposition methods.

It is demonstrated that variation of international crude oil prices causes changes of domestic refined oil prices with a weak feedback effect. Secondly, international crude oil prices and domestic refined oil prices are affected by their lag terms from positive and negative directions in different degrees. Thirdly, an international crude oil price shock has a significant positive impact on domestic refined oil prices while the impulse response of the international crude oil price variable to the domestic refined oil price shock is negatively insignificant. Furthermore, international crude oil prices and domestic refined oil prices have strong historical inheritance. In addition, the international crude oil price is significantly affected by its own distributions. However, a domestic refined oil price shock has a slight impact on international crude oil price changes. The domestic refined oil price variance is mainly caused by international crude oil price distributions and slightly affected by its own distributions. When international crude oil prices change, domestic refined oil prices do not respond in a timely manner to the international crude oil price change, and there exists a time lag.

This study not only effectively complements existing research but also provides important suggestions for policy makers in China. Firstly, the ultimate goal of the refined oil price reform is to establish a market-oriented pricing mechanism in China. For instance, on January 3, 2016, the National Development and Reform Commission issued a report on the further improvement of the oil price formation mechanism to improve the refined oil pricing mechanism. The policy has a certain effect on the stability of national oil prices in the short term. However, for the long-term analysis,

fundamentally it cannot solve the pricing problem of the refined oil market in China. Therefore, oil price reform must stick to the principle of integrating the international market price direction. Meanwhile, the establishment of an oil price formation mechanism not only reflects the change of oil prices in the international market but also takes many factors into consideration, such as the domestic market supply and demand, production costs and other social elements. Secondly, it is significant to promote the common development of the oil spot and future market actively. The future market is an open, centralized and unified market, and the oil future price can reflect the market supply and demand to the maximum extent. The development of oil future trading is significant for China, and thus participation in the pricing process of the international oil price is crucial in striving for oil pricing. Chinese enterprises can effectively avoid the risk of oil price fluctuation through oil future trading. Thirdly, the government should improve the system of petroleum reserves in China in the low-price environment, which would be helpful to solve the problem of domestic and international oil price inversion phenomenon and ensure national energy security.

Acknowledgements The authors gratefully acknowledge support from the Key Project of National Social Science Foundation of China (NO. 13&ZD159). We are also grateful to colleagues from the research laboratory for helpful suggestions that improve this paper.

References

Asche F, Gjølberg O, Völker T. Price relationships in the petroleum market: an analysis of crude oil and refined product prices. Energy Econ. 2003;25(3):289–301. doi:10.1016/S0140-9883(02) 00110-X.

Blanco JM, Vazquez L, Peña F, et al. New investigation on diagnosing steam production systems from multivariate time series applied to thermal power plants. Appl Energy. 2013;101: 589–99. doi:10.1016/j.apenergy.2012.06.060.

Bondia R, Ghosh S, Kanjilal K. International crude oil prices and the stock prices of clean energy and technology companies: Evidence from non-linear cointegration tests with unknown structural breaks. Energy. 2016;101:558–65. doi:10.1016/j.energy.2016.02.031.

Brahmasrene T, Huang JC, Sissoko Y. Crude oil prices and exchange rates: causality, variance decomposition and impulse response. Energy Econ. 2014;44:407–12. doi:10.1016/j.eneco.2014.05.011.

Broadstock DC, Cao H, Zhang DY. Oil shocks and their impact on energy related stocks in China. Energy Econ. 2012;34(6): 1888–95. doi:10.1016/j.eneco.2012.08.008.

Christopher A. Macroeconomics and reality. Econometrica. 1980;48(1):1–48. doi:10.2307/1912017.

Dickey DA, Fuller WA. Distribution of the estimators for autoregressive time series with a unit root. J Am Stat Assoc. 1979;74(366):427–31. doi:10.1080/01621459.1979.10482531.

Du LM, He YN, Chu W. The relationship between oil price shocks and China's macro-economy: an empirical analysis. Energy Policy. 2010;38(8):4142–51. doi:10.1016/j.enpol.2010.03.042.

Engle RF, Granger CWJ. Co-integration and error correction: representation, estimation, and testing. Econom Soc. 1987;55(2): 251–76. doi:10.2307/1913236.

Gallagher CM, Fisher TJ, Shen J. A Cauchy estimator test for autocorrelation. J Stat Comput Simul. 2015;85(6):1264–76. doi:10.1080/00949655.2013.874424.

Granger CWJ. Investigating causal relations by econometric models and cross- spectral methods. Econometrica. 1969;37(3):424–38. doi:10.2307/1912791.

Granger CWJ. Testing for causality: a personal viewpoint. J Econ Dyn Control. 1980;2:329–52. doi:10.1016/0165-1889(80)90069-X.

Hoang EC, Hoxha I. Corporate payout smoothing: a variance decomposition approach. J Empir Finance. 2015;35:1–13. doi:10.1016/j.jempfin.2015.10.011.

Jiang CH (2013) An empirical study on the transmission mechanism of the price of crude oil and refined oil retail price in China based on the VAR model during the period of 2003–2011. Macroecon Res. 2013;(4):28–38. doi:10.16304/j.cnki.11-3952/f.2013.04. 001. (in Chinese).

Jiang ZF, Jiang H. China's oil security strategy under the shadow of high oil prices. Modern Manag Sci. 2005;(8):69–70 (in Chinese). doi:10.3969/j.issn.1007-368X.2005.03.030.

Jiao JL, Fan Y, Wei YM. VECM based analysis of gasoline/diesel price anti-symmetry. Manag Sci China. 2006;14(3):97–102. doi:10.16381/j.cnki.issn1003-207x.2006.03.018 (in Chinese).

Jiao JL, Fan Y, Zhang JT, et al. Study of the interactive relationship between Chinese crude oil price and international crude oil price. Manag Forum. 2004;16(7):48–54. doi:10.14120/j.cnki.cn11-5057/f.2004.07.009 (in Chinese).

Johansen S, Juselius K. Maximum likelihood estimation and inference on cointegration-with applications to the demand for money. Oxford Bull Econ Stat. 1990;52(2):169–210. doi:10.1111/j.1468-0084.1990.mp52002003.x.

Johansen S. Statistical analysis of cointegration vectors. J Econ Dyn Control. 1988;12:231–54. doi:10.1016/0165-1889(88)90041-3.

Li ZG, Guo JG. Asymmetry between gasoline and crude oil prices in China based on asymmetric ECM modeling. Resour Sci. 2013;35(1):66–73 (in Chinese).

Liao SJ, Wang FX, Wu T, et al. Crude oil price decision under considering emergency and release of strategic petroleum reserves. Energy. 2016;102:436–43. doi:10.1016/j.energy.2016. 02.043.

Liu L, Ma GF. Cross-correlation between crude oil and refined product prices. Phys A Stat Mech Appl. 2014;413(1):284–93. doi:10.1016/j.physa.2014.07.007.

Mackinnon JG, Haug AA, Michelis L. Numerical distribution functions of likelihood ratio tests for cointegration. J Appl Econ. 1999;14(5):563–577. doi:10.1002/(sici)1099-1255(199909/10) 14:5<563::aid-jae530>3.3.co;2-i.

Moore MJ, Copeland LS. A comparison of Johansen and Phillips-Hansen cointegration tests of forward market efficiency Baillie and Bollerslev revisited. Econ Lett. 1995;47(2):131–5. doi:10. 1016/0165-1765(94)00547-F.

Ouyang XL, Lin BQ. An analysis of the driving forces of energy-related carbon dioxide emissions in China's industrial sector. Renew Sustain Energy Rev. 2015;45:838–49. doi:10.1016/j.rser. 2015.02.030.

Sørensen BE, Yosha O. International risk sharing and European monetary unification. J Int Econ. 1998;45(2):211–38. doi:10. 1016/S0022-1996(98)00033-6.

Timilsina GR. Oil prices and the global economy: a general equilibrium analysis. Energy Econ. 2015;49:669–75. doi:10. 1016/j.eneco.2015.03.005.

Wang X. Statistical test analysis on the international crude oil prices and domestic refined oil prices to study the interaction relation. Finance Econ. 2014;(4):162–4. doi:10.13546/j.cnki.tjyjc.000153. (in Chinese).

Xu B, Lin BQ. Assessing CO_2 emissions in China's iron and steel industry: a dynamic vector autoregression model. Appl Energy. 2016;161:375–86. doi:10.1016/j.apenergy.2015.10.039.

Zhang J, Xie MJ. China's oil product pricing mechanism: What role does it play in China's macroeconomy? China Econ Rev. 2016;38:209–21. doi:10.1016/j.chieco.2016.02.002.

Zhang T, Ma GF, Liu GS. Nonlinear joint dynamics between prices of crude oil and refined products. Phys A Stat Mech Appl. 2015;419:444–56. doi:10.1016/j.physa.2014.10.061.

3

Thermo-sensitive polymer nanospheres as a smart plugging agent for shale gas drilling operations

Wei-Ji Wang[1] · Zheng-Song Qiu[1] · Han-Yi Zhong[1] · Wei-An Huang[1] · Wen-Hao Dai[1]

Abstract Emulsifier-free poly(methyl methacrylate–styrene) [P(MMA–St)] nanospheres with an average particle size of 100 nm were synthesized in an isopropyl alcohol–water medium by a solvothermal method. Then, through radical graft copolymerization of thermo-sensitive monomer N-isopropylacrylamide (NIPAm) and hydrophilic monomer acrylic acid (AA) onto the surface of P(MMA–St) nanospheres at 80 °C, a series of thermo-sensitive polymer nanospheres, named SD-SEAL with different lower critical solution temperatures (LCST), were prepared by adjusting the mole ratio of NIPAm to AA. The products were characterized by Fourier transform infrared spectroscopy, transmission electron microscopy, thermogravimetric analysis, particle size distribution, and specific surface area analysis. The temperature-sensitive behavior was studied by light transmittance tests, while the sealing performance was investigated by pressure transmission tests with Lungmachi Formation shales. The experimental results showed that the synthesized nanoparticles are sensitive to temperature and had apparent LCST values which increased with an increase in hydrophilic monomer AA. When the temperature was higher than its LCST value, SD-SEAL played a dual role of physical plugging and chemical inhibition, slowed down pressure transmission, and reduced shale permeability remarkably. The plugged layer of shale was changed to being hydrophobic, which greatly improved the shale stability

Keywords Nanoparticle plugging agent · Polymer microspheres · Thermo-sensitive polymer · Wellbore stability · Shale gas · Drilling fluid

1 Introduction

At present, shale gas exploration and production has attracted much attention. Given its accumulation characteristics, extended-reach horizontal wells and cluster horizontal wells were drilled to produce shale gas. Because of the existence of micro-fissures and strong water sensitivity in shale formations, severe wellbore instability often occurs in the long horizontal sections, which seriously restricts the process of shale gas exploration and development (Cui et al. 2011; Dong et al. 2012; Wang et al. 2013). Shale formation is mainly composed of hard brittle shale, mainly of illite and mixed layer illite/smectite. For hard brittle shales, pore pressure transmission is the primary cause of wellbore instability. Therefore, the key to maintaining wellbore stability is to prevent pore pressure transmission. The effective sealing of micropores and micro-fissures is of great significance for preventing pore pressure transmission. Traditional plugging agents are difficult to form effective mud cake to prevent liquid penetration into shale matrix which has extremely low permeability and tiny pore throats. In recent years, nanoparticles are found to effectively plug shale pore throats to prevent liquid penetration into the formation, thus maintaining wellbore stability and protecting the reservoir (Roshan and Aghighi 2012; Rafieepour et al. 2013; Wen et al. 2014). According to previous experimental results, silica nanoparticles could significantly improve the densification of mud cakes, slow down pressure transmission and reduce shale permeability, while the rheology and lubrication of water-based drilling

✉ Zheng-Song Qiu
   zsqiu63@sina.com

1   School of Petroleum Engineering, China University of Petroleum, Qingdao 266580, Shandong, China

Edited by Yan-Hua Sun

fluid were improved (Cai et al. 2012; Hoelscher et al. 2012; Al-Baghli et al. 2015). Bai and Pu (2010) synthesized PMMA latex nanoparticles with an average size of 73 nm, which can be used as a lubricant in drilling fluids based on their "ball bearing" function to prevent pipe sticking. They can also be used as a filtration reducer based on its deformability under temperature and pressure, forming a tough filter cake and sealing the micro-fissures in the formations drilled. Qu et al. (2007) synthesized intercalated or exfoliated nanocomposite poly(styrene-b-acrylamide)/bentonite using reversible addition-fragmented chain transfer (RAFT) polymerization. Experiments showed that these products had high-temperature tolerance and were good filtration control agents. In the last 20 years, the investigation into nanomaterials has greatly developed in many fields. Great progress has been made in basic theory and application of nanooptical materials, nanosemiconductor materials, nanobiomedical materials, nanoenhanced materials, nanomodified surface, etc. (Lin et al. 2012; Cormick and Hunter 2014; Kearnes et al. 2014). The combination of smart polymers with environmental response behavior (temperature, pH value, electrolyte concentration, magnetic field strength, electric field strength, etc.) and nanoparticles to realize the potential of nanoparticles is the most common research (Wu et al. 2013; Gulfam and Chung 2014; Lian et al. 2015). In this study, nanomaterials technology, smart polymers, and drilling fluid technology were combined. Emulsifier-free poly(methyl methacrylate–styrene) [P(MMA–St)] nanospheres with an average particle size of about 100 nm were synthesized in an isopropyl alcohol–water medium by the solvothermal method. Then, the thermo-sensitive smart polymer P(NIPAm–AA) was modified onto the surface of P(MMA–St) nanospheres and thermo-sensitive smart nanoparticles were obtained. With the change in temperature, the hydrophilicity and hydrophobicity of nanoparticle surface would change accordingly. Moreover, we would adjust the transformation temperature of NIPAm by an introduction of hydrophilic monomer or hydrophobic monomer, getting smart nanoparticles with different transformation temperatures to adapt to shale formations with different temperatures.

2 Experimental

2.1 Materials

Methyl methacrylate (MMA), styrene (St), and N-isopropylacrylamide (NIPAm) were purchased from the Aladdin Industrial Corporation and used after vacuum distillation. Acroleic acid (AA), potassium persulfate (KPS, $K_2S_2O_8$), and tetrahydrofuran (THF) were purchased

from the Sinopharm Chemical Reagent Co. Ltd and used without further purification.

2.2 Preparation of thermo-sensitive poly(methyl methacrylate–styrene) nanoparticles

2.2.1 Preparation of poly(methyl methacrylate–styrene) latex nanoparticles

The latex particles prepared by emulsifier-free emulsion polymerization have good adhesion and good resistance to water. These latex particles are evenly distributed in a narrow size range with clear surfaces and relatively large particle sizes. Emulsifier-free emulsion polymerization was carried out by using a solvothermal method. A cosolvent-water mixture was used as the dispersed medium, and monomer polymerization was initiated in a closed system. The solvothermal method can improve the reaction temperature and pressure at the same time, so that the size of particles prepared in the medium decreased significantly, and the stability of the emulsion was improved (Hoa and Huyen 2013; Farooq et al. 2013; Mishra et al. 2014).

A total of 9.44 mmol MMA, 8.65 mmol St, and 0.3 mmol $K_2S_2O_8$ were dissolved in a 38-mL isopropanol-water mixture, and the pH value adjusted to 7 by adding 1 mol/L NaOH solution. Then, the mixture was loaded into a PTFE lined hydrothermal synthesis reactor. After that, this mixed solution was stirred vigorously for 30 min and heated up to 90 °C. After heating for 1.5 h at 90 °C, the reaction mixture was diluted in 100 mL benzene. The mixture was precipitated and washed with methyl alcohol to remove the residual monomers and homopolymers. After drying for 8 h at 90 °C in a vacuum drying oven and grinding in a ball mill, poly(methyl methacrylate–styrene) [P(MMA-St)] nanoparticles were obtained.

2.2.2 Synthesis of thermo-sensitive polymer nanoparticles

A mixture of NIPAm and AA with a given mole ratio (no AA, 90/10, 80/20, 70/30, 74/26, 66/34, 52/48) was dissolved in an H_2O/THF mixed solvent (the volume ratio of H_2O and THF was 2:1), and then P(MMA–St) was added. The mixture was ultrasonically dispersed for 30 min. 0.2 mmol $K_2S_2O_8$ was added dropwise, and the mixture was heated up to 80 °C and deoxygenated with N_2 for 9 h. The synthetic route of the thermo-sensitive poly(methyl methacrylate–styrene) nanoparticles is shown in Fig. 1. The obtained product was centrifuged at 10,000 rpm for 30 min and washed with absolute ethyl alcohol to remove residual monomers. After centrifugation, the precipitates were collected, dried for 8 h at 90 °C in a vacuum drying

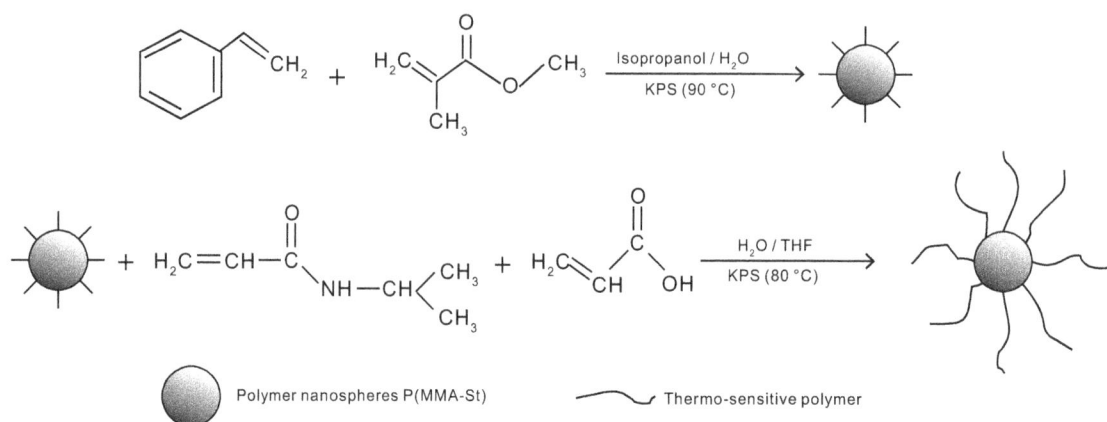

Fig. 1 Synthetic route of SD-SEAL nanoparticles

oven and ground in a ball mill for characterization. The product was abbreviated as SD-SEAL.

2.3 Structural characterization

The molecular structure of SD-SEAL was characterized by infrared spectroscopy which was recorded with a Nicolet 6700 FT-IR spectrometer (NEXUS, USA), scanning from 400 to 4000 cm^{-1} with a resolution of 4 cm^{-1} in transmission using KBr pellets. The KBr pellets were prepared by pressing mixtures of 1 mg of SD-SEAL powder and 100 mg of KBr. Transmission electron microscopy (TEM) measurements of SD-SEAL were acquired with a JEM-2100UHR electron microscope (JEOL, Japan). SD-SEAL solution with a concentration of 0.1 g/mL was dropped onto carbon-coated copper grids and dried in air. The microscopic morphology of shale was observed with an S-4800 field emission scanning electron microscope (Hitachi, Japan). The thermogravimetric analysis (TGA) of the SD-SEAL was performed on an SDT Q600 instrument (TA Instrument, USA). The sample was heated at a rate of 20 °C/min in nitrogen flow of 50 mL/min.

2.4 Performance characterization

2.4.1 Temperature-sensitive behavior

There are a lot of methods to measure the temperature sensitivity of smart polymers. The most simple and commonly used method is to determine the light transmittances of a polymer solution at different temperatures (Feng et al. 2005; Kokufuta et al. 2012; Rwei and Nguyen 2014). When the temperature is lower than its lower critical solution temperature (LCST) value, the smart polymer is strongly hydrophilic. Its water solution is almost transparent, and the light transmittance is high. However, when the temperature is higher than its LCST value, the

hydrophilicity of the smart polymer will be changed into hydrophobicity. At this time, micro-phase separation and turbidity will occur, so the light transmittance is almost zero. The curve of light transmittance as a function of temperature can be obtained after testing the light transmittances of the polymer solution at different temperatures. The temperature value corresponding to the inflection point of the curve is the LCST value of the polymer. That is the temperature corresponding to the light transmittance obviously declining. The temperature-sensitive behavior of polymer nanospheres (used as smart plugging agents) with different LCST values were acquired with an UV–Vis spectrophotometer (UV-1750, SHIMADZU International Trading Co., Ltd.).

2.4.2 Sealing performance evaluation

The pore pressure transmission test was used to measure the sealing performance of SD-SEAL using the simulation equipment for hydro-mechanics coupling of shale shown in Fig. 2 (van Oort 1994, 1997; Xu et al. 2005; Yuan et al. 2012). During pore pressure transmission tests, shale cores were installed in a core holder, and test fluids were pumped into the core holder from its upstream inlet to interact with the core. The confining pressure and the axial pressure were maintained at 5 MPa, the upstream pressure was maintained at 2.1 MPa, and the initial downstream pressure was 1.0 MPa. The pore pressure was determined by measuring the variation of the downstream pressure. Permeability of shale cores was calculated as follows (Xu et al. 2005):

$$K = \frac{\mu \beta V L}{A} \frac{\ln\left(\frac{P_m - P_o}{P_m - P(L, t_2)}\right) - \ln\left(\frac{P_m - P_o}{P_m - P(L, t_1)}\right)}{t_2 - t_1} \quad (1)$$

where K is the permeability of the shale core, μm^2; μ is the viscosity of fluids, mPa s; β is the static compression ratio

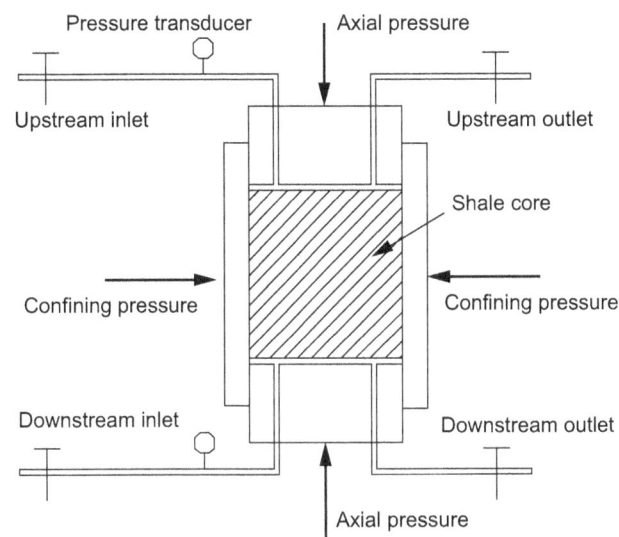

Fig. 2 Schematic of pressure penetration test apparatus (Xu et al. 2005)

of fluids, MPa^{-1}; V is the enclosed volume of downstream fluids, cm^3; L is the length of the shale core, cm; A is the cross-sectional area, cm^2; t is the total experimental time, s; P_m is the upstream pressure, MPa; P_o is the pore pressure, MPa; and $P(L, t)$ is the real-time downstream pressure, MPa.

2.4.3 Characterization of the core sealing surface

The microscopic morphology of the core sealing surface was observed with an S-4800 field emission scanning electron microscope (Hitachi, Japan). The wettability of the core sealing surface was measured with a JC2000D5M contact angle meter (Shanghai Zhongchen Digital Technic Apparatus Co., Ltd, China).

3 Results and discussion

3.1 Structural characterization of SD-SEAL

3.1.1 FT-IR

FT-IR spectra of P(MMA–St) and SD-SEAL were shown in Fig. 3. For P(MMA–St), the absorption bands at 3093, 3065, 3020, and 2996 cm^{-1} were characteristic absorption peaks of C–H bond from monosubstituted benzene rings. The strongest absorption band at 1730 cm^{-1} was due to carbonyl stretching vibration. The absorption bands at 1236 and 1142 cm^{-1} could correspond to the symmetric stretching vibration of C–O–C bond. The absorption bands at 754 and 700 cm^{-1} were the characteristic bending vibrations of C–H from monosubstituted benzene rings.

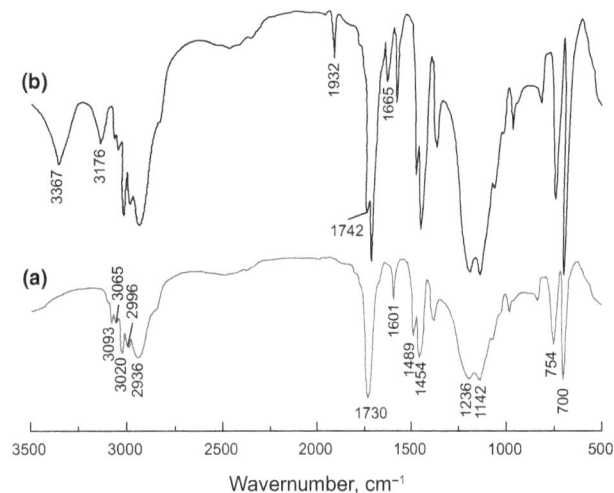

Fig. 3 IR spectra of P(MMA–St) (**a**) and SD-SEAL (**b**)

No absorption peak due to stretching vibration of C=C bond (1640 cm^{-1}) was observed in Fig. 3a, meanwhile P(St) and P(MMA) homopolymers were separated before testing. Therefore, the above discussion confirmed that the newly synthesized latex particles were copolymers of St and MMA.

For SD-SEAL, except for characteristic peaks of P(MMA–St) copolymers, the absorption bands at 3367 and 3176 cm^{-1} were attributed to the stretching vibration of N–H bonds. The absorption bands at 1742 and 1655 cm^{-1} were characteristic absorption peaks of amide I (C–O bond) and amide II (N–H bond). An absorption band at 1932 cm^{-1} was the association absorption peak of –COOH. Since the synthesized products had been extracted with acetone, the homopolymers of AA and NIPAm were separated from the products, and the characteristic absorption peaks of NIPAm, AA and P(MMA–St) were obviously observed in the FT-IR spectra. So there were chemical bonds between P(MMA–St) particles and P(NIPAm–AA) polymers rather than a simple physical mixture. Namely, under certain reaction conditions copolymerization took place between P(MMA–St) latex particles and P(NIPAm–AA) polymers.

3.1.2 TEM

TEM tests on P(MMA–St) and SD-SEAL were conducted as follows: A small amount of P(MMA–St) or SD-SEAL was put into a dialysis bag for 24 h to remove the electrolyte ions in the products which allows only water molecules, ions and small molecules to pass through. The P(MMA–St) or SD-SEAL solution with a concentration of 0.1 g/mL was dropped onto carbon-coated copper grids and dried in air. TEM images (Fig. 4) of P(MMA–St) and SD-SEAL were acquired with a JEM-2100UHR electron

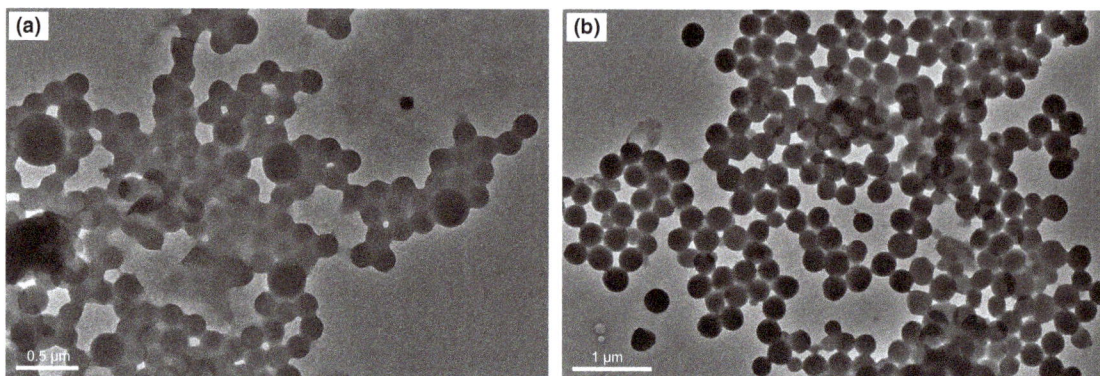

Fig. 4 TEM images of P(MMA–St) (**a**) and SD-SEAL (**b**)

microscope (JEOL, Japan). P(MMA–St) is a hydrophobic polymer, which is poorly dispersed in the aqueous solution, with irregular shapes and uneven particle sizes, and may form sticky agglomerates. SD-SEAL was well dispersed in the aqueous solution with regular shapes (mainly spherical) and uniform particle sizes (about 250 nm). Black spheres were observed in the center of the SD-DEAL particles, and the particle surfaces were covered with a thick gray polymer shell, indicating that thermo-sensitive polymer chains were successfully coated on the surfaces of P(MMA–St) nanospheres and products with core–shell structure were obtained.

3.1.3 Particle size distribution

The particle size distribution and specific surface area of 0.001 wt% SD-SEAL solution were measured. As can be seen from Fig. 5, SD-SEAL had a narrow particle size distribution, mainly 90−360 nm, and had a D_{50} value of 252 nm, a D_{10} value of 179 nm, and a D_{90} value of 312 nm. Also SD-SEAL had a very large specific surface area, reaching 25,450 m^2/kg, leading to a strong adsorption

capacity. The particle size distribution of SD-SEAL was consistent with its TEM characterization results, further verifying that the synthesized products were successful. After flowing into pores and micro-cracks of shales, the coarse particles were prone to bridge and seal the larger openings of shales, and the finer particles were prone to fill the gaps between coarse particles. Finally, a dense sealing layer with low permeability was formed.

3.1.4 TGA

The thermal decomposition of SD-SEAL was investigated by TGA (Fig. 6). The mass loss curve indicated two major stages. The first stage of mass loss occurred at around 200 °C corresponding to the evaporation of a small amount of adsorbed water and solvent (Mao et al. 2015; Zhong et al. 2015), while the second stage was the decomposition of SD-SEAL structures at around 380 °C, indicating that the newly synthesized products were highly temperature resistant, which was attributed to the presence of benzene in the SD-SEAL (Luo et al. 2016; Hu et al. 2016).

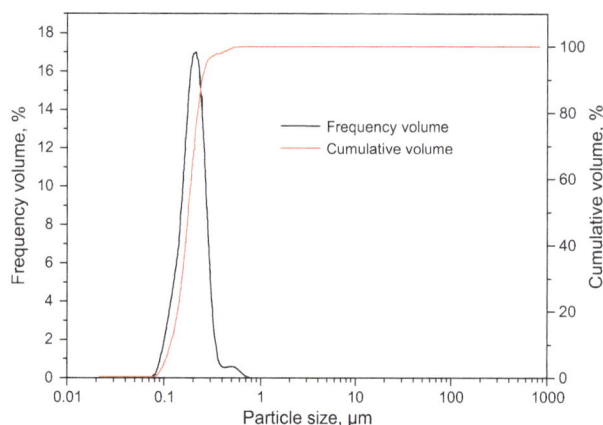

Fig. 5 Particle size distribution of SD-SEAL

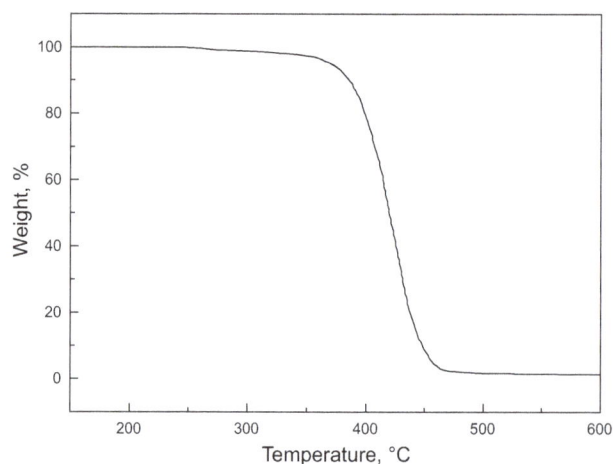

Fig. 6 TG curve of SD-SEAL

3.2 Temperature-sensitive behavior

The temperature-sensitive behavior of SD-SEAL was investigated by measuring the light transmittance of the SD-SEAL solution at different temperatures (Fig. 7). The experimental results showed that the light transmittance of the SD-SEAL solution dropped sharply when the temperature reached its LCST value, so SD-SEAL is temperature-sensitive. The LCST values of SD-SEAL increased with an increase in hydrophilic monomer AA. The LCST values were 53, 63, 81, 93, 106, 125, and 158 °C when the mole ratio of NIPAm to AA was no AA, 90/10, 80/20, 70/30, 74/26, 66/34, and 52/48.

The main driving force of the phase transition of SD-SEAL in aqueous solution was the hydrogen bond effect and hydrophobic effect (Huynh and Lee 2012; Chen et al. 2013; Xu et al. 2013). When temperature was lower than its LCST value, SD-SEAL had a high solubility in water due to its polar groups ($-CONH-$ of NIPAm and $-COOH$ of AA) on molecular chains. These polar groups interacted with surrounding water molecules to form strong hydrogen bonds. Because of the effect of hydrogen bonds and van der Waals force, the water molecules around the macromolecular chains would form solvation shells with high ordering degrees, which were connected by hydrogen bonds. So SD-SEAL can dissolve in water, and then its molecular chains can stretch in water, showing hydrophilic properties. When the temperature was above its LCST value, the hydrogen bonds formed between polar groups and water molecules were destroyed. Also the solvation shells of the hydrophobic parts of the molecular chain were destroyed, leading to an entropy increase in the dispersion system. The hydrophobic association of nonpolar isopropyl groups was dominant, showing hydrophobic properties of the whole molecule. Water molecules were expelled from solvation shells, causing phase separation. Therefore, with

an increase in temperature, the regularity of hydrogen bonds was destroyed and the molecules were changed from hydrophilic to hydrophobic.

3.3 Sealing performance

3.3.1 SEM observation of shale samples

The sealing performance of SD-SEAL was evaluated by the pressure transmission tests of shale samples collected from the Lungmachi Formation, Sichuan Basin. The microstructural characteristics of shale samples were observed by SEM (scanning electron microscopy). As shown in Fig. 8, the shale matrix is developed with parallel bedding planes, which are mainly formed by the dark organic layer and the organic-lean silicon layer. These bedding planes generally have a short horizontal extension and an intermittent development. Besides, their thicknesses range approximately from 200 to 500 μm. The SEM investigation results show that some micro-cracks and bedding planes are filled with the organic matter. The well-developed micro-cracks, which have widths about 0.5–3 μm, are significantly extended, bent, and partly show reticular distributions. The cracks are mainly distributed in the interior and edges of the rich organic matter layers, generally being parallel and vertical to these layers. Nanoscale pores (pore diameter, 200-800 nm) with poor connectivity are extensively observed in the shale matrix and organic matter.

3.3.2 Pore pressure transmission tests

In pore pressure transmission tests, the downstream fluid was 4 wt% sodium chloride (NaCl) solution, and the upstream fluids were 4 wt% NaCl solution, or a mixture of 4 wt% NaCl solution and 2 wt% SD-SEAL. The sealing performance of SD-SEAL (ratio of NIPAm to AA, 66/34; LCST, 125 °C) was tested at room temperature and temperatures above their LCST values (Fig. 9). The permeability of shale cores (Table 1) was calculated using Eq. (1).

As can be seen in Fig. 9 and Table 1, the pressure transmission rate of brine increased very quickly and reached a steady state after testing for 6.5 min. At room temperature, SD-SEAL slowed down the pressure transmission rate and then reduced the shale permeability remarkably. Under the action of pressure, nanoparticles were pressed into micropores and micro-fractures in the shale surface, forming a physical sealing layer. The shale permeability was reduced from 3.34×10^{-7} μm^2 to 0.268×10^{-7} μm^2. When the temperature was above the LCST value of SD-SEAL, the downstream pressure change was small. 4 h later, the pressure transmission curve was

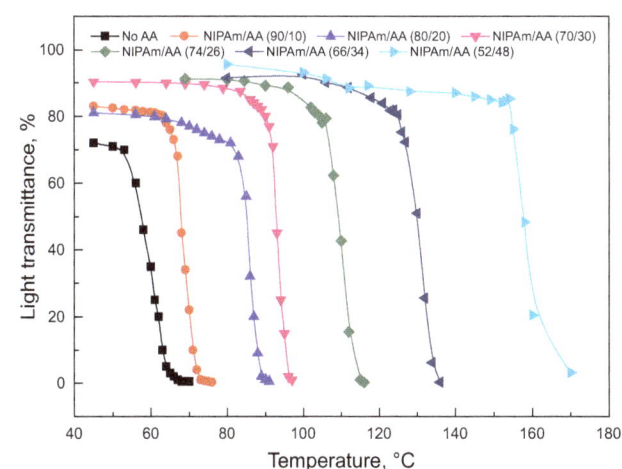

Fig. 7 Transmittance of SD-SEAL as a function of temperature

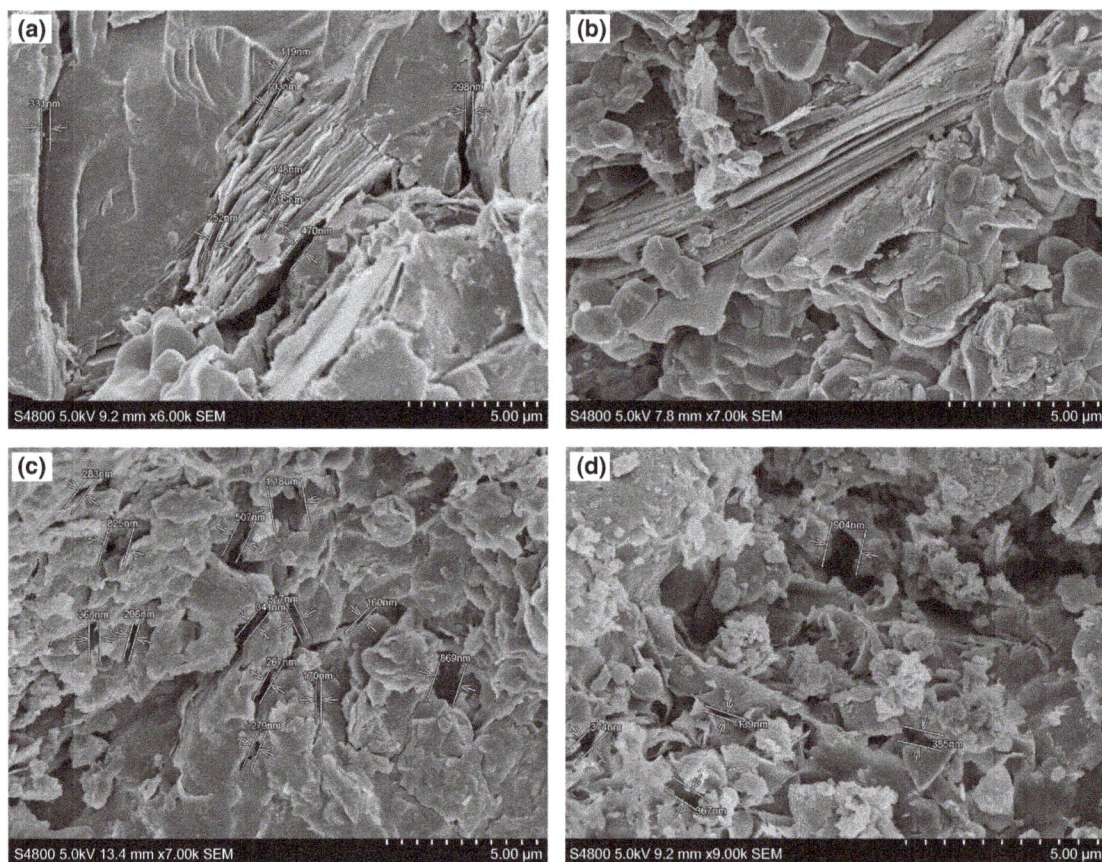

Fig. 8 SEM images of shale samples from the Lungmachi Formation. **a** Micro-fissures. **b** Lamellae development. **c** Micro-fractures and micropores. **d** Micropores

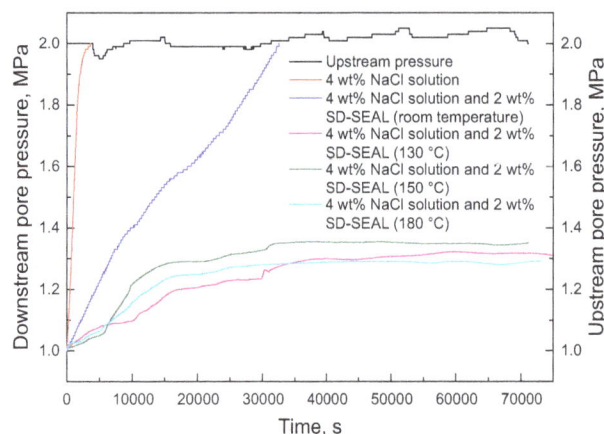

Fig. 9 Pore pressure transmission tests of shale samples

close to a horizontal line. The effect of SD-SEAL slowing down pressure transmission and reducing shale permeability was much better. At this moment, the surface of SD-SEAL (thermo-sensitive polymers) changed from hydrophilic to hydrophobic. A hydrophobic layer was formed on the shale surface with an effect of water resistance. The shale permeability was reduced from 3.34×10^{-7} μm^2 to less than 0.072×10^{-7} μm^2, indicating that the SD-SEAL had a good temperature resistance. Therefore, SD-SEAL played a dual role of physical plugging and chemical inhibition when the temperature was higher than its LCST value, which greatly improved the shale stability (Fig. 10).

The microstructural characteristics of the plugged layer of shale were observed with a scanning electron

Table 1 Permeability of shale cores

Test conditions	Shale permeability, 10^{-7} μm^2
4 wt% NaCl solution	3.340
4 wt% NaCl solution and 2 wt% SD-SEAL (room temperature)	0.268
4 wt% NaCl solution and 2 wt% SD-SEAL (130 °C)	0.061
4 wt% NaCl solution and 2 wt% SD-SEAL (150 °C)	0.072
4 wt% NaCl solution and 2 wt% SD-SEAL (180 °C)	0.048

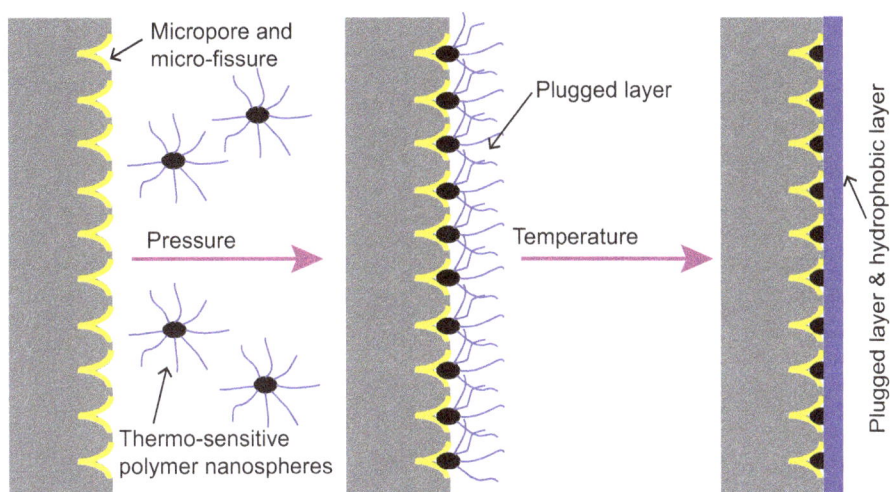

Fig. 10 Schematic diagram of physical plugging and chemical inhibition of SD-SEAL

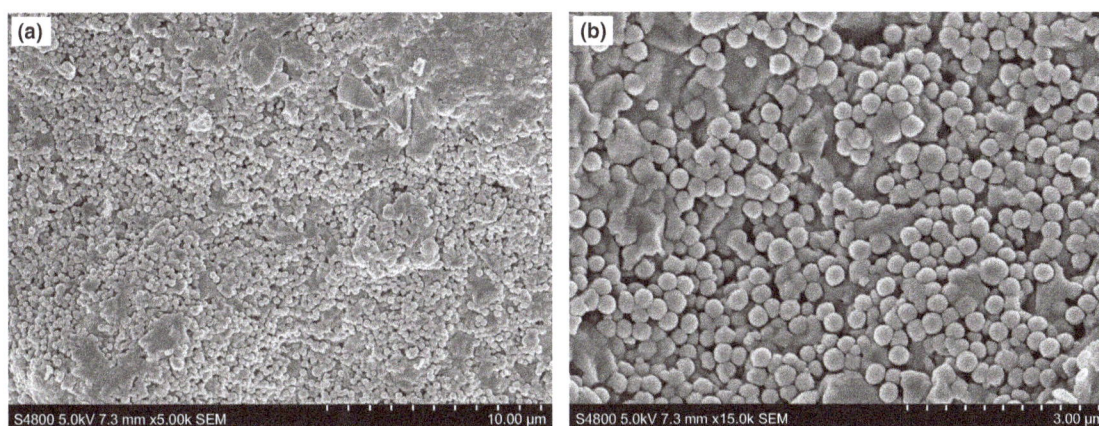

Fig. 11 SEM images (**a**, **b**) of the surface of the plugged shale sample at different locations

microscope. As can be seen in Fig. 11, after being sealed with SD-SEAL the shale surface was smooth and dense, SD-SEAL nanoparticles were tightly packed in micropores and micro-fractures in the shale sample, and spherical particles were clearly visible on the shale surface. This would significantly improve the core compaction and effectively reduce the core permeability.

3.4 Wettability tests

The wettability of the plugged layer of shale was measured with a contact angle measurement. The wettability test results (Table 2) showed that the shale surface was strongly hydrophilic with a wetting angle of 12° before sealing. At room temperature, the plugged layer of shale was hydrophilic with a wetting angle of 38°. When the testing temperature was higher than the LCST value of SD-SEAL, the

Table 2 Wettability test of the plugged layer of shale

Test condition	Wetting angle, °
Before sealing, room temperature	12
After sealing, room temperature	38
After sealing, $T = 130$ °C	136
After sealing, $T = 150$ °C	142
After sealing, $T = 180$ °C	139

plugged layer of shale was changed to being hydrophobic. The wetting angle was 136°, 142°, and 139°, respectively, when the testing temperature was 130, 150, and 180 °C. The wettability test results further verified that SD-SEAL had a hydrophobic effect when the testing temperature was higher than its LCST value.

4 Conclusions

P(MMA–St) nanospheres with an average particle size of 100 nm were synthesized by the solvothermal method. Then, through radical graft copolymerization of thermo-sensitive monomer NIPAm and hydrophilic monomer AA onto the surface of P(MMA–St) nanospheres, a series of thermo-sensitive polymer nanospheres with different LCST values were prepared by adjusting the mole ratio of NIPAm to AA. The light transmittance of the SD-SEAL solution dropped sharply when the temperature reached its LCST value, so SD-SEAL is temperature-sensitive. The LCST values of SD-SEAL increased with an increase in hydrophilic monomer AA. The LCST values were 53, 63, 81, 93, 106, 125, and 158 °C when the mole ratio of NIPAm to AA was no AA, 90/10, 80/20, 70/30, 74/26, 66/34, and 52/48. When temperature was higher than its LCST value, SD-SEAL played a dual role of physical plugging and chemical inhibition, slowed down pressure transmission, and reduced shale permeability remarkably, which greatly improved the shale stability.

Acknowledgements We would like to thank the financial support from the National Science Foundation of China (Nos. 51374233, 51474235), the Postdoctoral Innovative Project Foundation of Shandong Province (No. 201602027), the Qingdao Postdoctoral Applied Research Project (No. 2015242), the Fundamental Research Funds for the Central Universities (No. 15CX06021A), and the Graduate Student Innovation Project from China University of Petroleum (East China) (No. YCX2015011).

References

Al-Baghli W, Saradhi V, Anas M, et al. Nanotechnology improves wellbore strengthening and minimizes differential sticking problems in highly depleted formation. In: SPE annual technical conference and exhibition; 2015. doi:10.2118/174859-MS.

Bai XD, Pu XL. The performance of PMMA nano-latex in drilling fluids. Drill Fluid Complet Fluid. 2010;27(1):8–10 (**in Chinese**).

Cai JH, Chenevert ME, Sharma MM, et al. Decreasing water invasion into Atoka shale using unmodified silica nanoparticles. SPE Drill Complet. 2012;27(1):103–12. doi:10.2118/146979-PA.

Chen SL, Liu MZ, Jin SP, et al. pH-/temperature-sensitive car-boxymethyl chitosan/poly(N-isopropylacrylamide-co-methacrylic acid) IPN: preparation, characterization and sustained release of riboflavin. Polym Bull. 2013;71(3):719–34. doi:10.1007/s00289-013-1088-8.

Cormick C, Hunter S. Valuing values: better public engagement on nanotechnology demands a better understanding of the diversity of publics. NanoEthics. 2014;8(1):55–71. doi:10.1007/s11569-014-0188-8.

Cui SH, Ban FS, Yuan GJ. Status quo and challenges of global shale gas drilling and completion. Nat Gas Ind. 2011;31(04):72–5 (**in Chinese**).

Dong DZ, Zou CN, Yang H, et al. Progress and prospects of shale gas exploration and development in China. Acta Pet Sin. 2012;33(S1):107–14 (**in Chinese**).

Farooq MH, Xu XG, Yang HL, et al. Room temperature ferromagnetism of boron-doped ZnO nanoparticles prepared by solvother-mal method. Rare Met. 2013;32(3):264–8. doi:10.1007/s12598-013-0058-5.

Feng X, Chen L, Dong J, et al. Fast responsive temperature-sensitive hydrogel and its application on bioseparation. Acta Sci Nat (Univ Nankaiensis). 2005;38(6):34–40 (**in Chinese**).

Gulfam M, Chung BG. Development of pH-responsive chitosan-coated mesoporous silica nanoparticles. Macromol Res. 2014;22(4):412–7. doi:10.1007/s13233-014-2063-4.

Hoa NTQ, Huyen DN. Comparative study of room temperature ferromagnetism in undoped and Ni-doped TiO_2 nanowires synthesized by solvothermal method. J Mater Sci Mater Electron. 2013;24(2):793–8. doi:10.1007/s10854-012-0811-9.

Hoelscher KP, De Stefano G, Riley M, et al. Application of nanotechnology in drilling fluids. In: SPE international oilfield nanotechnology conference and exhibition; 2012. doi:10.2118/157031-MS.

Hu XL, Hou GM, Zhang MQ, et al. Studies on solid-state polymer composite electrolyte of nano-silica/hyperbranched poly(amine-ester). J Solid State Electrochem. 2016;20(7):1845–54. doi:10.1007/s10008-015-3073-7.

Huynh CT, Lee DS. Controlling the properties of poly(amino ester urethane)–poly(ethylene glycol)–poly(amino ester urethane) tri-block copolymer pH/temperature-sensitive hydrogel. Colloid Polym Sci. 2012;290(11):1077–86. doi:10.1007/s00396-012-2624-z.

Kearnes M, Macnaghten P, Davies SR. Narrative, nanotechnology and the accomplishment of public responses: a response to Thorstensen. NanoEthics. 2014;8(3):241–50. doi:10.1007/s11569-014-0209-7.

Kokufuta MK, Sato S, Kokufuta E. LCST behavior of copolymers of N-isopropylacrylamide and N-isopropylmethacrylamide in water. Colloid Polym Sci. 2012;290(16):1671–81. doi:10.1007/s00396-012-2706-y.

Lian Q, Zheng XF, Wang DJ. Synthesis of magnetic $Co_{0.5}Zn_{0.5}Fe_2O_4$-chitosan nanoparticles as pH responsive drug delivery system. Russ J Gen Chem. 2015;85(1):152–4. doi:10.1134/S1070363215010260.

Lin SF, Lin HS, Wu YY. Validation and exploration of instruments for assessing public knowledge of and attitudes toward nanotechnology. J Sci Educ Technol. 2012;22(4):548–59. doi:10.1007/s10956-012-9413-9.

Luo JJ, Ersen O, Chu W, et al. Anchoring and promotion effects of metal oxides on silica supported catalytic gold nanoparticles. J Colloid Interface Sci. 2016;482:135–41. doi:10.1016/j.jcis.2016.08.001.

Mao H, Qiu ZS, Shen ZH, et al. Hydrophobic associated polymer based silica nanoparticles composite with core–shell structure as a filtrate reducer for drilling fluid at ultra-high temperature. J Pet Sci Eng. 2015;129:1–14. doi:10.1016/j.petrol.2015.03.003.

Mishra D, Arora R, Lahiri S, et al. Synthesis and characterization of iron oxide nanoparticles by solvothermal method. Prot Met Phys Chem Surf. 2014;50(5):628–31. doi:10.1134/S2070205114050128.

Qu YZ, Sun JS, Su YN, et al. A nano composite material poly (styrene-b-acrylamide)/bentonite: the laboratory synthesis and laboratory research on its filtration control performance. Drill Fluid Complet Fluid. 2007;24(4):15–8 (**in Chinese**).

Rafieepour S, Jalayeri H, Ghotbi C, et al. Simulation of wellbore stability with thermo-hydro-chemo-mechanical coupling in troublesome formations: an example from Ahwaz oil field, SW Iran. Arab J Geosci. 2013;8(1):379–96. doi:10.1007/s12517-013-1116-x.

Roshan H, Aghighi MA. Chemo-poroelastic analysis of pore pressure and stress distribution around a wellbore in swelling shale: effect of undrained response and horizontal permeability anisotropy. Geomech Geoeng. 2012;7(3):209–18. doi:10.1080/17486025.2011.616936

Rwei SP, Nguyen TA. Formation of liquid crystals and behavior of LCST upon addition of xanthan gum (XG) to hydroxypropyl cellulose (HPC) solutions. Cellulose. 2014;22(1):53–61. doi:10.1007/s10570-014-0469-y.

van Oort E. A novel technique for the investigation of drilling fluid induced borehole instability in shales. In: SPE rock mechanics in petroleum engineering; 1994. doi:10.2118/28064-MS.

van Oort E. Physics–chemical stabilization of shales. In: SPE international symposium on oilfield chemistry; 1997. doi:10.2118/37263-MS.

Wang WF, Liu P, Chen C, et al. The study of shale gas reservoir theory and resources evaluation. Nat Gas Geosci. 2013;24(3):429–38 **(in Chinese)**.

Wen H, Chen M, Jin Y, et al. A chemo-mechanical coupling model of deviated borehole stability in hard brittle shale. Pet Explor Dev. 2014;41(6):817–23. doi:10.1016/S1876-3804(14)60099-9.

Wu YP, He S, Guo ZR, et al. Preparation and stabilization of silver nanoparticles by a thermo-responsive pentablock terpolymer. Polym Sci Ser B. 2013;55(11):634–42. doi:10.1134/S15600904 13130058.

Xu JF, Qiu ZS, Lv KH. Pressure transmission testing technology and simulation equipment for hydro-mechanics coupling of Shale. Acta Pet Sin. 2005;26(6):2115–8 **(in Chinese)**.

Xu X, Wang K, Gu YC, et al. Synthesis and characterization of pH and temperature sensitive hydrogel based on poly(N-isopropy-lacrylamide), poly(ε-caprolactone), methylacrylic acid, and methoxyl poly(ethylene glycol). Macromol Res. 2013;21(8):870–7. doi:10.1007/s13233-013-1098-2.

Yuan JL, Deng JG, Tan Q, et al. Borehole stability analysis of horizontal drilling in shale gas reservoirs. Rock Mech Rock Eng. 2012;46(5):1157–64. doi:10.1007/s00603-012-0341-z.

Zhong HY, Qiu ZS, Sun D, et al. Inhibitive properties comparison of different polyetheramines in water-based drilling fluid. J Nat Gas Sci Eng. 2015;26:99–107. doi:10.1016/j.jngse.2015.05.029.

4

Investigation of methane adsorption on chlorite by grand canonical Monte Carlo simulations

Jian Xiong[1] · Xiang-Jun Liu[1] · Li-Xi Liang[1] · Qun Zeng[2]

Abstract In this paper, the methane adsorption behaviours in slit-like chlorite nanopores were investigated using the grand canonical Monte Carlo simulation method, and the influences of the pore sizes, temperatures, water, and compositions on methane adsorption on chlorite were discussed. Our investigation revealed that the isosteric heat of adsorption of methane in slit-like chlorite nanopores decreased with an increase in pore size and was less than 42 kJ/mol, suggesting that methane adsorbed on chlorite through physical adsorption. The methane excess adsorption capacity increased with the increase in the pore size in micropores and decreased with the increase in the pore size in mesopores. The methane excess adsorption capacity in chlorite pores increased with an increase in pressure or decrease in pore size. With an increase in temperature, the isosteric heats of adsorption of methane decreased and the methane adsorption sites on chlorite changed from lower-energy adsorption sites to higher-energy sites, leading to the reduction in the methane excess adsorption capacity. Water molecules in chlorite pores occupied the pore wall in a directional manner, which may be related to the van der Waals and Coulomb force interactions and the hydrogen bonding interaction. It was also found that water molecules existed as aggregates. With increasing water content, the water molecules occupied the adsorption sites and adsorption space of the methane, leading to a reduction in the methane excess adsorption capacity. The excess adsorption capacity of gas on chlorite decreased in the following order: carbon dioxide > methane > nitrogen. If the mole fraction of nitrogen or carbon dioxide in the binary gas mixture increased, the mole fraction of methane decreased, methane adsorption sites changed, and methane adsorption space was reduced, resulting in the decrease in the methane excess adsorption capacity.

Keywords Chlorite · Methane · Nanopores · Grand canonical Monte Carlo · Adsorption capacity

1 Introduction

The study "Technically Recoverable Shale Oil and Shale Gas Resources: An Assessment of 137 Shale Formations in 41 Countries outside the United States" conducted by the US DOE's Energy Information Administration (EIA) in 2013, indicated that technically the shale gas resource in the world was approximately 220×10^{12} m^3, suggesting that there was a significant developmental potential for shale gas resources in the world (EIA 2013). Free, adsorbed, and dissolved gases exist in shale formations. Adsorbed gas is found on the surface of the mineral grains or in the micropore structure of organic matter in shale gas reservoirs. However, free gas is mainly contained in microfractures or larger pores in organic matter as well as in mineral grains in shale gas reservoirs. In 2002, Curtis studied the characteristics of American shale gas reservoirs, drawing the conclusion that adsorbed gas accounts for approximately 20%–85% of the total gas content and suggesting that adsorbed gas played an important role in the shale gas resource. Therefore, it is important to

✉ Jian Xiong
 361184163@qq.com

[1] State Key Laboratory of Oil and Gas Reservoir Geology and Exploitation, Southwest Petroleum University, Chengdu 610500, Sichuan, China

[2] Institute of Chemical Materials, Engineering Physical Academy of China, Mianyang 621999, Sichuan, China

Edited by Jie Hao

investigate the methane adsorption capacity of organic-rich shales to evaluate shale gas resources. Both the physico-chemical properties of shales and environmental factors could have an impact on the methane adsorption capacity of shales, illustrating that the mineralogical compositions are key factors that affect the methane adsorption capacity of shales. According to previous research (Liang et al. 2015; Liu et al. 2015; Xiong et al. 2015a), clay minerals are the essential mineralogical components of the shales from the Yanchang Formation of the Ordos Basin as well as the Longmaxi Formation and Wufeng Formation of the Sichuan Basin, the contents of which were comparatively higher. Therefore, it is important to investigate the methane adsorption capacity on chlorite, which is an important type of clay mineral. Currently, studies aimed at evaluating the methane adsorption capacity on chlorite mainly focused on isothermal adsorption experiments. Ji et al. (2012a, b) investigated the influences of pressure, temperature, and grain size on the methane adsorption capacity of chlorite. Fan et al. (2014) studied the influences of pressure and temperature on the methane adsorption capacity of chlorite. Tang and Fan (2014) studied the methane adsorption capacity of chlorite at different temperatures under a pressure of 20 MPa. Liang et al. (2016) investigated the methane adsorption capacity of chlorite under a pressure of 20 MPa. All of the above studies were based on isothermal adsorption experiments, that is, the value of the adsorption amount under equilibrium pressure can be used to evaluate the methane adsorption capacity of chlorite. However, this value comprehensively reflects the specific surface area of chlorite and the value of the adsorption amount per unit surface area and cannot fundamentally reflect the essence of microscopic adsorption mechanisms of the methane adsorption on the chlorite owing to the results obtained for the macroscopic behaviour.

Computational molecular simulations have recently attracted much attention as a theoretical research method that can be used to study the adsorption properties of the adsorbent and could therefore be used to investigate the adsorption mechanism of fluid molecules on porous material. Titiloye and Skipper (2005) used the grand canonical Monte Carlo (GCMC) and molecular dynamics (MD) methods to study the adsorption behaviours and structural properties of methane in slit-like montmorillonite pores. Using MD simulations, the microscopic structural properties and diffusion behaviours of carbon dioxide in slit-like montmorillonite pores were studied by Yang and Zhang (2005). Jin and Firoozabadi (2013, 2014) used GCMC to investigate the adsorption behaviours of methane and carbon dioxide in slit-like montmorillonite pores as well as the influence of water on the adsorption behaviours of methane and carbon dioxide. Using the GCMC method, Sun et al. (2015) performed research on the methane adsorption behaviours of different

types of clay minerals (montmorillonite, illite and kaolinite) and also studied the effects of different temperatures on the methane adsorption behaviours. Sui et al. (2015) studied the microscopic structural properties and diffusion behaviours of methane in slit-like montmorillonite pores using the GCMC and MD methods. Xiong et al. (2016) studied the microscopic adsorption mechanism of methane in slit-like montmorillonite pores using the GCMC method. These studies generated knowledge on methane adsorption on montmorillonite, illite and kaolinite. However, the detailed microscopic adsorption mechanism of methane in chlorite pores has not been well studied.

Hence, this article regarded chlorite as an object of study and used the computer molecular simulation technique to construct skeleton patterns of slit-like chlorite pores. Then, the impacts of pore sizes, temperatures, water and gas compositions on the methane adsorption behaviours in slit-like chlorite pores and the microscopic adsorption mechanism of methane in chlorite pores were studied using the GCMC simulations. Finally, the influence of the temperatures, water contents and compositions on the adsorption behaviours of methane on chlorite and their interaction mechanisms were discussed, which can provide important theoretical and instructional significance for the exploration and development of shale gas reservoirs.

2 Molecular model

The parameters of the chlorite crystal cell can be found in the literature (Joswig et al. 1980). The following parameters from this crystal cell are listed: $a = 0.5327$ nm, $b = 0.9232$ nm, $c = 1.440$ nm, $\alpha = \gamma = 90°$, $\beta = 97.16°$. According to the $9a \times 4b$ super-cell structure constructed in the x and y directions of the chlorite unit crystal cell structure, the size of this super-cell structure in the $x \times y$ direction is 4.794 nm $\times$ 3.693 nm. Based on this, a space can be added in the z direction between the two super-cell structures to construct pores with different sizes in the chlorite super-cell structure. Figure 1 shows the configuration of the slit-like chlorite pore, and Table 1 presents their basic parameters.

The Lennard–Jones (L–J) potential parameters and charges of the sites in the unit cell of chlorite are presented in Table 2 and are taken from the work of Cygan et al. (2004) and Jin and Firoozabadi (2013, 2014). Methane and nitrogen molecules were modelled using a TraPPE force field (Martin and Siepmann 1998; Potoff and Siepmann 2001), the water molecule was modelled using an SPC-E force field (Berendsen et al. 1987), and the carbon dioxide molecule was simulated by using the EPM2 model (Harris and Yung 1995). All fluid molecules retain electric neutrality. The L–J potential parameters and charges of each atom in the liquids

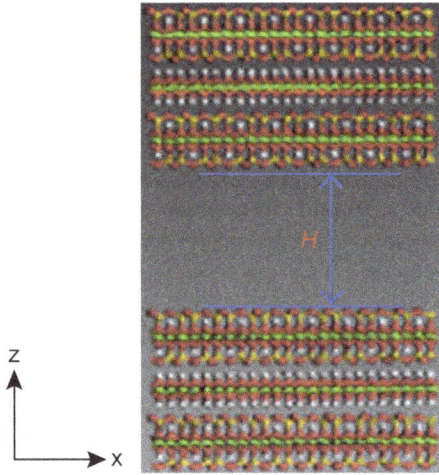

Fig. 1 Schematic representation of the slit-like chlorite pore (*H* represents different pore sizes) (*red circle* is oxygen atom, *white circle* is hydrogen atom, *yellow circle* is silicon atom, *purple circle* is aluminium atom, *green circle* is magnesium atom)

dielectric constant, 8.854×10^{-12} F/m; and σ_{ij} and ε_{ij} are the L–J potential parameters. According to the Lorentz–Botherlot mixed rules these are set as:

$$\sigma_{ij} = (\sigma_i + \sigma_j)/2 \qquad \varepsilon_{ij} = \sqrt{\varepsilon_i \times \varepsilon_j} \qquad (2)$$

where σ_i, σ_j are the collision diameters of the atoms or molecules i, j in nm and ε_i, ε_j are the potential well depths in kJ/mol.

3 Simulation method

3.1 Grand canonical Monte Carlo (GCMC)

Monte Carlo simulations have been widely used to study the adsorption properties of materials, while GCMC has been widely applied in the investigation of the adsorption behaviours of an adsorbate on an adsorbent. In this work, we use GCMC simulation to investigate the adsorption behaviours of methane in a slit-like chlorite pore. In the grand canonical ensemble, the chemical potential, volume, and temperature are the independent variables. Among these, the chemical potential is a function of the fugacity instead of the pressure. In this research, the Soave, Redlich and Kwong (SRK) state equation was used to calculate the fugacity (Soave 1972). The fugacity coefficients of pure methane at different temperatures and pressures in the simulations are shown in Fig. 2a, and the fugacity coefficients of the mole fraction of methane in the binary gas mixture at different pressures in the simulations are described in Fig. 2b, c. Simulation of the methane isothermal adsorption by the GCMC method was performed mainly using Sorption Module of the Materials Studio 6.0. The temperature in this simulation varied from 313 to 373 K, and the temperature interval was 20 K. The maximum simulated pressure was 40 MPa, and the simulation was under constant pressure, point by point, divided into a total of 15 points. The force field type used in this

are also shown in Table 2 and can be found in the above references. During the simulation, the force fields are based on the Dreiding force field, and chlorite is supposed to be a rigid body. Furthermore, owing to the lack of force fields for magnesium, we assign the same L–J parameters for magnesium as for aluminium in the Dreiding force field (Zeng et al. 2003; Jin and Firoozabadi 2013). In addition, the charges of magnesium and aluminium are +2 and +3, respectively. In the simulation, the interactions consist of the van der Waals force and Coulomb force. The L–J (12–6) potential model was used to describe the van der Waals force. The model to represent the Coulomb force and van der Waals force is given by:

$$E = \varepsilon_{ij}\left[\left(\frac{\sigma_{ij}}{r_{ij}}\right)^{12} - \left(\frac{\sigma_{ij}}{r_{ij}}\right)^{6}\right] + \frac{q_i q_j}{4\pi\varepsilon_0 r_{ij}}. \qquad (1)$$

where q_i, q_j are the charges of atoms in the system in C; r_{ij} is the distance between the atoms i and j in nm; ε_0 is the

Table 1 Parameters of the slit-like chlorite pore of different pore sizes

H, nm	x, nm	y, nm	z, nm	α	β	γ	Surface area, × 10⁻¹⁷ m²	Volume, × 10⁻²⁰ cm³	Density, g/cm³	Mass, × 10⁻¹⁹ g
1	4.794	3.693	5.227	90	97.16	90	1.77	9.727	2.292	2.23
1.5	4.794	3.693	5.727	90	97.16	90	1.77	10.61	2.101	2.23
2	4.794	3.693	6.227	90	97.16	90	1.77	11.50	1.940	2.23
3	4.794	3.693	7.227	90	97.16	90	1.77	13.27	1.681	2.23
4	4.794	3.693	8.227	90	97.16	90	1.77	15.04	1.483	2.23
6	4.794	3.693	10.227	90	97.16	90	1.77	18.58	1.200	2.23
8	4.794	3.693	12.227	90	97.16	90	1.77	22.12	1.008	2.23
10	4.794	3.693	14.227	90	97.16	90	1.77	25.66	0.869	2.23
15	4.794	3.693	19.227	90	97.16	90	1.77	34.51	0.646	2.23
20	4.794	3.693	24.227	90	97.16	90	1.70	43.36	0.514	2.23

Table 2 L–J potential parameters and charges of each atom

	Atoms	ε/k_B, K	σ, nm ·	q, e	References
Chlorite	O(t)	78.18	0.3166	−0.800	Cygan et al. (2004), Jin and Firoozabadi (2013, 2014)
	O(a)	78.18	0.3166	−1.000	
	O(o)	78.18	0.3166	−1.7175	
	O(OH)	78.18	0.3166	−1.7175	
	H(OH)	0.00	0.0000	0.7175	
	Si	9.26×10^4	0.3302	1.200	
	Al	4.54×10^4	0.5086	3.000	
	Mg	4.54×10^4	0.5086	2.000	
Methane	C	148.10	0.3730	0	Martin and Siepmann (1998)
Water	O	78.18	0.3166	−0.8476	Berendsen et al. (1987)
	H	0	0	0.4238	
Carbon dioxide	C	28.129	0.2757	0.6512	Harris and Yung (1995)
	O	80.507	0.3033	−0.3256	
Nitrogen	N	36	0.331	−0.482	Potoff and Siepmann (2001)
	COM	0	0	0.964	

Notes: (t) tetrahedron oxygen, (o) octahedral oxygen, (a) terminal oxygen, COM at the centre of the nitrogen–nitrogen bond

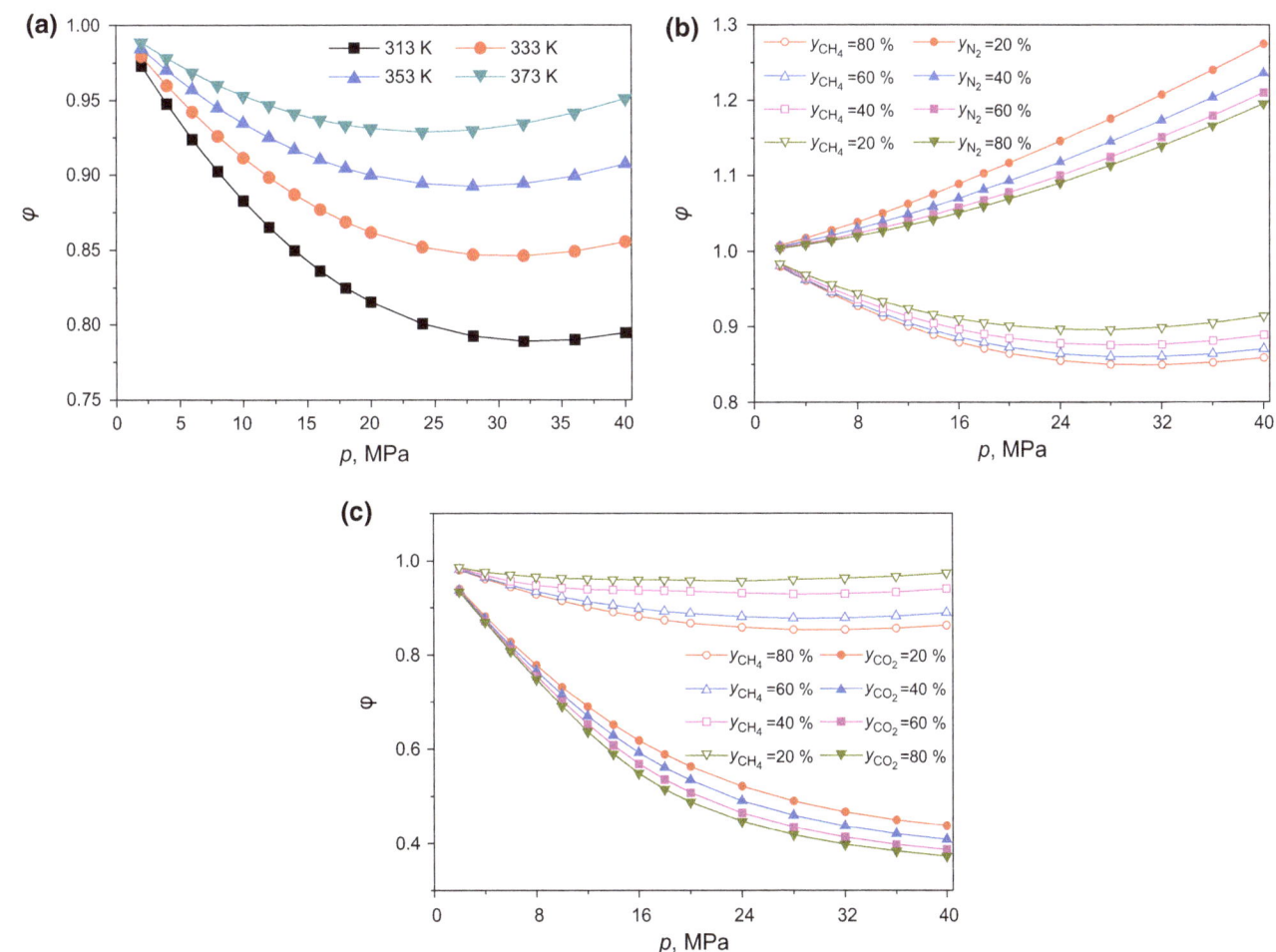

Fig. 2 Methane fugacity coefficient at different temperatures and pressures

simulation was the Dreiding force field, with the Coulomb force and van der Waals force interactions calculated by the Ewald & Group method and the atom interaction-based method with an L–J potential cutoff distance of 1.55 nm. The maximum number of load steps in each simulation was 3×10^6, including 1.5×10^6 balance steps and 1.5×10^6 process steps. The related statistics were obtained using the later 1.5×10^6 configurations.

3.2 Excess adsorption amount

The isothermal adsorption experiments of the chlorite exceed the critical temperature of methane (191 K), suggesting that the methane adsorption behaviour on chlorite belongs to supercritical adsorption. For supercritical adsorption, Gibbs proposed that an adsorbate molecule in the adsorbed phase on the surface of the adsorbent cannot be used as the total adsorption amount. Additionally, the distribution of adsorbate molecules in the adsorbed phase based on the gas phase density was independent of the gas/solid molecule inter-atomic forces (Xiong et al. 2015b). According to this view, Gibbs introduced the concept of the excess adsorption amount:

$$n_{ex} = n_{ab} - \rho_g V_a, \tag{3}$$

where n_{ex} is the excess adsorption amount in mol/g, n_{ab} is the absolute adsorption amount in mol/g, V_a is the adsorbed phase volume in cm^3, and ρ_g is the vapour density in g/cm^3 calculated by the SRK state equation (Soave 1972).

Figure 3 shows a schematic representation of the excess adsorption amount and absolute adsorption amount in which the area of a represents the excess adsorption amount and the total area of a and b expresses the absolute adsorption amount. We assume that the total amount of adsorbate in the adsorption system is N, corresponding to the total area of a, b, and c, as shown in Fig. 3, which is equal to the expression $(n_{ab} + \rho_g V_g)$. Therefore,

$\rho_g (V_a + V_g)$ represents the total area of b and c, as shown in Fig. 2. Then, the excess adsorption amount can be expressed as follows:

$$n_{ex} = N - \rho_g(V_a + V_g) = N - \rho_g V_p \tag{4}$$

where N is the total amount of gas in mol/g, V_g is the gas phase volume in g/cm^3, and V_p is the free volume in g/cm^3. The free volume in the pore can be determined by the method that uses He as the probe (Talu and Myers 2001). Therefore, the gas amount obtained from the results of the simulation is the total amount of gas, and based on the free volume in the pore, the total amount of gas can be converted into the excess adsorption amount of the gas according to Eq. (4).

4 Results and discussions

4.1 Influences of different pore sizes

Figure 4 presents the total amount and excess adsorption capacity isotherms of methane in chlorite pores for different pore sizes. Examination of Fig. 4a shows that the total amount of methane increased with the increment of the pore size and first increased rapidly and then increased slowly with the increase in pressure. At the same time, from Fig. 4b, it can be seen that the differences among the excess adsorption capacity of methane on chlorite in micropores were small. However, it also showed that the methane excess adsorption capacity gradually decreased as the pore size increased in the mesopores, and the excess adsorption capacity of methane in mesopores is significantly smaller than that in micropores. This may be because the potential superimposed effect of the pore wall can significantly affect the adsorption of methane molecules in micropores, and the methane adsorption capacity in the pore would therefore be limited by the pore volume, that is, the pore volume increases with the increase in the pore size and methane adsorption capacity. However, the adsorption of methane molecules in mesopores was mainly affected by the surface potential effect of the two sides of the pore wall; the interactions between the methane molecules and the chlorite decrease, and movement space of the methane molecules increases, which makes the force to escape from the chlorite pore wall easy to overcome. Then, the methane adsorption capacity decreases with increasing pore size. In addition, the excess adsorption capacity of methane first increased after the pressure drop. That is, there is a maximum value of the excess adsorption capacity $(n_{exc-max})$, and the corresponding pressure is known as the maximum pressure (p_{max}). This conclusion is in line with that of previous studies of organic-rich shales (Rexer et al. 2013; Gasparik et al. 2014; Yang et al. 2015).

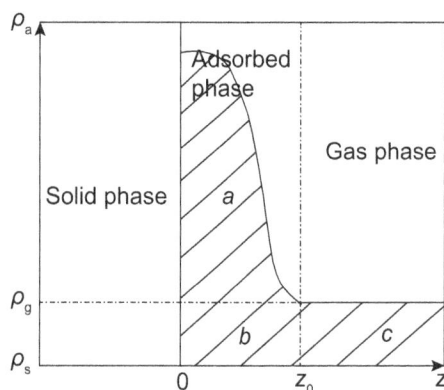

Fig. 3 Schematic representation of the excess adsorption amount and absolute adsorption amount

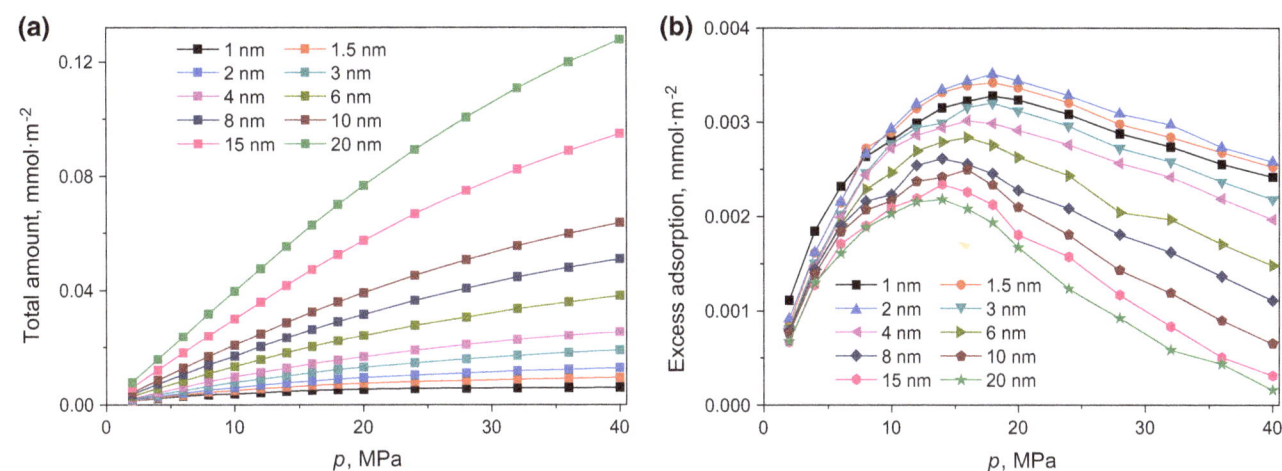

Fig. 4 Total amount isotherms of methane (**a**) and the excess adsorption isotherms of methane (**b**) in chlorite pores for different pore sizes (temperature of 333 K)

Table 3 shows the maximum values of the methane excess adsorption capacity and its corresponding pressure for different pore sizes. Examination of the data in Table 3 shows that the maximum pressure corresponding to the maximum value of the excess adsorption capacity was different and that the range of the maximum pressure was between 14 MPa and 18 MPa. This finding is in agreement with previous studies on organic-rich shales (Rexer et al. 2013; Gasparik et al. 2014; Yang et al. 2015), suggesting that the maximum pressure (p_{max}) was between 10 and 19 MPa, as was concluded from the experimental data. This result indicates that, to a certain extent, our simulation results are in reasonable agreement with the experimental results. Meanwhile, the maximum value of the excess adsorption capacity decreased with an increase in pore size in mesopores. The maximum value of the excess adsorption capacity reached a peak value of 0.00372 mmol/m^2 when the pore size was 2 nm, while the minimum value of the excess adsorption capacity was 0.00239 mmol/m^2 when the pore size was 20 nm. The conclusions illustrate that the methane adsorption capacity in chlorite micropores increased with an increase in pore size, whereas that in chlorite mesopores decreased with an increase in the pore size.

The average isoteric heat of methane in chlorite pores with different pore sizes is shown in Fig. 5. It is seen that the methane isosteric heat decreased gradually with an increase in pore size. The isosteric heat for a pore size of 1 nm was the maximum value (13.6 kJ/mol), and the average isosteric heat for a pore size of 20 nm was the minimum value (6.12 kJ/mol). Experimentally, Ji et al. (2012a, b) found that the average isosteric heat of methane adsorption on chlorite was 9.4 kJ/mol. Although there are differences between the simulated and experimental results, they also showed similarities to a certain extent. This result may be related to the differences of research methods and samples. The pore size in the experiments is distributed continuously between 20 nm and 100 nm (Ji et al. 2012a, b), and the methane isosteric heat obtained from the experiment reflects the synthesis results obtained for the sample with a continuous distribution of pore sizes. However, the pore skeletons of chlorite in the simulation have a single pore size, and the methane isosteric heat obtained from the simulation reflects the results for a single pore and changes with pore size. In addition, the isosteric heat of adsorption of methane in chlorite pores with different pore sizes was less than 42 kJ/mol, demonstrating that the methane adsorption on chlorite is of the physical adsorption type. This conclusion is in accord with previous studies that suggested the methane is adsorbed on chlorite by physical adsorption (Ji et al. 2012a, b; Fan et al. 2014; Tang and Fan 2014; Liang et al. 2016).

Table 3 Simulation results of methane adsorption on chlorite for different pore sizes

H, nm	$n_{exc-max}$, mmol/m^2	p_{max}, MPa	H, nm	$n_{exc-max}$, mmol/m^2	p_{max}, MPa
1	0.003276	18	6	0.002839	16
1.5	0.003417	18	8	0.002614	14
2	0.003505	18	10	0.002496	16
3	0.003199	18	15	0.002340	14
4	0.003016	16	20	0.002178	14

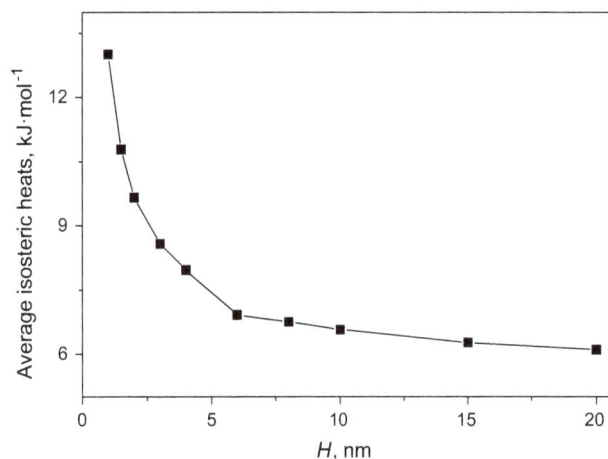

Fig. 5 Average methane isosteric heat in a chlorite pore with different pore sizes

According to the simulation results, we obtained the potential energy distribution of methane and chlorite. The methane and chlorite potential energy distribution curves for different pressures (pore size of 4 nm) are presented in Fig. 6a. It can be noted that the curve transforms the twin peaks into a unimodal distribution with the increase in the pressure. At the same time, as the pressure is increased, the potential energy distribution curves of methane and chlorite gradually moved to the left. Additionally, the most probable potential energy of methane and chlorite gradually decreased, that is, the most probable potential energy changed from -0.209 to -6.485 kJ/mol as the pressure increased from 2 to 36 MPa. This indicates that methane adsorption occurring in chlorite pores gradually changes from higher-energy adsorption sites to lower-energy adsorption sites with the increase in pressure and that the adsorption state of methane molecules in chlorite pores

under low pressure is not as stable as that under high-pressure conditions. In addition, the potential energy distribution curves of methane and chlorite with different pore sizes under a pressure of 20 MPa are presented in Fig. 6b. Examination of Fig. 6b shows that the potential energy distribution curves of methane and chlorite gradually moved to the right with the increase in pore size and the most probable potential energy of methane and chlorite gradually increased, that is, the most probable potential energy changed from -11.92 to -3.56 kJ/mol when the pore size increased from 1 nm to 20 nm. This suggests that methane adsorption occurring in chlorite pores gradually changed from lower-energy adsorption sites to higher-energy adsorption sites as the pore size increased, and the methane adsorption capacity in chlorite micropores was stronger than that in macropores.

4.2 Influence of different temperatures

The excess adsorption isotherms of methane for different temperatures (pore size of 4 nm) are listed in Fig. 7. It can be seen that the methane excess adsorption capacity decreased with increasing temperature under the same pressure; this may be due to methane adsorption on chlorite being of the physical adsorption type. When the temperature increases, the thermal motion of methane molecules would increase, resulting in an increase in the mean kinetic energy of methane molecules, generating a sufficiently large force to escape from the chlorite pore wall easily, thus causing a reduction in the methane adsorption capacity. This conclusion is in accord with the results of the isothermal adsorption experiments performed by Ji et al. (2012a, b), suggesting that the methane adsorption capacity on chlorite decreased with increasing temperature.

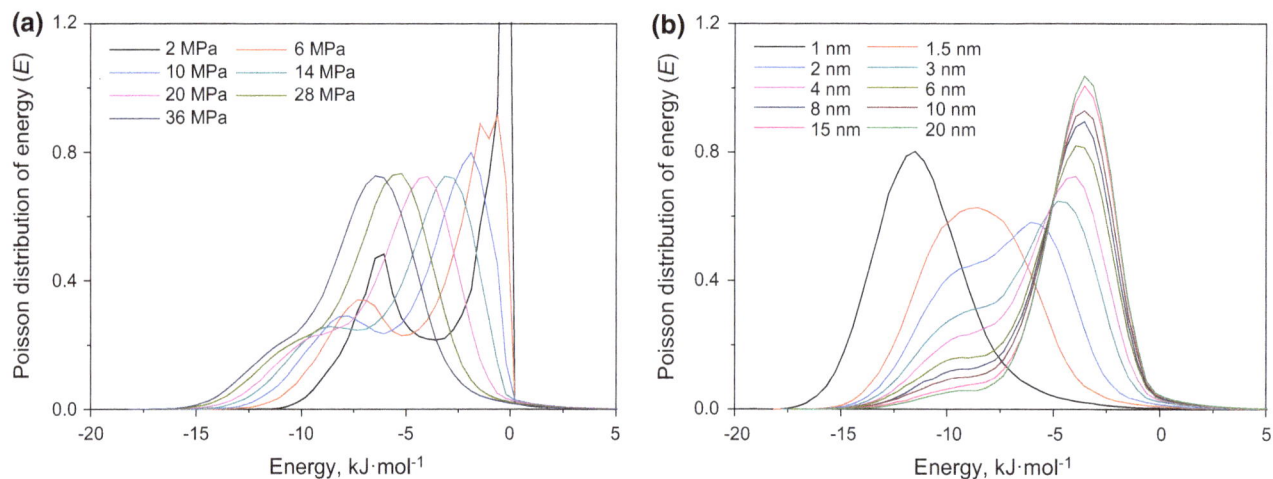

Fig. 6 Potential energy distribution curves of methane and chlorite for different pressures (pore size of 4 nm) (**a**) and different pore sizes (pressure of 20 MPa) (**b**)

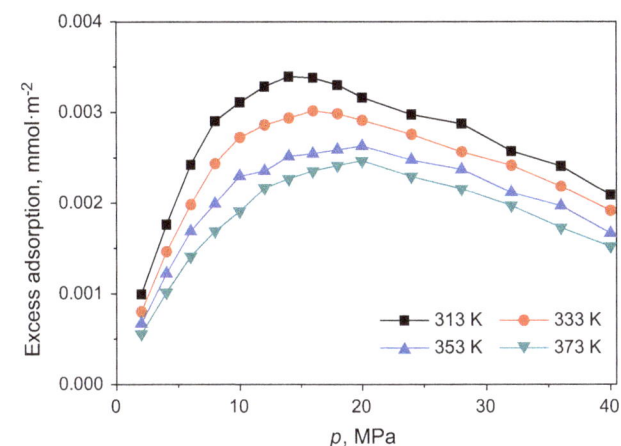

Fig. 7 Excess adsorption isotherms of methane for different temperatures (pore size of 4 nm)

Figure 8 shows the average methane isosteric heat in chlorite pores for different temperatures (pore size of 4 nm). We observe that the average isosteric heat of methane decreased with increasing temperature, indicating that the interactions between methane molecules and chlorite became weaker with increasing temperature, resulting in a decrease in the methane adsorption capacity. In the range of the simulated temperatures, the value of the average isosteric heat of methane in a chlorite pore with the pore size of 4 nm is between 7.59 and 8.21 kJ/mol (less than 42 kJ/mol), illustrating that the adsorption of methane in the chlorite pores is due to physical adsorption. These findings indicate that methane adsorption on chlorite is exothermic and the increase in temperature is not conducive for methane adsorption on chlorite. The potential energy distribution curves between methane and chlorite at different temperatures (pore size of 4 nm) are shown in Fig. 9. When the temperature increased, the potential energy distribution curve of methane and chlorite gradually

moved to the right. Furthermore, the most probable potential energy of methane and chlorite gradually increased, that is, the most probable potential energy changed from -4.393 to -3.138 kJ/mol when the temperature increased from 313 to 373 K. This finding suggests that the adsorption sites of methane molecules in chlorite pores gradually change from lower-energy adsorption sites to higher-energy adsorption sites with increasing temperature, causing the reduction in the methane adsorption capacity.

4.3 Influence of different water contents

To investigate the influence of water on the methane adsorption in chlorite pores, three simulation projects considering three water contents (wt% = the water molecules mass/the chlorite mass) were carried out. First the adsorption sites of water molecules in the slit-like chlorite pores need to be determined by using the annealing simulation method. The distribution of the different water contents in the chlorite pores is given in Fig. 10. In addition, the size of the chlorite pores is 4 nm and the temperature is 333 K in simulation.

Examination of Fig. 10 shows that water molecules occupied the chlorite pore walls in a directional manner, and the oxygen atoms of the water molecules were close to or pointed to the surface of the chlorite pore wall or hydrogen atoms of the surrounding water molecules, with the hydrogen atoms located at a distance from the surface of the chlorite pore wall. This may be due to the positive charges of the aluminium and silicon atoms on the surface of the chlorite pore wall and the negative charge of the oxygen atoms of the water molecule, causing a pattern in which the oxygen atoms of the water molecules are close to or point to the surface of the chlorite pore wall. This phenomenon arises from the Coulomb and van der Waals

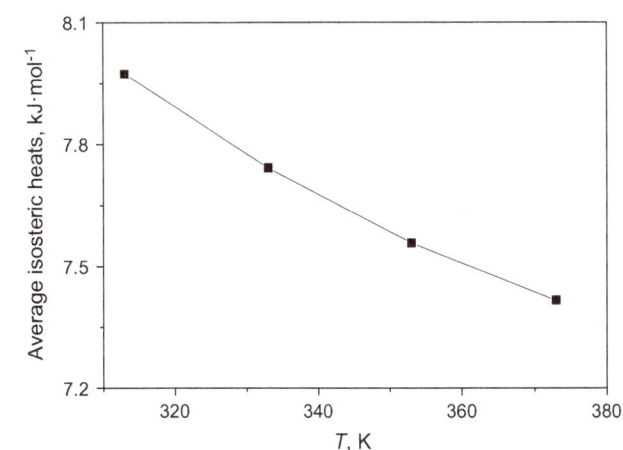

Fig. 8 Average isosteric heats of methane for different temperatures (pore size of 4 nm)

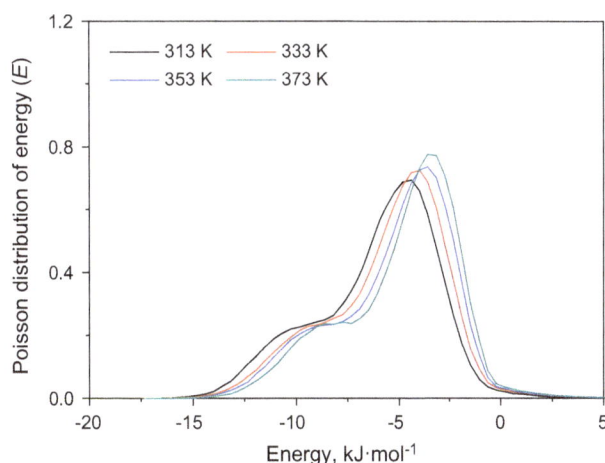

Fig. 9 Potential energy distribution curves of methane and chlorite at different temperatures (pore size of 4 nm)

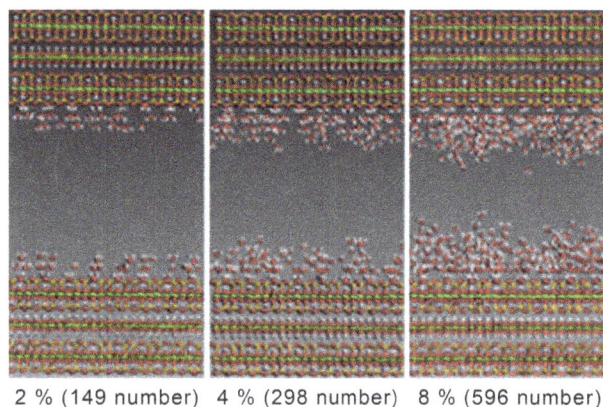

2 % (149 number)   4 % (298 number)   8 % (596 number)

Fig. 10 Distribution of different water contents in chlorite pores (the number in parentheses represents the number of water molecules)

force interactions between water molecules and chlorite, resulting in the aggregation of water molecules in the chlorite pore. In addition, due to the hydrogen bonding interaction, the oxygen atoms of the water molecules point to the hydrogen atoms of surrounding water molecules. All of the data indicate that the water molecules are adsorbed on the surface of the pore walls and occupy the adsorption space of the methane molecules in the form of aggregation.

Figure 11 shows the methane excess adsorption isotherms for different water contents. It can be seen that the methane excess adsorption capacity on chlorite is reduced when the water contents increased under the same temperature and pressure, implying that water molecules inhibit methane adsorption on chlorite. This conclusion is in agreement with a previous study of methane adsorption on montmorillonite (Jin and Firoozabadi 2013, 2014), indicating that water reduced the methane adsorption capacity on montmorillonite. The potential energy distribution curves of methane and chlorite for different water contents are shown in Fig. 12. Inspection of Fig. 12 shows

that the curves have two peaks, with the main peak lying in the higher-energy area and the secondary peak located in the lower-energy area. The most probable potential energy of methane and chlorite did not change significantly with the increase in the water contents, indicating that the methane molecules in the higher-energy adsorption sites could not be occupied with the change of water contents. However, the secondary peak of the potential energy distribution curve gradually became broad, implying that the water molecules occupy the lower-energy adsorption sites of methane molecules. It can be deduced that water molecules mainly occupied lower-energy adsorption sites on the chlorite pore walls instead of higher-energy adsorption sites, illustrating that the water molecules and methane molecules compete with each other for adsorption space and adsorption sites. Therefore, the adsorption space and adsorption sites occupied by water molecules decreased the adsorption space and adsorption sites of methane molecules, leading to a decrease in the methane adsorption capacity.

4.4 Influence of different mole fractions of nitrogen

To investigate the influence of mole fraction of nitrogen on competitive adsorption of nitrogen and methane in the chlorite pores, five simulation projects considering five mole fractions of nitrogen in the methane/nitrogen binary gas mixture ($y_{CH4} = 80\%$ means that the mole fraction of methane is 80% while the mole fraction of nitrogen is 20%) would be carried out. The size of the chlorite pores is 4 nm and the temperature is 333 K in the simulation.

The excess adsorption isotherms of methane for different mole fractions of nitrogen are shown in Fig. 13. The methane excess adsorption capacity decreased with the increase in the nitrogen mole fraction at the same temperature and pressure, indicating that a lower mole fraction

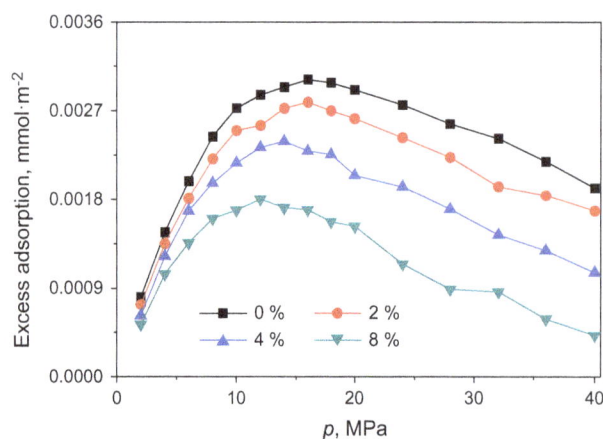

Fig. 11 Excess adsorption isotherms of methane for different water contents

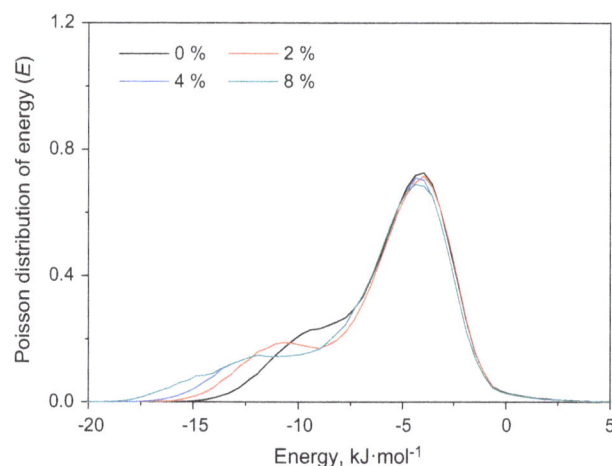

Fig. 12 Potential energy distribution curves of methane and chlorite for different water contents

of the methane in the methane/nitrogen binary gas mixture leads to a smaller methane adsorption capacity on chlorite.

In the adsorption system of the methane/nitrogen binary gas mixture, the potential energy distribution curves for different mole fractions of nitrogen are presented in Fig. 14. It can be seen that the most probable potential energy of methane and chlorite under different nitrogen mole fractions was smaller than that of the nitrogen, demonstrating that the potential energy distribution between methane and chlorite was different from that between nitrogen and chlorite, that is, the methane adsorption occurring on the chlorite pore walls was due to the lower-energy adsorption sites, whereas nitrogen adsorption occurred in higher-energy adsorption sites, illustrating that nitrogen adsorption on chlorite in the adsorption system of the methane/nitrogen binary gas mixture was less stable than that of methane. Figure 15 (the distributions of methane and nitrogen on the surface of chlorite) also illustrates this conclusion. Figure 15 shows that the methane molecules and nitrogen molecules on the chlorite surface were distributed among different adsorption sites.

Furthermore, we found that the interactions between methane and nitrogen led to a change of the potential energy distribution curves of methane and chlorite. At the same time, the potential energy distribution curve of methane and chlorite gradually moved to the right and the most probable potential energy of methane and chlorite gradually increased with increasing nitrogen mole fraction. Namely, the methane adsorption gradually changed from lower-energy adsorption sites to higher-energy adsorption sites with the increase in the nitrogen mole fraction, resulting in a decrease in the methane adsorption capacity on chlorite, implying that the nitrogen adsorption occurring in the chlorite pores led to the change of the adsorption site of methane molecules and the reduction of the adsorption

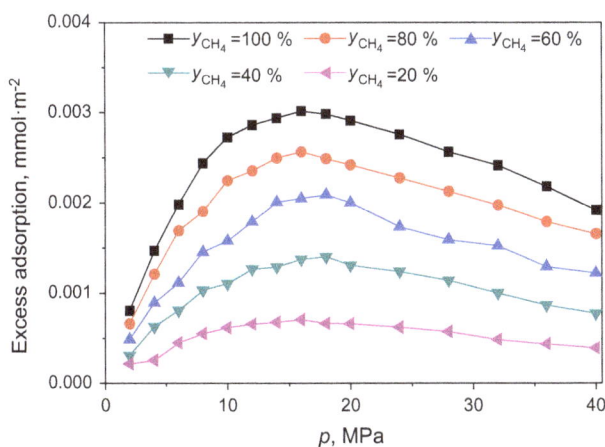

Fig. 14 Potential energy distribution curves for different mole fractions of nitrogen

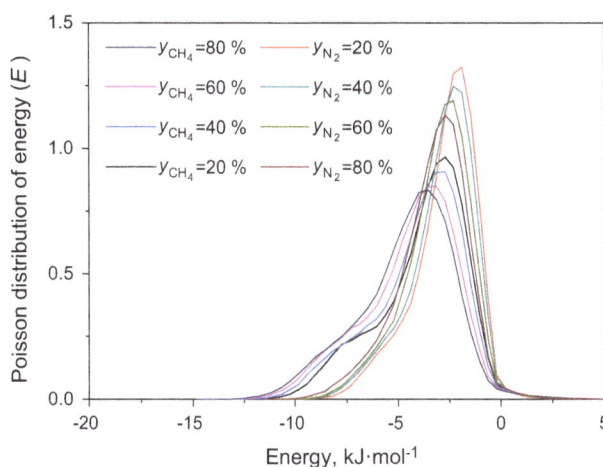

Fig. 15 Distributions of methane and nitrogen on the surface of chlorite (right does not include the chlorite cell)

space for methane molecules. Hence, in the adsorption system of the methane/nitrogen binary gas mixture, the methane adsorption capacity on chlorite is greater than that of nitrogen. According to the previous analysis, the methane adsorption capacity decreased with the increase in the mole fraction of nitrogen due to the decrease in the methane mole fraction in the gas phase, the change of the adsorption sites of the methane molecules and the reduction in the adsorption space of the methane molecules.

4.5 Influence of different mole fractions of carbon dioxide

To investigate the influence of the mole fraction of carbon dioxide on competitive adsorption of carbon dioxide and methane in the chlorite pores, simulations with five mole fractions of carbon dioxide in the carbon dioxide/methane binary gas mixture ($y_{CH4} = 80\%$ means that the mole fraction of methane is 80% while the mole fraction of carbon dioxide is 20%) were carried out. The size of the chlorite pores is 4 nm, and the temperature is 333 K in these simulations.

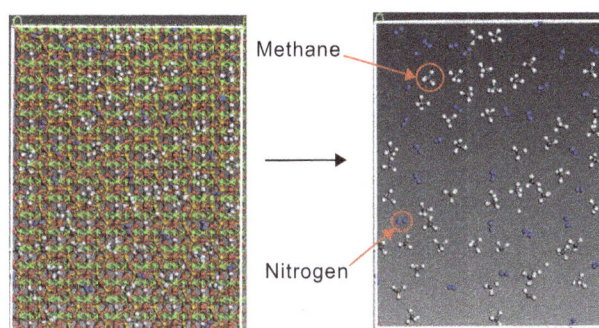

Fig. 13 Excess adsorption isotherms of methane on chlorite for different mole fractions of nitrogen

Excess adsorption isotherms of methane for different mole fractions of carbon dioxide are shown in Fig. 16. It can be seen that the methane excess adsorption capacity decreased as the carbon dioxide mole fraction increased at the same pressure and temperature, indicating that the smaller methane mole fraction in the carbon dioxide/methane binary gas mixture led to lower methane adsorption capacity.

In the adsorption system of the carbon dioxide/methane binary gas mixture, the potential energy distribution curves for different mole fractions of carbon dioxide are shown in Fig. 17. It can be seen that the most probable potential energy of methane and chlorite for different carbon dioxide mole fractions was higher than that of carbon dioxide, suggesting that the potential energy distribution between methane and chlorite was different from that between carbon dioxide and chlorite. That is, carbon dioxide adsorption on the chlorite pore walls occurred in the lower-energy adsorption sites, whereas methane adsorption was located in higher-energy adsorption sites, implying that methane adsorption on chlorite in the adsorption system of the carbon dioxide/methane binary gas mixture was less stable than that of carbon dioxide. Figure 18 (the distributions of methane and carbon dioxide on the chlorite surface) also illustrates this conclusion. Figure 18 shows that methane and carbon dioxide molecules on the chlorite surface were distributed among different adsorption sites.

Furthermore, we also observed that the interactions between methane and carbon dioxide could change the potential energy distribution curves of methane and chlorite. At the same time, the potential energy distribution curve of methane and chlorite gradually moved to the right and the most probable potential energy of methane and chlorite gradually increased with increasing carbon dioxide mole fraction, meaning that the methane adsorption sites gradually changed from lower-energy adsorption sites to

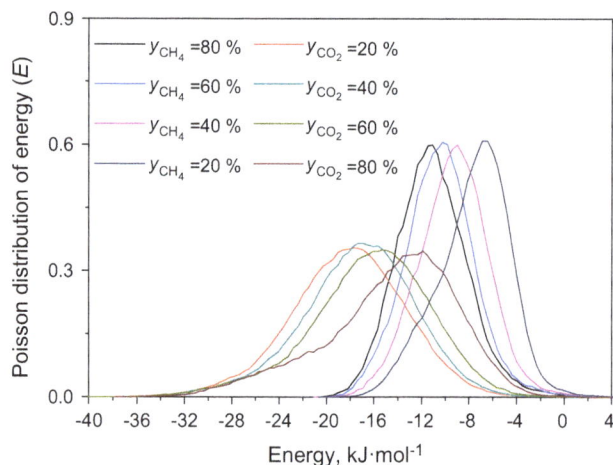

Fig. 17 Potential energy distribution curves for different mole fractions of carbon dioxide

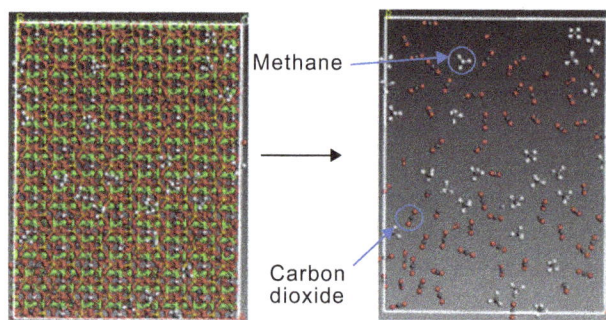

Fig. 18 Distributions of methane and carbon dioxide on the surface of chlorite (right does not include the chlorite cell)

higher-energy adsorption sites with the increasing carbon dioxide mole fraction, leading to a reduction in the methane adsorption capacity on chlorite. This phenomenon demonstrates that carbon dioxide adsorption occurring in chlorite pores results in a change of the adsorption site of methane molecules and a reduction in their adsorption space. Thus, the carbon dioxide adsorption capacity on chlorite is greater than that of methane in the carbon dioxide/methane binary gas mixture adsorption system. Based on the previous analysis, the methane adsorption capacity decreased with the increase in the carbon dioxide mole fraction, resulting in a reduction in the methane mole fraction in the gas phase, a change of the adsorption sites of the methane molecules and a reduction in the methane adsorption space.

5 Conclusions

First, the average methane isosteric heat decreased with increasing pore size, which was smaller than 42 kJ/mol in the chlorite–methane adsorption system, suggesting that

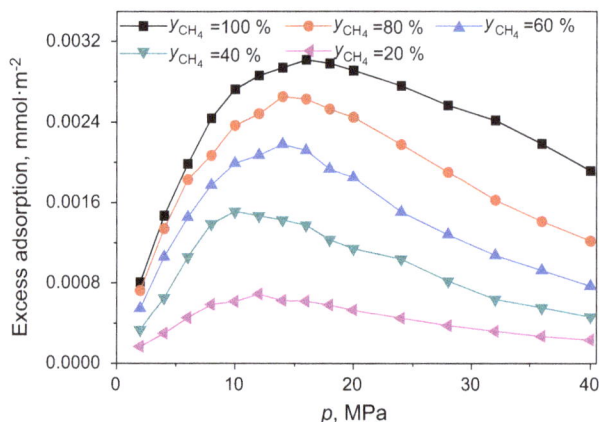

Fig. 16 Excess adsorption isotherms of methane in chlorite pores for different carbon dioxide mole fractions

methane adsorbed on the chlorite through a physical adsorption.

Second, the methane adsorption capacity in chlorite pores increased with the increase in pressure or decrease in the pore size. In chlorite micropores, the methane adsorption capacity increased with the increase in pore size, showing an opposite trend to the findings for mesopores.

In addition, the isosteric heat of methane decreased with the increase in temperature. At that time, the methane adsorption site changed from lower-energy adsorption sites to higher-energy adsorption sites, leading to the decline of the methane adsorption capacity. Water molecules in the chlorite pores occupied the pore wall in a directional manner and occupied the adsorption sites and adsorption space of methane molecules, causing a decrease in the methane adsorption capacity.

Finally, in a system of methane with nitrogen or carbon dioxide, the order of the potential energy between gas and chlorite is as follows: nitrogen > methane > carbon dioxide, implying that the adsorption capacity of carbon dioxide on chlorite is larger than that of methane and that the adsorption capacity of methane is greater than that of nitrogen. An increase in the mole fraction of nitrogen or carbon dioxide would lead to a change of the methane adsorption sites as well as a reduction in the methane adsorption space and the mole fraction of methane in the gas phase, resulting in a decrease in the methane adsorption capacity.

Acknowledgements This research was supported by the United Fund Project of National Natural Science Foundation of China (Grant No. U1262209) and the National Natural Science Foundation of China (Grant No. 41602155), and the Young Scholars Development Fund of SWPU (No. 201599010137).

References

Berendsen HJC, Grigera JR, Straatsma TP. The missing term in effective pair potentials. J Phys Chem. 1987;91(24):6269–71. doi:10.1021/j100308a038.

Curtis JB. Fractured shale gas systems. AAPG Bull. 2002;86(11):1921–38. doi:10.1306/61eeddbe-173e-11d7-8645000102c1865d.

Cygan RT, Liang JJ, Kalinichev AG. Molecular models of hydroxide, oxyhydroxide, and clay phases and the development of a general force field. J Phys Chem B. 2004;108(4):1255–66. doi:10.1021/jp0363287.

Fan EP, Tang SH, Zhang CL, Guo QL, Sun CG. Methane sorption capacity of organics and clays in high-over matured shale-gas systems. Energy Explor Exploit. 2014;32(6):927–42. doi:10.1260/0144-5987.32.6.927.

Gasparik M, Bertier P, Gensterblum Y, et al. Geological controls on the methane storage capacity in organic-rich shales. Int J Coal Geol. 2014;123:34–51. doi:10.1016/j.coal.2013.06.010.

Harris JG, Yung KH. Carbon dioxide's liquid-vapor coexistence curve and critical properties as predicted by a simple molecular model. J Phys Chem. 1995;99(31):12021–4. doi:10.1021/j100031a034.

Ji LM, Qua JL, Zhang TW, Xia YQ. Relationship between methane adsorption capacity of clay minerals and micropore volume. Earth Sci (J China Univ Geosci). 2012a;37(5):1043–50. doi:10.3799/dqkx.2012.111 (**in Chinese**).

Ji LM, Zhang TW, Milliken KL, Qua JL, Zhang XL. Experimental investigation of main controls to methane adsorption in clay-rich rocks. Appl Geochem. 2012b;27:2533–45. doi:10.1016/j.apgeochem.2012.08.027.

Jin ZH, Firoozabadi A. Methane and carbon dioxide adsorption in clay-like slit pores by Monte Carlo simulations. Fluid Phase Equilib. 2013;360:456–65. doi:10.1016/j.fluid.2013.09.047.

Jin ZH, Firoozabadi A. Effect of water on methane and carbon dioxide sorption in clay minerals by Monte Carlo simulations. Fluid Phase Equilib. 2014;382:10–20. doi:10.1016/j.fluid.2014.07.035.

Joswig W, Fuess H, Rothbauer R, et al. A neutron diffraction study of a one-layer triclinic chlorite (penninite). Am Mineral. 1980;65(3–4):349–52.

Liang LX, Luo DX, Liu XJ, Xiong J. Experimental study on the wettability and adsorption characteristics of Longmaxi Formation shale in the Sichuan Basin, China. J Nat Gas Sci Eng. 2016;33:1107–18. doi:10.1016/j.jngse.2016.05.024.

Liang LX, Xiong J, Liu XJ. Mineralogical, microstructural and physiochemical characteristics of organic-rich shales in the Sichuan Basin, China. J Nat Gas Sci Eng. 2015;26:1200–12. doi:10.1016/j.jngse.2015.08.026.

Liu XJ, Xiong J, Liang LX. Investigation of pore structure and fractal characteristics of organic-rich Yanchang Formation shale in central China by nitrogen adsorption/desorption analysis. J Nat Gas Sci Eng. 2015;22:62–72. doi:10.1016/j.jngse.2014.11.020.

Martin MG, Siepmann JI. Transferable potentials for phase equilibria.1. United-atom description of n-alkanes. J Phys Chem B. 1998;102(14):2569–77. doi:10.1021/jp972543+.

Potoff JJ, Siepmann JI. Vapor–liquid equilibria of mixtures containing alkanes, carbon dioxide, and nitrogen. AIChE J. 2001;47(7):1676–82. doi:10.1002/aic.690470719.

Rexer TF, Benham MJ, Aplin AC, Thomas KM. Methane adsorption on shale under simulated geological temperature and pressure conditions. Energy Fuels. 2013;27(6):3099–109. doi:10.1021/ef400381v.

Soave G. Equilibrium constants from a Modified Redlich–Kwong equation of state. Chem Eng Sci. 1972;27(6):1197–203. doi:10.1016/0009-2509(72)80096-4.

Sui HG, Yao J, Zhang L. Molecular simulation of shale gas adsorption and diffusion in clay nanopores. Computation. 2015;3(4):687–700. doi:10.3390/computation3040687.

Sun RY, Zhang YF, Fan KK, Shi YH, Yang SK. Molecular simulation of adsorption characteristics of clay minerals in shale. CIESC Journal. 2015;66(6):2118–22. doi:10.11949/j.issn.0438-1157.20141766 (**in Chinese**).

Talu O, Myers AL. Molecular simulation of adsorption: Gibbs dividing surface and comparison with experiment. AIChE J. 2001;47(5):1160–8. doi:10.1002/aic.690470521.

Tang SH, Fan EP. Methane adsorption characteristics of clay minerals in organic-rich shales. J China Coal Soc. 2014;39(8):1700–6. doi:10.13225/j.cnki.jccs.2014.9015 (**in Chinese**).

Titiloye JO, Skipper NT. Monte Carlo and molecular dynamics simulations of methane in potassium kaolinite clay hydrates at elevated pressures and temperatures. J Colloid Interface Sci. 2005;282(2):422–7. doi:10.1016/j.jcis.2004.08.131.

U.S. Energy Information Administration (EIA). Technically Recoverable Shale Oil and Shale Gas Resources: An Assessment of 137 Shale Formations in 41 Countries Outside the United States. http://www.eia.gov/analysis/studies/worldshalegas. 24 Sep 2013.

Xiong J, Liu XJ, Liang LX. Experimental study on the pore structure characteristics of the Upper Ordovician Wufeng Formation shale in the southwest portion of the Sichuan Basin, China. J Nat Gas Sci Eng. 2015a;22:530–9. doi:10.1016/j.jngse.2015.01.004.

Xiong J, Liu XJ, Liang LX, Lei M. Improved Dubibin–Astakhov model for shale gas supercritical adsorption. Acta Pet Sinica. 2015b;36(7):849–57. doi:10.7623/syxb2015.07.009 **(in Chinese)**.

Xiong J, Liu XJ, Liang LX. Molecular simulation on the adsorption behaviors of methane in montmorillonite slit pores. Acta Pet Sinica. 2016;37(8):1021–9. doi:10.7623/syxb201608008 **(in Chinese)**.

Yang F, Ning ZF, Zhang R, Zhao HW, Krooss BM. Investigations on the methane sorption capacity of marine shales from Sichuan Basin, China. Int J Coal Geol. 2015;146:104–17. doi:10.1016/j.coal.2015.05.009.

Yang X, Zhang C. Structure and diffusion behavior of dense carbon dioxide fluid in clay-like slit pores by molecular dynamics simulation. Chem Phys Lett. 2005;407(4):427–32. doi:10.1016/j.cplett.2005.03.118.

Zeng QH, Yu AB, Lu GQ, Standish RK. Molecular dynamics simulation of organic-inorganic nanocomposites: layering behavior and interlayer structure of organoclays. Chem Mater. 2003;15(25):4732–8. doi:10.1021/cm0342952.

5

A parametric study of the hydrodynamic roughness produced by a wall coating layer of heavy oil

S. Rushd[1] · R. S. Sanders[1]

Abstract In water-lubricated pipeline transportation of heavy oil and bitumen, a thin oil film typically coats the pipe wall. A detailed study of the hydrodynamic effects of this fouling layer is critical to the design and operation of oil–water pipelines, as it can increase the pipeline pressure loss (and pumping power requirements) by 15 times or more. In this study, a parametric investigation of the hydrodynamic effects caused by the wall coating of viscous oil was conducted. A custom-built rectangular flow cell was used. A validated CFD-based procedure was used to determine the hydrodynamic roughness from the measured pressure losses. A similar procedure was followed for a set of pipe loop tests. The effects of the thickness of the oil coating layer, the oil viscosity, and water flow rate on the hydrodynamic roughness were evaluated. Oil viscosities from 3 to 21300 Pa s were tested. The results show that the equivalent hydrodynamic roughness produced by a wall coating layer of viscous oil is dependent on the coating thickness but essentially independent of oil viscosity. A new correlation was developed using these data to predict the hydrodynamic roughness for flow conditions in which a viscous oil coating is produced on the pipe wall.

Keywords Pipeline transportation · Heavy oil · Wall fouling · Lubricated pipe flow · CFD simulation

List of symbols

A_{eff}	Effective cross-sectional area (m^2)
D	Internal diameter of the pipeline (m)
D_h	Hydraulic diameter (m)
D_{eff}	Effective diameter (m)
f	Friction factor
H	Nominal height of the test cell (m)
h_{eff}	Effective height (m)
h_{tp}	Height of the test plate (m)
k	Turbulence kinetic energy (m^2/s^2)
k_s	Nikuradse sand grain equivalent hydrodynamic roughness (m)
m_w	Mass flow rate of water (kg/s)
p	Static (thermodynamic) pressure (Pa)
ΔP	Pressure loss (Pa)
$\Delta P/$	Pressure gradient (Pa/m)
L	
P_{ij}	Reynolds stress production tensor
Re_w	Water Reynolds number
S_i	Sum of body forces
T	Temperature (°C)
t_c	Average thickness of wall fouling/coating layer (mm)
U_i	Velocity vector
V	Average water velocity (m/s)
V_{eff}	Effective velocity (m/s)
δ_{ij}	Identity matrix or Kronecker delta function
τ_{ij}	Stress tensor
ρ_w	Water density (kg/m^3)
μ	Viscosity (Pa s)
μ_o	Viscosity of coating oil (Pa s)
μ_t	Turbulent viscosity (Pa s)
μ_w	Water viscosity (Pa s)
Φ_{ij}	Pressure–strain tensor
ω	Turbulence eddy frequency

✉ R. S. Sanders
   sean.sanders@ualberta.ca

[1] Department of Chemical and Materials Engineering, University of Alberta, Edmonton, AB, Canada

Edited by Yan-Hua Sun

1 Introduction

The reserves of non-conventional oils, such as heavy oil and bitumen, form a substantial part of the known global petroleum resources (IEA 2014; CAPP 2014). These oil reserves are asphaltic, dense, and highly viscous, with bitumen being more dense and viscous than heavy oil (Saniere et al. 2004). Densities of these oils are nearly the same as that of water, whereas their viscosities are higher than that of water by orders of magnitude (McKibben et al. 2000). Therefore, the production of these non-conventional oils requires extraordinary techniques that are not needed to recover traditional petroleum deposits. Various mining and in situ production technologies are being used to extract non-conventional oils in Canada. After extraction, the oil is transported from a production site to an upgrading facility. Since pipeline transportation is a cost-effective technology, the non-conventional oil industry is keen to use this technology for transporting both bitumen and heavy oil (Nunez et al. 1998; Saniere et al. 2004; Hart 2014).

Water-lubricated pipeline transportation of non-conventional oils, known as lubricated pipe flow (LPF), is one option for transporting these viscous fluids. It is more economical when compared with other technologies, such as heating, solvent dilution, emulsification, and partial upgrading (Nunez et al. 1998; Saniere et al. 2004). In LPF, a thin water annulus prevents continuous contact between the pipe wall and the viscous oil core, resulting in much lower energy requirements than would be needed to transport the viscous oil alone in the pipeline (Arney et al. 1993; Joseph et al. 1999; McKibben et al. 2000; Rodriguez et al. 2009; de Andrade et al. 2012). The water could be naturally present in the oil or could be injected for the purpose of producing LPF. A concern for the application of LPF is the formation of an oil film on the wall (Nunez et al. 1998; Saniere et al. 2004). This oil layer is usually identified as "wall fouling." The probable flow regime in LPF is schematically presented in Fig. 1. In this figure, a fouling oil layer is shown to surround a thin water annulus that lubricates the oil-rich core. The mechanism of wall fouling has not previously been studied in any detail. Previous

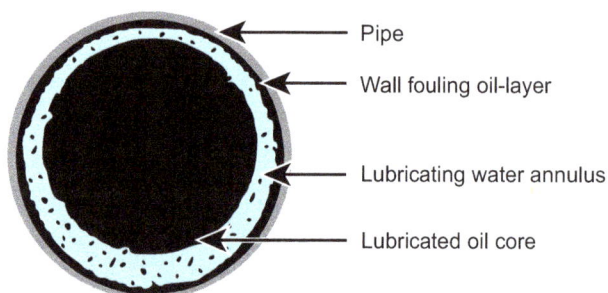

Fig. 1 Illustration of the flow regime in a LPF pipeline

experimental works suggest the fouling layer is a natural consequence of the lubrication process (Joseph et al. 1999; Vuong et al. 2009). Frictional pressure losses in a fouled pipe are much higher (say, 8–15 times) than those measured in an unfouled pipe (Arney et al. 1996), but still orders of magnitude lower than those would be expected for transporting only heavy oil or bitumen. It has been found in repeated tests that the formation of this wall coating is practically unavoidable in the industrial application of LPF technology (McKibben et al. 2000, 2016). Different degrees of wall fouling occur depending on the conditions of LPF, e.g., water cut, oil viscosity, and superficial velocity (Joseph et al. 1999; Schaan et al. 2002; Rodriguez et al. 2009; Vuong et al. 2009).

The wall fouling layer in a water-lubricated pipeline can be considered as a stationary coating film of viscous oil adhered on the pipe wall. This is because the relative velocity of this layer is negligible compared to the average mixture velocity (Joseph et al. 1999; McKibben et al. 2000, 2016; Shook et al. 2002; Schaan et al. 2002; Vuong et al. 2009). This wall coating layer can produce a large equivalent hydrodynamic roughness value: The typical equivalent roughness of a commercial steel pipe is about 0.045 mm, while the hydrodynamic roughness (inferred from pressure loss measurements) of a pipeline with a viscous oil layer on the pipe wall can be greater than 1 mm (Brauer 1963; Shook et al. 2002). The roughness is produced primarily through contact between the viscous oil coating and the turbulent water layer that flows over the film while lubricating the oil core. The result is a rippled/rough wall that is associated with very large hydrodynamic roughness values (Brauer 1963; Picologlou et al. 1980; Shook et al. 2002). While the presence of the coating reduces somewhat the cross-sectional area available for flow, which also causes an increase in pressure loss for a given throughput, the increased hydrodynamic roughness plays a much more important role in this increase.

The conventional method for describing the hydrodynamic roughness produced by a rough solid wall is the Nikuradse sand grain equivalent (Perry and Green 1997; Whyte 1999). This definition of equivalent roughness is extensively used for commercial metal pipes or channels and has also been used to describe the hydrodynamic roughness caused by a biofilm on a solid wall (Picologlou et al. 1980). Much like an oil fouling layer, the biofilm is conformable and can substantially increase the hydrodynamic roughness, in turn increasing power requirements for pumping water through bio-fouled pipes and channels (Barton et al. 2008; Andrewartha et al. 2008).

Previous studies of equivalent hydrodynamic roughness involved either a rectangular flow cell or a pipe for experiments (Barton et al. 2008; Andrewartha et al. 2008). In rectangular flow cells, one wall is typically "roughened" (e.g., through the formation of a biofilm) while the other three walls are kept hydrodynamically smooth. The time-averaged velocity profile perpendicular to the rough wall is then measured to determine the hydrodynamic roughness on the basis of correlations, such as the law of the wall (Andrewartha et al. 2008). The reliability of the measurement was subject to the type of instrumentation selected for the measurements and also the size of the flow cell. Typically, a large channel was used to ensure that the measured velocity profile would not be affected by the presence of the walls.

Pressure loss measurements have been typically used to determine the hydrodynamic roughness for the pipeline tests using some basic equations of fluid dynamics, such as the Darcy–Weisbach equation or the Colebrook correlation (e.g., Barton et al. 2008). A basic analytical approach such is appropriate when the hydrodynamic roughness can be represented by a single, constant value. In other words, this approach is not applicable for the flow cells with asymmetric wall roughness.

In the present study, a customized rectangular flow cell was used to perform a parametric investigation of the equivalent hydrodynamic roughness produced by a wall coating of viscous oil. A relatively small test cell was chosen because the goal was to test a number of different oils under a wide range of flow conditions. As a result, it was very difficult to make accurate measurements of the velocity profile of the flow above the coated surface and instead the pressure loss (under fully developed flow conditions) was measured for the different flow conditions tested. The asymmetry of wall roughness in the flow cell (one rough wall and three smooth walls), however, meant that a simple analytical approach to relate pressure loss to hydrodynamic roughness could not be used. Therefore, the flow conditions in the experimental cell were modeled using a commercially available CFD package (ANSYS CFX 13.0) and simulations were conducted to determine the hydrodynamic roughness that would give a predicted pressure loss equal to that measured during an experiment. The validated CFD-based procedure has been described in detail elsewhere (Rushd et al. 2016). Based on the results presented here, a new correlation is proposed for the equivalent hydrodynamic roughness produced by a viscous layer of wall coating in terms of the coating thickness. This correlation can be used to estimate the hydrodynamic roughness from a measured or a known value of the physical wall coating thickness.

2 Experimental setup and procedure

A 2.5-m-long rectangular flow cell was designed and fabricated for the present study. The flow cell consisted of a channel whose lower surface was comprised of segmented steel plates. These plates were coated with a measured, constant thickness (t_c) of oil prior to the start of each flow experiment. The effective cross section of the flow channel without a wall coating was 15.9 mm × 25.4 mm. Its entrance length was 1.5 m, which was more than $60D_h$; where $D_h = 19.6$ mm is the hydraulic diameter defined as $4A/P$, where A is the cross-sectional area and P is the wetted perimeter of the cross-sectional area. The flow cell had two Plexiglas windows so that it was possible to observe the shape of oil–water interface. This custom-built cell was placed in a 25.4-mm pipe loop as shown in Fig. 2a. A cross-sectional view of the flow cell is presented as Fig. 2b. A photograph of the cell under actual flow conditions, after the rough/rippled topology was developed on the wall coating layer, is given in Fig. 2c.

The steady-state pressure loss across the flow cell was measured with a differential pressure transducer (Validyne P61). The experiments were conducted for a range of water flow rates, coating thickness, and oil viscosities. Typical pressure gradient measurements (30 s averages) are presented for a specific flow condition in Fig. 3. These results demonstrate that the change in pressure loss for a given water flow rate is negligible when the thickness of the wall coating layer is constant.

For experiments, each of the segmented plates comprising the bottom wall of the flow visualizing section was coated separately with a specific thickness of the viscous oil and placed in the flow cell to form a coating layer of uniform thickness. The average thickness of the coating layer (t_c) was determined by weighing the test plates without and with coating oil. It should be noted that the coated plates were also weighed before and after each experiment. The difference between the measured weights was negligible, i.e., t_c was unaffected by the flow rate and thus was taken as a controlled parameter.

Please refer to Rushd (2016) for more detailed descriptions of the experimental apparatus, procedures, and results.

3 Experimental parameters

The rectangular flow cell was used to study the hydrodynamic effect of different viscous wall coatings. The measured variable was the pressure loss (ΔP). The controlled parameters are listed in Table 1. The most important of these parameters are the average thickness (t_c) and the viscosity (μ_o) of the coating oil. Recall that only the bottom wall of the rectangular flow cell was coated with oil. The experimental value of t_c for an oil was selected depending on oil viscosity (μ_o) and flow rate of water (m_w). The thickness (t_c) that could be maintained under the highest flow rate for the lower viscosity oils ($\mu_o \sim 65$ and

(a)

(b)

(c)

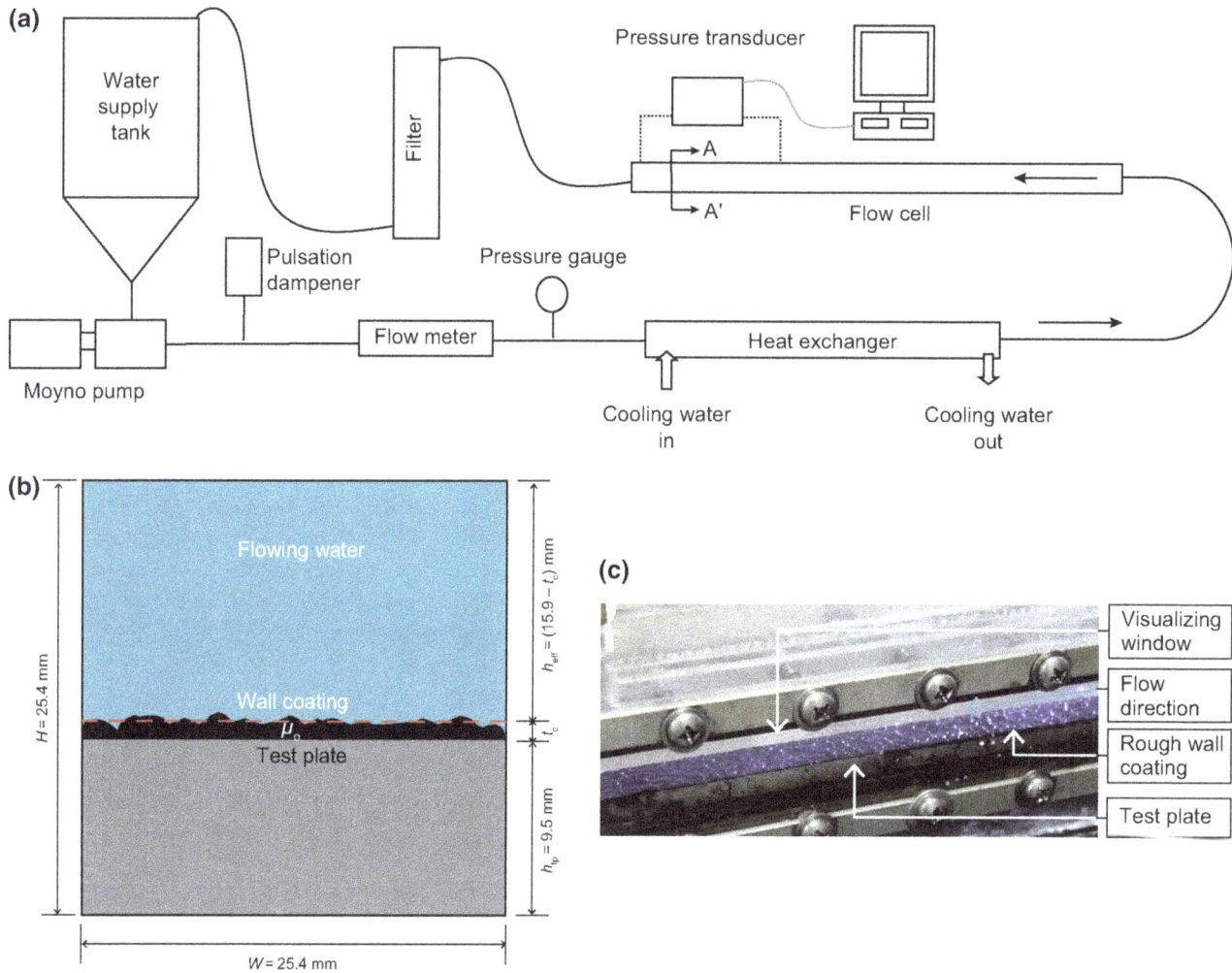

Fig. 2 Illustration of the experimental facility. **a** Schematic presentation of the flow loop. **b** Cross-sectional view (section A–A') of the flow cell. **c** Photograph showing a test with a rough wall coating of viscous oil (μ_o = 21300 Pa s)

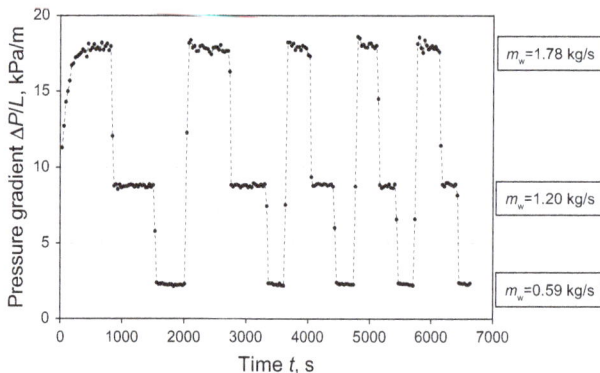

Fig. 3 Illustration of pressure gradients ($\Delta P/L$) measured over time (t) for different water flow rates (m_w) (t_c = 0.2 mm; μ_o = 21300 Pa s)

Table 1 Controlled parameters for the rectangular flow cell experiments

Controlled parameter	Value(s)
Thickness of wall coating t_c, mm	0.2, 0.5, and 1.0
Viscosity of coating oil μ_o, Pa s	65, 320, 2620, and 21300
Mass flow rate of water m_w, kg/s	0.59, 0.91, 1.20, 1.52, and 1.78
Water Reynolds numbers Re_w	$29000 < Re_w < 87000$
Flow temperature T, °C	20

for these oils were 0.2, 0.5, and 1.0 mm. The overall uncertainty associated with the measurement of t_c in the flow cell was 10%. Thus, the coating thickness (t_c) for the first phase of experiments was selected so that it would not change significantly with water flow rate. The purpose of these tests was to evaluate the effects of the flow rate and the viscosity on the hydrodynamic roughness while keeping the coating thickness constant.

320 Pa s) in the flow cell was 0.2 mm. Similarly, the maximum t_c for the higher-viscosity oils (μ_o ~ 2620 and 21300 Pa s) was 1.0 mm. Coating thickness values tested

4 CFD simulations

As mentioned previously, the CFD simulations were used to determine the unknown equivalent sand grain roughness (k_s) of the oil-covered bottom wall of the flow cell. This was done by modeling the water flow through the cell over the viscous coating. The CFD software package, ANSYS CFX 13.0, was used for simulation. The software solves the governing differential equations that include Reynolds-averaged Navier–Stokes (RANS) continuity and momentum equations. The Reynolds stress term in RANS was modeled using an omega-based Reynolds stress model, ω-RSM. Full details of the governing equations are given in Appendix 1.

The geometry of the 3D computational domain used for the simulation was identical to the rectangular flow cell; however, two different flow cell lengths ($l = 1.0$ m; $l = 2.0$ m) were used for computations even though the actual flow cell was 1.0 m in length. This was done to ensure the length independence of the simulations. The width (w) was equal to that of the flow cell (25.4 mm). The height ($h = 15.9 - t_c$ mm) was varied depending on the average thickness (t_c) of oil coating on the bottom wall. The values of t_c tested during the present study are shown in Table 1.

The flow geometry was created and meshed with ANSYS ICEM CFD. The software was used to discretize the flow domain into structured grids, one for the bulk of the flow and one for the near-wall region. Coarse, intermediate, and fine mesh grids were tested. The mesh resolution was based on the number of nodes, n, in each mesh. In the present study, the mesh resolution is classified as follows: coarse ($n < 50000$), intermediate ($50000 < n < 500000$), and fine ($n > 500000$). The total number of nodes found to be sufficient for grid independence was $n = 670200$. An example of the fine mesh used here is shown in Fig. 4. The number of nodes in the near-wall region was selected so that $y^+ > 11.06$. At $y^+ = 11.06$, ANSYS CFX 13.0 uses the logarithmic law of the wall (i.e., the wall function). For these simulations, typically, $y^+ = 25$. All computations were performed to obtain steady-state solutions. Double precision was used in the computations, and solutions were considered converged when the normalized sum of the absolute dimensionless residuals of the discretized equations was less than 10^{-6}. The typical computational time required for the convergence of a simulation was 45 min.

At the inlet of the flow domain, the experimental mass flow rate of water and a turbulent intensity of 5% were prescribed as the boundary condition. A zero pressure condition was specified at the outlet. The no-slip condition was used at the boundaries representing walls. The two side walls and the upper wall in the rectangular domain were taken as hydrodynamically smooth ($k_s = 0$) based on the results of simulations conducted for clean walls (Rushd et al. 2016). Flow conditions where the bottom wall was coated with oil required one to specify the k_s value for this wall. However, the values of k_s were unknown for the oil-coated bottom wall of the flow cell for any given flow condition. A trial and error procedure was adopted to determine the appropriate k_s value. Starting from a low value, k_s was increased in steps and the simulation was repeated until a reasonable agreement between the measured and predicted pressure loss (maximum 5% difference) was observed. The final value of k_s at which this condition was met was considered to be the equivalent hydrodynamic roughness of the corresponding rough wall. The trial and error approach is summarized in Fig. 5. This CFD-based trial and error approach of estimating k_s was validated using data on biofilms taken from the literature and from flow cell tests using materials of known roughness (Rushd et al. 2016).

5 Results and discussion

As mentioned earlier, two effects were produced by the wall coating layer: a slight reduction in the effective flow area and a drastically increased hydrodynamic roughness (k_s). The reduction in the flow area was taken into account through the average thickness of the wall coating layer (t_c), which is a physical parameter that can be measured directly. The hydrodynamic roughness (k_s) value corresponding to each combination of viscous wall coating thickness (t_c), and water Reynolds number was determined using the CFD-based procedure describe above. The results were used to develop a correlation between k_s and t_c.

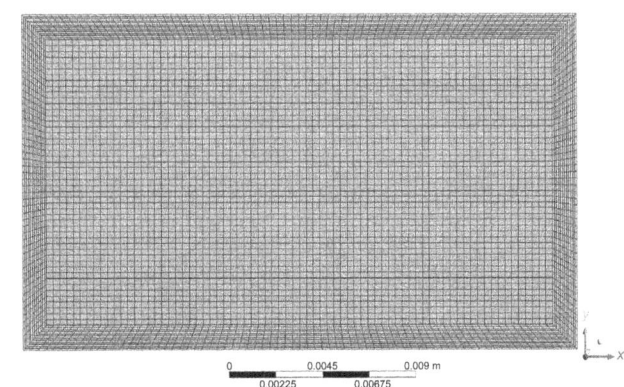

Fig. 4 Two-dimensional illustration of the fine mesh used for simulations

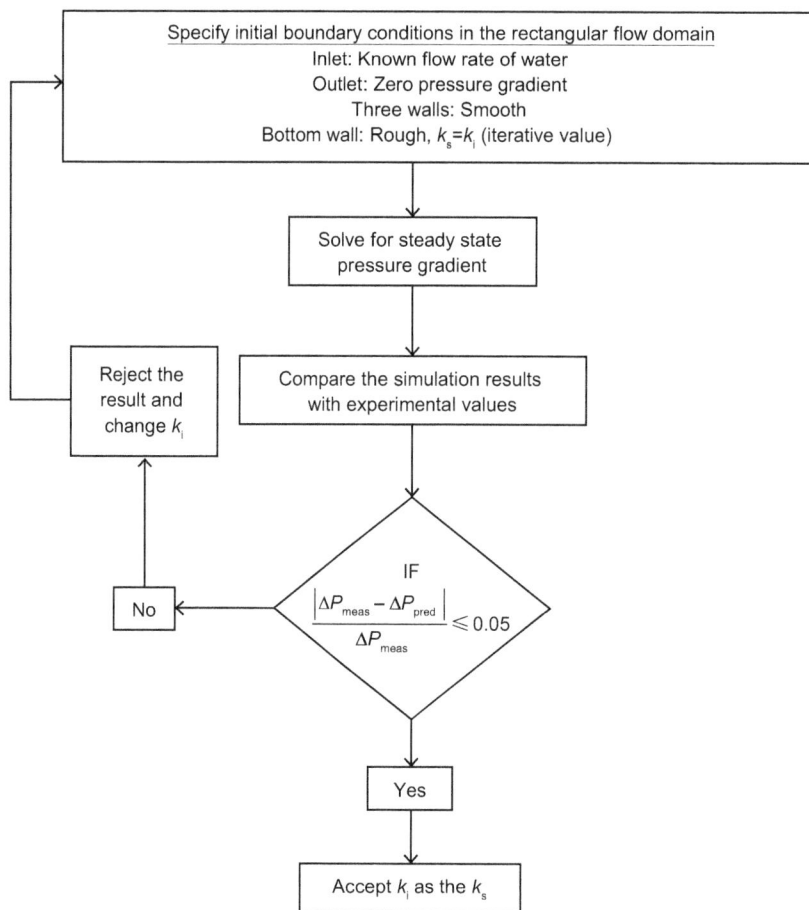

Fig. 5 Flowchart describing the steps involved in the simulation procedure for computing the equivalent sand grain roughness (k_s)

5.1 Rectangular flow cell results

The effect of wall coating thickness (t_c) on the measured pressure gradient, for tests involving a specific oil ($\mu_o = 2620$ Pa s), is demonstrated in Fig. 6a. It can be seen from the figure that operation at higher average velocities ($V = m_w/(\rho_w A_{eff})$) causes the pressure gradients ($\Delta P/L$) to increase approximately with V^2, as would be expected for the turbulent flow of water through a channel or pipe. Note, however, that compared to the clean wall condition, the measured pressure gradients are significantly higher when the wall is coated with oil ($t_c > 0$). Clearly, the primary contributor to the measured pressure loss at any velocity is the presence of the oil coating in the flow cell. Although four different oils with viscosities ranging from 65 to 21300 Pa s were tested (see Table 1), the results for any given oil were almost identical to those presented in Fig. 6a (Rushd 2016). In other words, oil viscosity played a negligible role over the range of viscosities tested here. The observation that viscosity of the coating layer (μ_o) had no appreciable effect on the measured pressure gradients ($\Delta P/L$) is demonstrated in Fig. 6b.

As can be observed from Fig. 6a, a small increase in coating thickness (t_c) causes a significant increase in pressure gradient ($\Delta P/L$). The cause of this substantial increase is related primarily to the increase in hydrodynamic roughness of the oil coating layer produced through its interaction with the turbulent water flow through the channel. The coating thickness t_c reduces D_h by 0.5%–4% (depending on the value of t_c tested). If the wall coating layers behaved hydrodynamically as "smooth" surfaces ($k_s = 0$), the reduced D_h would cause a 4%–20% increment in $\Delta P/L$. The range can be calculated on the basis of Blasius law for a rectangular flow cell (Jones 1976). As Figs. 6a and 7 show, the measured increase in pressure loss with increasing t_c is in the range of 50%–200%. The substantial increase in $\Delta P/L$ indicates the importance of the roughness of the coating layer.

In Fig. 7, the measured values are shown in comparison with the predictions obtained from CFD simulations. The simulated pressure gradients agree well with the corresponding measurements when the rectangular flow cell was clean, i.e., when the bottom wall was not coated with oil ($t_c = 0$). For these simulations, all four walls of the

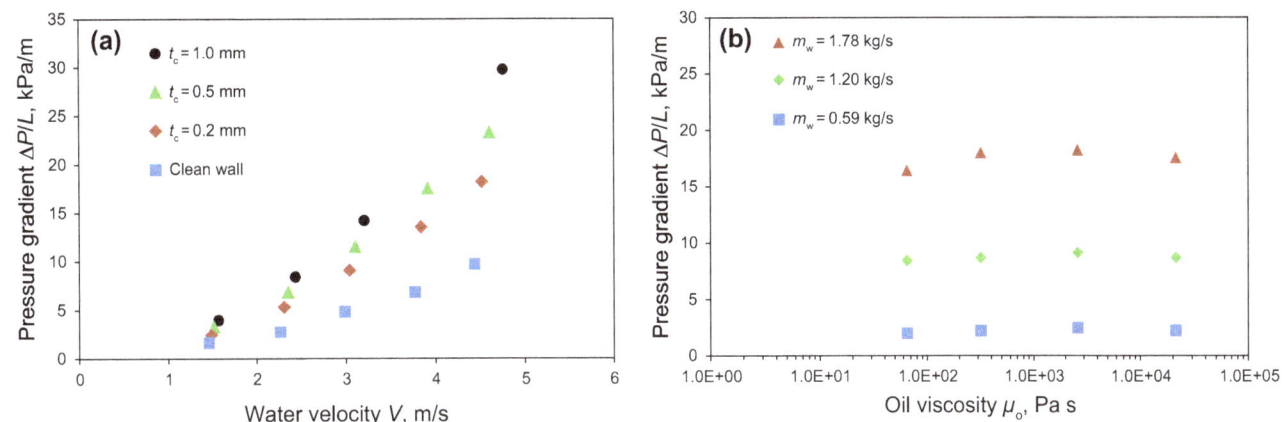

Fig. 6 Experimental results for the rectangular flow cell. **a** Pressure gradient ($\Delta P/L$) as a function of bulk water velocity (V) and oil coating thickness (t_c) ($\mu_o = 2620$ Pa s). **b** Effect of water flow rate (m_w) and oil viscosity for a fixed coating thickness ($t_c = 0.2$ mm)

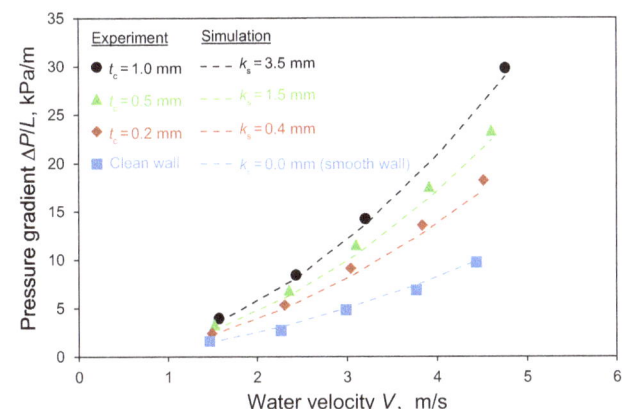

Fig. 7 Comparison of simulation and experimental results for the rectangular flow cell ($\mu_o = 2620$ Pa s)

rectangular flow cell were considered "smooth," i.e., $k_s = 0$. The agreement between the experimental and simulation results indicates the clean walls of the rectangular flow cell to be hydrodynamically smooth. The figure also shows that when the bottom wall was coated with oil, the measured values of $\Delta P/L$ could be accurately predicted. Another important point to note is that the hydrodynamic roughness (k_s) produced by a constant coating thickness (t_c) was not dependent on velocity, for the range of velocities tested here.

5.2 Pipe loop results

The CFD-based methodology of determining the equivalent hydrodynamic roughness that was developed for the flow cell experiments was then applied to determine the values of k_s for comparable tests carried out with a recirculating pipe loop. A 103.3-mm (ID) pipe having an internal wall fouled/coated with two different heavy oils ($\mu_o \sim 3$ and 27 Pa s) was used in experiments. The wall coatings were developed in the course of testing LPF. After

completing a set of LPF tests, water at 20 °C was pumped through the pipeline, replacing the oil core. The flow scenario for the pipeline testing is shown schematically in Fig. 8. Pressure loss and wall coating thickness measurements were made simultaneously at mean (bulk) water velocities of $V = 0.5$, 1.0, 1.5, and 2.0 m/s. A custom-built double pipe heat exchanger (Schaan et al. 2002) and a hot film probe were used to obtain wall coating thickness measurements. The wall coating thickness for the pipeline tests decreased with increasing velocity, i.e., t_c values were dependent on V because the coating was partially stripped from the wall as the water velocity was increased. The pressure measurements always reached a steady-state condition at each velocity, which allowed for the calculation a steady-state value of k_s. A more detailed description of the apparatus and test procedure is provided by McKibben et al. (2016).

As was done for the rectangular flow cell tests, CFD simulations of the pipe loop tests were conducted to

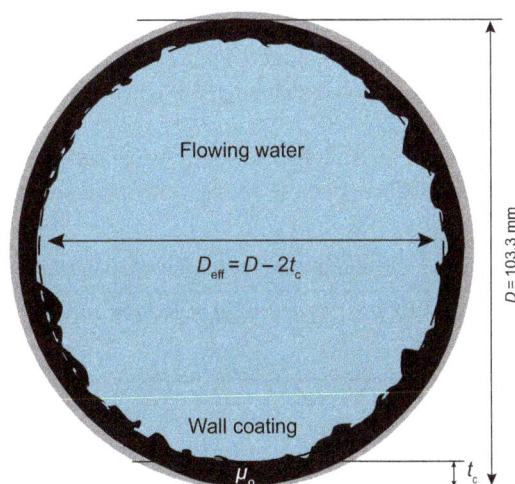

Fig. 8 Schematic cross-sectional view of test section in the pipeline

determine the equivalent hydrodynamic roughness (k_s). A typical comparison of the measured pressure gradients and those obtained from simulations for the pipeline tests are shown in Table 2. Only the results for the higher-viscosity oil ($\mu_o \sim 27$ Pa s) are shown, as the trend is similar for the other oil ($\mu_o \sim 3$ Pa s). Because of the way the pipe loop experiments were conducted, different values of t_c were tested at different water velocities. As the results in Table 2 show, the agreement between the measured pressure gradients and the values determined using the CFD methodology (where k_s is set by trial and error) is excellent.

The values of k_s for the pipeline tests were corroborated by estimating the same values on the basis of the Colebrook correlation:

$$\frac{1}{\sqrt{f}} = -2\log_{10}\left(\frac{k_s}{3.7 D_{\text{eff}}} + \frac{2.51}{Re_w\sqrt{f}}\right), \qquad (1)$$
$$4 \times 10^3 < Re < 1 \times 10^8$$

The values of k_s for these tests were determined using Eq. (1) and the CFD methodology. These are two completely different approaches for determining k_s. The values calculated on the basis of the Colebrook formula agree reasonably well with those obtained with the CFD method. The results from the two calculation methods are presented in Table 2 for the higher-viscosity oil (~ 27 Pa s). Similar agreement was found when comparing the two calculation methods for the 3 Pa s oil coating as well.

5.3 Correlation development

A correlation between k_s and t_c is proposed here, on the basis of the rectangular flow cell data and the pipe flow tests described previously:

$$k_s = 2.76 t_c, \quad 0.2 \text{ mm} \le t_c \le 2.0 \text{ mm} \qquad (2)$$

The correlation is illustrated in Fig. 9. The proportionality constant of the equation is determined with a regression analysis for which $R^2 = 0.96$. The average uncertainty associated with the predictions of this correlation is $\pm 14\%$. The data set used for developing the correlation set presented in Table 3.

As shown in Fig. 9, eight data points were used to develop the correlation. Three of these points were obtained from the

experiments conducted with the flow cell, and five points were obtained from pipeline tests. Multiple combinations of oil viscosity, water flow rate, and coating thickness were used for the flow cell experiments. Therefore, the three data points for the rectangular flow cell actually correspond to 24 different flow conditions, meaning the correlation is based on 29 distinct flow conditions.

The relationship between k_s and t_c proposed in this work is the first of its kind. To the best of our knowledge, a similar correlation is not available in the literature. An example of its application to predict pressure losses in a fouled/coated pipeline is presented in Appendix 2. Using the correlation in its current form is subject to the knowledge of coating thickness (t_c), which is possible to measure for a LPF system (Schaan et al. 2002). Additional efforts are currently underway to correlate t_c with flow parameters so that its direct measurement will no longer be necessary.

6 Summary

The objective of the present study is to provide detailed information about the hydrodynamic roughness that a wall coating of viscous oil produces. The results reported here can be summarized as follows:

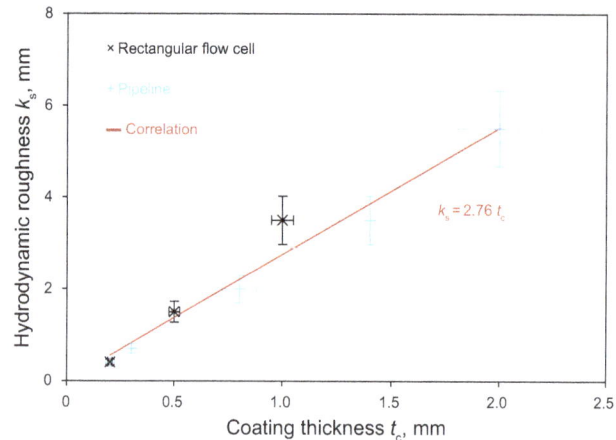

Fig. 9 Correlation between hydrodynamic roughness (k_s) and coating thickness (t_c)

Table 2 Comparison of equivalent hydrodynamic roughness for pipeline tests ($\mu_o \sim 27$ Pa s)

Water velocity V, m/s	Wall coating thickness t_c, mm	Pressure gradient $\Delta P/L$, kPa/m		Hydrodynamic roughness k_s, mm	
		Measured	CFD method	Colebrook correlation	CFD method
1.0	2.0	0.45	0.45	5.9	5.5
1.5	1.4	0.81	0.84	4.1	3.5
2.0	0.8	1.10	1.17	2.5	2.0

Table 3 Data used to develop the correlation between k_s and t_c (Eq. 2)

Apparatus	Hydraulic diameter D_h, mm	Oil viscosity μ_o, Pa s	Average velocity V, m/s	Coating thickness t_c, mm	Hydrodynamic roughness k_s, mm
Rectangular flow cell	20	65, 320, 2 620, 21 300	1.5	0.2	0.4
			3.1		
			4.5		
		2 620, 21 300	1.5	0.5	1.5
			3.1		
			4.6		
			1.6	1.0	3.5
			3.2		
			4.8		
Pipeline	100	3	1.0	0.2	0.4
			1.5	0.3	0.7
		27	1.0	0.8	2.0
			1.5	1.4	3.5
			2.0	2.0	5.5

1. A film of viscous wall coating substantially increases the measured pressure loss, primarily as a result of the rough/rippled structure that forms on the surface of the coating, which produces a very large value of the equivalent hydrodynamic roughness, k_s.
2. Experiments were conducted using two different geometries—a rectangular flow cell and a 100-mm (diameter) pipeline loop—and using different oils to produce the wall coating layer. The oil viscosities spanned four orders of magnitude ($3 \leq \mu_o \leq 2.1 \times 10^4$ Pa s). Water alone (i.e., no oil in the bulk flow) was circulated through both test cells under highly turbulent conditions ($2.9 \times 10^4 < Re_w < 2 \times 10^5$).
3. The results obtained from the two test geometries were in very close agreement and showed that the hydrodynamic roughness produced by a wall coating of viscous oil was essentially independent of oil viscosity. In fact, the thickness of the coating layer directly determines its hydrodynamic roughness.
4. A new correlation was proposed to relate the hydrodynamic roughness produced by a viscous wall to the thickness of the coating layer. The correlation will be a critical component of any advance model of lubricated pipe flow with wall fouling.

Acknowledgements The research was conducted through the support of the *NSERC Industrial Research Chair in Pipeline Transport Processes* (held by RS Sanders). The contributions of Canada's Natural Sciences and Engineering Research Council (NSERC) and the Industrial Sponsors (Canadian Natural Resources Limited, Fort Hills LLP, Nexen Inc., Saskatchewan Research Council Pipe Flow Technology Centre[TM], Shell Canada Energy, Syncrude Canada Ltd., Total E&P Canada Ltd., Teck Resources Ltd. and Paterson & Cooke Consulting Engineers Ltd.) are recognized with gratitude. We are also grateful to Dr. Adane and Dr. Islam for their kind assistance in preparing the manuscript.

Appendix 1: description of ω-RSM

Most important features of the ω-RSM are described here on the basis of Fletcher et al. (2009) and ANSYS CFX-Solver Theory Guide (2010). In this narrative, the differential equations are presented with index notation[1]. The Reynolds-averaged Navier–Stokes (RANS) equations of continuity and momentum transport for an incompressible fluid can be presented with following equations.

Continuity:

$$\frac{D\rho}{Dt} = 0 \tag{3}$$

Momentum transport:

$$\rho \frac{DU_i}{Dt} = -\frac{\partial}{\partial x_i}\left[p + \frac{2}{3}\mu \frac{\partial U_k}{\partial x_k}\right] + \frac{\partial}{\partial x_j}\left[\mu\left(\frac{\partial U_i}{\partial x_j} + \frac{\partial U_j}{\partial x_i}\right)\right] - \rho \frac{\partial}{\partial x_j}\left(\tau_{ij}\right) + S_i \tag{4}$$

[1] In Cartesian coordinates, for example, U_i represents all three components (x, y, z) of the vector U. Likewise, τ_{ij} stands for the six components $(xx, xy, xz, yx, yy, yz, zx, zy, zz)$ of the tensor τ. The differential operators are denoted similarly. Also, the summation convention is implied.

In Eq. (4), p is the static (thermodynamic) pressure; S_i is the sum of body forces, and τ_{ij} is the fluctuating Reynolds stress contributions.

A number of models are available in ANSYS CFX 13.0 for the Reynolds stresses (τ_{ij}) in the RANS equations. Among the available models, ω-RSM is selected as the most suitable for the current work. In this model, τ_{ij} is made to satisfy a transport equation. A separate transport equation is solved for each of the six Reynolds stress components of τ_{ij}. The differential transport equation for Reynolds stress is as follows:

$$\rho \frac{D\tau_{ij}}{Dt} = -\rho P_{ij} - \rho \Phi_{ij} + \frac{2}{3} \beta' \rho \omega k \delta_{ij} + \frac{\partial}{\partial x_k} \left[\left(\mu + \frac{\mu_t}{\sigma_k} \right) \frac{\partial \tau_{ij}}{\partial x_k} \right]$$

(5)

The Reynolds stress production tensor P_{ij} is given by:

$$P_{ij} = \tau_{ik} \frac{\partial U_j}{\partial x_k} + \tau_{jk} \frac{\partial U_i}{\partial x_k}, P = \frac{1}{2} P_{kk}$$

(6)

The constitutive relation for the pressure–strain term Φ_{ij} in Eq. (5) is expressed as follows:

$$\Phi_{ij} = \beta' C_1 \rho \omega \left(\tau_{ij} + \frac{2}{3} k \delta_{ij} \right) - \hat{\alpha} \left(P_{ij} - \frac{2}{3} P \delta_{ij} \right)$$
$$- \hat{\beta} \left(D_{ij} - \frac{2}{3} P \delta_{ij} \right) - \hat{\gamma} \rho k \left(S_{ij} - \frac{1}{3} S_{kk} \delta_{ij} \right)$$

(7)

In Eq. (7), the tensor D_{ij} and the model coefficients are

$$D_{ij} = \tau_{ik} \frac{\partial U_k}{\partial x_j} + \tau_{jk} \frac{\partial U_k}{\partial x_i}$$

(8)

Model coefficients:

$\beta' = 0.09$

$\hat{\alpha} = (8 + C_2)/11$

$\hat{\beta} = (8C_2 - 2)/11$

$\hat{\gamma} = (60C_2 - 4)/55$

$C_1 = 1.8$

$C_2 = 0.52$

In addition to the stress equations, the ω-RSM uses the following equations with corresponding coefficients for the turbulent eddy frequency ω and turbulent kinetic energy k.

$$\rho \frac{D\omega}{Dt} = \alpha \rho \frac{\omega}{k} P_k - \beta \rho \omega^2 + \frac{\partial}{\partial x_k} \left[\left(\mu + \frac{\mu_t}{\sigma} \right) \frac{\partial \omega}{\partial x_k} \right]$$

(9)

$$\rho \frac{Dk}{Dt} = P_k - \beta' \rho k \omega + \frac{\partial}{\partial x_j} \left[\left(\mu + \frac{\mu_t}{\sigma_k} \right) \frac{\partial k}{\partial x_j} \right]$$

(10)

$$P_k = \mu_t \left(\frac{\partial U_i}{\partial x_j} + \frac{\partial U_j}{\partial x_i} \right) \frac{\partial U_i}{\partial x_j} - \frac{2}{3} \frac{\partial U_k}{\partial x_k} \left(3\mu_t \frac{\partial U_k}{\partial x_k} + \rho k \right)$$

(11)

Coefficients:

$\sigma^* = 2$

$\sigma = 2$

$\beta = 0.075$

$$\alpha = \frac{\beta}{\beta'} - \frac{\kappa^2}{\sigma (\beta')^{0.5}} = \frac{5}{9}$$

$\kappa = 0.41$

$\sigma_k = 2$

In the previously mentioned transport equations, the turbulent viscosity μ_t is defined as

$$\mu_t = \rho \frac{k}{\omega}.$$

(12)

Along with the basic differential equations, the flow near a stationary wall is significant for the turbulent model, ω-RSM. Usually a wall is treated with "no-slip" boundary condition for CFD simulations. Mesh-insensitive automatic near-wall treatment is available for the ω-RSM implementation in ANSYS CFX 13.0. The treatment is meant to control the smooth transition from the viscous sub-layer to the turbulent layer through the logarithmic zone. Important features of the near-wall treatment for ω-RSM are outlined as follows.

(1) In case of a hydrodynamically smooth wall, the viscous sub-layer is connected to the turbulent layer with a log-law region. Velocity profiles for the near-wall regions are as follows: Viscous sub-layer:

$$u^+ = y^+$$

(13)

Log-law region:

$$u = (1/k) \ln(y) + B - \Delta B$$

(14)

Here, $u^+ = U_t/u_\tau$, $y^+ = \rho \Delta y u_\tau / \mu = \Delta y u_\tau / v$, and $u_\tau = (\tau_w/\rho)^{0.5}$. In the log-law, B and ΔB are constants. The value of B is considered as 5.2 and that of ΔB is dependent on the wall roughness. For a smooth wall, $\Delta B = 0$. The term Δy, in the definition of y^+, is calculated as the distance between the first and the second grid points off the wall. Special treatment of y^+ in CFX allows one to arbitrarily refine the mesh.

(2) For a hydrodynamically rough wall, the roughness is scaled with Nikuradse sand grain equivalent (k_s). Dimensional roughness k_s^+ is defined as $k_s^+ u_\tau / v$. A wall is treated hydrodynamically rough when k_s^+ is greater than 70. The value of ΔB is empirically correlated to k_s^+ as follows.

$$\Delta B = \frac{1}{\kappa} \ln \left(1 + 0.3 k_s^+ \right)$$

(15)

ΔB represents a parallel shift of logarithmic velocity profile compared to the smooth wall condition.

(3) At the fully rough condition ($k_s^+ > 70$), the viscous sub-layer is assumed to be destroyed. Effect of viscosity in the near-wall region is neglected.

(4) The equivalent sand grains are considered to have a blockage effect on the flow. This effect is taken into account by virtually shifting the wall by a distance of $0.5\,k_s$.

Appendix 2: sample calculation illustrating application of the proposed correlation

An example illustrating the application of the proposed correlation, i.e., Equation (3) is presented here. The correlation is used to predict the frictional pressure loss for a specific pipe flow case, and then, the predicted value is compared with the measured value. The data for this example are reported by McKibben et al. (2016).

Measured/known parameters

Internal diameter of the pipeline (D): 103.3 mm
Average water velocity (V): 1.0 m/s
Density of water (ρ_w): 997 kg/m^3
Viscosity of water (μ_w): 0.001 Pa s
Average thickness of wall coating/fouling (t_c) (measured): 2.0 mm

Calculations

Effective diameter (D_{eff}): $D_{eff} = D - 2t_c = 99.3$ mm

Effective velocity (V_{eff}): $V_{eff} = V\left(\dfrac{D}{D_{eff}}\right)^2 = 1.1$ m/s

Reynolds number (Re_w): $Re_w = \dfrac{D_{eff}V_{eff}\rho_w}{\mu_w} = 1.1 \times 10^5$

Equivalent hydrodynamic roughness (k_s): $k_s = 2.76t_c = 5.52$ mm

Darcy friction factor (f), obtained using the Swamee–Jain correlation: $f = 0.25\left[\log\left(\dfrac{k_s}{3.7D_{eff}} + \dfrac{5.74}{Re_w^{0.9}}\right)\right]^{-2} = 0.0756$

Predicted pressure gradient

$$\frac{\Delta P}{L} = f\frac{\rho_w V_{eff}^2}{2D_{eff}} = 0.44 \text{ kPa/m}$$

Measured pressure gradient

($\Delta P/L$): 0.45 kPa/m

References

Andrewartha JM, Sargison JE, Perkins KJ. The influence of freshwater biofilms on drag in hydroelectric power schemes. WSEAS Trans Fluid Mech. 2008;3(3):201–6.

ANSYS CFX-Solver Theory Guide. Release 13.0. ANSYS Inc., Southpointe, 275 Technology Drive, Canonsburg, PA 15317. 2010. http://www.ansys.com.

Arney MS, Bai R, Guevara E, Joseph DD, Liu K. Friction factor and holdup studies for lubricated pipelining—I. Experiments and correlations. Int. J. Multiph Flow. 1993;19(6):1061–76. doi:10.1016/0301-9322(93)90078-9.

Arney MS, Ribeiro GS, Guevara E, Bai R, Joseph DD. Cement-lined pipes for water lubricated transport of heavy oil. Int. J. Multiph Flow. 1996;22(2):207–21. doi:10.1016/0301-9322(95)00064-X.

Barton AF, Wallis MR, Sarglson JE, Bula A, Walker GJ. Hydraulic roughness of biofouled pipes, biofilm character, and measured improvements from cleaning. J Hydraul Eng. 2008;134:852–7.

Brauer H. Flow resistance in pipes with ripple roughness. Chemische Zeitung (Chemist Review Eng). 1963;87:199–210.

Canadian Association of Petroleum Producers (CAPP). Crude oil, forecast, market and transportation. 2014. http://www.capp.ca.

de Andrade THF, Crivelaro KCO, Neto SRdeF, de Lima AGB. Numerical study of heavy oil flow on horizontal pipe lubricated by water. In: Ochsner A, da Silva LFM, Altenbach H, editors. Materials with complex behaviour II, vol. 16., Advanced Structured MaterialsHeidelberg: Springer Berlin; 2012. p. 99–118.

Fletcher DF, Geyer PE, Haynes BS. Assessment of the SST and omega-based Reynolds stress models for the prediction of flow and heat transfer in a square-section U-bend. Comput Therm Sci. 2009;1:385–403. doi:10.1615/ComputThermalScien.v1.i4.20.

Hart A. A review of technologies for transporting heavy crude oil and bitumen via pipelines. J Pet Explor Prod Technol. 2014;4:327–36. doi:10.1007/s13202-013-0086-6.

International Energy Agency (IEA). World energy outlook 2014. www.iea.org.

Jones OC Jr. An improvement in calculation of turbulent friction in rectangular ducts. J Fluid Eng-T ASME. 1976;98:173–80.

Joseph DD, Bai R, Mata C, Sury K, Grant C. Self-lubricated transport of bitumen froth. J Fluid Mech. 1999;386:127–48. doi:10.1017/S0022112099004413.

McKibben MJ, Gillies RG, Shook CA. Predicting pressure gradients in heavy oil-water pipelines. Can J Chem Eng. 2000;78:752–6. doi:10.1002/cjce.5450780418.

McKibben M, Gillies R, Rushd S, Sanders RS. A new technique for the scale-up of pilot plant tests of water-lubricated pipeline flows of heavy oil and bitumen. J Petrol Explor. Prod. Technol. 2016. Manuscript submitted for publication (under revision)

Nunez GA, Rivas HJ, Joseph DD. Drive to produce heavy crude prompts variety of transportation methods. Oil Gas J. 1998;96(43):59–63.

Perry RH, Green DW. Perry's chemical engineers' handbook. Section 6. 7th ed. New York: McGraw-Hill International Editions; 1997.

Picologlou BF, Characklis WG, Zelver N. Biofilm growth and hydraulic performance. J Hydraul Div. 1980;106(5):733–46.

Rodriguez OMH, Bannwart AC, de Carvalho CHM. Pressure loss in core-annular flow: modeling, experimental investigation and full-scale experiments. J Pet Sci Eng. 2009;65(1–2):67–75. doi:10.1016/j.petrol.2008.12.026.

Rushd S, Islam A, Sanders RS. A CFD methodology to determine the hydrodynamic roughness of a surface with application to viscous oil coatings. J. Hydraul. Eng. 2016. Manuscript submitted for publication (under revision).

Rushd S. A new approach to model friction losses in the water-assisted pipeline transportation of heavy oil and bitumen. PhD thesis, University of Alberta, Canada. 2016.

Saniere A, Henaut I, Argillier JF. Pipeline transportation of heavy oils, a strategic, economic and technological challenge. Oil Gas Sci Technol—Rev. IFP. 2004;59:455–66.

Schaan J, Sanders RS, Litzenberger C, Gillies RG, Shook CA. Measurement of heat transfer coefficients in pipeline flow of Athabasca bitumen froth. In: Proceedings of 3rd North American conference on multiphase technology, Banff, Canada. BHR Group, Cranfield, UK. 2002:25–38.

Shook CA, Gillies RG, Sanders RS. Pipeline hydrotransport with applications in the oil sand industry. SRC Publication No. 11508-1E02. Saskatchewan Research Council Pipeflow Technology Centre, Saskatoon, SK, Canada. 2002.

Vuong DH, Zhang HQ, Sarcia C, Li M. Experimental study on high viscosity oil/water flow in horizontal and vertical pipes. In: SPE annual technical conference and exhibition. 4–7 October, New Orleans, Louisiana, USA. 2009. doi:10.2118/124542-MS.

Whyte FM. Fluid mechanics. 4th ed. Boston: McGraw-Hill; 1999.

6

Origin and depositional model of deep-water lacustrine sandstone deposits in the 7th and 6th members of the Yanchang Formation (Late Triassic), Binchang area, Ordos Basin, China

Xi-Xiang Liu[1,2] · **Xiao-Qi Ding**[3] · **Shao-Nan Zhang**[1,2] · **Hao He**[4]

Abstract Sandstones attributed to different lacustrine sediment gravity flows are present in the 7th and 6th members of the Yanchang Formation in the Ordos Basin, China. These differences in their origins led to different sandstone distributions which control the scale and connectivity of oil and gas reservoirs. Numerous cores and outcrops were analysed to understand the origins of these sandstones. The main origin of these sandstones was analysed by statistical methods, and well logging data were used to study their vertical and horizontal distributions. Results show that the sandstones in the study area accumulated via sandy debris flows, turbidity currents and slumping, and sandy debris flows predominate. The sandstone associated with a single event is characteristically small in scale and exhibits poor lateral continuity. However, as a result of multiple events that stacked gravity flow-related sandstones atop one another, sandstones are extensive overall, as illustrated in the cross section and isopach maps. Finally, a depositional model was developed in which sandy debris flows predominated and various other types of small-scale gravity flows occurred

✉ Xiao-Qi Ding
xiaoqiding@qq.com

[1] State Key Laboratory of Oil and Gas Reservoir Geology and Exploitation, Southwest Petroleum University, Chengdu 610500, Sichuan, China

[2] School of Geoscience and Technology, Southwest Petroleum University, Chengdu 610500, Sichuan, China

[3] College of Energy, Chengdu University of Technology, Chengdu 610059, Sichuan, China

[4] Second Production Plant, PetroChina Changqing Oilfield Company, Qinyang 745000, Gansu, China

Edited by Jie Hao

frequently, resulting in extensive deposition of sand bodies across a large area.

Keywords Sediment gravity flows · Sandy debris flows · Binchang area · Yanchang Formation · Ordos Basin

1 Introduction

Discussion of the origin of deep-water sandstones began with "Turbidity currents as a cause of graded bedding" (Kuenen and Migliorini 1950). Then, Bouma (1962) proposed the "Bouma sequence", which summarized the vertical sequence of sedimentary structures in turbidites and became the criterion for recognizing turbidites. Based on these criteria, other scholars contributed valuable insights related to this type of deposit and improved turbidity current theory (Yang et al. 2015a). In the same time frame, many depositional models were developed, such as the ancient fan model, the modern submarine fan model and the general fan model (Shanmugam 2000). All these models suggest that turbidity-related sand bodies feature fan-shaped deposits. This consensus has been widely applied in the exploration of deep-water sandstone reservoirs. However, as more deep-water sandstone reservoirs have been discovered, the turbidity current theory has begun to fall short. Shanmugam (1996) noted that high particle concentrations (>30% by volume) lead the support mechanism in the flow change to disperse pressure and viscosity of clay minerals which essentially represent the physical properties of the sandy debris flow. Other scholars noted that sandy debris flows tend to be deposited in the form of lumps, which modelled as interrupted sand bodies (Shanmugam and Moiola 1997) and tongue-like sand bodies (Zou et al. 2012). Distribution patterns of these deposits are entirely different from those of turbidites. Recently, scholars described a special type of turbidity current called hyperpycnal flow (Mulder et al. 2003; Zavala et al. 2011). Unlike turbidity currents, hyperpycnal flow is of significantly long duration because it is caused by seasonal floods, snow-melt floods and melting glaciers (Huneke and Mulder 2011). Therefore, the resulting sandstones are interrupted less than those related to waning turbidity currents or sandy debris flows and show gradational changes in the vertical and the horizontal direction (Zavala and Arcuri 2016).

The above analyses suggest that reservoir characteristics such as the sand distribution pattern, scale and continuity of these various deep-water sandstones differ significantly because of variations in transport and depositional models. The 7th and 6th members of the Yanchang Formation in the Binchang area of the Ordos Basin were deposited in a deep-water environment (Lei et al. 2015) and are the main production formation of tight oil in China (Yang et al. 2012; Ding et al. 2013; Yuan et al. 2015). Understanding the origin, distribution pattern, scale and continuity of this sandstone is important, and these factors must be considered during exploration and decision-making in the development of the oil field.

This paper aims to discuss the main origin of sand bodies by summarizing the sedimentary characteristics observed in cores and outcrops and then to determine the scale and distribution pattern based on well logging information. Finally, it aims to construct a depositional model for these deep-water lacustrine sandstones and provide a reference for further exploration and development of this area.

2 Geological background

The Ordos Basin is a multi-episodic cratonic basin in central China that covers an area of 370,000 km^2. The basin is bordered by the Yin Mountain on the north, the Liupan Mountain on the west, the Lüliang Mountain to the east and the Qin Mountain on the south. The study area is located around Xunyi and Binxian, near the Weibei uplift, on the southern Yi-Shan slope in the Ordos Basin (Fig. 1a–c). The basement of the Ordos Basin is Archaeozoic and lower Proterozoic metamorphic rocks. The basin has accumulated sediments since the Mesoproterozoic, resulting in a sediment column that is 5,000–10,000 m thick. The sedimentary rocks in the basin that predate the Permian were deposited under marine conditions (He 2003). By the middle Permian to Late Triassic, the northern China Plate and Yangtze Plate began to collide and merge (Deng et al. 2013; Yang and Deng 2013). This collision caused the Qin Mountains to be uplifted and the residual marine basin to

close. The collision also resulted in southward migration of the basin centre and steepened the bottom topography in the southern portion of the basin. By the Late Triassic, when deposition of the Yanchang Formation began, the basin was a deltaic–lacustrine depositional system (Fig. 1b) with a 900–1600-m-thick set of clastic rocks that recorded a complete cycle of lacustrine basin initiation, development and cessation (Fu et al. 2013; Liu et al. 2015a). The formation can be divided into 10 members (Fig. 2). Member 10 was deposited during the initial creation of the lake. Members 9 and 8 were deposited during an episode of major transgression. Member 7 was deposited during an episode of accelerated subsidence, when the water depth increased from 50 to 120 m, causing significant lake expansion. Laterally extensive black shale was deposited during this time (Lei et al. 2015). Deep-water sandstones of Member 7 were deposited on top of the black shale. The lake began to shrink during the deposition of Member 6. Members 5–1 were deposited during a period of major lake contraction, after which the lake gradually disappeared. Adequate coring data from the Yanchang Formation and outcrops around the study area make it an ideal place to study the origin of deep-water lacustrine sandstones.

3 Origin analysis of deep-water sandstones

3.1 Facies and interpretation

Based on the observation of 42 cored wells and outcrops near the study area, 7 typical facies were identified according to their lithologies and sedimentary characteristics (Table 1).

3.1.1 Facies A

Facies A is composed of black-dark grey mudstone and shale. Silty laminaes are ubiquitous in the mudstone (Fig. 3a). The thickness of the mudstone (or shale) layer reaches 40 m in the study area.

The thick black mudstone reflects a reducing and weak hydrodynamic condition in the deposition area. This facies, therefore, represents a deep-water environment.

3.1.2 Facies B

Facies B is composed of light to dark grey fine-grained sandstone with an individual bed thickness ranging from 2 to 3 m. No significant grading is observed. Abrupt contacts with mudstones are present at both the bottom and top surface of sandstones. Most of the top contact surfaces are irregular (Fig. 3b). Dark muddy clasts are ubiquitous and

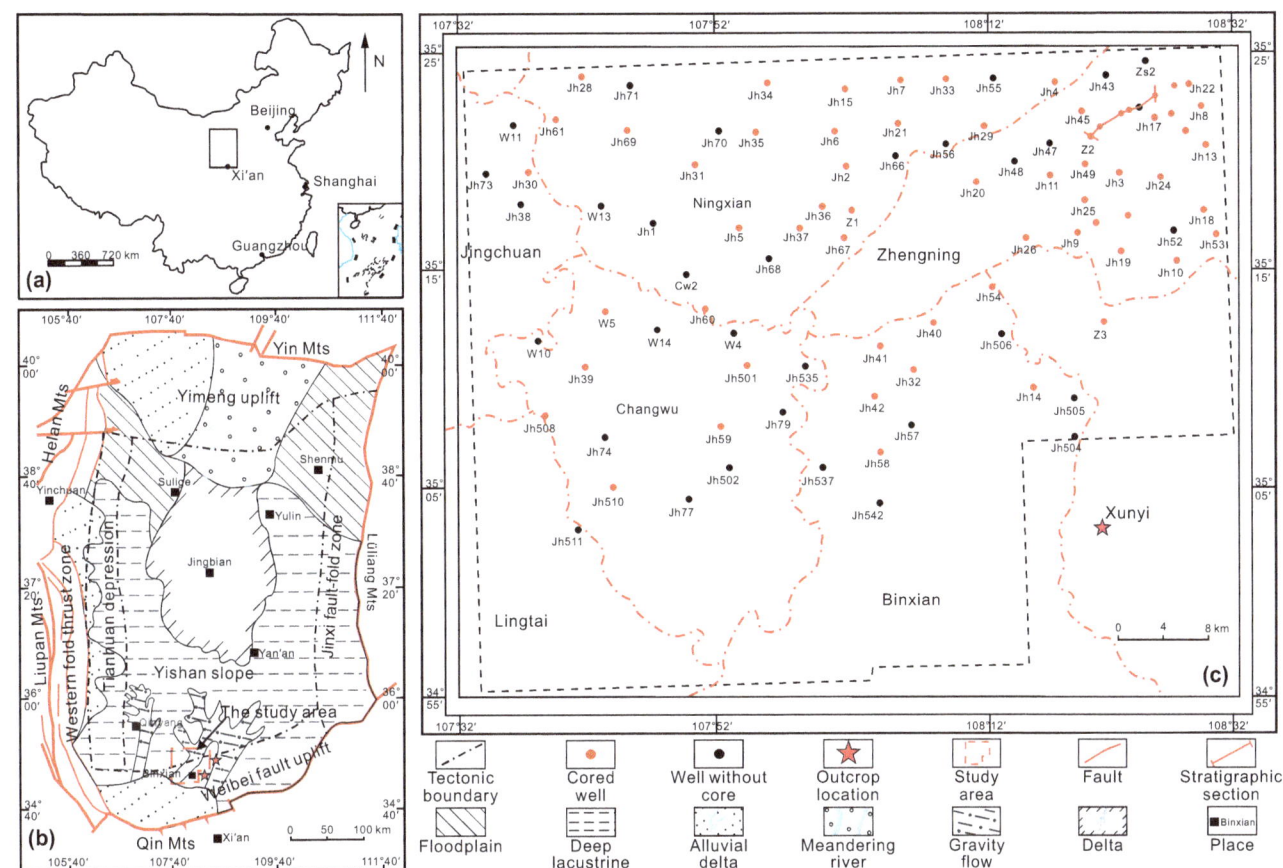

Fig. 1 Geologic setting and well distribution of the study area. **a** Map showing the Ordos Basin in central China; **b** sketch map of the Ordos Basin showing the location of the study area and the lithofacies palaeo-geographic map of the 7th member of the Yanchang Formation; and **c** detailed map showing the study area, wells and outcrops

usually exhibit a planar fabric (Fig. 3c). Yellow muddy clasts are also present, some of which are coated with black mudstone (Fig. 3d). Unlike clasts commonly found at the bottom of a layer in a fluvial environment, these clasts are primarily found in the middle or upper parts of sandstones.

The presence of clasts in the middle or upper parts of sandstones indicates the viscosity and density of the transporting fluid, which was able to support mudstone clasts and allow grains to deposit via freezing en masse and without grading. This is typical of sediments transported by plastic sandy debris flows (Zou et al. 2012; Shanmugam 2013; Xian et al. 2013). The planar fabric of the clasts further supports the hypothesis that the sediment was transported by plastic flows (Talling et al. 2012a; Yang et al. 2014). The presence of mudstone coats can be attributed to muddy slurry adhering to the clasts as they were rolled or spun by the shear stresses along their upper and lower surfaces (Li et al. 2014, 2016). These features also indicate that the deposits did not result from a turbidity current (with water between grains). Therefore, it can be concluded that this facies is the product of sandy debris flow deposition.

3.1.3 Facies C

Facies C is composed of light grey sandstone lacking grading, sedimentary structures or muddy clasts. The bottom and top of the sandstone are typically abrupt contacts with mudstones. However, the bottom contact surfaces are very smooth and the top contact surfaces are rough and irregular (Fig. 4). In outcrops, this facies displays abrupt lateral terminations with tongue-like shapes (Fig. 5a).

Massive sandstones such as these could have been deposited by grain flow, sandy debris flow or hyperpycnal flow (Bagnold 1956; Zavala 2006; Li et al. 2011a; Gao et al. 2012). Hampton (1975) concluded, based on experimentation, that a clay content of less than 2% is required for the development of grain flow. However, the thin section observation and X-ray diffraction analysis show that the average clay content in these sandstones is as high as 8.6%, which suggests that these sandstones did not form via grain flow. Hyperpycnal flow, however, is quasi-steady flow that produces sand bodies with good lateral continuity from the source to the depositional district (Tan et al. 2015; Yang et al. 2015b). However,

Series	Formation	Member	Thickness (m)	Submember	Lithology	Depositional environment
Quaternary						
Lower Cretaceous	Zhidan		0–1200			Eolian and alluvial
Upper Jurassic	Anding		0–340			Lacustrine
Middle Jurassic	Zhiluo		0–600			Fluvial
	Yanan		0–300			Fluvial-lacustrine
Lower Jurassic	Fuxian		0–156			Fluvial
Upper Triassic	Yanchang	$Chang_1$	0–240	$Chang_1{}^1$ $Chang_1{}^2$ $Chang_1{}^3$		Swamp and shallow lacustrine
		$Chang_2$	120–150	$Chang_2{}^1$ $Chang_2{}^2$ $Chang_2{}^3$		Fluvial and delta
		$Chang_3$	90–110	$Chang_3{}^1$ $Chang_3{}^2$ $Chang_3{}^3$		
		$Chang_{4+5}$	80–100	$Chang_{4+5}{}^2$ $Chang_{4+5}{}^1$		Shallow lacustrine
		$Chang_6$	110–130	$Chang_6{}^1$ $Chang_6{}^2$ $Chang_6{}^3$		Delta and shallow lacustrine
		$Chang_7$	100–120	$Chang_7{}^1$ $Chang_7{}^2$ $Chang_7{}^3$		Delta and deep lacustrine
		$Chang_8$	75–90	$Chang_8{}^1$ $Chang_8{}^2$		Delta and shallow lacustrine
		$Chang_9$	80–110	$Chang_9{}^1$ $Chang_9{}^2$		
		$Chang_{10}$	240–280	$Chang_{10}{}^1$ $Chang_{10}{}^2$ $Chang_{10}{}^3$		Fluvial

Legend: Conglomerate, Coarse sandstone, Medium sandstone, Fine sandstone, Silt sandstone, Mudstone, Shale, Loess

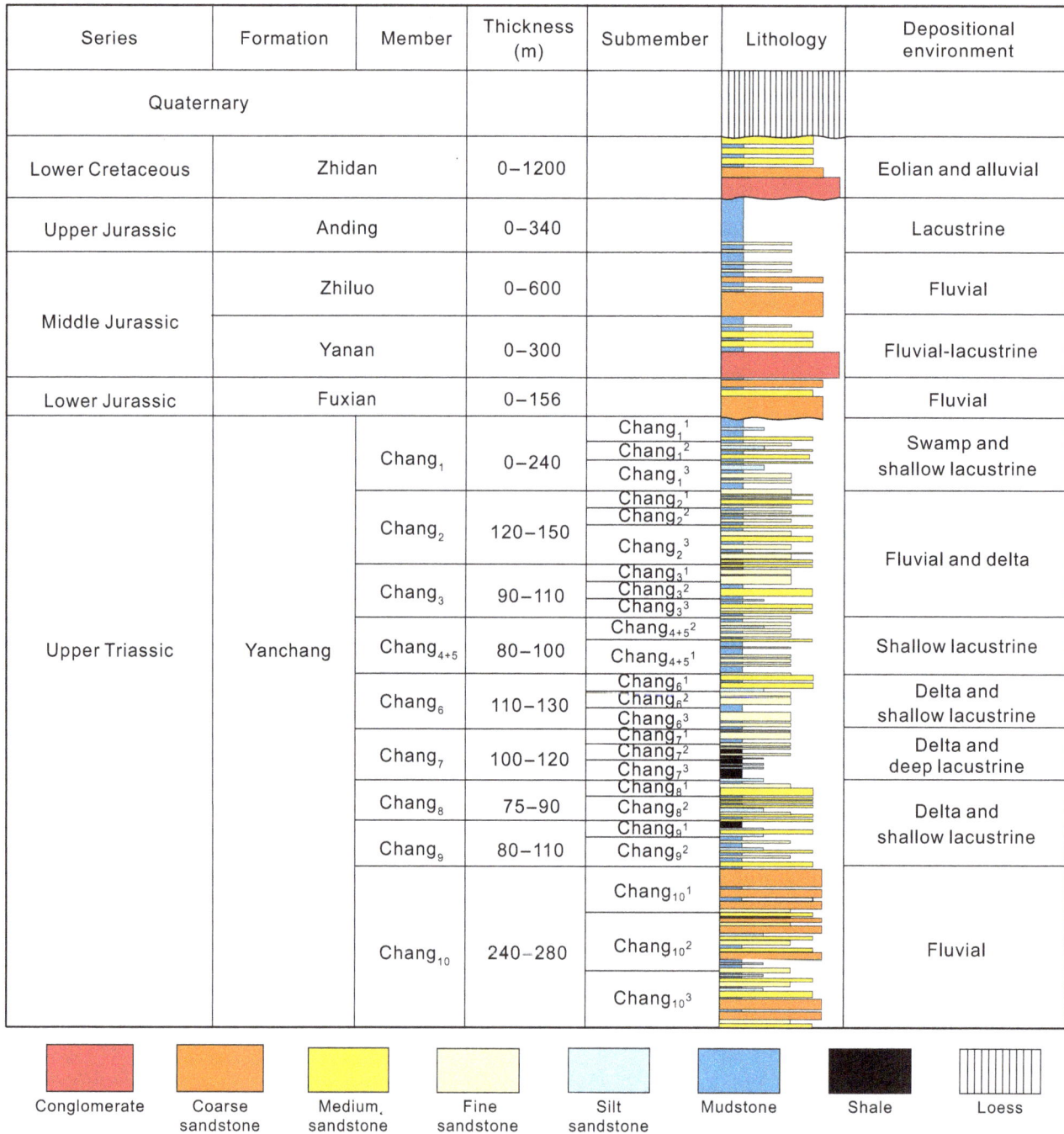

Fig. 2 Stratigraphic column showing the Mesozoic stratigraphy, lithology and depositional environments in the study area, southern part of the Ordos Basin, central China

the sand bodies in the study area are generally characterized by small scales and poor continuity (see chapter 4.2). Hyperpycnal flows significantly erode the underlying layer, but sandstones observed in cores and outcrops in the study area display typically smooth contacts with underlying layers. Smooth bottom contacts are more likely produced by hydroplaning associated with sandy debris flows (Marr et al. 2001). The abrupt termination of sandstones and the irregular contact surfaces with overlying layers also indicate that the sediment was deposited by the freezing en masse of a plastic sandy debris flow (Talling et al. 2012b; Shanmugam 2016). Therefore, these massive sandstones are more likely the result of sandy debris flow deposition.

Table 1 Characteristics of different facies in the study area

Facies	Lithology and sedimentary structure
A	Black-dark grey mudstone or shale. Silty laminae are ubiquitous
B	Light to dark grey fine grained sandstones. Dark muddy clasts are ubiquitous and exhibit a planar fabric. Mud-coated clasts can be observed
C	Light grey massive sandstones. Their bottom and top often are in abrupt contact with mudstones. The bottom contact surface usually very smooth and the top contact surface rough and irregular
D	Grey fine-to-silt sandstone with small scale parallel bedding or cross bedding. Weak normal grading can be observed
E	Light grey fine-to-silt sandstones or mudstones with lenticular, flaser or wavy bedding and normal grading can be observed. Their bottom often has abrupt contact with mudstones
F	Grey sandstones with deformation structures, such as glide planes, steep fabric and slump folds
G	Black mudstone or silty mudstone that is mainly characterized by steeply dipping mud layers (steep fabric) and sandstone injection

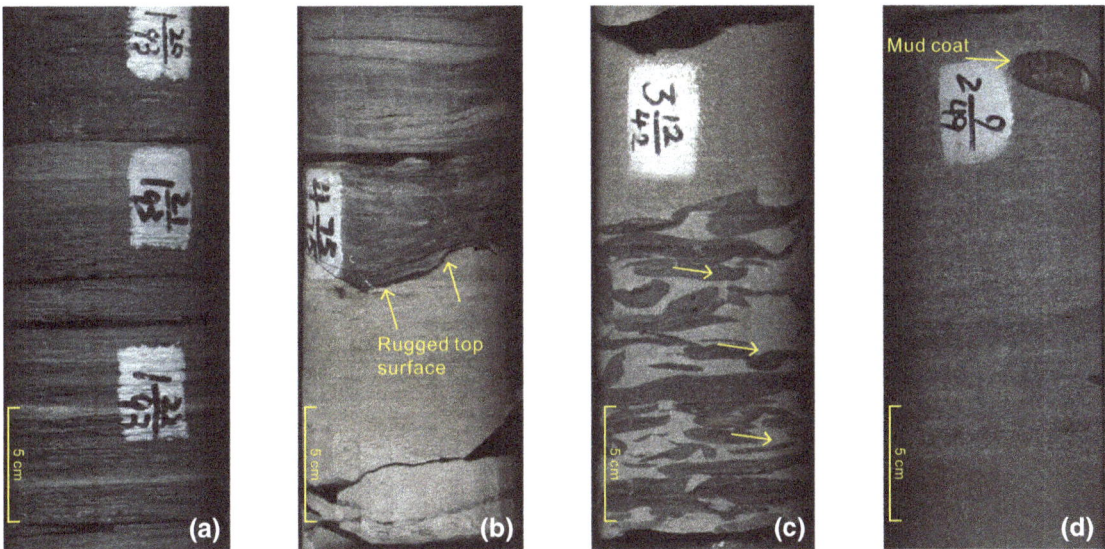

Fig. 3 Photographs of sedimentary structures in facies A and facies B. **a** Black mudstone with silty laminae, well Jh4, 1319.57–1319.81 m. **b** Sandstone with irregular abrupt top contact surface, well Jh4, 1370.89–1371.23 m. **c** Mud clasts with planar fabric, well Jh9, 937.9–938.14 m. **d** Yellow mud clast with a black mud coat, well Jh2, 1317.11–1317.35 m

Fig. 4 Photographs of facies C showing massive sandstone with an irregular top surface and a smooth bottom surface, well Jh25, 1277.51–1280.10 m

Fig. 5 Photographs of an outcrop of the Chang 7 member. **a** Upper sandstone, facies C, featuring an abrupt termination with a tongue-like shape. **b** Photograph of facies D. **c** Detailed photograph of the lower sandstone showing a typical vertical assemblage of facies B, C and D. **d** Planar fabric, which is ubiquitous in facies B

3.1.4 Facies D

Facies D is composed of grey fine to silty sandstone with parallel bedding and small-scale cross-bedding. Weak normal grading can be observed. The thickness of facies D is no more than 10 cm. It typically overlies facies B or facies C (Fig. 5b–d).

The parallel bedding or cross-bedding is related to the traction process. However, the scale of the traction structure is small which reflects the weak hydrodynamic condition and a small impact area. Additionally, normal grading usually indicates that this facies is associated with turbidity currents (Mulder and Alexander 2001). Therefore, we conclude that as a debris flow advanced through the water, shear stress generated along the upper boundary of the water–slurry interface. Then, the shear stress leads to erosion and entrainment of sediment from the surface of the debris flow into the clear water above (Marr et al. 2001). The entrainment of sediment into the overriding clear water resulted in the formation of a dilute subsidiary

turbidity current and a traction ability in the surrounding water. Thus, this facies with weak normal grading and small traction structure is generally observed above facies B and facies C and could be a criterion for identify sandy debris flow of different stages. This phenomenon has been observed by other scholars (Yang et al. 2014; Liu et al. 2015b).

3.1.5 Facies E

Facies E is composed of light grey fine to silty sandstone or mudstone with lenticular (Fig. 6a), flaser (Fig. 6b) or wavy bedding (Fig. 6c). Most of the sand beds are no more than 1 m thick. A transition from flaser bedding at the bottom to lenticular bedding at the top can be observed where the sand body is thick enough. Normal grading can also be observed, and the bottom of the sandstone has abrupt contacts with mudstone (Fig. 6).

The wavy, flaser and lenticular bedding are all related to traction processes (Mulder and Alexander 2001), and the

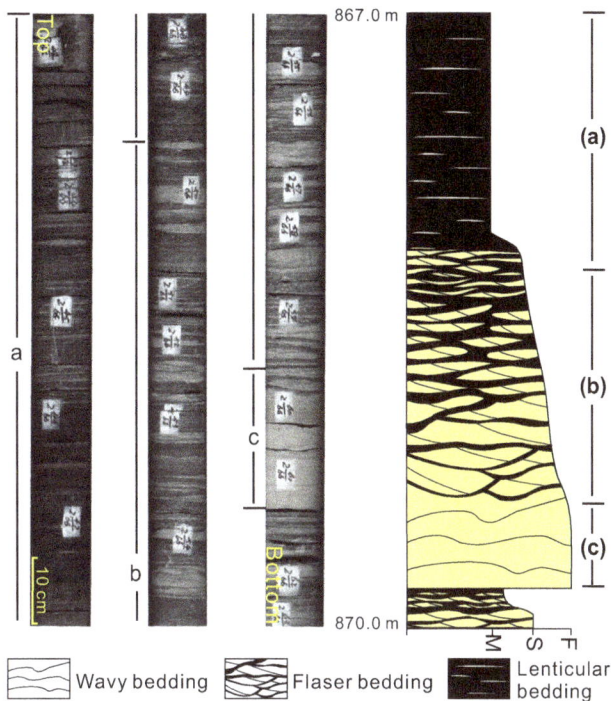

Fig. 6 Photographs of facies E showing normal grading and different kinds of bedding. **a** Mudstone with lenticular bedding. **b** Silty to fine sandstone with flaser bedding and an upward-increasing mud content. **c** Fine sandstone with wavy bedding. Jh9, 867–870 m

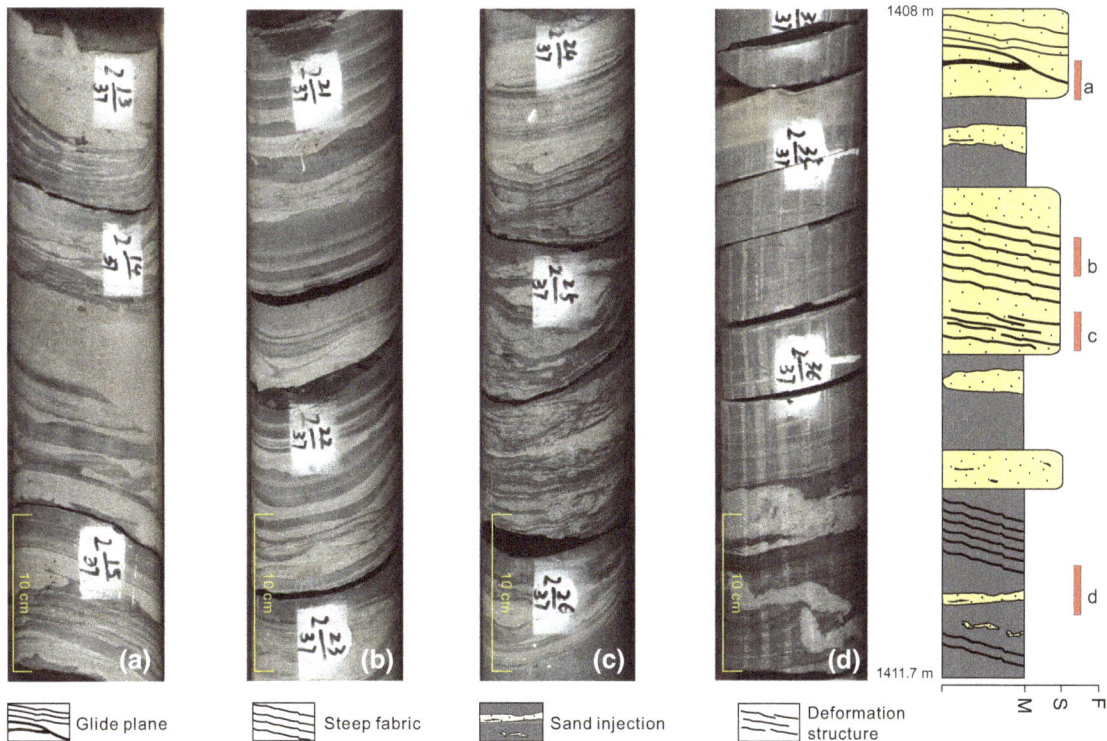

Fig. 7 Typical association of facies F and G, which are related to sand slumping. **a** Glide plane. **b** Steep fabric in sandstone. **c** Slight deformation of a pelitic layer in sandstone **d** Sand injection and steeply dipping mud layer in mudstone; well Jh7, 1408–1411.7 m

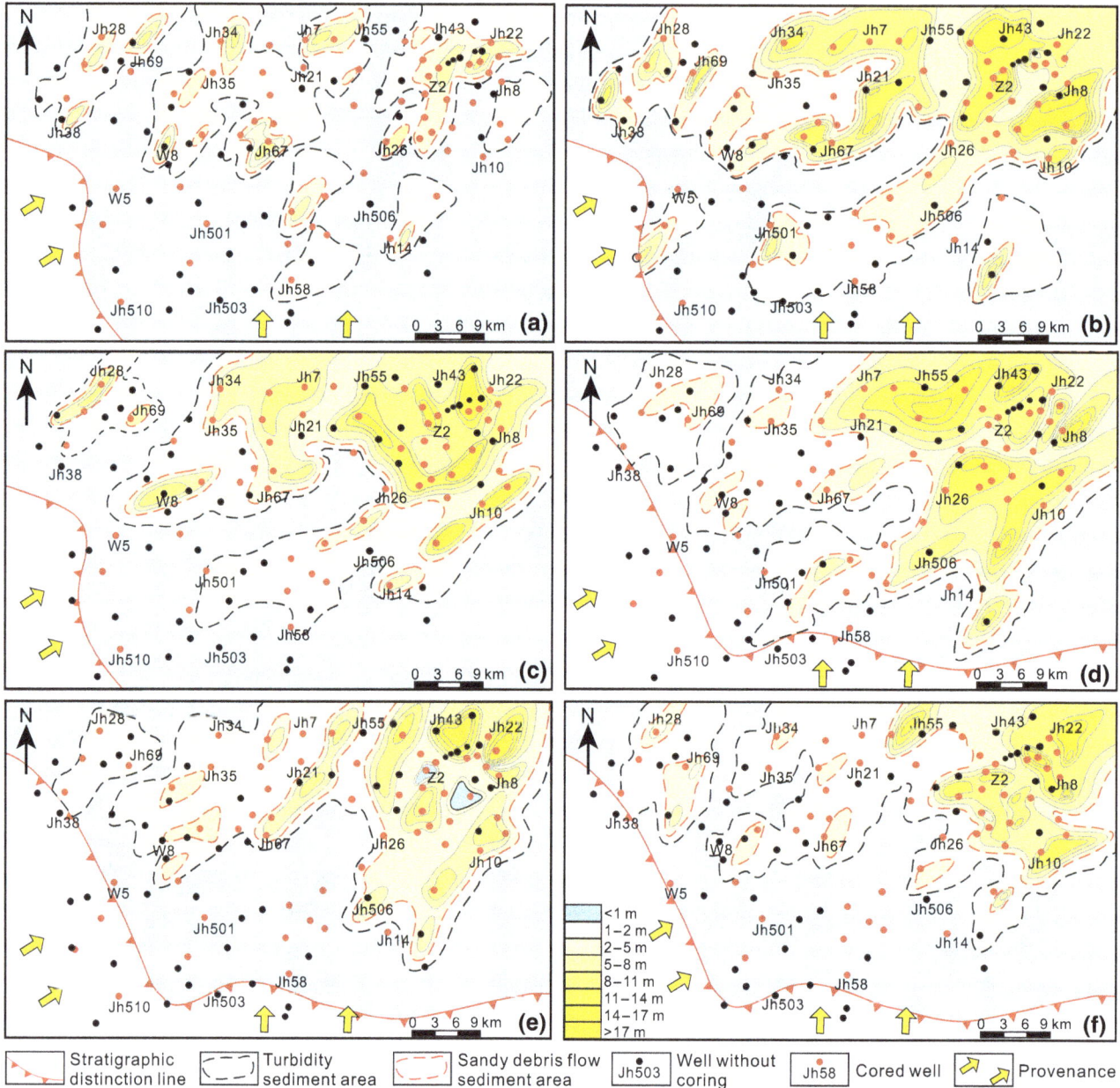

Fig. 8 Isopach maps of sandstone thicknesses; note that abrupt thickness variations and the thick sandstones are generally distributed in the northeast portion of the study area. **a** Isopach map of the Chang$_7^3$ submember. **b** Isopach map of the Chang$_7^2$ submember. **c** Isopach map of the Chang$_7^1$ submember. **d** Isopach map of the Chang$_6^3$ submember. **e** Isopach map of the Chang$_6^2$ submember. **f** Isopach map of the Chang$_6^1$ submember

normal grading suggests that the sediments were deposited by waning turbidity currents that deposited coarse-grained material first followed by fine-grained material (Li et al. 2011b, c; Pu et al. 2014). However, as in facies D, these beds are thin, which indicates that the energy of the currents was low. Therefore, we conclude that the deposits may have been formed by subsidiary turbidity currents transformed from sandy debris flows.

3.1.6 Facies F

Facies F is composed of grey sandstone with deformation structures such as glide planes (Fig. 7a), steep fabric (Fig. 7b) and slump folds (Fig. 7c). The bottoms and tops are typically abrupt contacts with mudstones, which also display deformation structures (Fig. 7). The deformed layers could reach 4 m in thickness.

Table 2 Characteristics of sandstones of various origins in the Binchang area, Ordos Basin, China

Origin	Slumping	Sandy debris flow	Turbidity current
Lithology	Silty fine-grained sandstone or mudstone	Silty fine-grained sandstone	Mainly silty sandstone
Colour	Light greyish-black	Light grey	Light to dark grey
Sedimentary structure	Glide planes	Mud clasts with planar fabric	Wavy bedding
	Steep fabric	Irregular upper contacts	Flaser bedding
	Slump folds		Lenticular bedding
Grading	No distinct grading	No distinct grading	Normal grading
Facies	Facies F	Facies B	Facies D
	Facies G	Facies C	Facies E
Bed geometry	–	Tongue-shaped	Fan-shaped

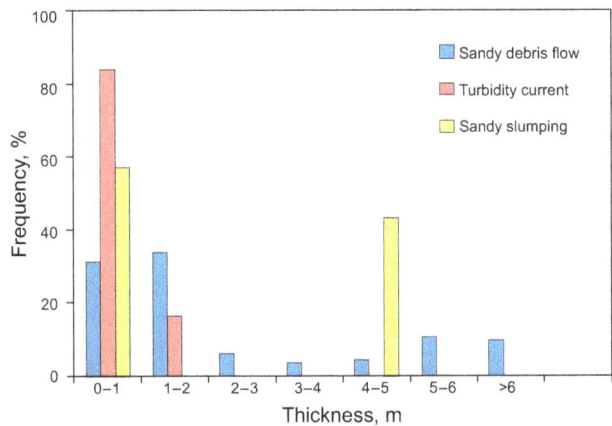

Fig. 9 Thicknesses of sandstone bodies of various origins

The slump folding suggests that the sediment experienced soft-sediment deformation. The glide plane is a dislocation in the sedimentary structure or bedding, which suggest internal soft-sediment deformation. The steep fabric reflects that the sandstone experienced slumping (Herrington et al. 1991). Thus, this facies is attributed to synsedimentary sand slumping.

3.1.7 Facies G

Facies G is composed of black mudstone and light grey silty mudstone that is characterized primarily by steeply dipping mud layers (steep fabric) and sandstone injection (Fig. 7d). This facies is usually adjacent to facies F (Fig. 7).

The black mudstone indicates deposition in a deep-water environment. In this environment, the steeply dipping mud layers can only be attributed to mud slumping. The injected sand is also related to loading induced by slumping (Shanmugam 2012). This facies is therefore the product of slumping.

3.2 Spatial distribution of sandstones

The sedimentary facies control the distribution of the sand bodies, and the origin of the sandstones can be deduced from the geometry of the sandstone beds. Based on the analysis of logging data in cored wells, a logging identification standard has been established. With the identification standard, sand bodies with a thickness of at least 1 m were distinguished. Then, sand-thickness contour maps of the various submembers were prepared.

The isopach map illustrates the localized thickening and thinning that is common in these beds. Sand bodies thicker than 5 m exist in the form of lumps and display rapid lateral thickness change (Fig. 8). Amy et al. (2005) noted that this horizontal distribution of sand bodies is a typical characteristic of sandy debris flows. Moreover, the thick sand bodies distributed primarily in the north-eastern part of the study area, far from the sedimentary sources. This suggests that the sandstone may not have been transported by waning traction flows whose sediment thickness decreases with the transporting distance (Shanmugam 2012).

3.3 Origins of deep-water sandstones

From the above analyses, black mudstones (shales) are widely distributed, and the coarsest grains in the 7th and 6th members of the Yanchang Formation are fine sand, which indicate weak hydrodynamic conditions. The rocks lack large-scale bedding, indicating weak traction processes. The lateral continuity of the sandstone bodies is poor, which suggests that the sand was not transported by flows of long duration. The thick sandstone beds display abundant characteristics of plastic flow transport such as massive sandstones (facies B), mud clasts with planar fabrics, irregular upper contacts and mud coatings. Additionally, previous investigators noted that during the deposition of the Yanchang Formation, strong tectonic movements such as earthquakes and volcanic eruptions occurred frequently; these events could trigger gravity flows (Deng et al. 2013; Zhang et al. 2014). Thus, we conclude that the deep-water sandstones of the 7th and 6th members of the Yanchang Formation are the product of sediment gravity flows including sandy debris flows, subsidiary turbidity currents and sandy slumping (Table 2).

4 Discussion

4.1 The main origin of deep-water sandstone

Based on the above-mentioned standard, the facies of the deposits in the study area were identified by observing 715.3 m of cores collected from 42 cored wells. In total, the black mudstones are 273.1 m, sandstones attributed to sandy debris flows are 354.5 m, turbidite sandstones are 65 m, and the sandstones attributed to sand slumping are 22.7 m. In terms of thickness, the 6th and 7th members of the Yanchang Formation are dominated by sandy debris flow sandstones followed by turbidite sandstones. The slump sand bodies are present in only a few wells.

A statistical analysis of the thicknesses of single-event sand bodies attributed to sandy debris flows shows that the layers are generally less than 2 m thick (Fig. 9, blue bars).

However, some thick sandstone bodies of this type are developed in local areas.

The total thickness of the turbidite sandstones is small, and most of the single-event sand bodies of this type are less than 1 m thick (Fig. 9, brown bars), which suggests that the scale of turbidity currents in the study area was small. Thus, turbidity currents were not the main origin of the deep-water sandstone in the study area.

Some thick sand bodies can be attributed to slumping (Fig. 9, green bars). However, the total thickness of this type of sandstone is small, and this type is observed in only a few wells. Overall, this type of sand body is not abundant in the study area.

Therefore, the thick sand bodies in the study area were deposited primarily by sandy debris flows. The sand bodies attributed to turbidity currents come second, and the slumping-related sandstone bodies are present only in local areas.

4.2 Scale of deep-water sandstones

A cross section through 6 wells, with an average well spacing of approximately 1 km, was made (Fig. 10) to further investigate the scale of the sand bodies in the study area. Based on the lateral distribution characteristics of the sand bodies in the cross section, the sand bodies in study area are suggested to have the following features:

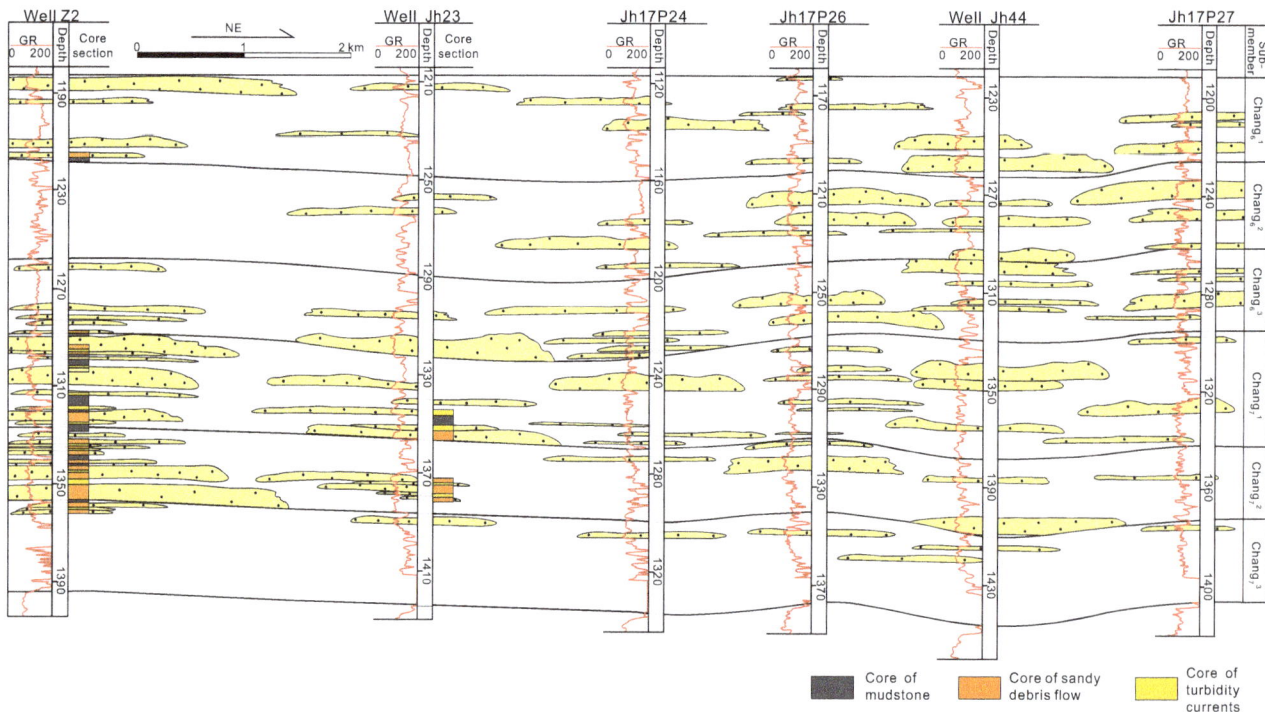

Fig. 10 Cross section of Chang 7 and Chang 6 members. See Fig. 1c for the location of the section

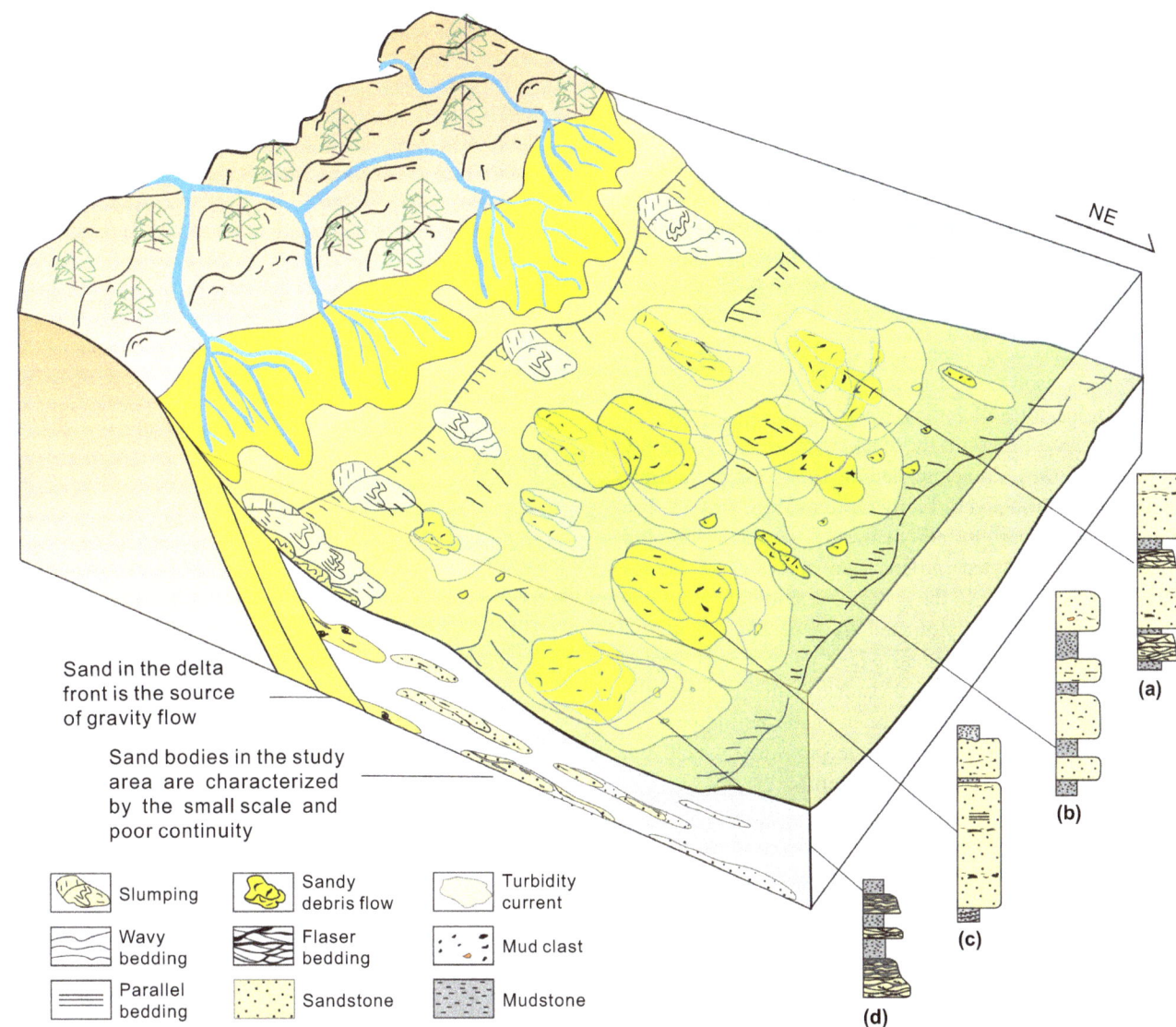

Fig. 11 Depositional model of sediment gravity flows in the 6th and 7th members of the Yanchang Formation in the Binchang area, Ordos Basin, China. **a** Interbedded sandy debris flows and turbidity currents; **b** interbedded sandy debris flows and mudstones; **c** stacking of multiple sandy debris flows; and **d** interbedded turbidity currents and mudstones

(1) The scale of single-event sand bodies is small, and individual sand bodies are usually less than 2 km in size.

(2) The single-event sandstone bodies are thin, although the frequency of gravity flow deposition resulted in a large cumulative thickness.

(3) The frequency of gravity flow events varied significantly from one neighbouring well to the next, and thick sandstones are developed only in small area.

The above features can be observed directly in the outcrops. One particular outcrop in the study area exhibits two sets of sandy debris flow sandstones at the top and bottom (Fig. 5). The sand bodies at the top are approximately 6 m thick but pinch out laterally within 15 m

(Fig. 5a). The sand bodies at the bottom are only 2 m thick but display parallel bedding and planar fabric (Fig. 5b–d), which indicates that the sand bodies are the result of two stages of sandy debris flows. All the above evidence adequately supports the conclusion that the sand bodies in the study area are characterized by small individual events that rapidly thin out laterally, which leads to poor continuity. However, a large cumulative thickness resulted from the stacking of multiple gravity flow deposits.

4.3 Sedimentary models

Based on the above analysis, a sedimentary model of the deep-water sandstones in the Binchang area was developed

(Fig. 11). A significant number of deposits in the study area were re-mobilized by sediment gravity flow and re-transported from a gently sloping front delta area to the adjacent lake plain due to an earthquake disturbance or the naturally unstable nature of the delta front area (i.e. sand deposited on water-bearing muddy deposits) (Zhang et al. 2014). Thus, the sand in the delta front area is the source of the deep-water sandstones. Re-transportation occurred mainly in the form of sandy debris flows and turbidity currents. Due to the short travel distance of the slumping-related sand (Shanmugam 2012), these deposits are believed to be present near the delta front zone, outside the study area.

The thick sand bodies in the study area are generally present in the form of lumps. These thick sand bodies, in most cases, resulted from the stacking of sediment gravity flows, in the form of stacking and interfingering of multiple sandy debris flows and turbidity currents (Fig. 11a–c). The individual subsidiary turbidity current deposits are thin but extensively distributed around the thick sand bodies (Fig. 11d). According to the drilling situation and previous research, the thick layers of sand are generally located far from the material source areas where palaeo-water depths were greater (Fu et al. 2013). This distribution is due to the nature of sandy debris flows, which, under the influence of gravity, are likely to move into deeper water where the gravitational potential energy is lower.

5 Conclusions

1. Three different types of sandstone were identified in the study area. The massive sandstone (facies C) and the sandstone containing muddy clasts and "mud coatings" (facies B) are the products of sandy debris flows. Facies D and facies E, which display traction structures and normal grading, represent deposition by turbidity currents. These deposits are thin, which indicates that the energy of the turbidity currents was low. Therefore, the deposits are interpreted as being the products of subsidiary turbidity currents generated by sandy debris flows. Facies F and G, which are characterized by soft-sediment deformation, are the product of sand slumping.
2. The sandstones in the study area were deposited mainly by sandy debris flows. This type of sandstone is characterized by small-scale single events, rapid lateral pinch outs and multi-stage development. The turbidite sandstones are widespread, but the scale of beds represented by single events was too small to form thick sandstones.
3. A depositional model was developed in which sandy debris flows predominated, turbidity currents were secondary and sandy slumping was localized. In this

model, the sandstones developed far from their source at substantial palaeo-water depths. The thick sandstones resulted from the accumulation of multiple gravity flow deposits, which led to lump plan-form geometry. The turbidites generally overlie the sandy debris flow deposits or are interbedded with mudstone near the sandy debris flow deposits. And lumping-related sandstone bodies are present only in local areas.

Acknowledgements This work was supported by the Science Foundation Programs (41302115). We are grateful to two reviewers who provided useful comments to help improve the manuscript and the senior engineer He Pumin, North China Company of SINOPEC, who assisted in the observation of cores.

References

Amy LA, Talling PJ, Peakall J, et al. Bed geometry used to test recognition criteria of turbidites and (sandy) debrites. Sed Geol. 2005;179(s 1–2):163–74. doi:10.1016/j.sedgeo.2005.04.007.

Bagnold RA. The flow of cohesionless grains in fluids. Philos Trans R Soc Math Phys Eng Sci. 1956;249(964):235–97. doi:10.1098/rsta.1956.0020.

Bouma AH. Sedimentology of some Flysch deposits: a graphic approach to facies interpretation. Amsterdam: Elsevier Press; 1962. p. 1–5.

Deng XQ, Luo AX, Zhang ZY, et al. Geochronological comparison on Indosinian tectonic events between Qinling Orogeny and Ordos Basin. Acta Sedimentol Sin. 2013;31(6):939–53. doi:10.14027/j.cnki.cjxb.2013.06.015 **(in Chinese)**.

Ding XQ, Han MM, Zhang SN. The role of provenance in the diagenesis of siliciclastic reservoirs in the Upper Triassic Yanchang Formation, Ordos Basin, China. Pet Sci. 2013;10(2):149–60. doi:10.1007/s12182-013-0262-9.

Fu JH, Deng XQ, Chu MJ, et al. Features of deepwater lithofacies, Yanchang Formation in Ordos basin and its petroleum significance. Acta Sedimentol Sin. 2013;31(5):928–38. doi:10.14027/j.cnki.cjxb.2013.05.011 **(in Chinese)**.

Gao HC, Zheng RC, Wei Q, et al. Reviews on fluid properties and sedimentary characteristics of debris flows and turbidity currents. Adv Earth Sci. 2012;27(8):815–27. doi:10.11867/j.issn.1001-8166.2012.08.0815 **(in Chinese)**.

Hampton MA. Competence of fine-grained debris flows. J Sediment Res. 1975;45(4):834–44. doi:10.1306/212F6E5B-2B24-11D7-8648000102C1865D.

He ZX. The evaluation of the Ordos Basin and oil and gas. Bejing: Peking: Petroleum Industry Press; 2003. p. 88–105 **(in Chinese)**.

Herrington PM, Pederstad K, Dickson JAD. Sedimentology and diagenesis of resedimented and rhythmically bedded chalks from the Eldfisk Field, North Sea central Graben. AAPG Bull. 1991;75(11):1661–74. doi:10.1306/0c9b29cf-1710-11d7-8645000102c1865d.

Huneke H, Mulder T. Deep-sea sediments. London: Elsevier Press; 2011. p. 46–54.

Kuenen PH, Migliorini CI. Turbidity currents as a cause of graded bedding. J Geol. 1950;58(2):91–127. doi:10.1086/625710.

Lei Y, Luo XR, Wang X, et al. Characteristics of silty laminae in Zhangjiatan shale of southeastern Ordos Basin, China: implications for shale gas formation. AAPG Bull. 2015;99(4):661–87. doi:10.1306/09301414059.

Li XB, Chen Q, Liu HQ, et al. Features of sandy debris flows of the Yanchang Formation in the Ordos basin and its oil and gas exploration significance. Acta Geologica Sinica (English edition). 2011a;85(5):1187–202. doi:10.1111/j.1755-6724.2011.00550.x.

Li XB, Fu JH, Chen QL, et al. The concept of sandy debris flow and its application in the Yanchang Formation deep water sedimentation of the Ordos basin. Adv Earth Sci. 2011b;26(3):286–94. doi:10.11867/j.issn.1001-8166.2011.03.0286 **(in Chinese)**.

Li XB, Liu HQ, Zhang ZY, et al. Argillaceous parcel structure: a direct evidence of debris flow origin of deep-water massive sandstone of Yanchang Formation, Upper Triassic, the Ordos Basin. Acta Sedimantol Sinica. 2014;32(4):611–22. doi:10.14027/j.cnki.cjxb.2014.04.005 **(in Chinese)**.

Li XB, Zl Yang, Wang J, et al. Mud-coated intraclasts: a criterion for recognizing sandy mass-transport deposits–deep-lacustrine massive sandstone of the Upper Triassic Yanchang Formation, Ordos Basin, Central China. J Asian Earth Sci. 2016;. doi:10.1016/j.jseaes.2016.06.007.

Li Y, Zheng RC, Zhu GJ, et al. Reviews on sediment gravity flow. Adv Earth Sci. 2011c;26(2):157–65. doi:10.11867/j.issn.1001-8166.2011.02.0157 **(in Chinese)**.

Liu F, Zhu XM, Li Y, et al. Characteristics of the late Triassic deep-water slope break belt in southwestern Ordos Basin and its control on sandbodies. Geol J China Univ. 2015a;21(4):674–84. doi:10.16108/j.issn1006-7493.2015051 **(in Chinese)**.

Liu F, Zhu XM, Li Y, et al. Sedimentary characteristics and facies model of gravity flow deposits of Late Triassic Yanchang Formation in southwestern Ordos Basin, NW China. Pet Explor Dev. 2015b;42(5):633–45. doi:10.1016/S1876-3804(15)30058-6.

Marr JG, Harff PA, Shanmugam G, et al. Experiments on subaqueous sandy gravity flows: the role of clay and water content in flow dynamics and depositional structures. GSA Bull. 2001;113(11):1377–86. doi:10.1130/0016-7606(2001)113%3C1377:EOSSGF%3E2.0.CO;2.

Mulder T, Alexander J. The physical character of subaqueous sedimentary density currents and their deposits. Sedimentology. 2001;48(2):269–99. doi:10.1046/j.1365-3091.2001.00360.x.

Mulder T, Syvitski JPM, Migeon S, et al. Marine hyperpycnal flows: initiation, behavior and related depositsA review. Mar Pet Geol. 2003;20(6–8):861–82. doi:10.1016/j.marpetgeo.2003.01.003.

Pu XG, Zhou LH, Han WZ, et al. Gravity flow sedimentation and tight oil exploration in lower first member of Shahejie Formation in slope area of Qikou Sag, Bohai Bay Basin. Pet Explor Dev. 2014;41(2):153–64. doi:10.1016/S1876-3804(14)60018-5.

Shanmugam G. High-density turbidity currents: are they sandy debris flows? J Sediment Res. 1996;66(1):2–10. doi:10.1306/D426828E-2B26-11D7-8648000102C1865D.

Shanmugam G. 50 years of the turbidite paradigm (1950s—1990s): deep-water processes and facies models—a critical perspective. Mar Pet Geol. 2000;17(2):285–342. doi:10.1016/S0264-8172(99)00011-2.

Shanmugam G. New perspectives on deep-water sandstones: origin, recognition, initiation, and reservoir quality. Oxford: Elsevier Press; 2012. p. 75–85.

Shanmugam G. New perspectives on deep-water sandstones: implications. Pet Explor Dev. 2013;40(3):316–24. doi:10.1016/b978-0-444-56335-4.00009-6.

Shanmugam G. Submarine fans: a critical retrospective (1950–2015). J Palaeogeogr. 2016;5(2):110–84. doi:10.1016/j.jop.2015.08.011.

Shanmugam G, Moiola RJ. Reinterpretation of depositional processes in a classic flysch sequence in the Pennsylvanian Jackfork Group, Ouachita Mountains: Reply. AAPG Bull. 1997;81:476–91. doi:10.1306/522b43a7-1727-11d7-8645000102c1865d.

Talling PJ, Giuseppe M, Fabrizio F. Can liquefied debris flows deposit clean sand over large areas of sea floor? field evidence from the Marnoso-arenacea formation, Italian Apennines. Sedimentology. 2012a;60(3):720–62. doi:10.1111/j.1365-3091.2012.01358.x.

Talling PJ, Masson DG, Sumner EJ, et al. Subaqueous sediment density flows: depositional processes and deposit types. Sedimentology. 2012b;59(7):1937–2003. doi:10.1111/j.1365-3091.2012.01353.x.

Tan MX, Zhu XM, Zhu SF. Research on sedimentary process and characteristics of hyperpycnal flows. Geol J China Univ. 2015;21(1):94–104. doi:10.16108/j.issn1006-7493.2014095 **(in Chinese)**.

Xian BZ, Wan JF, Dong YL, et al. Sedimentary characteristics, origin and model of lacustrine deep-water massive sandstone: an example from Dongying Formation in Nanpu Depression. Acta Pet Sinica. 2013;29(9):3287–99 **(In Chinese)**.

Yang H, Deng XQ. Deposition of Yanchang Formation deep-water sandstone under the control of tectonic events, Ordos Basin. Pet Explor Dev. 2013;40(5):549–57. doi:10.1016/S1876-3804(13)60072-5.

Yang H, Fu JH, He HQ, et al. Formation and distribution of large low-permeability lithologic oil regions in Huaqing, Ordos Basin. Pet Explor Dev. 2012;39(6):683–91. doi:10.1016/S1876-3804(12)60093-7.

Yang RZ, He ZL, Qiu GQ, et al. Late Triassic gravity flow depositional systems in the southern Ordos Basin. Pet Explor Dev. 2014;41(6):724–33. doi:10.1016/S1876-3804(14)60086-0.

Yang T, Cao YC, Wang YZ, et al. Status and trends in research on deep-water gravity flow deposits (English edition). Acta Geol Sinica. 2015a;89(2):610–31. doi:10.1111/1755-6724.12451.

Yang T, Cao YC, Wang YZ, et al. Sediment dynamics process and sedimentary characteristics of hyperpycnal flows. Geol Rev. 2015b;61(1):23–33. doi:10.16509/j.georeview.2015.01.006 **(in Chinese)**.

Yuan XJ, Lin SH, Liu Q, et al. Lacustrine fine-grained sedimentary features and organic rich shale distribution pattern: a case study of Chang 7 Member of Triassic Yanchang Formation in Ordos Basin, NW China. Pet Explor Dev. 2015;42(1):37–47. doi:10.1016/S1876-3804(15)60004-0.

Zavala C. Ancient lacustrine hyperpycnites: a depositional model from a case study in the Rayoso Formation (Cretaceous) of West-Central Argentina. J Sedim Res. 2006;76(1–2):41–59. doi:10.2110/jsr.2006.12.

Zavala C, Arcuri M. Intrabasinal and extrabasinal turbidites: origin and distinctive characteristics. Sediment Geol. 2016;337:36–54. doi:10.1016/j.sedgeo.2016.03.008.

Zavala C, Marcano J, Carvajal J, et al. Genetic Indices in hyperpycnal systems: a case study in the late Oligocene–Early Miocene Merecure Formation, Maturin subbasin, Venezuela. Sediment transfer from shelf to deep water —revisiting the delivery system. Oklahoma: SEPM Press; 2011. p. 31–51.

Zhang H, Peng PA, Zhang WZ. Zircon U-Pb ages and Hf isotope characterization and their geological significance of Chang 7 tuff of Yanchang Formation in Ordos Basin. Acta Pet Sinica. 2014;30(2):565–75 **(in Chinese)**.

Zou CN, Wang L, Li Y, et al. Deep-lacustrine transformation of sandy debrites into turbidites, Upper Triassic, Central China. Sediment Geol. 2012;265–266:143–55. doi:10.1016/j.sedgeo.2012.04.004.

7

Towards the development of cavitation technology for upgrading bitumen: Viscosity change and chemical cavitation yield measurements

Deepak M. Kirpalani[1] · Dipti Prakash Mohapatra[1]

Abstract Among the different methods used for reducing viscosity of bitumen, acoustic cavitation during sonication is well recognised. Several chemical methods were used to detect the production of reactive species such as hydroxyl radicals and hydrogen peroxide during acoustic cavitation processes. However, quantification of cavitation yield in sonochemical systems is generally limited to low frequencies and has not been applied to bitumen processing. An empirical determination of the cavitation yield in mid- to high-frequency range (378, 574, 850, 992, and 1173 kHz) was carried out by measuring the amount of iodine liberated from the oxidation of potassium iodide (KI). Further, cavitation yield and the effects of different sonic operating conditions such as power input (16.67%–83.33%) and solute concentration on cavitation yield were carried out in KI solution and sodium carboxymethyl cellulose–water mixture to obtain benchmark changes in rheology and chemistry using these two model fluids. The findings were then applied to bitumen upgrading through sonication. Through this study, it was found that the chemical cavitation yield peaked with a sonication frequency of 574 kHz. It was also found that cavitation yield and viscosity change were correlated directly in bitumen and a 38% lower bitumen viscosity could be obtained by acoustic cavitation.

Keywords Bitumen · Cavitation yield · CMC–water · KI solution · Viscosity correlation

1 Introduction

Bitumen is a complex mixture of hydrocarbons of different families (aromatics, naphthenes, and paraffin among others), oxygen, nitrogen, and sulphur compounds as well as trace metals. The recovery of bitumen from oil-bearing rocks, including tar sand (also called oil sands or bituminous sand), has become increasingly important for energy security of the continent (Speight 2007). Due to the growing world oil demand and scarcity of the conventional oil reserves, increasing attention is turning towards huge unconventional resources such as heavy oil and oil sands deposits due to their enormous volume and worldwide distribution. Production from these reservoirs is challenging owing to the immobile nature of bitumen, and reducing the in situ viscosity of the oil is considered as the main objective of any recovery process. In order to efficiently produce bitumen, the viscosity of the oil must be substantially reduced. Furthermore, efficient transportation of bitumen (e.g. by pipeline) can also be difficult unless the viscosity of the oil is first reduced (Kariznovi et al. 2013).

Among the different methods used for reducing viscosity of bitumen without the usage of excess energy or changing the properties of oil, acoustic cavitation during sonication is well recognised (Castaneda et al. 2014; Gogate et al. 2003). It causes heating, and intense agitation of a liquid medium or suspension, and activates chemical processes and enhancement of heat and mass transfer processes. Ultrasound is capable of producing extraordinarily high transient temperature and pressure in a localised spot within bitumen by the occurrence and collapse of

✉ Deepak M. Kirpalani
Deepak.Kirpalani@nrc-cnrc.gc.ca

[1] Energy Mining and Environment Portfolio, National Research Council of Canada, 1200 Montreal Road, Ottawa, ON K1A 0R6, Canada

Edited by Xiu-Qin Zhu

acoustic cavitation (Suslick 1990). Yan and Yaping (1996) have studied the change in viscosity of heavy oil from the Gudao oil field by using ultrasonic and surface active agents. They reported that ultrasonic waves can effectively decrease the viscosity of heavy oil and increase fluid flow ability, which aids in producing additional oil and in transporting heavy oil over long distances. Najafi and Amani (2011) studied the asphaltene flocculation inhibition with ultrasonic radiation (45 kHz frequency and 75 W power intensity) and observed that the viscosity of heavy oil decreases due to the disintegration of asphaltene flocs under ultrasonic irradiation. Further, Mohammadreza et al. (2012) studied the effect of ultrasonic irradiation on rheological properties of asphaltenic crude oil and concluded that ultrasonic irradiation increases the value of the yield stress required for the flow of crude oil samples. However, in order to understand the action of ultrasound and its effect on the change in heavy oil viscosity, it is required to study the effect of sonication frequency, solute concentrations, and acoustic power intensity on cavitation yield.

Cavitation yield measurement by determining the amount of iodine liberated during ultrasound of potassium iodide (KI) solutions is a simple and widely accepted method (Koda et al. 2003). The oxidation of KI is widely regarded as a standard to calibrate sonochemical efficiency. In addition, preparation and handling of KI solutions is simple and easy (Koda et al. 2003). During the sonication process, cavitation activates the generation of hydroxide and peroxy radicals which act as oxidising agents for the solute (KI) leading to release of iodine (Weissler et al. 1950). Entezari and Kruus (1996) reported the amount of oxidation of iodide to iodine at sonication frequency of 20 and 90 kHz. They observed that the cavitation yield using a 900-kHz transducer was 20 times greater than that of the 20-kHz transducer. Kirpalani and McQuinn (2006) determined the cavitation yield in a high-frequency ultrasound system (1.7 and 2.4 MHz) by measuring the amount of iodine liberated from the oxidation of KI solution. They observed that the concentration of KI and temperature affect the cavitation yield of the system such that the iodine production is proportional to both conditions. Furthermore, Ebrahiminia et al. (2013) studied the efficacy of different exposure parameters on cavitation production by 1 MHz ultrasound using iodide dosimetry and reported that with increasing time of sonication or intensity, the absorbance is increased.

Therefore, the following study was carried out to determine the cavitation yield under mid- to high-range sonication frequencies of 378, 574, 850, 992, and 1173 kHz and to investigate the effect of sonication frequency, power input, and solute concentration on yield. Further experiments were carried out with bitumen in order to determine the effect of cavitation yield on the viscosity change in heavy oil.

2 Materials and methods

2.1 Reagents

Bitumen was collected from a facility at Mildred Lake near Fort McMurray, Alberta, Canada. All the chemicals used were of analytical grade. HPLC-grade toluene, heptane, methanol (MeOH), dichloromethane (DCM), and acetone used for cleaning and extraction purposes were purchased from Fisher Scientific (Ontario, Canada). HPLC-grade water was prepared in the laboratory using a Milli-Q/Milli-RO Millipore system (Milford, MA, USA). Nitric acid, hydrochloric acid, and hydrogen peroxide were supplied by Fisher Scientific (Ontario, Canada).

2.2 Sonication experiment

A mid- to high-frequency sonochemical processing system was assembled using a broadband transducer (Ultraschall-technik-Meinhardt GMBH). The transducer was installed at the bottom of a coolant-jacketed glass column reactor with a diameter of 5 and 100 cm in height.

The ultrasound was supplied by a power amplifier (HM8001-2) through a function generator (HM 8030-5 and HM 8032). Sonication experiments were carried out at five different frequencies, 378, 574, 850, 992, and 1173 kHz, using two different broadband transducers with the same effective diameter. The reactor was supplied with different power inputs starting from 16.67% to 83.33%. The cooling system was operated to maintain a constant temperature. The experiment was carried out with a sample volume of 100 mL held within a jacked glass cooling column. A cooling jacket with ethylene glycol as a coolant was set up to maintain near-isotropic conditions to minimise the effect of acoustic energy conversion to heat. The local fluid temperature was monitored with a K-type thermocouple and maintained between 20 ± 1 °C for the duration of the experiments.

2.3 Acoustic intensity measurements based on calorimetric determination

Experiments were carried out to determine the acoustic intensity based on calorimetric determination at 378, 574, 860, 992, and 1173 kHz in water. Acoustic intensity based on calorimetric determination was calculated by measuring the temperature increase in water under ultrasound irradiation and using Eqs. (1) and (2):

$$\text{Actual power input} = \frac{dT}{dt} C_p M \tag{1}$$

$$\text{Acoustic intensity } (I) = \frac{\text{Actual power input}}{\text{Area of transducer tip}} \tag{2}$$

where $\frac{dT}{dt}$ is the rate of increase in the liquid temperature, C_p is the specific heat capacity of the liquid, and M is the mass of the liquid.

2.4 Measurement of cavitation yield

Experiments were carried out to determine the cavitation yield at 378, 574, 860, 992, and 1173 kHz in water. Further studies were carried out to observe the effect of these sonication frequencies in heavy oil. Cavitation yield measurements were taken to determine the amount of iodine liberated when potassium iodide solutions at different concentrations were subjected to ultrasound (Kirpalani and McQuinn 2006). Cavitation yield was defined as the grams of iodine liberated per unit power density during the oxidation of potassium iodide by hydroxide and peroxy radicals (Gedanken 2004). The amount of liberated iodine was measured using a UV/VIS spectrophotometer at 350 nm and quantified using a calibration curve ranging from 0 to 4×10^{-3} M of iodine for determining the cavitation yield at various intervals of time. A calibration curve for an iodine–water solution is presented in Fig. 1.

2.5 Measurement of viscosity

Viscosity of unsonicated and sonicated samples was measured by using a rotational viscometer (Brookfield DVII PRO + (Brookfield Engineering Laboratories, Inc., Stoughton, MA, USA)) equipped with Rheocalc32 software. The viscosity data acquisition and analysis was carried out using Rheocalc V2.6 software (B.E.A.V.I.S.—Brookfield Engineering Advanced Viscometer Instruction Set). Viscosity was measured at 50 rpm with 60% power amplitude at controlled temperature.

3 Results and discussion

3.1 Effect of sonication frequency

A series of experiments were conducted using 0.1 and 1 wt% KI aqueous solutions in the sonication vessel in order to establish the effect of sonication frequencies (378, 574, 850, 992, and 1173 kHz) and time (0–30 min) on cavitation yield. The cavitation yield for oxidation of potassium iodide solutions under different sonication frequencies and sonication time is presented in Fig. 2 (0.1 wt% KI) and Fig. 3 (1.0 wt% KI). The results showed that the sonication frequencies can significantly affect the cavitation yield, and higher yield was observed at mid-frequencies as compared with high frequencies. Note that other studies also reported the enhancement of cavitation yield under different frequencies (Kirpalani and McQuinn 2006; Seymour et al. 1997). A comparison of two studies, Entezari and Kruus (1996) performed at 20 and 900 kHz and Seymour et al. (1997) at 640 kHz, showed a higher cavitation yield at sonication frequency of 640 kHz compared to 20 and 900 kHz. Furthermore, a study by Feng et al. (2002) reported that when using the low MHz frequency range, the lower the frequency used, the higher the cavitation yield. They observed that the sound intensity of low MHz frequency ultrasound is above 6 W cm^{-2}, and the cavitation yield of the combined irradiation (0.87 MHz and 28 kHz) is more than 1.6 times of that of combined

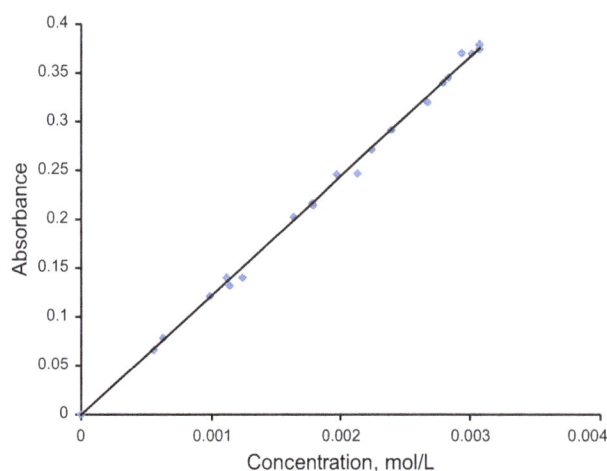

Fig. 1 Iodine calibration curve obtained using UV–Vis spectrophotometry

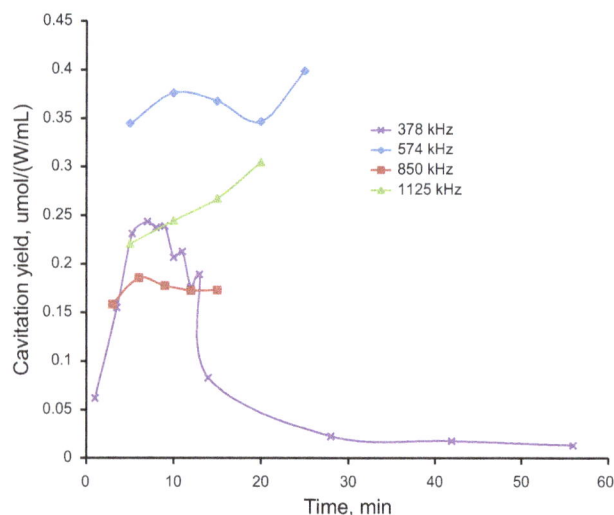

Fig. 2 Cavitation yield obtained under different conditions of sonication frequency and time using 0.1 wt% KI solution

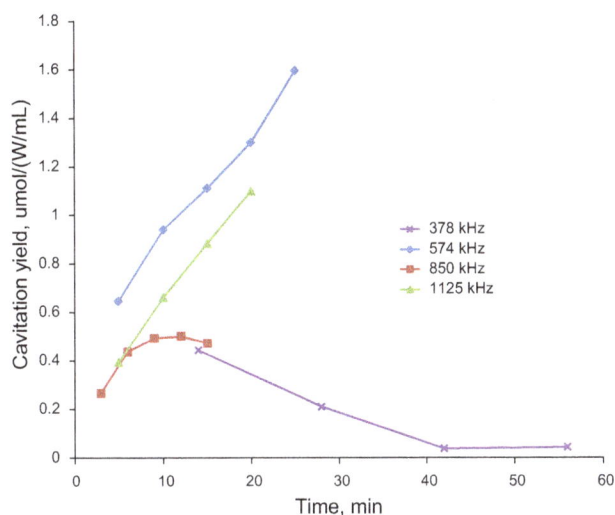

Fig. 3 Cavitation yield obtained under different conditions of sonication frequency and time using 1.0 wt% KI solution

irradiation using a higher MHz frequency (1.7 MHz and 28 kHz). Formation of highly reactive free radicals due to the formation, growth, and implosive collapse of bubbles in a liquid is the primary mechanism of a sonochemical reaction. The extent of radical formation in a single cavitation bubble is a function of the following parameters: amount of water vapour trapped in the bubble and the temperature and pressure peak reached in the bubble during the collapse. A mechanistic approach to the enhancement of the yield of a sonochemical reaction showed that the collapse of cavitation bubbles and sonochemical yield is a complicated function of several interdependent physical processes such as rectified diffusion, water vapour transport, and entrapment in cavitation bubbles. Further, it was reported that the degassing of the reaction medium intensifies the collapse of the cavitation bubbles, resulting in higher production of OH and other radicals, which enhance the yield of the sonochemical reaction (Sivasankar et al. 2007).

Furthermore, higher oxidation of iodide to iodine was observed at a sonication frequency of 574 kHz compared to the other four frequencies 378, 860, 992, and 1173 kHz (Figs. 2, 3). It was also observed that in case of 1 wt% KI aqueous solutions, the cavitation yield increases with increase in sonication time for a sonication frequency of 574 and 1173 kHz. Higher oxidation of iodide to iodine during sonication frequency of 574 kHz was due to the presence of more dissolved air in the medium. The presence of dissolved air in the solution reduces the threshold pressure during cavitation resulting in significant rise in the number of cavities formed and consequent increase in reaction rates. Lida et al. (2005) showed that the degree of gas saturation, the type of gas, and the temperature of a sonicated dosimeter solution at different sonication

frequencies play an important role in determining the extent of inertial cavitation occurrence. Further, Ebrahiminia et al. (2013) observed that sonication frequency plays an important role for reactive radical generation in the medium and showed that with increasing the sonication frequency, the cavitation yield increased when the frequency was greater than 250 kHz.

3.2 Effect of power supplied and KI concentration

The effect of power supplied on the oxidation of KI at sonication frequencies of 378, 574, 992, and 1173 kHz was observed. Figure 4 presents the cavitation yield obtained over a power input range of 16.67% to 83.33% with different sonication frequencies carried out for 30-min sonication treatments on 1.0 wt% KI solutions. The results showed that the cavitation yield increased by increasing power input up to 50% in all the frequency level tested. The effect of power supplied on the oxidation of KI within a sonication frequency of 20 and 900 kHz has been studied previously (Henglein and Gutierrez 1993; Weissler et al. 1950). A study by Entezari and Kruus (1996) showed that at 900-kHz sonication frequency, increase in power supply (between 8 and 76 W) leading to linearly increased production of iodine with a cavitation threshold at 0.14 W cm^{-2}. Merouani et al. (2010) studied the influence of several operational parameters on the sonochemistry dosimetry approaches such as KI oxidation, H_2O_2 production, and Fricke reaction using 300-kHz ultrasound. They observed that the main experimental parameters that showed significant effect in KI oxidation dosimetry were initial KI concentration, pH, and acoustic power. Further, Lim et al. (2014) studied the effects of liquid height/

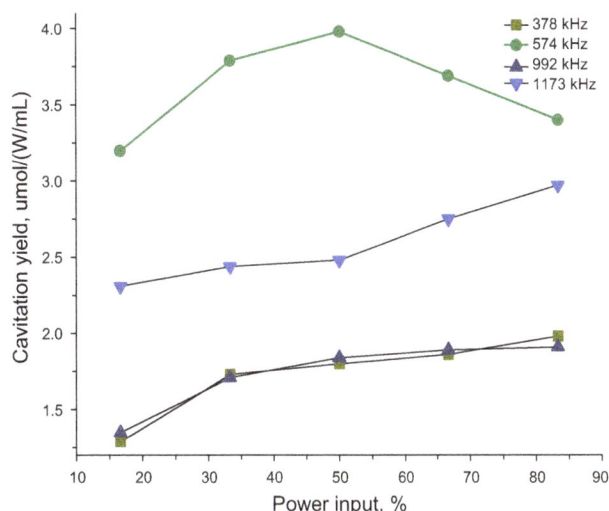

Fig. 4 Cavitation yield obtained with different power inputs at different sonication frequencies sonicated for 30 min in 1.0 wt% KI solutions

volume, initial concentration of reactant, and acoustic power (23, 40, and 82 W) on sonochemical oxidation. They observed that as the liquid height/volume and the input power changed, the power density varied from 23 to 1640 W L^{-1} and the maximum cavitation yields of triiodide ion for 23, 40, and 82 W were observed as 0.05, 0.1, and 0.2/0.3 L, respectively. They also reported that low power was more effective for the small volume and the large volume required high power level. However, a previous study by Henglein and Gutierrez (1993) showed that at large volume of KI solution, the iodine production tends to show a nonlinear increase with an increase in power input. Kirpalani and McQuinn (2006) established a relationship between acoustic power and the amount of iodine liberated by varying the power supply of 3–21 W and 8–18 W to 1.7 and 2.4 MHz transducers, respectively. They reported that for 100 mL of KI solution, with increase in power supply, the amount of iodine liberated increased. Further, experiments were performed to measure the power inputs (W) under four different sonication frequency conditions of 378, 574, 992, and 1173 kHz with different power inputs starting from 16.67% to 83.33%. Table 1 presents different power input settings and their measured power output under four different sonication frequency conditions: 378, 574, 992, and 1173 kHz. It is clear that the power intensity and voltage at each power input level is not constant and does not follow a particular pattern which could be used to determine the effects of alternating

intensity levels at a different frequency. Acoustic intensity measurements based on calorimetric determination under different frequencies and power input are presented in Table 2. The results showed increased cavitation yield with increasing power input under 378 and 574 kHz sonication frequency conditions. However, it was observed that using a higher power input increased the amount of iodine that is liberated, but due to the increase in power required, the cavitation yield decreased in the system.

Furthermore, experiments were carried out to determine the effect of KI concentration on the production of iodine in an ultrasonic system. The yield of KI oxidation reaction can be defined as the number of moles of iodine liberated per unit time per unit reaction volume per unit mole of KI per unit power input to the system (Sivasankar et al. 2007). A decrease in cavitation yield as a result of decreasing KI concentration was observed for all types of sonication frequencies tested (Figs. 2, 3). Naidu et al. (1994) also observed an increase in iodine liberation in solutions with higher KI concentrations sonicated at 25 kHz. However, a higher cavitation yield was observed with sonication frequency of 574 kHz in both cases of KI concentration. Increase in cavitation yield in case of 574-kHz sonication frequency can be attributed to the amount of hydroxyl radicals produced due to rapid collapse of bubbles and also due to the consumption of hydroxyl radicals by iodide ions that increase with an increase in KI concentration.

Table 1 Measured power inputs for a range of intensity settings under different conditions of sonication frequency

Sonication frequency, kHz	Power input, %	Power intensity, W	Voltage, V
378	16.67	210	118
	33.33	248	118
	50	304	118
	66.66	360	117
	83.33	426	116
574	16.67	209	118
	33.33	252	118
	50	304	118
	66.66	376	118
	83.33	446	118
990	16.67	208	116
	33.33	249	116
	50	295	117
	66.66	346	118
	83.33	413	118
1173	1173	199	116
	1173	256	116
	1173	303	118
	1173	369	118
	1173	452	117

Table 2 Acoustic power delivered to the reaction system at different frequencies and power input determined using calorimetric method

Sonication frequency, kHz	Power input, %	Actual power, W	Acoustic intensity, W/cm^2
378	16.67	6.671	0.218
	50	12.644	0.395
	83.33	25.820	0.817
574	16.67	5.834	0.184
	50	10.826	0.342
	83.33	27.652	0.876
990	16.67	5.638	0.178
	50	10.522	0.334
	83.33	22.981	0.726
1173	16.67	10.447	0.331
	50	11.675	0.369
	83.33	32.733	1.034

3.3 Effect of cavitation yield on viscosity

In order to establish the effect of cavitation yield under different sonication frequencies, 378, 574, 992, and 1173 kHz, on viscosity change, first experiments were carried out in KI solution and sodium carboxymethyl cellulose (CMC)–water mixture followed by application in bitumen. CMC is used as a model fluid to describe the behaviour of bitumen at 1000 mPa s approximately since it is a shear-thinning fluid similar to heavy oil (Muller 1994). Effect of cavitation yield on viscosity change under different conditions of sonication frequency (378, 574, 992, and 1173 kHz) and power input (16.67%–83.33%) in 0.7 wt% CMC–water sonicated for 30 min is presented in Fig. 5. Higher viscosity changes were observed with sonication frequencies such as 378 and 574 kHz as compared to frequencies such as 992 and 1173 kHz. Higher viscosity changes observed in 378- and 574-kHz sonication frequencies were due to higher cavitation yield (Fig. 6). It was

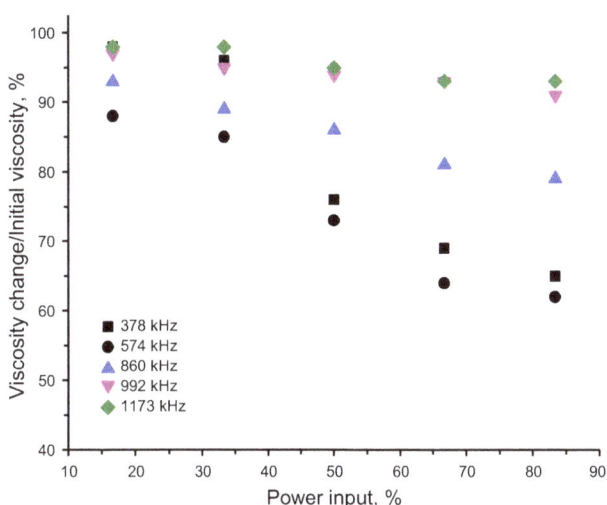

Fig. 6 Viscosity change at different sonication frequencies for different power inputs in bitumen

also observed that with increasing power intensities leading to higher cavitation yield results in higher viscosity change. Chemical effects are recognised to be dominant at low frequency (Suslick et al. 1999), while physical effects are dominant at high frequency as it requires very high energy input to generate cavitation at high frequencies. Change in viscosity occurs due to intense shear or tensile force within the fluid (Mohapatra and Kirpalani 2016). This shear is caused by physical effects such as microconvection, and microstreaming or microstirring. The principal physical effect of ultrasonic cavitation is the formation of fine emulsion droplets between immiscible phases that eliminates the mass transfer resistance, while principal chemical effect is the production of radicals through collapse of cavitation bubbles, which accelerate the reaction (Kuppa and Moholkar 2010). For all the four types of sonication frequencies, increase in viscosity was observed with increasing power intensity except for 378 kHz where a

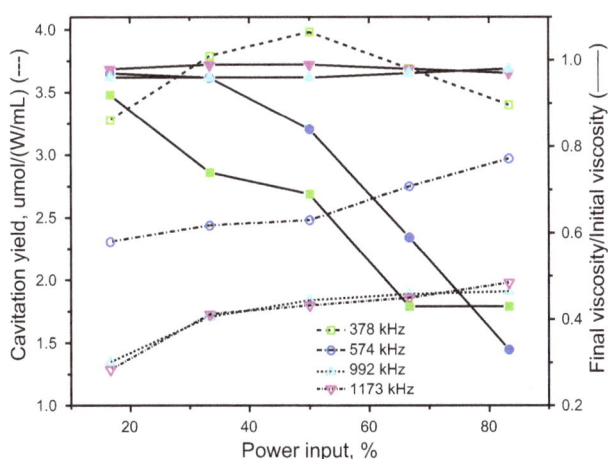

Fig. 5 Cavitation yield and viscosity change under different conditions of sonication frequency and power inputs in CMC–water sonicated for 30 min

decrease in viscosity change was observed after 50% power intensity.

Furthermore, the effect of five different sonication frequency conditions, 378, 574, 860, 992, and 1173 kHz, with different power inputs from 16.67% to 83.33% on change in viscosity of bitumen was investigated. Typically, in oil sands operations, bitumen is transported between unit operations by dilution with naphtha or a paraffinic solvent (C6–C8 paraffinics). In this work, bitumen was diluted with naphtha to viscosity of 480 mPa s to evaluate the benefits of acoustic cavitation for improved transportability and reduced pumping (energy) costs. Bitumen was first diluted with naphtha at the N/B ratio of 0.5. The diluted bitumen sample was then centrifuged at 20 000 rpm for over 20 min to remove the fine solids. All the experiments were carried out at room temperature. The supernatant was then used for preparation of a series of diluted bitumen samples by adding more naphtha to get a viscosity of 480 mPa s.

The viscosity change in bitumen for different sonication frequencies with different power inputs is presented in Fig. 6. The major viscosity changes were observed to occur at sonication frequencies of 378 and 574 kHz with increasing power input. The highest viscosity change of 38% was observed at a sonication frequency of 574 kHz with 83.33% power input, and lowest viscosity changes (5%) were observed at sonication frequency of 1173 kHz with a power input of 83.33%. However, it was observed that the change in viscosity increases in all the five sonication frequency conditions tested with increasing power input. The increase in viscosity change (lower viscosity) with increasing the power input was due to higher disintegration of asphaltene flocs. Higher cavitation yield causes more acoustic cavitation in the medium, which then causes floc disintegration and cell breakage, leading to release of intracellular materials to the aqueous phase. Cavitation yield during different sonication frequencies and power input conditions is a result of a series of energy conversions (electrical energy → mechanical energy → acoustic energy → cavitation energy) that occur in the ultrasonic processor. In case of varying frequencies, such chain of energy conversion strongly depends on the total bubble volume fraction in the medium which further depends on the number of bubbles and their size distribution (Moholkar and Warmoeskerken 2003).

Some studies report that the viscosity of bitumen depends upon the asphaltene content (Mohammadreza et al. 2012; Najafi and Amani 2011; Luo and Gu 2007). Hence, during the sonication treatment of bitumen, the dissolution of asphaltenic components occurs resulting in the breakdown of asphaltene molecules to lighter molecules leading to decrease in viscosity. A study by Mack (1932) concluded that the significant viscosity increase with the asphaltene content was due to strong aggregation of asphaltene particles. Further, many studies also observed ultrasonic technology as a method for reducing asphaltene flocculation rates by changing the kinetics of aggregation as well as for removing deposits (Mohammadreza et al. 2012; Mousavi-Dehghani et al. 2004).

4 Conclusions

Optimisation of cavitation technology including different sonication frequencies and power intensities was carried out in KI solution and CMC–water mixture and further applied in bitumen heavy oil processing. This study examines the effects of mid- to high sonication frequencies, solute concentrations, and power intensity on cavitation yield and viscosity changes in the medium. This study showed that mid-range frequency levels such as 378 and 574 kHz are more effective in terms of increase in cavitation yield and viscosity changes in bitumen heavy oil. The highest viscosity decrease of 38% was observed at a sonication frequency of 574 kHz with 83.33% power input, and the lowest viscosity decrease (5%) was observed at a sonication frequency of 1173 kHz with a power input of 83.33%. The higher viscosity change observed with the 574-kHz sonication frequency was due to a higher cavitation yield which showed a direct relationship between cavitation yield and viscosity change.

Different factors such as power inputs and solute concentration affect the cavitation yield in the medium. An optimum condition of 574 kHz sonication frequency and 83.33% power input was observed as the best method for decreasing the viscosity of bitumen compared to other frequencies tested.

References

Castaneda LC, Munoz JAD, Ancheyta J. Current situation of emerging technologies for upgrading of heavy oils. Catal Today. 2014;220–222:248–73.

Ebrahiminia A, Mokhtari-Dizaji M, Toliyat T. Correlation between iodide dosimetry and terephthalic acid dosimetry to evaluate the reactive radical production due to the acoustic cavitation activity. Ultrason Sonochem. 2013;20:366–72.

Entezari MH, Kruus P. Effect of frequency on sonochemical reactions II. Temperature and intensity effects. Ultrason Sonochem. 1996;3:19–24.

Feng R, Zhao Y, Zhu C, Mason TJ. Enhancement of ultrasonic cavitation yield by multi-frequency sonication. Ultrason Sonochem. 2002;9:231–6.

Gedanken A. Using sonochemistry for the fabrication of nanomaterials. Ultrason Sonochem. 2004;11:47–55.

Gogate PR, Mujumdar S, Pandit AB. Large-scale sonochemical reactors for process intensification: design and experimental validation. J Chem Technol Biotechnol. 2003;78:685–93.

Henglein A, Gutierrez M. Sonochemistry and sonoluminescence: effects of external pressure. J Phys Chem. 1993;97:158–62.

Kariznovi M, Nourozieh H, Guan JG, Abedi J. Measurement and modeling of density and viscosity for mixtures of Athabasca bitumen and heavy n-alkane. Fuel. 2013;112:83–95.

Kirpalani DM, McQuinn KJ. Experimental quantification of cavitation yield revisited: focus on high frequency ultrasound reactors. Ultrason Sonochem. 2006;13:1–5.

Koda S, Kimura T, Kondo T, Mitome H. A standard method to calibrate sonochemical efficiency of an individual reaction system. Ultrason Sonochem. 2003;10:149–56.

Kuppa R, Moholkar VS. Physical features of ultrasound-enhanced heterogeneous permanganate oxidation. Ultrason Sonochem. 2010;17:123–31.

Lida Y, Yasui K, Tuziuti T, Sivakumar M. Sonochemistry and its dosimetry. Microchem J. 2005;80:159–64.

Lim M, Ashokkumar M, Son Y. The effects of liquid height/volume, initial concentration of reactant and acoustic power on sonochemical oxidation. Ultrason Sonochem. 2014;21:1988–93.

Luo P, Gu Y. Effects of asphaltene content on the heavy oil viscosity at different temperatures. Fuel. 2007;86:1069–78.

Mack C. Colloid chemistry of asphalts. J Phys Chem. 1932;36(12):2901–14.

Mousavi-Dehghani SA, Riazi MR, Vafaie-Sefti M, Mansoori GA. An analysis of methods for determination of onsets of asphaltene phase separations. J Pet Sci Eng. 2004;42:145–56.

Mohammadreza MS, Ahmad R, Iman N, Mohammad DS. Effect of ultrasonic irradiation on rheological properties of asphaltenic crude oils. Pet Sci. 2012;9:82–8.

Mohapatra DP, Kirpalani D. Bitumen heavy oil upgrading by cavitation processing: effect on asphaltene separation, rheology and metal content. Appl Petrochem Res. 2016;. doi:10.1007/s13203-016-0146-1.

Moholkar VS, Warmoeskerken MMCG. Integrated approach to optimization of an ultrasonic processor. AIChE J. 2003;49(11):2918–32.

Merouani S, Hamdaoui O, Saoudi F, Chiha M. Influence of experimental parameters on sonochemistry dosimetries: KI oxidation, Fricke reaction and H_2O_2 production. J Hazard Mater. 2010;178:1007–14.

Muller FL. Rheology of shear thinning polymer solutions. Ind Eng Chem Res. 1994;33:2364–7.

Naidu DVP, Ranjan R, Kumar R, Gandhi KS, Arakeri VH, Chandrasekheran S. Modelling of a batch sonochemical reactor. Chem Eng Sci. 1994;49:877–88.

Najafi I, Amani M. Asphaltene flocculation inhibition with ultrasonic wave radiation: a detailed experimental study of the governing mechanisms. Adv Pet Explor Dev. 2011;2:32–6.

Seymour JD, Wallace HC, Gupta RB. Sonochemical reactions at 640 kHz using an efficient reactor: oxidation of potassium iodide. Ultrason Sonochem. 1997;4:289–93.

Sivasankar T, Paunikar AW, Moholkar VS. Mechanistic approach to enhancement of the yield of a sonochemical reaction. AIChE J. 2007;53(5):1132–43.

Speight JG. The chemistry and technology of petroleum. 4th ed. Boca Raton: CRC Press, Taylor & Francis Group; 2007.

Suslick KS. Sonochemistry. Science. 1990;247:1439–45.

Suslick KS, Didenko Y, Fang MM, Hayon T, Kolbeck KJ, et al. Acoustic cavitation and its chemical consequences. Philos Trans R Soc Lond A. 1999;357:335–53.

Weissler A, Cooper H, Snyder S. Chemical effect of ultrasonic waves: oxidation of potassium iodide solution by carbon tetrachloride. J Am Chem Soc. 1950;72:1769–75.

Yan X, Yaping Z. Experiment of reduction viscosity by ultrasonic wave. Oil Gas Surf Eng. 1996;15:20–1.

8

The surface properties of aluminated meso–macroporous silica and its catalytic performance as hydrodesulfurization catalyst support

Zhi-Gang Wang[1] · Jia-Ning Pei[1] · Sheng-Li Chen[1] · Zheng Zhou[1] ·
Gui-Mei Yuan[1] · Zhi-Qing Wang[2] · Guo-Qiang Ren[2] · Hong-Jun Jiang[2]

Abstract Aluminated mesoporous silica was prepared by
multiple post-grafting of alumina onto uniform mesoporous
SiO_2, which was assembled from monodisperse SiO_2
microspheres. Hydrodesulfurization (HDS) catalyst was
prepared by loading Ni and Mo active components onto the
aluminated uniform mesoporous SiO_2, and its HDS catalytic performance was evaluated using hydrodesulfurization of dibenzothiophene as the probe reaction at 300 °C
and 6.0 MPa in a tubular reactor. The samples were characterized by N_2 physisorption, scanning electronic microscopy, Fourier transform infrared spectrum, X-ray
diffraction (XRD), temperature-programmed desorption of
ammonia (NH_3-TPD), ^{27}Al nuclear magnetic resonance
(^{27}Al-NMR) and high-resolution transmission electron
microscopy (HRTEM). The results showed that the Si–OH
group content of SiO_2 was mainly dependent on the pretreatment conditions and had significant influence on the
activity of the NiMo catalyst. The surface properties of the
aluminated SiO_2 varied with the Al_2O_3-grafting cycles.
Generally after four cycles of grafting, the aluminated SiO_2
behaved like amorphous alumina. In addition, plotting of
activity of NiMo catalysts supported on aluminated meso–
macroporous silica materials against the Al_2O_3-grafting
cycle yields a volcano curve.

Keywords Aluminum grafting · Hydrodesulfurization ·
Surface properties · Catalyst support · SiO_2

1 Introduction

The demand for more environmentally friendly petroleum
products with lower sulfur content is growing due to
environmental problems caused by SO_x emissions. Therefore, development of new catalysts with high hydrodesulfurization (HDS) activity is required. Previous
investigations (Breysse et al. 2003, 2008) indicate that the
nature and characteristics of catalyst support have significant influence on the performance of HDS catalyst. Generally, for industrial NiMo and CoMo HDS catalysts, the
supports are usually γ-Al_2O_3 or aluminosilicates rather than
pure SiO_2, mainly because of the stronger support–metal
interaction and subsequent better HDS activity of Al_2O_3-
supported catalyst than that of SiO_2-supported catalyst
(Scheffer et al. 1988). Ordered mesoporous Al_2O_3 is particularly suited as catalyst support due to its suitable surface and textural properties (Venezia et al. 2010; Morris
et al. 2008). Therefore, a number of researchers have aimed
to directly synthesize ordered mesoporous alumina (Bagshaw and Pinnavaia 1960; Yada et al. 1997; Cabrera et al.
1999; Márquez-Alvarez et al. 2008). However, the ordered
mesoporous alumina obtained by direct synthesis has some
shortcomings like the complexity of the synthesis process,
lack of reproducibility, low thermal stability at high temperature, structural non-uniformity and wide pore distribution (Cheralathan et al. 2008). Fortunately, one
alternative route to prepare ordered mesoporous alumina is
available, that is, grafting alumina onto ordered SiO_2
mesoporous materials. This alternative method has distinct
advantages over direct synthesis with respect to

✉ Sheng-Li Chen
slchen@cup.edu.cn

[1] State Key Laboratory of Heavy Oil Processing, College of
Chemical Engineering, China University of Petroleum,
Beijing 102249, People's Republic of China

[2] Shanghai Petrochemical Company of Sinopec,
Jinshan, Shanghai 200540, People's Republic of China

Edited by Xiu-Qin Zhu

reproducibility, structural ordering and thermal stability (Mokaya and Jones 1999; Mokaya 1997).

Up to now, various pure silica mesoporous materials, such as MCM-41 (Goldbourt et al. 2002; Landau et al. 2001), SBA-15 (Zukal et al. 2008; Baca et al. 2008) and KIT-1 (Ryoo and Kim 1997; Jun and Ryoo 2000), have been employed as templates in many investigations concerning alumination of SiO_2 surfaces through grafting technology. Several researchers have investigated the HDS performance of catalysts supported on alumina-grafted materials. Klimova et al. (2003) found that a catalyst with a higher content of alumina has lower HDS activity, and concluded that over strong metal–support interaction, caused by incorporation of alumina in MCM-41, made the reduction and sulfidation of the active component more difficult. However, in the study of the effect of incorporation of alumina on HDS activity of NiMo/SBA-15 catalysts, it was found that the blockage of catalyst pores by alumina could lead to significant decline in activity in the case of high content of alumina (Rayo et al. 2009). On the one hand, owing to the small pore size of these molecular sieves, the pore structure probably suffered from the incorporation of alumina. On the other hand, the surface property is significantly affected by the incorporation of alumina. Therefore, it is difficult ascertain which is the main cause for the decrease in HDS activity. Fortunately, our research group has prepared SiO_2 opal by ordered face-centered cubic packing of monodisperse silica microspheres (Zhou et al. 2012a, b). The ordered compact of SiO_2 opal with a high coordination number has very uniform pore size distribution, and the pore size can be tuned according to the needs of experiments by changing the diameter of the monodisperse SiO_2 microspheres. Hence, SiO_2 opals with larger pore size are more suitable materials for the grafting of multilayer alumina, since larger pore sized opal has better pore structure stability against the amount of grafted alumina, than MCM-41 or SBA-15. In addition, the content of silanol groups on the surface of template porous SiO_2 opal is supposed to have a great influence on the post-synthesis of alumination (Mokaya and Jones 1999; Li et al. 2001; Zhao et al. 1997), and the content of the silanol groups mainly depends on pretreatment conditions, such as calcination temperature and whether or not there has been a hydrothermal treatment. However, very few researchers have aimed to investigate the effect of the content of silanol groups of template porous SiO_2 on HDS activity of catalyst using aluminated SiO_2 as catalyst support.

In the present work, the effects of pretreatment conditions of the SiO_2 materials and the Al_2O_3-grafting cycle on the surface properties of aluminated SiO_2 opal materials were investigated. In order to ascertain the relationship between the HDS activity and the surface property of the aluminated SiO_2 used as catalyst support, the catalytic performance of NiMo catalysts supported on these aluminated SiO_2 opal materials was tested using the HDS of dibenzothiophene (DBT) as probe reaction.

2 Experimental

2.1 Preparation of SiO_2 opal materials and reference alumina

Monodisperse SiO_2 microspheres were synthesized through hydrolysis and condensation of tetraethyl orthosilicate (98%, J&K) in alcohol (99.7%, Sinopharm Chemical Reagent) and in the presence of water and ammonia by the seed particle growth method. Detailed descriptions of the synthesis procedures were reported in the previous papers of our research group (Chen et al. 1996; Liu et al. 2009; Chen 1998). In this work, monodisperse SiO_2 microspheres with diameters of 100, 250 and 500 nm were prepared. First, SiO_2 opals were obtained by ordered packing of SiO_2 microspheres, then the assembled SiO_2 opals were calcined at three temperatures in the range of 500–900 °C for 2 h, and finally these SiO_2 opals were hydrothermally treated at 220 °C for 5 h to recover the surface silanol groups which were lost during the calcination. The pretreatment of the SiO_2 opal includes the calcination and hydrothermal treatment. In addition, a reference alumina, denoted as $(Al(NO_3)_3$-C), was prepared by calcination of $Al(NO_3)_3$·9H$_2$O (99%, Sinopharm Chemical Reagent) in static air at 500 °C for 5 h.

2.2 Alumination of SiO_2 opal materials

In the alumination procedure, hydrothermally treated SiO_2 opals were added to 0.67 M aluminum nitrate ($Al(NO_3)_3$) solution at 80 °C, kept in the solution for 12 h and then filtered to remove the solution. To remove the $Al(NO_3)_3$ solution in the pores of the SiO_2 opals, the filtered SiO_2 opals were washed three times with distilled water. The SiO_2 opals were then dried at 100 °C for 6 h and calcined in static air at 500 °C for 5 h. The above steps compose one cycle of Al_2O_3 grafting, and the Al_2O_3-grafting process can be repeated several times. The sample is denoted as m-SiO_2-Al-n, where m is the diameter of the monodisperse SiO_2 microspheres, from which the SiO_2 opal was made of, and n is the number of Al_2O_3-grafting cycles. These aluminated SiO_2 opal materials were used as supports for the hydrodesulfurization catalyst.

2.3 Catalyst preparation

NiMo catalysts were prepared by incipient wetness co-impregnation of aqueous solutions of $(NH_4)Mo_7O_2\cdot4H_2O$ (>99%, Sinopharm Chemical Reagent) and $Ni(NO_3)_2\cdot6H_2O$ (>98%, Sigma-Aldrich) into catalyst support. After impregnation, the catalysts were dried at 100 °C for 6 h and then calcined at 500 °C for 5 h. The active metallic component loadings in the catalysts were 5.42 μmol MoO_3/m^2 support and 1.81 μmol NiO/m^2 support. For comparison, the reference alumina was loaded with the same loading amount of MoO_3 and NiO by incipient wetness co-impregnation of $(NH_4)Mo_7O_2\cdot4H_2O$ and $Ni(NO_3)_2\cdot6H_2O$ aqueous solutions. This catalyst was dried and calcined as described above. All catalysts were crushed and sieved to a size of about 0.23 mm before catalytic performance testing.

2.4 Characterization of support and catalyst

The supports and catalysts were characterized by N_2 physisorption, scanning electron microscopy (SEM), X-ray diffraction (XRD), temperature-programmed desorption (TPD), ^{27}Al nuclear magnetic resonance (NMR) and high-resolution transmission electron microscopy (HRTEM). N_2 adsorption/desorption isotherms were measured with a Micromeritics ASAP 2020 automatic analyzer (ASAP 2020, Micromeritics, USA) at −196 °C. Specific surface areas were calculated by the BET method. The total pore volume and pore size distributions were measured by the mercury penetration method on an Autopore II 9220 mercury porosimeter using a contact angle of 140°. The morphology of catalysts was observed on a scanning electron microscope (Quanta 200F, FEI, USA) using an accelerating voltage of 20 kV. HRTEM micrographs were obtained by using a transmission electron microscope (JEM-2100, JEOL, Japan) operated at 200 kV. The sulfided catalysts were crushed and then ultrasonically dispersed in heptane, and the suspension was collected on carbon-coated grids. HRTEM micrographs were taken from different parts of the same sample dispersed on the microscope grid. Wide-angle XRD patterns were recorded in the range of $10° < 2\theta < 85°$ on a Bruker D8 Advance diffractometer (D8 Advance, Bruker, German), using Cu Kα radiation and a goniometer speed of 1° (2θ) min^{-1}. The acid sites amount and acid strength distribution were determined by the NH_3-TPD. The ammonia in the effluent gas was detected by a thermal conductivity detector (TCD). Before NH_3-TPD experiments, the samples were pretreated in situ at 550 °C for 50 min in a N_2 flow in order to remove water and other contaminants. The samples were then cooled to 110 °C, and a N_2/NH_3 mixture (30/10 mol/mol) at a flow rate of 40 ml/min was fed for 30 min. The desorption step was performed in a N_2 stream (30 mL/min) at a heating rate of 10 °C/min. ^{27}Al MAS NMR was carried out on a Bruker AVANCE III 600 spectrometer at a resonance frequency of 156.4 MHz using a 4 mm HX double-resonance MAS probe at a sample spinning rate of 15 kHz. The ^{27}Al chemical shift was referenced to 1 M aqueous $Al(NO_3)_3$. ^{27}Al MAS NMR spectra were recorded by the small flip angle technique with a pulse length of 0.5 μs (<π/12), a 1 s recycle delay and 4000 scans. Fourier transform infrared spectroscopy (FTIR) was recorded on a Bruker IFS 66 V spectrometer in the range of 3800–600 cm^{-1} (4 cm^{-1} resolution, 256 scans/spectrum) using a Thermo Spectra-Tech high-temperature cell. All the spectra were recorded at 150 °C in argon after 2 h of pretreatment at 450 °C in argon.

2.5 HDS activity of NiMo catalyst

The HDS activity tests of the NiMo catalysts were carried out in a bench-scale stainless-steel tubular reactor at 300 °C and 6.0 MPa using a sulfur-free lube base oil solution of DBT (1000 ppm of S) as HDS feedstock. The lube base oil was provided by Sinopec Shanghai Gaoqiao Petrochemical Corporation (China). Prior to catalytic HDS, the catalysts were sulfided by a mixture of 2 wt% CS_2 and cyclohexane with the following temperature program: reactor temperature was raised to 230 °C from room temperature at a heating rate of 7 °C/min and kept at 230 °C for 3 h; then the temperature was raised to 340 °C at a heating rate of 3 °C/min and kept at 340 °C for 3 h. The presulfiding conditions were as follows: liquid hourly space velocity (LHSV), 5 h^{-1}; H_2/Oil (v/v) ratio, 200; operation pressure, 6.0 MPa. After sulfidation, the stream was switched from sulfiding agent to HDS feedstock and maintained for 20 h to achieve catalyst stability, and the hydrodesulfurized sample was collected at the appropriate time (the time interval depended on the LHSV). Then, these samples were washed with 5 wt% sodium hydroxide solution three times to remove the dissolved H_2S. The sulfur content in the samples was measured by using a THA-2000S UV-induced fluorescence sulfur analyzer (Taizhou Jinhang Analytical Instruments Co. Ltd., China).

3 Results and discussion

3.1 Support and catalyst characterization

The pore structure characterization of supports and their NiMo catalysts is shown in Tables 1 and 2, respectively. It can be seen that the three SiO_2 opal supports without pretreatment had almost the same total pore volume (about 0.23 cm^3/g), but their average pore diameter and specific

Table 1 Pore structure characterization of catalyst supports

Samples	D_m, Nm	CT, °C	HT	GC	Alumina content, g/100 m² SiO₂	SSA, m²/g	V_p, cm³/g	d_p, nm
A	100	–	–	–	–	34.8	0.22	25.2
A-500-H-2	100	500	Y	2	0.053	32.5	0.21	25.8
A-700-2	100	700	N	2	0.042	30.1	0.21	27.9
A-700-H-2	100	700	Y	2	0.052	29.8	0.21	28.1
A-900-H-2	100	900	Y	2	0.031	27.3	0.19	27.8
A-700-H-4	100	700	Y	4	0.088	30.4	0.20	26.3
B	250	–	–	–	–	18.8	0.23	48.9
B-700-H-4	250	700	Y	4	–	17.8	0.22	49.4
C	500	–	–	–	–	9.8	0.23	93.9
C-700-H-4	500	700	Y	4	–	9.9	0.23	93.0

D_m is the diameter of microspheres, nm; CT is the calcination temperature, °C; HT is hydrothermal treatment; GC is the grafting cycles of alumina; d_p is the average pore diameter, nm; V_p is the pore volume, cm³/g; SSA is the specific surface area, cm²/g. Alumina content was measured by EDX

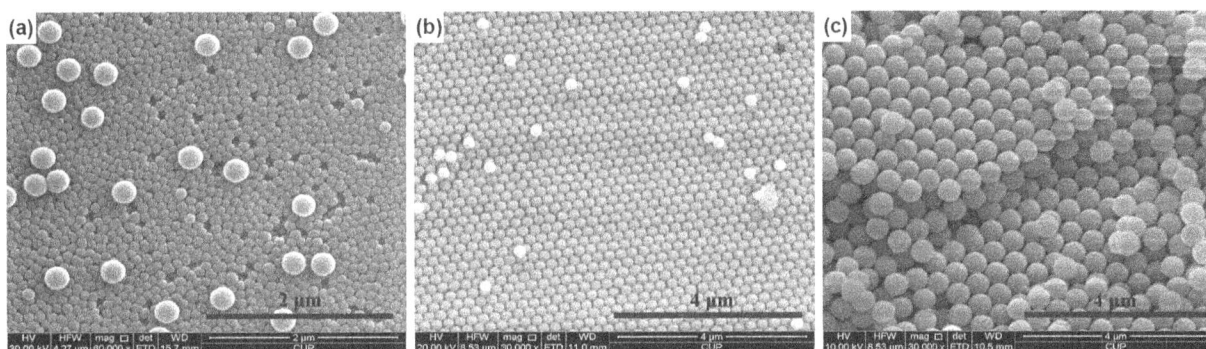

Fig. 1 SEM images of NiMo catalysts with microspheres diameters of 100 nm (**a**), 250 nm (**b**), and 500 nm (**c**)

Table 2 Pore structure properties of NiMo catalysts

Sample	d_p, nm	V_p, cm³/g	SSA, m²/g
NiMo/A-700-H-4	27.5	0.19	26.2
NiMo/B-700-H-4	49.4	0.20	16.2
NiMo/C-700-H-4	100.3	0.23	9.2
NiMo/Al(NO₃)₃-C	4.9	0.12	95.2

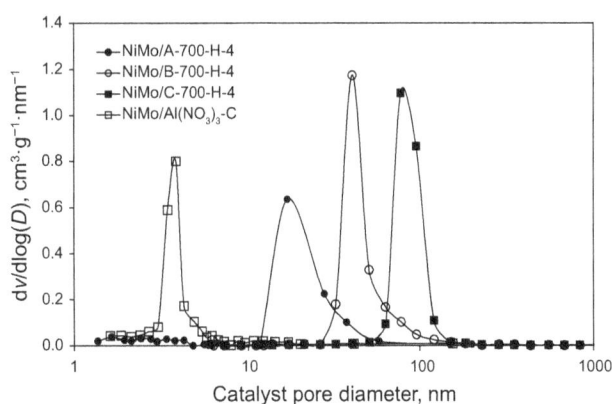

Fig. 2 Pore size distribution of four NiMo catalysts

surface area were different from each other (Table 1). Due to the shrinkage of the SiO₂ microspheres during calcination and sealing of the micropores on the surface of the SiO₂ microspheres during hydrothermal treatment, the specific surface area of SiO₂ opal which was calcined and hydrothermally treated had declined slightly. Additionally, the pore structure did not change significantly after successive Al₂O₃-grafting steps and NiMo impregnation (Tables 1, 2; Fig. 1), indicating that the well-defined pore structure of the original SiO₂ opals was still well maintained after several cycles of Al₂O₃ grafting and NiMo impregnation. Figure 2 shows that catalysts were opal-like materials with face-centered cubic packing of monodisperse nonporous spheres.

Clearly, the well-defined pore structure was not damaged by NiMo impregnation and Al₂O₃-grafting treatment.

The silanol group of the samples pretreated at different conditions was evaluated by FTIR spectroscopy (Fig. 3). According to the report by Brandriss and Margel (1993) the absorbance at approximately 3750 cm⁻¹ and at the range approximately from 3265 to 3645 cm⁻¹ was assigned to

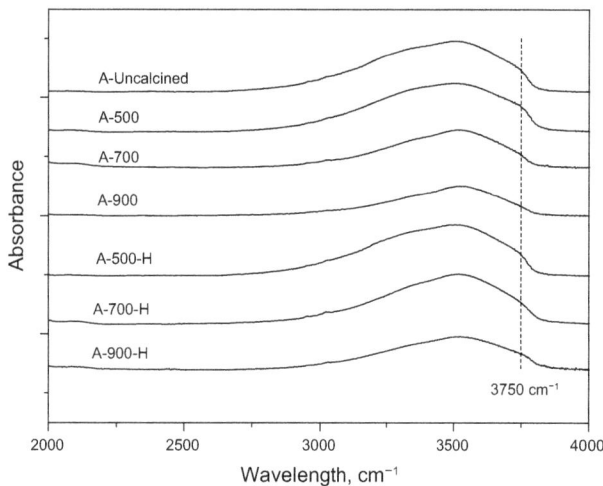

Fig. 3 Infrared spectra for SiO_2 opal, assembled from 100-nm microspheres pretreated at different conditions

free silanol and chemisorption of water through silanol group, respectively. Upon heating, the intensity of the absorbance belonging to free silanol and the chemisorption of water began to decrease, indicating that the surface silanol group started to condense and eliminate water. After hydrothermal treatment, the silanol group can be almost completely recovered if the calcination temperature is below 500 °C; the silanol group could be partly recovered if the calcination temperature above 500 °C.

The ^{27}Al NMR results of this work showed that the pretreatment conditions of the SiO_2 opals and the alumination procedure had a strong effect on the Al coordination (Fig. 4). It was reported that three distinct signals with maximum values centered at 3, 34 and 56 ppm should be assigned to octahedral Al (oct), pentahedral Al (pen) and tetrahedral Al (tet) atoms, respectively (Zukal et al. 2008).

It can be seen that the chemical shift of the Al (tet) of the grafted materials was closer to that of tetrahedral Al in zeolite and amorphous aluminum silicate reported in literature (Klimova et al. 2003; Hensen et al. 2010; Góra-Marek et al. 2005), indicating the formation of Al–O–Si bonds resulted from the grafting of Al atoms on the surface of SiO_2 microspheres. It is noteworthy that the spectra of these Al_2O_3-grafted materials were different from those of γ-Al_2O_3, which consists mostly of Al(tet) and Al(oct) with a Al(tet)/Al(oct) atomic ratio of 3/7, and the Al(pen) accounts for very small amount. According to the report by De Witte (De Witte et al. 1995), the Al(pen) usually exists in the interface between silica and alumina or in the amorphous silica–alumina, indicating that the local arrangement of aluminum atoms in Al_2O_3-grafted materials was different from that in bulk γ-Al_2O_3. In addition, the effect of calcination temperature on Al coordination is also found from the ^{27}Al NMR spectrum of A-700-H-2 and A-900-H-2 in Fig. 4. The Al(tet) was formed through the silanol group on the surface of the SiO_2 microspheres; therefore, the content of the Al(tet) mainly depended on the amount of silanol group which was largely affected by the calcination temperature. Based on the results of infrared spectra (see Fig. 2), the content of silanol group of A-700-H is higher that of A-900-H, so more Al(tet) could exist after two Al_2O_3-grafting cycles.

Further characterization of the supports was undertaken by NH_3-TPD to investigate the acidity of support with different amounts of alumina, and the result is shown in Fig. 5. NH_3 desorption peaks at 120–220, 220–400 and 400–550 °C represent weak, medium and strong acid, respectively, and the acidity data of them are summarized in Table 3. Table 3 shows that the pure silica support presented mostly weak acid sites originating from the

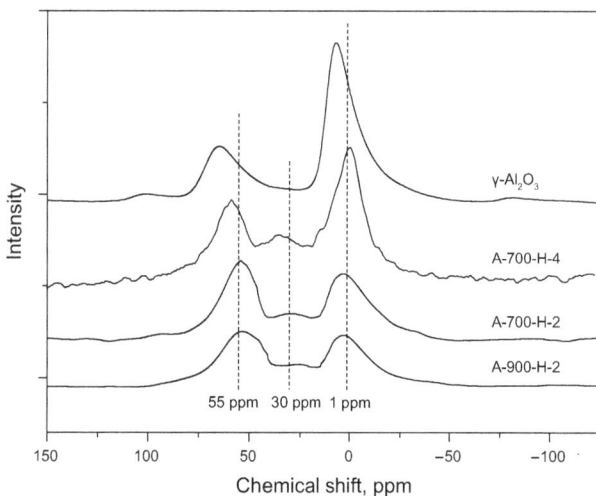

Fig. 4 ^{27}Al NMR spectra of SiO_2 opal grafted with different amounts of alumina and γ-Al_2O_3

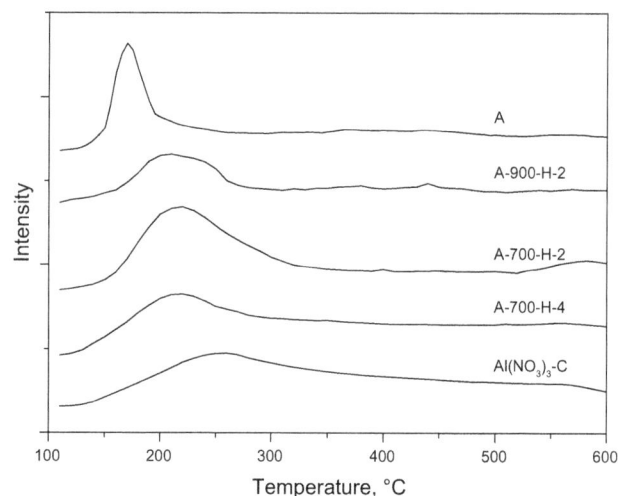

Fig. 5 NH_3-TPD spectra of alumina-grafted materials, parent SiO_2 opal and reference alumina

Table 3 Acidity density of the parent SiO_2 opal, grafted materials and reference alumina

Samples	Acidity density, µmol NH_3/m^2			
	Total acid	Weak acid 110–220 °C	Medium acid 220–400 °C	Strong acid 400–600 °C
A	1.23	0.82	0.23	0.18
A-900-H-2	1.51	0.51	0.68	0.32
A-700-H-2	2.99	0.88	1.36	0.75
A-700-H-4	2.29	0.61	1.10	0.58
$Al(NO_3)_3$-C	2.22	0.23	1.48	0.61

silanol group on the surface, and only a small amount of medium and strong acid sites. However, the amorphous alumina (Sample $Al(NO_3)_3$-C) presented mostly medium and strong acid sites. After two cycles of alumina grafting (Sample A-700-H-2), the alumina-grafted materials showed a significant decrease in the amount of weak acid sites, indicating that the silanol group was consumed in the reaction with aluminum species, and medium and strong acid sites were created by Al_2O_3 grafting. After four cycles of alumina grafting (Sample A-700-H-4), the total acidity decreased, possibly due to the formation of more amorphous alumina like the sample $Al(NO_3)_3$-C. Based on the result of the ^{27}Al NMR (Fig. 4), it can be concluded that the total acidity increased first and then decreased with the grafting cycle, confirming that some of the grafted alumina atoms would be bonded with silica atoms of the silica microspheres after two-cycle grafting, but the acid sites would be covered by the alumina overlayer during the subsequent Al_2O_3 grafting. By comparing the NH_3-TPD spectra (Fig. 5) of the A-700-H-2 and A-900-H-2 and infrared spectra (Fig. 3) of the A-700-H and A-900-H, it can be found that higher content of silanol group is more conducive to forming acid sites during the grafting procedure.

The supports and their corresponding NiMo catalysts were analyzed by XRD, and the results are shown in Fig. 6. No distinct diffraction peaks are observed in the XRD patterns of $Al(NO_3)_3$-C (Fig. 6a), indicating that amorphous alumina was formed during calcination of $Al(NO_3)_3$ at 500 °C. The XRD patterns of SiO_2 opal showed a broad signal in the 2θ range between 15° and 35° (samples A in Fig. 6a), and this was attributed to the amorphous silica (Zepeda et al. 2008; Nava et al. 2007). It is suggested that the alumina unbonded with silicon atoms of SiO_2 microspheres can be regarded as amorphous alumina similar to the $Al(NO_3)_3$-C sample. As shown in Fig. 6b, no reflections belonging to molybdenum and nickel oxides are observed in the XRD patterns of the NiMo/$Al(NO_3)_3$-C catalyst. This result indicated a good dispersion of deposited Ni and Mo oxide species on the support surface. For NiMo/SiO_2 catalyst, besides the peaks from the support, some peaks of molybdenum oxide appeared at 2θ of 12.7°,

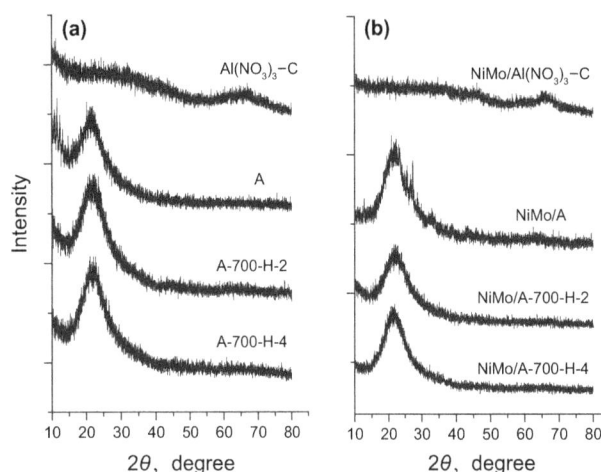

Fig. 6 XRD patterns for supports (**a**) and corresponding NiMo catalysts (**b**)

22.7° and 33.4° (JCPDS (Joint Committee on Powder Diffraction Standards) card 35-609). An additional reflection peak appeared at 2θ of 25.6°, indicating the presence of β-$NiMo_4$ phase (JCPDS card 21-0868). However, in the case of NiMo/Al_2O_3-grafted SiO_2 catalyst, no reflections of MoO_3 and β-$NiMo_4$ are found, indicating that the incorporation of alumina in the supports enhanced the dispersion of active components of the catalysts.

3.2 Catalytic activity in hydrodesulfurization of dibenzothiophene

In the present study, the catalytic activity of NiMo catalyst using alumina-grafted SiO_2 as support was tested using the HDS of dibenzothiophene (DBT) in a fixed-bed tubular reactor as the probe reaction. The HDS catalytic activity of NiMo catalysts supported on parent SiO_2 opals, and amorphous alumina was also tested for comparison purposes. Surface area hourly space velocity (SHSV) was applied to describe the ratio of flow rate of liquid feedstock to the catalyst surface area. According to the literature (Chen et al. 2005; Chen and Ring 2004), HDS reactions of individual compounds follow first-order kinetics, so the material balance of the isothermal and plug-flow reactor is given by Eq. (1).

$$\ln\left(\frac{C_0}{C_t}\right) = k \cdot SHSV^{-1} \tag{1}$$

$$SHSV = \frac{Q}{S} \tag{2}$$

where C_0 and C_t is the sulfur concentration in the feedstock and product stream, respectively, ppm; k is the pseudo-first-order reaction rate constant, m s^{-1}; Q is the feedstock flow rate, m^3 s^{-1}; S is the surface area of the catalyst loaded in the reactor, m^2.

For the HDS catalytic performance testing, $SHSV$ was varied through changing the Q, the liquid effluents from the reactor were collected, and their sulfur content was measured. The relationship between $\ln(C_0/C_t)$ and $SHSV^{-1}$ is shown in Fig. 7. The best fit straight lines of $\ln(C_0/C_t)$ versus $SHSV^{-1}$ were obtained, and the slope of the straight line was the k. Three SiO$_2$ opals, prepared with 100, 250 and 500 nm SiO$_2$ microspheres, were employed as templates to fabricate NiMo catalysts. The k of various NiMo catalysts supported on SiO$_2$ opal grafted with different amounts of Al$_2$O$_3$ is calculated and presented in Fig. 8, and the n in the legend at the top left corner represent the Al$_2$O$_3$-grafting cycles. It can be found that the catalytic activities of all the NiMo catalysts supported on aluminated SiO$_2$ opals were significantly higher than those of NiMo catalysts supported on parent SiO$_2$ opals. In addition, the activity of NiMo catalysts supported on the aluminated SiO$_2$ yielded a volcano curve as a function of the Al$_2$O$_3$-grafting cycles. According to the results of ^{27}Al NMR in Fig. 4, a certain amount of Al was implanted within the outer surface of the silica microspheres during the grafting

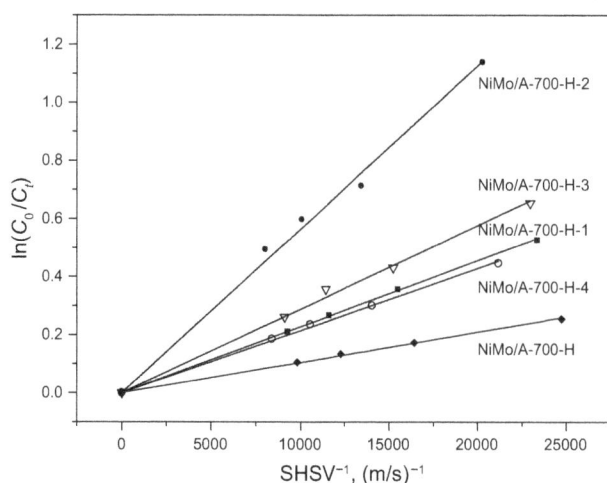

Fig. 8 Reaction rate constant k of various NiMo catalysts supported on Al-grafted materials with different amounts of alumina (300 °C, 6.0 MPa, H$_2$/Oil (v/v) = 800)

process, and amorphous aluminum silicate was formed. Then the content of amorphous aluminum silicate gradually increased with the grafting cycles. Therefore, the activities of the catalysts supported on aluminated SiO$_2$ began to increase till it reached a maximum. The maximum k of the three NiMo catalysts was 5.64 m s^{-1} for NiMo/A-700-H-2, 6.32 m s^{-1} for NiMo/B-700-H-2 and 7.60 m s^{-1} for NiMo/C-700-H-2, respectively. Excessive Al grafting resulted in the formation of amorphous alumina that caused a significant decrease in k. As a result, the k of the three NiMo catalysts supported on the SiO$_2$ with four Al$_2$O$_3$-grafting cycles was almost equal to that of NiMo catalyst

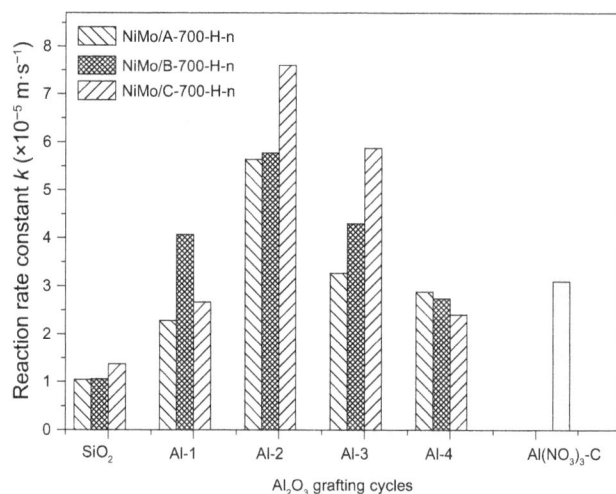

Fig. 7 Relationship between $SHSV^{-1}$ and $\ln(C_0/C_t)$ (300 °C, 6.0 MPa, H$_2$/Oil (v/v) = 800)

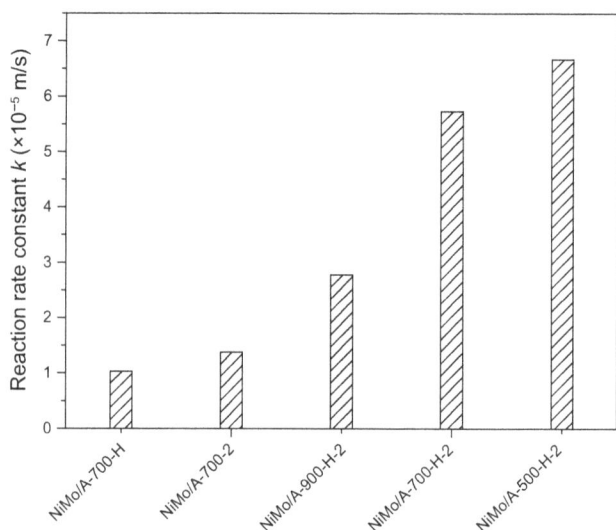

Fig. 9 Effect of pretreatment conditions on the reaction rate constant k of NiMo catalysts (300 °C, 6.0 MPa, H$_2$/Oil (v/v) = 800)

supported on Al(NO$_3$)$_3$-C. Several researchers (Landau et al. 2001; Rayo et al. 2009) have reported the catalytic activity of catalysts supported on Al$_2$O$_3$-grafted materials, and they found that when the content of the alumina exceeds a certain level, the activity of the catalysts began to decrease significantly. In their studies, molecular sieves were used, and the pore textural properties of molecular sieves were prone to being damaged when a high content of alumina is introduced. Therefore, the decrease in the activity may be attributed to this damage of the textural properties. In the present research work, the textural properties of the aluminated SiO$_2$ changed only slightly with the grafting cycles (see Tables 1, 2), suggesting that the decrease in the catalyst activity was resulted from the difference in surface property rather than textural structure.

The supports pretreated at different conditions were used to prepare NiMo catalysts, and the effect of pretreatment conditions on the catalytic activity was evaluated. It can be found that the NiMo/A-700-H-2 catalyst showed significantly higher HDS activity compared to the NiMo/A-700-2, indicating hydrothermal treatment can increase the catalyst activity through the recovery of silanol groups of the SiO$_2$ spheres. As shown in Fig. 9, the catalyst activity increased with a decrease in calcination temperature and declined greatly when the calcination temperature reached 900 °C. This was because the silanol groups lost during calcinations, especially at higher temperature, were just partly recovered in the hydrothermal treatment procedure, and the higher the calcination temperature, the more the silanol groups on the silica spheres were lost.

As shown in Fig. 10, the size and the dispersions of sulfided Mo species (MoS$_2$) of different supports were observed by using HRTEM. The size of active component MoS$_2$ on SiO$_2$ support is large and its dispersion is bad in Fig. 10a, in comparison with that in Fig. 10b, due to the weak interaction between metal oxide and SiO$_2$ support (Scheffer et al. 1988), resulting in the low reaction rate constant k (see Fig. 8). With the alumination of SiO$_2$ supports, the dispersion of sulfided Mo species was improved significantly (Fig. 10b), and this is in accordance with the XRD results (see Fig. 6). It was suggested that the support, which is able to disperse the active components well, has relatively high catalytic activity.

4 Conclusions

NiMo catalysts supported on SiO$_2$ opals aluminated by grafting with Al$_2$O$_3$ were prepared and characterized, and their HDS catalytic activity was tested. It was shown with the increase in grafted alumina content on the surface of SiO$_2$, the surface properties changed from SiO$_2$ to amorphous aluminosilicate and then to amorphous alumina. In line with this, the HDS rate constant k of NiMo catalysts supported on the aluminated SiO$_2$ followed a volcano curve when plotted as a function of alumina content. The alumination process of SiO$_2$ opals by chemical grafting did not affect their well-defined pore structure, and alumination of the SiO$_2$ surface led to an increase in the metal–support interaction and improved the dispersibility of the active species on the support surface.

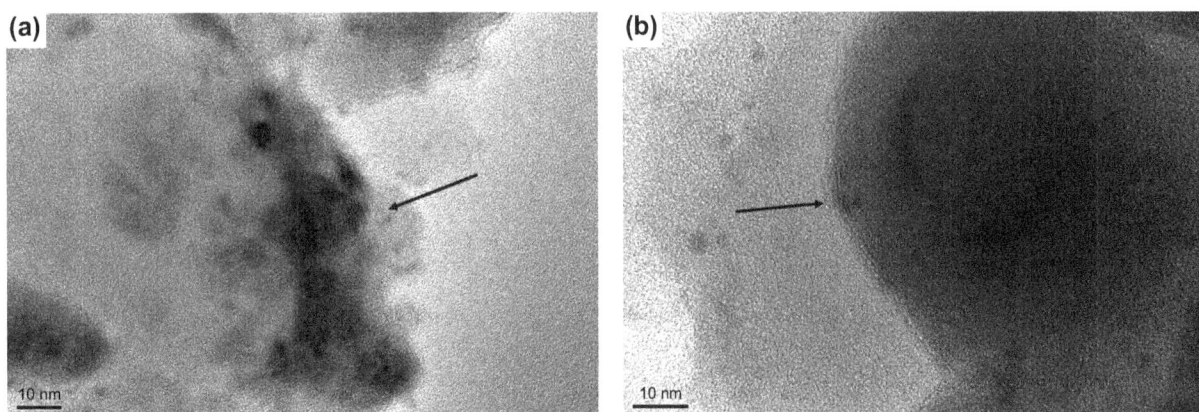

Fig. 10 HRTEM images of NiMo catalysts supported on **a** A-700-H, **b** A-700-H-2

Acknowledgements Financial support by the National Natural Science Foundation of China (Grant No. 91534120) and the Shanghai Petrochemical Company of Sinopec (under the contract number 30450127-13-ZC0607-0001) is greatly acknowledged.

References

Baca M, Rochefoucauld E, Ambroise E, et al. Characterization of mesoporous alumina prepared by surface alumination of SBA-15. Micropor Mesopor Mater. 2008;110:232–41. doi:10.1016/j.micromeso.2007.06.010.

Bagshaw SA, Pinnavaia TJ. Mesoporous alumina molecular sieves. Ang Chem Inter Ed Eng. 1960;35:1102–5. doi:10.1002/anie.199611021.

Brandriss S, Margel S. Synthesis and characterization of self-assembled hydrophobic monolayer coatings on silica colloids. Langmuir. 1993;9(5):1232–40. doi:10.1021/la00029a014.

Breysse M, Afanasiev P, Geantet C, et al. Overview of support effects in hydrotreating catalysts. Catal Today. 2003;86:5–16. doi:10.1016/S0920-5861(03)00400-0.

Breysse M, Geantet C, Afanasiev P, et al. Recent studies on the preparation, activation and design of active phases and supports of hydrotreating catalysts. Catal Today. 2008;130:3–13. doi:10.1016/j.cattod.2007.08.018.

Cabrera S, Haskouri JE, Alamo J, et al. Surfactant-assisted synthesis of mesoporous alumina showing continuously adjustable pore sizes. Adv Mater. 1999;11(5):379–81. doi:10.1002/(SICI)1521-4095(199903)11:5<379:AID-ADMA379>3.0.CO;2-6.

Chen J, Ring Z. HDS reactivities of dibenzothiophenic compounds in a LC-finer LGO and H_2S/NH_3 inhibition effect. Fuel. 2004;83(3):305–13. doi:10.1016/j.fuel.2003.08.009.

Chen J, Yang H, Ring Z. Study of intra-particle diffusion effect on hydrodesulphurization of dibenzothiophenic compounds. Catal Today. 2005;109:93–8. doi:10.1016/j.cattod.2005.08.006.

Chen SL, Dong P, Yang GH, et al. Characteristic aspects of formation of new particles during the growth of monosize silica seeds. J Colloid Interf Sci. 1996;180(1):237–41. doi:10.1006/jcis.1996.0295.

Chen SL. Preparation of monosize silica spheres and their crystalline stack. Colloids. Surf A Physicochem Eng Asp. 1998;142:59–63. doi:10.1016/S0927-7757(98)00276-3.

Cheralathan KK, Hayashi T, Ogura M. Post-synthesis coating of alumina on the mesopore walls of SBA-15 by ammonia/water vapour induced internal hydrolysis and its consequences on pore structure and acidity. Micropor Mesopor Mater. 2008;116:406–15. doi:10.1016/j.micromeso.2008.05.001.

De Witte BM, Grobet PJ, Uytterhoeven JB. Pentacoordinated aluminum in noncalcined amorphous aluminosilicates, prepared in alkaline acid mediums. J Chem Phys. 1995;99:6961–5. doi:10.1021/j100018a031.

Goldbourt A, Landau MV, Vega S. Characterization of aluminum species in alumina multilayer grafted MCM-41 using ^{27}Al FAM(II)-MQMAS NMR. J Phys Chem B. 2002;107(3):724–31. doi:10.1021/0217132.

Góra-Marek K, Derewiński M, Sarv P, et al. IR and NMR studies of mesoporous alumina and related aluminosilicates. Catal Today. 2005;101:131–8. doi:10.1016/j.cattod.2005.01.010.

Hensen EJM, Poduval DG, Magusin PCMM, et al. Formation of acid sites in amorphous silica-alumina. J Catal. 2010;269:201–18. doi:10.1016/j.jcat.2009.11.008.

Jun S, Ryoo R. Aluminum impregnation into mesoporous silica molecular sieves for catalytic application to Friedel-Crafts alkylation. J Catal. 2000;195(2):237–43. doi:10.1006/jcat.2000.2999.

Klimova T, Calderón M, Ramírez J. Ni and Mo interaction with Al-containing MCM-41 support and its effect on the catalytic behavior in DBT hydrodesulfurization. Appl Catal A Gen. 2003;240:29–40. doi:10.1016/S0926-860X(03)00503-9.

Landau MV, Dafa E, Kaliya ML, et al. Mesoporous alumina catalytic material prepared by grafting wide-pore MCM-41 with an alumina multilayer. Micropor Mesopor Mater. 2001;49:65–81. doi:10.1016/S1387-1811(01)00404-8.

Li Z, Gao L, Zheng S. Investigation of the dispersion of MoO_3 onto the support of mesoporous silica MCM-41. Appl Catal A Gen. 2001;236:163–71. doi:10.1016/S0926-860X(02)00302-2.

Liu Z, Chen SL, Dong P, et al. Diffusion coefficient of petroleum residue fractions in a SiO_2 model catalyst. Energ Fuels. 2009;23:2862–6. doi:10.1021/ef801100v.

Márquez-Alvarez C, Žilková N, Pérez-Pariente J, et al. Synthesis, characterization and catalytic applications of organized mesoporous aluminas. Catal Rev. 2008;50:222–86. doi:10.1080/01614940701804042.

Mokaya R, Jones W. Grafting of Al onto purely siliceous mesoporous molecular sieves. Phys Chem Chem Phys. 1999;1:207–13. doi:10.1039/A807919F.

Mokaya R. Post-synthesis grafting of Al onto MCM-41. Chem Commun. 1997;22:2185–6. doi:10.1039/A705340A.

Morris SM, Fulvio PF, Jaroniec M. Ordered mesoporous alumina-supported metal oxides. J Am Chem Soc. 2008;130:15210–6. doi:10.1021/ja806429q.

Nava R, Ortega RA, Alonso G, et al. CoMo/Ti-SBA-15 catalysts for dibenzothiophene desulfurization. Catal Today. 2007;127:70–84. doi:10.1016/j.cattod.2007.02.034.

Rayo P, Ramirez J, Rana MS, et al. Effect of the incorporation of Al, Ti, and Zr on the cracking and hydrodesulfurization activity of NiMo/SBA-15 catalysts. Ind Eng Chem Res. 2009;48:1242–8. doi:10.1021/ie800862a.

Ryoo R, Kim MJ. Generalised route to the preparation of mesoporous metallosilicates via post-synthetic metal implantation. Chem Commun. 1997;22:2225–6. doi:10.1039/A704745B.

Scheffer B, Arnoldy P, Moulijn JA. Sulfidability and hydrodesulfurization activity of Mo catalysts supported on alumina, silica, and carbon. J Catal. 1988;112:516–27. doi:10.1016/0021-9517(88)90167-4.

Venezia AM, Murania R, La Parola V, et al. Post-synthesis alumination of MCM-41: effect of the acidity on the HDS activity of supported Pd catalysts. Appl Catal A Gen. 2010;383:211–6. doi:10.1016/j.apcata.2010.06.001.

Yada M, Kitamura H, Machida M, et al. Biomimetic surface patterns of layered aluminum oxide mesophases templated by mixed surfactant assemblies. Langmuir. 1997;13:5252–7. doi:10.1021/la9704462.

Zepeda TA, Pawelec B, Fierro JLG, et al. Effect of Al and Ti content in HMS material on the catalytic activity of NiMo and CoMo hydrotreating catalysts in the HDS of DBT. Micropor Mesopor Mater. 2008;111:157–70. doi:10.1016/j.micromeso.2007.07.025.

Zhao XS, Lu GQ, Whittaker AK, et al. Comprehensive study of surface chemistry of MCM-41 using ^{29}Si CP/MAS NMR, FTIR, Pyridine-TPD, and TGA. J Phys Chem B. 1997;101:6525–31. doi:10.1021/jp971366+.

Zhou Z, Chen S-L, Hua D, et al. Preparation and evaluation of a well-ordered mesoporous nickel-molybdenum/silica opal hydrodesulfurization model catalyst. Transit Met Chem. 2012a;37:25–30. doi:10.1007/s11243-011-9552-5.

Zhou Z, Chen SL, Hua D, et al. Structure and activity of NiMo/alumina hydrodesulfurization model catalyst with ordered opal-like pores. Catal Comm. 2012b;19:5–9. doi:10.1016/j.catcom.2011.12.009.

Zukal A, Šiklová H, Čejka J. Grafting of alumina on SBA-15: Effect of surface roughness. Langmuir. 2008;24:9837–42. doi:10.1021/la801547u.

9

Use of community mobile phone big location data to recognize unusual patterns close to a pipeline which may indicate unauthorized activities and possible risk of damage

Shao-Hua Dong[1] · He-Wei Zhang[1] · Lai-Bin Zhang[1] · Li-Jian Zhou[2] · Lei Guo[2]

Abstract Damage caused by people and organizations unconnected with the pipeline management is a major risk faced by pipelines, and its consequences can have a huge impact. However, the present measures to monitor this have major problems such as time delays, overlooking threats, and false alarms. To overcome the disadvantages of these methods, analysis of big location data from mobile phone systems was applied to prevent third-party damage to pipelines, and a third-party damage prevention system was developed for pipelines including encryption mobile phone data, data preprocessing, and extraction of characteristic patterns. By applying this to natural gas pipelines, a large amount of location data was collected for data feature recognition and model analysis. Third-party illegal construction and occupation activities were discovered in a timely manner. This is important for preventing third-party damage to pipelines.

Keywords Pipeline · Big location data · Third-party damage · Model · Prevention

✉ Shao-Hua Dong
    shdong@cup.edu.cn

[1] The Pipeline Technology Research Center, China University of Petroleum (Beijing), Beijing 102249, China

[2] PetroChina R&D Center, Langfang 065000, Hebei, China

Edited by Yan-Hua Sun

1 Introduction

The risk caused by third-party damage is an important issue during the entire life of pipelines. During 2001–2015, 30%–40% of pipeline accidents in China were caused by third-party damage. According to European accident statistics, 52% of pipeline accidents in European were due to third-party external damage during 1984–1992 (Dong 2015); 40.4% in the USA and Europe according to the PHMAS latest statistics. Accidents caused by third-party construction accounted for ~20% in 1993–2010. More than 702 leakage accidents occurred during 2010–2016, and 177 of those accidents were caused by third-party damage (external force or excavation by third party), accounting for 25.21%.

Typical third-party accidents in China had a great impact and caused huge economic losses. Several accidents have been reported: On October 6, 2004, because of mechanical failure, pipeline leakage occurred during third-party construction on the Shaanxi–Beijing pipelines in Shenmu Town, Yulin City, Shaanxi Province. On December 30, 2009, the Lan-Zheng-Chang oil products pipeline leaked because of third-party construction, leading to diesel fuel being spilt into the Weihe River. On May 2, 2010, third-party construction caused pipeline rupture on No. 223 pile of the East-Huang oil pipeline in Jiulong Town, Jiaozhou City, leading to leakage of 240 tonnes of crude oil. On July 28, 2010, the propylene pipeline in the Qixia District in Nanjing City exploded because of third-party construction failure. More than 13 people were killed, 28 people were seriously injured, and more than 100 people were slightly injured. On June 30, 2014, because of third-party unauthorized excavation, a leakage accident occurred on 14# + 700 m of Xingang–Songgang pipe of Xin-Da pipeline, and the oil spilt into the municipal sewer network. On September 16, 2015, a medium-pressure gas PE pipeline leaked due to the construction in Xujiawan, Gansu Province, near the Lanya-qinn River.

At present, pipeline patrol is the main measure for monitoring third-party activities and preventing damage; however, because these activities are hidden and random, the patrol monitoring is not effective, especially for third-party mining on pipelines. Illegal activities such as oil and gas stealing are often carried out during the rest time of line patrol officers. For fiber optic early warning and third-party intrusion detection technologies with a high false alarm rate, a large number of databases should be built. This is because cable vibration caused by mining action on site is used to determine third-party activities. However, many similar activities take place, and it is difficult to accurately determine damage. At the same time, some places have different cable and pipeline trenches, thus limiting the applicability of the technology.

Big location data (BLD) have been widely utilized. BLD have become an important resource to observe human community activity and analyze geographical conditions. By analyzing the BLD of oil and gas transport vehicles, human social attribute and relationship with the environment can be extended from a simple positioning data, and a type of intelligent and social application is formed (Daggitt et al. 2016; Doornik and Hendry 2015; Duan et al. 2014; Ettinger-Dietzel et al. 2016; Hashem et al. 2016; Narayanan and Cherukuri 2016; Teli et al. 2016; Tsou 2015).

IBM used mobile phone signals and a signal tower to locate the specific personnel, thus timely accessing the information as to whether the specific personnel came to the region, and established models to perform complex analyses. Then, some information related to the specific personnel was obtained, including the mobile phone behavior of people together with their location, to determine future behavior and help to analyze their movement (Hashem et al. 2016).

Inspired by the above analysis, big location data were used to help prevent third-party damage in this study and to solve the problems in the current third-party damage identification such as real-time deficiencies and small monitoring scope. By establishing the location relationship between a specific cell phone signal and signal towers along the pipeline and obtaining the mobile phone GPS location information, the data of mobile phone signals were analyzed, and third-party damage behavior was evaluated. An area of about 10 km on a pipeline suffering from a higher third-party risk was selected for monitoring using the BLD to uninterruptedly determine the existing excavation and construction activities. A big data association

model of mobile phone signal position was developed to provide timely alarms.

2 Extraction of big location data

Big data are a combination of large complex datasets. The scale and complexity of these datasets exceeds the capabilities of current database management software and traditional data processing technology in acquisition, management, retrieval, analysis, mining, and visualization (Liu 2012).

2.1 Features of BLD

An important part of the big data is BLD. The location data are a combination of geographical data and human social information data containing the space position and time identification. Here, the space position can be accurate geographical coordinates and also can be a conventional place or position (Guo et al. 2013, 2014).

The features of BLD are as follows:

(1) BLD are multiple, heterogeneous, and rapidly changing with typical characteristics such as a large volume, rapid update speed, diversity, and low density.

(2) The common characteristic of BLD is space–time identification; this can be described by absolute location, coordinate, relative position, and language. In addition, the space–time identification of the location data should be accurate and reliable. Accuracy, reliability, and credibility are required in processing and analyzing the location data.

(3) This has a feature called "complex but sparse". Because of the constraint in data acquisition technology, BLD may not reflect the overall picture of the object.

Analysis of BLD means extraction of clues from the local research object and establishment of several characteristic patterns based on a single area r_i or moving object o_i. The extraction methods for a feature model can be divided into two categories as follows:

(1) First-order characteristics: this refers to characteristics that can be easily calculated from the location records, map data, or historical track of moving objects in the region, such as the mean and variance.

(2) Second-order characteristics: this refers to characteristics where the hybridity of original observation data can be eliminated to a certain extent. These features are processed by higher-order statistics.

2.2 Extraction features of mobility pattern in a bar area

Mobility pattern (MP) φ_{mp}: take one or two (peer) moving objects o as the observation target, and the aspects over a period of time include the mobility uniqueness feature, randomness and periodic features, metastatic nature, static and dynamic intermittence, and expectations of movement (Pan et al. 2013; Quinlan 1993a, b).

(1) Uniqueness feature, f_{uniq}

The mobility uniqueness feature can be used to distinguish moving objects and defined as the probability of a track $trai_i$ that can be determined according to the number of given regions $\|F\|$, average size of a region $\overline{F_{size}}$, and interval of statistical time $\overline{F_{time}}$:

$$P_F\{|trai_i| \leq 2|F_{size}, F_{time}|, \|F\|\} \qquad (1)$$

When $\overline{F_{size}}$ and $\overline{F_{time}}$ are relatively appropriate, the activities of the bar area are considered. For example, the probability to determine a unique path is very high in an area with a length of 200 m and width of 50 m on both sides of the pipeline ($\overline{F_{size}} = 0.02$ km^2, $\overline{F_{time}} = 0.5$ h), and it is only about 8 regions ($\|F\| = 8$) (De Montjoye et al. 2013) When $\|F\|$ is fixed, similar power-law relationships of probability with $\overline{F_{size}}$ and $\overline{F_{time}}$ are established.

$$\begin{aligned} f_{uniq} &= \alpha - \left(\overline{F_{size}}\right)^\beta \\ f_{uniq} &= \alpha - \left(\overline{F_{time}}\right)^\beta \end{aligned} \qquad (2)$$

β is a power exponent and linear with $\|F\|$:

$$\beta = \lambda_1 - \lambda_2\|F\| \qquad (3)$$

By observing a small number of regions with abnormal activities surrounding the pipeline, third-party damage by the relevant personnel or tracks of third-party construction users can only be determined. This shows that individual mobility has a high degree of regularity and also shows that the mobility behavior significantly differs among different populations.

(2) Periodic features, f_{peri}

For a moving object, o_i, a discrete Fourier transform was conducted for the binarization of its access region's sequence F_j (1 means visiting, 0 means not visiting). By observing the frequency of the largest Fourier transform coefficient, the cycle of position TP_j^i can be obtained (Liu et al. 2010)

It is supposed that a group of regions $A = \{F_1, F_2,..., F_{\|F\|}\}$ with the same access period $TP = \{T_1, T_2,..., T_Q\}$ is divided into Q time slots. Thus, the detailed probability distribution matrix of each individual mobility $P = [P_1, P_2,..., P_j]$ can be obtained.

Among them, $P_j = [P_r(F_1|T = T_j), P_r(F_2),..., P_r(F_{\|F\|})]$ represents the column probability vector. The location record of the T time period in BLD is generated into $[T/TP] = m$ probability distribution matrix $\{P_1, P_2,..., P_m\}$ according to the cycle of TP. Then, the periodic behavior of moving objects can be analyzed by calculating their Kullback–Leibler (KL) divergence (Yuan et al. 2013).

The more precise standard location entropy can be obtained:

$$H(P) = -\sum_{t_j=1}^{Q} \sum_{A} P_r\left(F_i|T = T_j\right) \log_2 P_r\left(F_i|T = T_j\right) \qquad (4)$$

Then, the entropy of relative distribution is:

$$KL(P_1\|P_2) = \sum_{t_j=1}^{Q} \sum_{A} P_{rP_1}\left(F_j\right) \log_2 \frac{P_{rP_1}\left(F_j\right)}{P_{rP_2}\left(F_j\right)} \qquad (5)$$

According to the order of relative entropy, hierarchical cluster, the probability distribution of n continuous or discontinuous location $\{P_1, P_2,..., P_n\}$, several clusters frequently matching with each other and having the same period (possibly maximum) could be obtained. This represented several typical periodic motion patterns of moving objects o_i (Song et al. 2010). During the clustering, the position probability distribution for associating two clusters C_i and C_j can be calculated as follows:

$$P^{New} = \frac{|C_i|}{|C_i| + |C_j|} P_i + \frac{|C_j|}{|C_i| + |C_j|} P_j \qquad (6)$$

3 Privacy protection for location data

Location information is generally formed by the identification and location information. Identification information is used to describe the user-specific attributes and characteristics that can be uniquely identified by the user. Location information represents a current specific location or track within a certain time of the user.

The privacy protection measures are as follows: When users submitted a service request to the server, accurate location information was provided by the mobile client, and the user's real identity was hidden at the same time. This method can provide high-quality location service to the user according to the location information (Wang 2015). The relationship is shown in Fig. 1.

4 Techniques used in the BLD detection of third parties along the pipeline

(1) Acquisition technology of third-party intrusion signal and GPS signal data

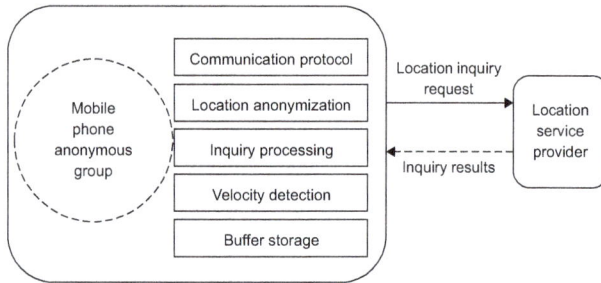

Fig. 1 Location privacy protection

The mobile data and GPS signals of third-party personnel activities along the pipeline were continually collected for 24 h. The signals were used to establish the location relationship between specific cell phone signals and signal towers along the pipeline and to obtain information related to mobile phone GPS location and cell phone towers. The data collected from the mobile equipment (including unique device ID, latitude, longitude, and time stamp) were stored in a database or loaded into the Hadoop platform.

(2) Storage technology for BLD

A computational framework model such as Hadoop, efficient space–time index and distributed analysis technology for flow media, map data, and track data were established. Because BLD are nonrelational, database storage technologies were used, such as Hbase, Big SQL, and Mango.

(3) Preprocessing technology of third-party mobile data

The filtering, integrity, reduction, and discretization methods for third-party communication mobile data were established as the pretreatment. Then, data mining, machine learning, and other methods were used for further processing and mining of location data to analyze the correlation.

By the pretreatment of map and location trace data, the plane map for continuous space was discretized and divided into several regions based on the BLD of map or road network data. The main methods include grid division, division according to road network, division according to position density, and division according to reference sites (Thiessen polygon) (Ester et al. 1996; Li et al. 2013; Pan et al. 2013; Liu et al. 2010; Yuan et al. 2012; Zheng et al. 2013; Zhu et al. 2013), as shown in Figs. 2, 3 and 4. In the analysis of BLD, especially the track data, the dataset should have a high sampling rate to make a simple linear interpolation in the track data. ST-matching, IVMM, Passby, and other algorithms and methods were used to relate the track data and map data (Lou et al. 2009; Liu et al. 2012; Tang et al. 2012; Yuan et al. 2010).

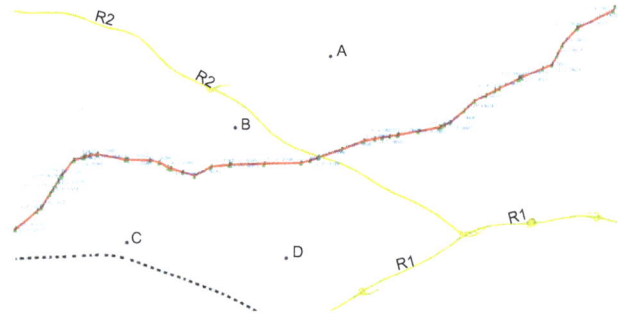

Fig. 2 Location distribution: road, traffic, and village network diagram near the long-distance transportation pipeline

(4) Technology for feature extraction of third-party damage risk.

The feature model between the mobile phone locations and risk of third-party damage was established according to the time feature, which was used to extract the valuable information and following three types of features: (a) Regional static characteristics. Taking a certain area as the observation object, the indexes related to the map were extracted, including the road network characteristics and change rate of concerned points. (b) Mechanical characteristics of regional position movement. The behavior of the moving group targets in the area such as the time evolution of the regional traffic mobility was extracted. (c) Movement patterns characteristics of individuals/groups in different periods. Taking the moving individual/group as the observation object, the mobile behavior characteristics of individual/group within a period of time were extracted. The second-order statistical characteristics and their application to the service calculation of the specific location were studied (Duan et al. 2014). By establishing the model, the signs of risk of third-party damage and destruction were identified.

With the acquisition of BLD, the data quantity gradually increased, and the pattern recognition methods were constantly updated. Logistic regression, support vector machine (SVM), random and uncertain analysis model, wavelet transform, and neural network model were used to analyze the BLD. Combining the behavior of third-party personnel with pipeline risk characteristics, the precision of the forecast warning model was improved.

(5) Visualization methods for third-party damage risk based on BLD.

A statistical chart was used to show the results or data trends in data processing. Based on the characteristics of large scale and diversity, visualization methods were developed to accurately simulate the development state and motional tendency of third-party intrusion along the pipeline.

Fig. 3 Personnel activities

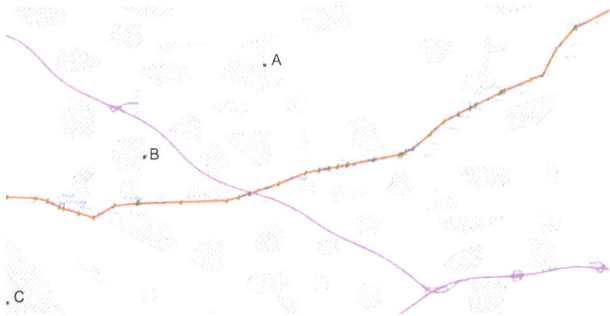

Fig. 4 Discrete reference point map along the pipeline

(6) Forecast warning system for third-party damage to pipeline.

Through the abovementioned research, a third-party forecasting and early warning system for pipeline were established, including data acquisition, data storage, data analysis and modeling, data risk visualization, and trend analysis.

5 Case study

5.1 Application steps

The length of the pipeline in this case is 9.8 km. By accessing the mobile phone signals, important results were obtained in the modeling of third-party damage prevention. Specific steps for mobile phone BLD analysis are as follows:

(1) Data acquisition

This is the first step. Wireless service providers are responsible for collecting location information. A mobile phone provides services using a group of mobile phone signal towers. Its specific location can be obtained by triangulation to the distance from the nearby towers, and the position accuracy is less than 20 m. Most smart phones can even provide more accurate GPS location information (the accuracy is about 20 m). Location data including latitude and longitude require 26 bytes if all this information are stored. If you are dealing with 2 million users and store their position information per minute, the size is about 0.1 TB per day.

In this case, the particular personnel can be three types of people: pipeline managers who have periodic and frequent activities on pipeline base, station, and line; planned construction personnel along the pipeline section, who report to the management. Their activities are clear to managers, illegal excavation, construction and sabotage persons are the focus of the monitoring data analysis.

In practical engineering applications, mobile signals within a distance of ± 50 m from the mobile tower to pipeline have been accessed from mobile companies. Mobile companies encrypt the data, changing mobile signals into specific codes. The movement of these codes is under analysis, not involving personal privacy and security.

(2) Big data storage and processing

Because of the nonrelational BLD, database storage technologies such as Hbase, Big SQL, Mango, and others were used to establish Hadoop analysis (Fig. 5).

(3) Dimension reduction analysis

For the dimension reduction treatment of a BLD network in a space scale, the core is to reduce the nodes (region) or edge (regional association) of the network and obtain global features by analyzing the key components. The main methods are dimensionality according to super betweenness and dimension reduction according to principal components. For the time scale, the dimension is

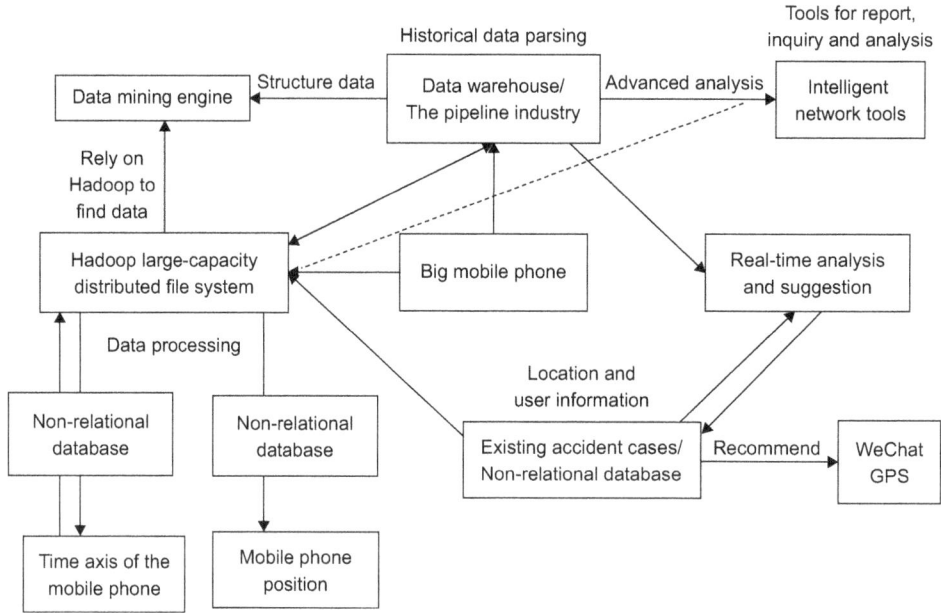

Fig. 5 Hadoop distributed storage hardware integration for big location data

mainly about time discretization, which reduces the similarity between different time periods.

According to the time dimension (determined by the maximum frequency of the occurrence of third-party damage to pipeline), the time periods were shortened to 20:00–22:00, 12:00–14:00, and 2:00–4:00 with a higher risk. For space dimension reduction, the location data in the range of 30 m around the pipeline showed the range of activity.

(4) Feature extraction and modeling of local location data

For the hybrid of BLD, extraction of the static data of mobile phone users should take a certain region as the object of observation and obtain some indicators related to landforms and maps of the area including the road network features, change rate of points, and other static characteristics. Based on the technology for extracting the mobility pattern features in a bar area, the trajectory of the relevant personnel of third-party damage risk or construction can only be determined through the feature probability extraction of individual location and two or more co-locations.

The model for feature probability extraction $H(P)$ is:

$$H_1(P) = -\sum_{t_j=1}^{Q} \sum_{A} P_r\left(F_i | T = T_j\right) \log_2 P_r\left(F_i | T = T_j\right) \quad (7)$$

$$H_2(P) = -\sum_{t_n=1}^{Q} \sum_{A} P_r\left(F_m | T = T_n\right) \log_2 P_r\left(F_m | T = T_n\right) \quad (8)$$

$$H(P) = H_1(P) \cap H_2(P) \quad (9)$$

where Q is equal to 3 (Time periods are 20:00–22:00, 12:00–14:00, and 2:00–4:00); A is the strip area for 9.8 km and 20 m within the scope of the pipeline; $H_1(P)$ is the location probability for individual 1; $H_2(P)$ is the location probability for individual 2; $H(P)$ is the intersection degree of the location probability for the two people in the same area. Generally, warning is needed when it is greater than 90%.

In this case, the model of a third-party damage critical region was developed according to the analysis of accident statistics. The accident statistics show that 85% of the third-party accidents have the same features: more than two people, more than two times, and each static time for 0.5 h. All these elements appeared in the same region.

(5) Data analysis

The mobile phone data were tested for 30 days, and 253,708 bar location data were collected. Then, all the data were screened as follows: in accordance with two or more people (not limited to the same person), at least arriving at the same place twice (with two) above, and each static time more than 0.5 h. After the screening and statistical analysis, the final statistical data were 232, as shown in Table 1.

The statistical analysis in Fig. 6 shows two high risk points of abnormal personnel situation during 22:00–24:00 and 2:00–4:00, and they are the highest risk. The level of personnel risk appearing at the wasteland, hills, and gullies is medium. The level of personnel risk appearing at the fields, railways, water conservancy project, and sites is low.

Table 1 Statistics of mobile phone location data

Time	Location										
	Gully	Field	Wasteland	Hill	Railway	Highway	Site	Water conservancy project	River	Forest	Statistics
6:00–8:00	1	13	0	0	0	1	0	0	0	1	16
8:00–10:00	2	25	0	0	1	2	3	2	3	2	40
10:00–12:00	0	37	1	0	1	2	4	3	2	0	50
12:00–14:00	1	11	2	2	1	1	0	1	1	1	21
14:00–16:00	0	13	1	0	0	3	3	2	2	1	25
16:00–18:00	0	25	1	0	2	5	4	3	2	2	44
18:00–20:00	0	21	0	0	1	2	2	3	3	1	33
22:00–24:00	0	0	0	1	1	0	0	0	0	0	2
24:00–2:00	0	0	0	0	0	0	0	0	0	0	0
2:00–4:00	0	0	1	0	0	0	0	0	0	0	1
4:00–6:00	0	0	0	0	0	0	0	0	0	0	0
Statistics	4	145	6	3	7	16	16	14	13	8	

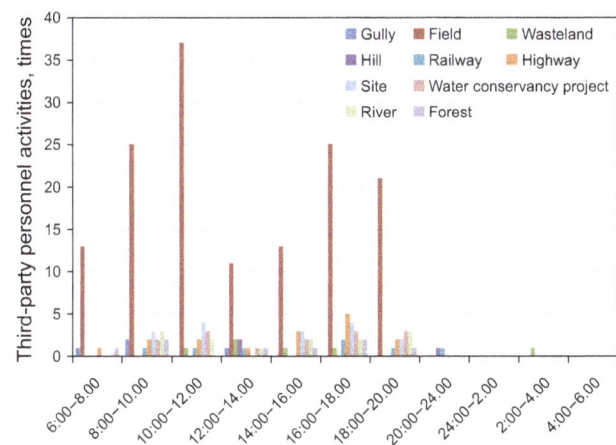

Fig. 6 Diagram of third-party personnel activities and time

After analysis, most people working in the fields around the pipeline, about 145, belong to normal production. The gully data were verified as returning farmland to forest plant operation; however, the abnormal data at 2:00–4:00 were verified as illegal construction for green houses near the pipeline and confirmed as not reporting to the pipeline protection department. An illegal earth borrowing occurred on the hill at 22:00–24:00, and the railway construction near the pipeline belongs to emergency inspection at night.

12:00–14:00 is lunch time, attributing 21 times to the model: 11 of them are involved in field farming; one of them on gully land is involved in forest operation. The wasteland, railway, highway, water conservancy, rivers, and woodland account for five times in total and belong to normal operation; however, the construction lacking normal monitoring on hills and wasteland work along the pipeline account for four times.

The data analysis shows one illegal construction on hills around the pipeline, one construction of a greenhouse at the edges of the wasteland, and other situations belong to normal work (fishing by the river). By analyzing the BLD, the cross-projects along the pipeline would be understood, and abnormal situations would be rapidly detected and monitored.

5.2 Brief summary

(1) Comparison for technologies

By studying these technologies, the following characteristics are given in Table 2.

Through comparison, several limitations were observed in the existing third-party prevention technologies. For example, the monitoring range of optical fiber early warning is small, and the prediction function is not present. The warning occurs after the occurrence of mining behavior. The big data have the features of forecast warning and protection. By collecting and analyzing the real-time data within 50 m of the pipeline, maintenance personnel can reach the scene to prevent third-party construction damage.

(2) Scientific problems to be solved

By analyzing big data, the early warning problem of the risk of third-party damage for bar area pipeline facilities was solved. With the established intersection degree model of location probability, the characteristics of the risk of third-party damage to pipelines can be accurately defined. Furthermore, the technology can also be extended to third-

Table 2 Comparison of prevention methods for preventing third-party damage to pipelines

Indicators	Optical fiber vibration	Remote sensing recognition	Actual patrol	BLD early warning
Monitoring distance	30 km	A wide range	5 km	100 km
Real-time performance	Real time	Not real time	Interval	Real time
False alarm rate	High	Middle	None	Low
Accuracy range	5 m	0.61 m	500 m	50 m
Features	(1) Vibration signal is complex	(1) Data acquisition time of remote sensing image recognition is long	Patrol is intermittent	(1) A full-time monitoring
	(2) False alarm rate is higher	(2)The analysis is difficult		(2) It can provide warning of the third-party construction work, damage, and destruction along the pipeline
	(3) Investment costs a lot	(3) Not applicable for third-party monitoring		(3) The disadvantage is that a lot of data is needed, and the model should be constantly improved
	(4) Characteristics of third-party activities are not obvious			

party monitoring for railways, highways, and electricity networks.

6 Conclusions

(1) For the first time, BLD technology was used to reduce the risk of third-party damage to pipelines. A set of BLD acquisition technologies was established, including encryption technology, data preprocessing technology, third-party damage pattern feature extraction technology, and third-party damage risk visualization methods. A prediction and warning system was developed for third-party damage to pipelines based on BLD.

(2) The case study shows that illegal third-party construction around the pipeline can be rapidly found using this technique. Early detection of risks and automatic classification of the system can help to control the third-party risk to pipelines.

(3) Through time and regional dimension reduction to reduce the nodes in the mobile data network, the periods with high third-party risk can be extracted, thus effectively solving the discretization problem of third-party location data.

(4) The developed method in this study has overcome the deficiency of other methods, such as the uncertainty and false alarm rate of optical fiber vibration and remote sensing image analysis. By analyzing the

data, a three-dimensional network of enterprise defense can be gradually established.

(5) The method can be used in pipeline safety management and increase the strength of research and application.

Acknowledgements This work was supported by Pipeline Management Data Analysis and Typical Model Research [Grant Number 2016B-3105-0501] and CNPC (China National Petroleum Corporation) project, Research on Oil and Gas Pipeline Safety and Reliability Operating [Grant Number 2015-B025-0628].

References

Daggitt ML, Noulas A, Shaw B, et al. Tracking urban activity growth globally with big location data. R Soc Open Sci. 2016;3(4):150688. doi:10.1098/rsos.150688.

De Montjoye YA, Hidalgo CA, Verleysen M, et al. Unique in the crowd: the privacy bounds of human mobility. Sci Rep. 2013;3:1376. doi:10.1038/srep01376.

Dong SH. Pipeline integrity management technology and practice. Beijing: Sinopec Press; 2015. p. 2–15 (**in Chinese**).

Doornik JA, Hendry DF. Statistical model selection with "big data". Cogent Econ Finance. 2015;3(1):1045216. doi:10.1080/23322039.2015.1045216.

Duan R, Hong O, Ma G. Semi-supervised learning in inferring mobile device locations. Qual Reliab Eng Int. 2014;30(6):857–66. doi:10.1002/qre.1701.

Ester M, Kriegel HP, Sander J, et al. A density-based algorithm for discovering clusters in large spatial databases with noise. In: Proceedings of the 2nd international conference on knowledge discovery and data mining (KDD-96). 1996, August. 96(34): p. 226–31.

Ettinger-Dietzel SA, Dodd HR, Westhoff JT, et al. Movement and habitat selection patterns of smallmouth bass Micropterus dolomieuin an Ozark river. J Freshw Ecol. 2016;31(1):61–75. doi:10.1080/02705060.2015.1025867.

Guo C, Fang Y, Liu JN, et al. Study on social awareness computation methods for location-based service. J Comput Res Dev. 2013;50(12):2531–42 **(in Chinese)**.

Guo C, Fang Y, Liu JN, et al. Analysis and processing of large data processing research. J Wuhan Univ (Information Science Edition). 2014;39(4):379–85. doi:10.13203/j.whugis20140210 **(in Chinese)**.

Hashem IAT, Chang V, Anuar NB, et al. The role of big data in smart city. Int J Inf Manage. 2016;36(5):748–58. doi:10.1016/j.ijinfomgt.2016.05.002.

Li Z, Ding B, Han J, et al. Mining periodic behaviors for moving objects. In: The 16th ACM SIGKDD international conference on knowledge discovery and data mining. ACM. 2013, April. p. 1099–108. doi:10.1145/1835804.1835942.

Liu JN. The recent progress on high precision applications of Beidou navigation satellite system. Report of the stanford's 2012 PNT challenges and opportunities symp. (SCPNT 2012), 2012. **(in Chinese)**.

Liu K, Li Y, He F, et al. Effective map-matching on the most simplified road network. In: The 20th international conference on advances in geographic information systems. ACM. 2012, November. p. 609–12. doi:10.1145/2424321.2424429.

Liu S, Liu Y, Ni LM, et al. Towards mobility-based clustering. In: The 16th ACM SIGKDD international conference on knowledge discovery and data mining. ACM. 2010, July. p. 919–28. doi:10.1145/1835804.1835920.

Lou Y, Zhang C, Zheng Y, et al. Map-matching for low-sampling-rate GPS trajectories. In The 17th ACM SIGSPATIAL international conference on advances in geographic information systems. ACM. 2009, November. p. 352–61. doi:10.1145/1653771.1653820.

Narayanan M, Cherukuri AK. A study and analysis of recommendation systems for location-based social network (LBSN) with big data. IIMB Manag Rev. 2016;28(1):25–30. doi:10.1016/j.iimb.2016.01.001.

Pan G, Qi G, Wu Z, et al. Land-use classification using taxi GPS traces. IEEE Trans Intell Transp Syst. 2013;14(1):113–23. doi:10.1109/TITS.2012.2209201.

Quinlan JR. Combining instance-based and model-based learning. In: The tenth international conference on machine learning. 1993a. p. 236–43. doi:10.1016/B978-1-55860-307-3.50037-X.

Quinlan JR. C4. 5: programs for machine learning. San Mateo: Morgan Kaufmann Publishers; 1993b.

Song C, Qu Z, Blumm N, et al. Limits of predictability in human mobility. Science. 2010;327(5968):1018–21. doi:10.1126/science.1177170.

Tang Y, Zhu AD, Xiao X. An efficient algorithm for mapping vehicle trajectories onto road networks. In: The 20th international conference on advances in geographic information systems. ACM. 2012, November. p. 601–4. doi:10.1145/2424321.2424427.

Teli P, Thomas MV, Chandrasekaran K. Big data migration between data centers in online cloud environment. Proced Technol. 2016;24:1558–65. doi:10.1016/j.protcy.2016.05.135.

Tsou MH. Research challenges and opportunities in mapping social media and big data. Cartogr Geogr Inf Sci. 2015;42(sup1):70–4. doi:10.1080/15230406.2015.1059251.

Wang XY. Analysis and processing method of location service data and privacy protection. J Jixi Univ (Integrated Edition). 2015;15(7):51–3 **(in Chinese)**.

Yuan J, Zheng Y, Xie X. Discovering regions of different functions in a city using human mobility and POIs. In: The 18th ACM SIGKDD international conference on knowledge discovery and data mining. ACM. 2012, August. p. 186–94. doi:10.1145/2339530.2339561.

Yuan J, Zheng Y, Xie X, et al. T-drive: enhancing driving directions with taxi drivers' intelligence. IEEE Trans Knowl Data Eng. 2013;25(1):220–32. doi:10.1145/1869790.1869807

Yuan J, Zheng Y, Zhang C, et al. An interactive-voting based map matching algorithm. In: The 2010 eleventh international conference on mobile data management. IEEE Computer Society. 2010, May. p. 43–52. doi:10.1109/MDM.2010.14.

Zheng Y, Liu F, and Hsieh HP. U-Air: when urban air quality inference meets big data. In: The 19th ACM SIGKDD international conference on knowledge discovery and data mining. ACM. 2013, August. 1436–44. doi:10.1145/2487575.2488188.

Zhu B, Huang Q, Guibas L, et al. Urban population migration pattern mining based on taxi trajectories. In: 3rd international workshop on mobile sensing: the future, brought to you by big sensor data, Philadelphia, USA. 2013, April.

10

Selecting China's strategic petroleum reserve sites by multi-objective programming model

Hui Li[1,2] · Ren-Jin Sun[1] · Kang-Yin Dong[1,3] · Xiu-Cheng Dong[1] · Zhong-Bin Zhou[4] ·
Xia Leng[5]

Abstract An important decision for policy makers is selecting strategic petroleum reserve sites. However, policy makers may not choose the most suitable and efficient locations for strategic petroleum reserve (SPR) due to the complexity in the choice of sites. This paper proposes a multi-objective programming model to determine the optimal locations for China's SPR storage sites. This model considers not only the minimum response time but also the minimum transportation cost based on a series of reasonable assumptions and constraint conditions. The factors influencing SPR sites are identified to determine potential demand points and candidate storage sites. Estimation and suggestions are made for the selection of China's future SPR storage sites based on the results of this model. When the number of petroleum storage sites is less than or equals 25 and the maximum capacity of storage sites is restricted to 10 million tonnes, the model's result best fit for the current layout scheme selected thirteen storage sites in four scenarios. Considering the current status of SPR in China, Tianjin, Qingdao, Dalian, Daqing and Zhanjiang, Chengdu, Xi'an, and Yueyang are suggested to be the candidate locations for the third phase of the construction plan. The locations of petroleum storage sites suggested in this work could be used as a reference for decision makers.

Keywords Strategic petroleum reserve · Storage site selection · Multi-objective modeling · China

1 Introduction

Oil, known as the lifeblood of industry, is a strategic raw material as well as a major source of energy supporting economic and social development (Karl 2007). The first of three oil crises began in 1970s, causing higher prices and a stagnant global economy (Helm 2002). Accordingly, western countries began to store emergency petroleum reserves (Fan and Zhang 2010). With the expansion in scale, a strategic petroleum reserve is considered as an effective tool to improve energy security and alleviate price fluctuations (Hubbard and Weiner 1985). According to the reserve agreement, 28 countries of the International Energy Agency were required to hold 90 days of net oil imports for their respective countries. As a second-largest oil consumer in the world, energy security in China is among the most serious challenges due to its increasing oil dependency (Dong et al. 2017). Additionally, the geopolitical impacts of OPEC on oil supply and prices are a significant factor for China to have its own SPR (Chen et al. 2016). Following other nations' establishment, China has built its own SPR to ensure that its oil supply will not be disrupted. The preliminary work on China's SPR was prepared in 1993, approved by the government in 2003, and commenced in 2004 (Jiao et al. 2014). The timeline of China's

✉ Ren-Jin Sun
sunrenjin@cup.edu.cn

[1] School of Business Administration, China University of Petroleum-Beijing, Beijing 102249, China

[2] Energy Systems Research Center, University of Texas at Arlington, Arlington, TX 76019, USA

[3] Department of Agricultural, Food and Resource Economics, Rutgers, State University of New Jersey, New Brunswick, NJ 08901, USA

[4] School of Management, Yangtze University, Hubei 434023, China

[5] Sinopec Offshore Oilfield Services Company, Shanghai 200000, China

Edited by Xiu-Qin Zhu

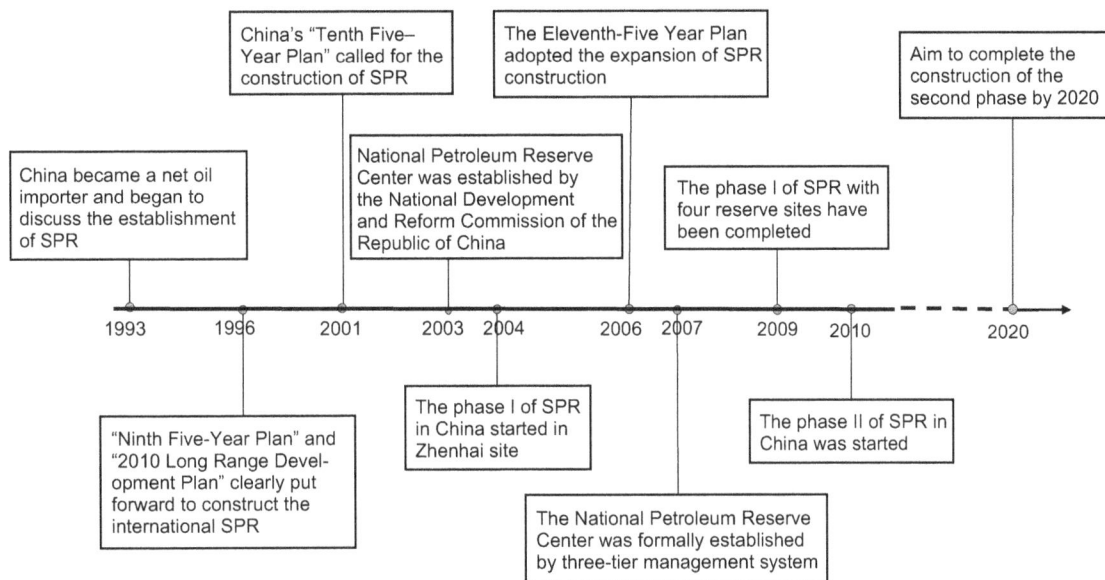

Fig. 1 Timeline of the China's SPR development

SPR program is shown in Fig. 1. According to China's stated policy on SPR, the country is expected to build storage capacity equivalent to 90 days of its net imports in three phases over 15 years. In reality, China has delayed completion of the second phase until 2020 (Park 2015). Generally, the establishment of an SPR in China has helped in dealing with oil crises and ensuring energy and economic security.

An SPR project is a complex system, and many practical questions need to be answered (Davis 1981). The selection of the petroleum reserve sites has a profound effect on SPR program operation. Not all locations are appropriate, so it is important for policy makers to identify the most suitable and efficient locations. Based on China's SPR construction status (Fig. 2), four reserve sites were selected in Phase I, locating in the coastal cities of Zhenhai, Zhoushan, Huangdao, and Dalian. Eight reserve sites in Phase II are planned for inland areas, including Tianjin, Jinzhou, Dushanzi, Xishan, Lanzhou, Jintan, Huizhou, and Zhanjiang. The third selection will likely be located in Wanzhou, Henan, Caofeidian, and Tianjian (Wu and Wang 2012). Three criteria affect the location selection at which China stores its purchased oil. The first is transportation convenience. Most sites are not only key petroleum import harbors but also near substantial demand centers, so the oil can be transported with low transportation costs and a short response time; the second factor is achieving maximum safety. The apparent move from coastal cities to inland is in consideration of increased safety for the SPR in underground tanks scattered around the country (Wu 2014). Third, the location distribution is closely associated with the layout of oil consumption and the routes of imports (Zhu 2007). Today, the preparatory work for the third phase has been launched and the sites selections are still

being determined. How to decide on the storage sites rationally and scientifically is a practical issue we need to consider to achieve the full strategic, economic, and social benefits of the SPR program.

SPR site location is categorized as an emergency facility location problem (Farahani et al. 2012). The p-median method, p-center method, and covering method are widely used in the literature to decide emergency facility locations (Toregas et al. 1971; Roth 1969; Li et al. 2011). However, these discrete methods are proposed based on single objective such as minimum time and distance. With regard to the SPR site selection, there is limited related research. Zhang et al. (2008) presented an uncertain planning model to determine the location of SPR sites with minimal cost of transport based on hybrid intelligent algorithms. Based on fuzzy comprehensive evaluation and maximal covering models, Chen suggested 14 candidate cities for China's second and third SPR construction (Chen 2010). Compared with these methods, the prominent advantage of the multi-objective model is that it takes full consideration of cost and time (Wilson et al. 2013). The comparisons between this study and other facilities location study are given in Table 1 (De Vos and Rientjes 2008). Accordingly, this paper presents a time and cost multi-objective programming model to identify the optimal SPR storage site locations and guide the construction of SPR in China.

2 Factors influencing SPR storage site location

Generally, the location decision for an SPR storage site should consider the economic considerations and other non-economic considerations, such as safety, efficiency, and

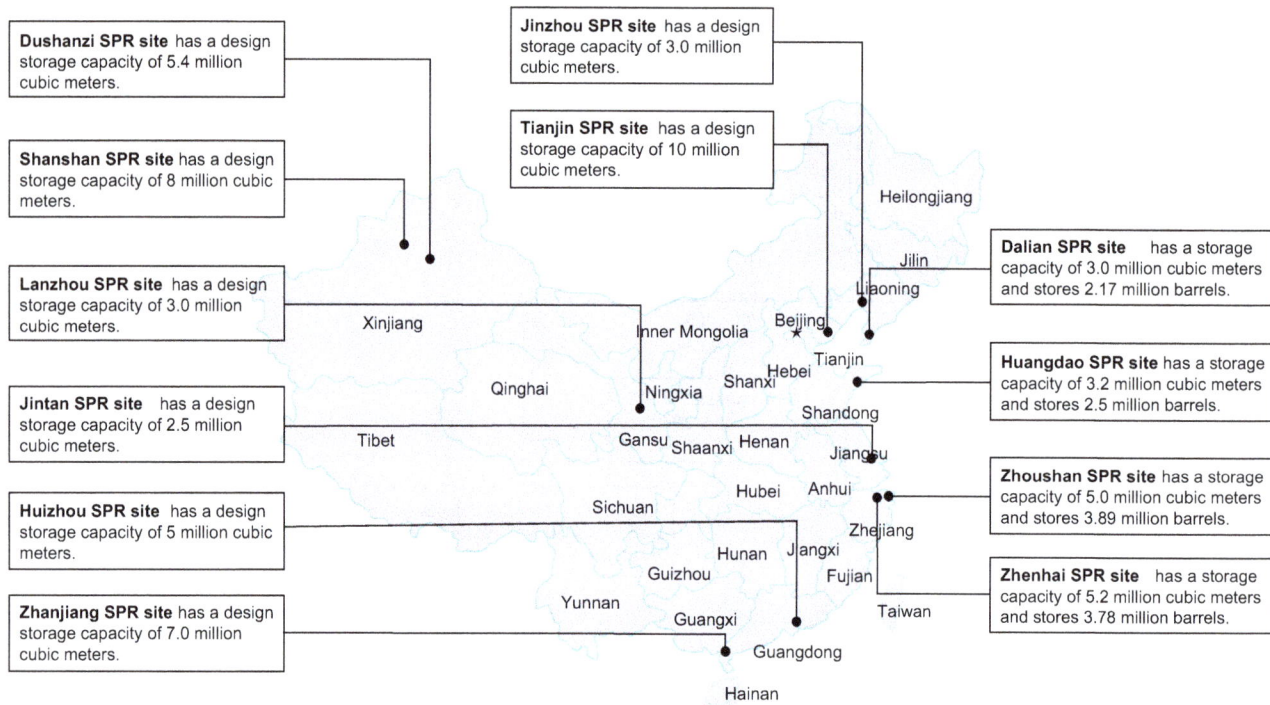

Fig. 2 Location of existing strategic oil reserve bases in China. *Red boxes* indicate SPR sites in Phase I; *green boxes* indicate SPR sites in Phase II

Table 1 Comparison between this study and other facilities location study

Items	Other facilities location studies	This study
Logistics network layer	Monolayer	Monolayer
Number of demand points	Certainty	Uncertainty
Number of facilities	Certainty	Uncertainty
Distance	Linear distance	Shortest distance
Objective function	Single objective	Multi-objective
Optional point capacity	Unlimited	Limited
Emergency response time	Unlimited	Limited
Purpose	Select location	Select location
		Determine reserve scale
		Distribute schedule

fairness. Among them, cost and time minimization is the two most important factors in the choice of a particular place to locate the SPR. In other words, the SPR site should be optimally located to achieve the minimum response time and construction cost (Dong et al. 2013). Because the SPR is characterized by providing emergency oil, the time to access the emergency oil supply must be as short as possible (Oregon 2003). Furthermore, as a construction project, reducing the construction cost is regarded as a basic goal in SPR site selection. With regard to the factors affecting SPR storage sites, it is necessary to take into consideration all the relevant factors. Niu indicated that five factors affected the location decision, including military benefit, economic

benefit, transportation convenience, sustainability, and compatibility (Niu et al. 2010). According to Li and Tan, SPR storage site location should satisfy two principles: absorbing the imported oil and quickly transporting it to a refining center (Li and Tan 2002). In addition, the accessibility of the oil resource, the convenience of transportation, the superiority of storage, and the effectiveness of releasing the emergency oil should be taken into account in location decisions. Based on the previous work, the major factors influencing location are discussed below:

- *Availability of petroleum resources* in determining the reserve location for the SPR, the availability of petroleum resources is of vital importance, because the availability can ensure emergency supply and reduce the cost of production. The basic requirement for a reserve site is that it can hold enough petroleum to address supply disruptions (Bai et al. 2016a, b).

- *Proximity to refining centers* policy makers must consider nearness to refining centers. Petroleum is difficult to transport over long distances so an SPR should be located in close proximity to refining centers. Locating reserve sites near refining centers can reduce transportation cost and reduce response time (Majid et al. 2016).
- *Convenience of transportation* convenience of transportation also influences the SPR storage sites. The four modes of petroleum transportation (pipeline, water, rail, and road) play a significant role. Thus, the junction points of these transport types become priority areas for SPR location (Niu et al. 2013).
- *Petroleum storage consideration* petroleum storage consideration refers to storing the petroleum safely in natural and climatic conditions. These factors can influence the location of an SPR storage site. Stable geological conditions with underground petroleum storage caverns can provide an added advantage over conventional storage tanks in ensuring the storage security of the SPR (Tillerson 1979).
- *Distance between reserve site and demand center* the distance between the reserve site and demand center is associated with the effectiveness of SPR release. A long distance not only causes response time delay but also increases transportation cost. Accordingly, the reserve site should be located as close as possible to the demand center (Williams 2008).
- *Strategic considerations* as petroleum is a strategic commodity, strategic consideration is important in determining the SPR storage site location. The location decision should be consistent with the strategic planning for development with coverage as complete as possible (Bai et al. 2016a, b).

As mentioned above, the influencing factors address three aspects: demand point information, reserve site information, and information related to both (Liu et al. 2014). The information about demand points refers to the geographic location and the amount of emergency petroleum demand. The reserve site information contains the location and storage capacity. The distance between demand point and reserve site and the transportation conditions also have an effect on SPR site selection.

3 Methodology

3.1 Data

3.1.1 Demand points information

Based on the analysis of factors affecting site location determination and following the principle of proximity, we collected the information about SPR emergency demand points in terms of geological location and demand amount. According to the Chinese Statistical Bureau report in 2014, there were over 260 refineries (CNPC 2015). As the largest refinery companies, CNPC and Sinopec accounted for 28% and 38%, respectively, of the refining capacity. Furthermore, CNOOC, Yanchang Shaanxi Petroleum, and local enterprises played a significant role in Chinese refining (The Oxford Institute for Energy Studies 2016). Therefore, we selected 64 refineries located in 52 cities as the SPR demand points, as shown in Table 2. The demand amount is calculated by indicators based on the petroleum self-sufficiency ratio, risk probability, and transportation distance (Martínez-Palou et al. 2011). To ensure the 90 days of SPR supply, the reserve amount is assumed to equal the demand amount. Due to the limitation of data availability, this demand amount for each point is calculated based on data from 2012 when 270 million tonnes of imported crude oil were imported. Furthermore, the distribution of emergency demand of each city should follow the industry production ratio. The results are given in Table 3.

3.1.2 Candidate reserve site location

According to the regulation of National Petroleum Reserve in China, the reserve location should ensure the supply of emergency SPR efficiently and safely. Based on Liu, the candidate reserve site location is determined by three indicators: the flow function, the spatial structure, and the flow track. Table 4 shows the function types of crude oil flow in China based on calculation of the oil self-sufficiency ratio and liquidity ratio. The spatial structure of oil flow is consistent with the petroleum distribution in China, including source system, transit system, and sink system, as shown in Fig. 3. Taking the above factors into consideration, 55 prefecture-level cities are listed as candidate SPR sites in China. The detailed information is shown in Table 5. Specifically, the candidate sites include 17 coastal harbors, 18 input origins, 5 terminal hinges, 6 inland ports, and 9 intersecting regions. The selected site locations cover nearly all the key areas of petroleum resource mobility.

Table 2 Information of strategic oil reserve demand (unit: 10^4 tonnes). *Source* CNPC, Sinopec, CNOOC, Yanchang Shaanxi Yanchang Petroleum, and local enterprise official websites

No.	City	Refinery name	Company	Capacity	No.	City	Refinery name	Company	Capacity
1	AQ	Anqing Petrochemical	Sinopec	500	28	JZ	Jinzhou Petrochemical	CNPC	700
2	BJ	Yanshan Petrochemical	Sinopec	1000	29	LN	Liaoyang Petrochemical	CNPC	1000
3	QZ	Fujian Petrochemical	Sinopec	1200	30	PJ	Liaohe Petrochemical	CNPC	500
4	LZ	Lanzhou Petrochemical	CNPC	1050			Huajin Petrochemical	CSGC	500
5	QY	Qingyang Petrochemical	CNPC	300	31	HHHT	Hohhot Petrochemical	CNPC	500
6	YM	Yumen Petrochemical	CNPC	300	32	YC	Pagoda Petrochemical	Local	350
7	GZ	Guangzhou Petrochemical	Sinopec	1320	33	GL	Golmud Petrochemical	CNPC	150
8	HZ	Huizhou Petrochemical	CNOOC	1200	34	BZ	Zhonghai Asphalt	Local	500
9	MM	Maoming Petrochemical	Sinopec	2550	35	DZ	Hengyuan Petrochemical	Sinopec	300
10	ZJ	Zhanjiang Dongxing Petrochemical	Sinopec	500	36	DM	Dongming Petrochemical	Local	1200
11	BH	Beihai Petrochemical	Sinopec	500	37	DY	Local Refinery	Local	1000
12	QZ	Guangxi Petrochemical	CNPC	1000	38	JN	Jinan Petrochemical	Sinopec	500
13	CZ	North China Petrochemical	CNPC	1000	39	QD	Qingdao Petrochemical	Sinopec	500
		Cangzhou Petrochemical	Sinopec	350			Qingdao Petrochemical	Sinopec	1000
14	SJZ	Shijiazhuang Petrochemical	Sinopec	500	40	WF	Changyi Petrochemical	ChemChina	500
15	LY	Luoyang Petrochemical	Sinopec	800	41	ZB	Qilu Petrochemical	Sinopec	1050
16	DQ	Daqing Petrochemical	CNPC	1000	42	XA	Xi'an Petrochemical	Sinopec	250
17	HRB	Harbin Petrochemical	CNPC	500	43	YA	Yan'an refinery	Yanchang	800
18	JM	Jinmen Petrochemical	Sinopec	600			Yong-Ping refinery	Yanchang	450
19	WH	Wuhan Petrochemical	Sinopec	850	44	YL	Yulin refinery	Yanchang	1000
20	YY	Changling Petrochemical	Sinopec	800	45	XY	Changqing Petrochemical	CNPC	500
21	SY	Jilin Petrochemical	CNPC	1000	46	SH	Gaoqiao Petrochemical	Sinopec	1250
22	NJ	Jinling Petrochemical	Sinopec	1350			Shanghai Petrochemical	Sinopec	1600
		Yangzi Petrochemical	Sinopec	900	47	CD	Sichuan Petrochemical	CNPC	1000
23	JJ	Jiujiang Petrochemical	Sinopec	500	48	TJ	Dagang Petrochemical	CNPC	500
24	DL	Dalian Petrochemical	CNPC	2050			Tianjin Petrochemical	Sinopec	1500
		West Pacific Petrochemical	CNPC	1000	49	DSZ	Dushanzi Petrochemical	CNPC	1600
25	FS	Fushun Petrochemical	CNPC	1000	50	KRMY	Karamay Petrochemical	CNPC	500
26	HLD	Jinxi Petrochemical	CNPC	1000	51	Kuqa	Talimu Petrochemical	Sinopec	500
27	JZ	Jinzhou Petrochemical	CNPC	700	52	Urumqi	Urumqi Petrochemical	CNPC	600

City abbreviations are given in "Appendix 1"

3.1.3 Distance calculation

Petroleum transport is a major component of an SPR system with a range of transportation options available, including railway, pipeline, highway, and water networks. Considering the wide range of cities involved and the incomplete information on pipeline networks, we calculated the shortest distance between the demand point and candidate site based on the shortest path theory of railway mileage. The result is shown in Table 6.

3.1.4 Selection criterion

In the SPR site selection model, four important parameters should be discussed. The first parameter refers to the time the reserve site needs to respond to an emergency demand.

In terms of economic and safety considerations, the candidate location can cover the demand point once. The storage capacity parameter of candidate reserve sites depends on the maximum and minimum capacity. Considering the limitation of SPR storage capacity, two maximum storage scenarios of 10 and 8 million tonnes are discussed based on Liu, which should outweigh the minimum capability (≥ 1 million tonnes). Then, the emergency response time should be evaluated on the basis of the average railway speed (90–100 km/h), so the response range is assumed to 900 km. The last parameter is the reserve numbers. Too few or too many reserve sites will affect the scheme. This paper will discuss 20 or 25 reserve sites considering the current status of and prospects for SPR construction.

Table 3 Strategic oil reserve demand for each demand point (unit: 10^4 tonnes)

No.	Demand point	Amount	No.	Demand point	Amount	No.	Demand point	Amount
1	AQ	66	19	WH	113	37	JN	133
2	BJ	133	20	YY	106	38	QD	199
3	QZ	159	21	SY	133	39	WF	66
4	LZ	139	22	NJ	299	40	ZB	139
5	QY	40	23	JJ	66	41	XA	33
6	YM	40	24	DL	405	42	YN	166
7	GZ	175	25	FS	133	43	YL	133
8	HZ	159	26	HLD	133	44	XY	66
9	MM	338	27	JZ	93	45	SH	378
10	ZJ	66	28	PJ	133	46	NC	133
11	BH	66	29	LY	66	47	TJ	265
12	QZ	133	30	HHHT	66	48	DSZ	212
13	CZ	179	31	YC	66	49	KLMY	66
14	SJZ	66	32	GM	20	50	KC	66
15	LY	106	33	BZ	20	51	WLMQ	80
16	DQ	86	34	DZ	66	52	ZH	305
17	HRB	66	35	DM	40			
18	JM	80	36	DY	159			

Table 4 Function types of crude oil flow in China. *Sources* China Energy Statistical Yearbook 2012

Flow function	Provinces
Source region	Tianjin, Heilongjiang, Shanxi, Qinghai, Xinjiang, Inner Mongolia
Sink region	Beijing, Zhejiang, Anhui, Fujian, Jiangxi, Jilin, Guangdong, Liaoning, Hubei, Jiangsu, Gansu, Sichuan, Hainan, Shanghai, Guangxi, Hunan, Ningxia, Henan, Shandong
Transit region	Hebei
Self-sufficient region	Yunnan

3.2 SPR site location model

3.2.1 Problem description

First, we define two sets, G and F, which are the function of potential demand points and reserve sites, respectively. G and F assume that the geographic location of the demand points and reserve sites, the distance between them, and the transportation costs are known. To provide emergency petroleum with minimal time and cost, the problem is to determine numbers of reserve sites from among the optional locations.

3.2.2 Assumptions

We have based our model on several assumptions as follows:

- The model is categorized to the discrete allocation problem, comprising various optional demand points and reserve sites.

- The demand points are evaluated by geographic location and amount of petroleum demand.
- Petroleum demand volume for each point is fixed.
- The storage capacity for each optional reserve site is limited by the maximum and minimum volume. The construction scale of reserve sites is restricted to the storage capacity.
- A demand point is supported by the reserve site under the assumption that the coverage frequency equals 1.
- The amount of petroleum supply from the reserve site should meet the demands under the coverage.
- The supply volume of the selected reserve site should be less than or equal to its maximum storage capacity.
- The transfer velocity is constant; that is, the transfer distance has an equivalence relation with transfer time.
- In reality, the location and construction of reserve sites satisfy the standard of technology that all the reserve sites are consistent with the technical standards.

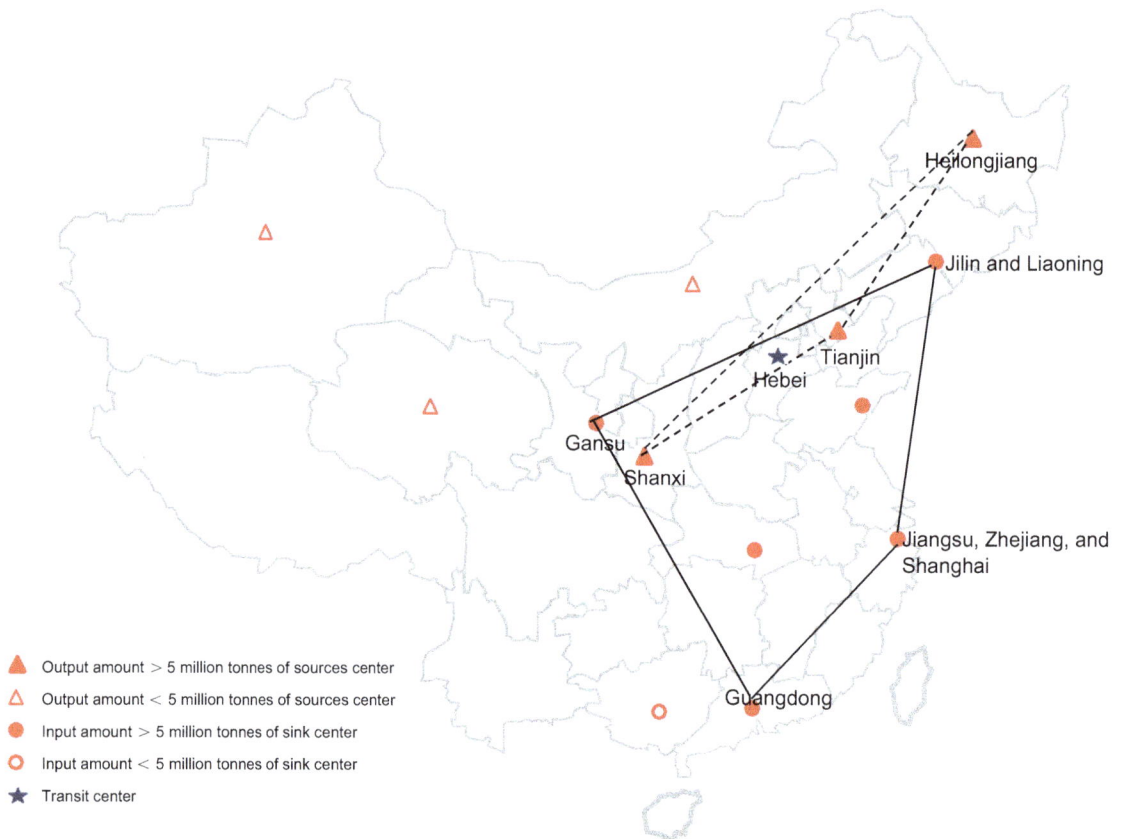

Fig. 3 Spatial structure of China's oil resources flow

Table 5 List of the candidate SPR storage sites

City	Function type	City	Function type	City	Function type
BJ	Terminal	XN	Input node	FZ	Coastal port
LZ	Terminal	YM	Input origin	GZ	Coastal port
CD	Terminal	RQ	Input origin	HZ	Coastal port
AS	Terminal	NY	Input origin	MM	Coastal port
SY	Terminal	PY	Input origin	ZZ	Coastal port
CZ	Transit	DQ	Input origin	QHD	Coastal port
SJZ	Transit	SY	Input origin	LYG	Coastal port
LY	Transit	PJ	Input origin	DL	Coastal port
ZZ	Transit	GLM	Input origin	DD	Coastal port
AQ	Inland port	HTG	Input origin	JZ	Coastal port
JM	Inland port	DY	Input origin	YK	Coastal port
WH	Inland port	XA	Input origin	QD	Coastal port
YY	Inland port	BKQ	Input origin	YT	Coastal port
NJ	Inland port	DSZ	Input origin	SS	Coastal port
JJ	Inland port	KMY	Input origin	TJ	Coastal port
TL	Input node	KC	Input origin	NB	Coastal port
BT	Input node	KL	Input origin	TS	Coastal port
HT	Input node	SS	Input origins		
YC	Input node	KM	Input origins		

3.2.3 Symbol

The SPR storage site selection model can be expressed as follows:

(1) Indices symbols

$G = \{G_i | i = 1, 2, \ldots, m\}$ is a set of SPR demand points;

$F = \{F_j | j = 1, 2, \ldots, n\}$ is a set of optional SPR storage sites;

(2) Parameters symbols

w_i is crude demand of SPR $i \in G$;

C_j^- and C_j^+ are the minimum and maximum storage capacity of reserve sites $j \in F$;

P is the constant parameters of the optional reserve sites;

d_{ij} is the distance between demand point i and candidate reserve site j;

λ is the emergency response time;

C is the total volume of SPR storage capacity in the country;

Table 6 Railway distance between demand cities (unit: miles)

	AQ	BJ	FZ	LZ	YM	GZ	HZ	MM	ZJ	CZ	SJZ	RQ	QHD	TS	LY	ZZ	NY	PY	TL	DQ	JM	WH	YY	SY	NJ	LYG	JJ
AQ	0	1199	784	1726	2534	1141	1053	1471	1547	996	1071	1083	1357	1216	815	746	655	798	1818	2336	605	357	513	2160	285	599	210
BJ	1199	0	1986	1485	2104	2149	2066	2485	2588	218	292	147	291	179	822	714	934	521	752	1323	1179	1156	1339	1148	1038	738	1338
QZ	859	1500	180	2409	3203	752	608	1047	1123	1883	1861	1920	2214	2101	1551	1507	1439	1559	2655	3193	1276	992	985	3018	1077	1361	772
LZ	1711	2087	2300	0	844	2330	2354	2653	2240	1473	1216	1400	1763	1622	1018	1124	1064	1307	2224	2836	1233	1382	1506	2567	1723	1672	1618
QY	1322	1160	1915	503	1326	2007	1965	2253	2371	1092	880	1060	1441	1300	631	737	677	881	1902	2420	843	993	117	2244	1335	1285	1229
YM	2534	2105	3128	848	0	3178	3178	3466	3119	2095	1838	2022	2385	2244	1852	1950	1890	2010	2846	3364	2057	2206	2330	3189	2549	2498	2442
GZ	1140	2150	927	2330	3132	0	150	340	416	2025	1952	2007	2404	2270	1489	1475	1323	1650	2862	3436	1108	1014	827	3208	1381	1676	933
HZ	1053	2065	786	2393	3177	151	0	458	534	1942	1865	1925	2318	2177	1555	1511	1389	1567	2779	3297	1224	977	930	3121	1295	1589	846
MM	1472	2484	1225	2658	3464	342	456	0	95	2361	2284	2343	2736	2596	1821	1807	1656	1982	3197	3716	1440	1346	1162	3540	1713	2008	1265
ZJ	1550	2593	1301	2576	3119	418	532	93	0	2471	2338	2453	2846	2610	1940	1926	1744	2092	3307	3832	1559	1465	1278	3650	1814	2084	1341
BH	272	2498	1926	1487	2106	577	690	253	178	2375	2242	2357	2750	2610	1891	716	936	522	755	3730	1463	1370	1182	3554	1998	747	1303
QZ	1561	2400	1499	2046	2924	616	730	291	213	2283	2150	2265	2658	2517	1752	1738	1586	1904	3119	3637	1371	1277	1090	3462	1811	2106	1385
CZ	997	215	1796	1477	2095	2030	1943	2362	2471	0	230	105	375	231	699	591	811	398	833	1351	1056	1033	1217	1176	841	580	1132
SJZ	1071	292	1789	1217	1835	1956	1868	2288	2323	230	0	200	568	427	519	411	663	324	1029	1598	908	959	1068	1372	917	663	1102
LY	815	822	1477	1014	1841	1492	1534	1823	1941	700	534	687	1075	934	0	138	225	321	1536	2054	473	563	687	1879	748	686	790
DQ	2337	1257	3021	2745	3364	3370	3282	3702	3810	1352	1546	1379	989	1129	2039	1931	2151	1738	556	0	2395	2372	2556	175	2117	1853	2470
HEB	2315	1237	2992	2740	3344	3350	3263	3683	3752	1332	1527	1359	970	1110	2019	1911	2131	1718	490	153	2376	2353	2537	225	2098	1816	2451
JM	605	1178	1166	1229	2055	1110	1212	1442	1560	1056	923	1038	1420	1292	472	511	259	677	1892	2410	0	246	306	2235	743	973	480
WH	357	1155	925	1380	2206	1019	977	1352	1470	1039	955	1015	1408	1267	563	515	398	654	1869	2431	255	0	235	2213	537	846	231
YY	513	1340	980	1495	2330	827	930	1160	1278	1218	1073	1200	1593	1452	687	673	521	839	2054	2572	306	229	0	2397	762	1057	333
SY	2146	1083	2847	2571	3190	3294	3108	3528	3636	1178	1372	1204	815	955	1865	1757	1977	1563	2054	175	2221	2198	2382	0	1943	1661	2296
NJ	286	1026	910	1747	2554	1387	1299	1719	1795	840	915	927	1150	1009	747	678	675	713	1611	2180	749	542	766	1594	0	324	463
JJ	210	1298	697	1611	2441	927	847	1267	1343	1132	1098	1158	1493	1352	789	745	623	797	1954	2472	478	230	334	2296	459	753	0
DL	1912	539	1786	2335	2933	2270	2192	2602	2678	921	1115	948	558	698	1189	1089	1272	903	458	1013	1516	1974	1672	844	917	1309	1346
FS	1800	745	2495	2219	2838	2834	2756	3176	3285	827	1021	853	463	603	1513	1405	1625	1212	99	653	1870	1847	2030	485	1582	1309	1954
HLD	1500	431	2152	1906	2525	2531	2444	2856	2972	514	708	540	150	290	1200	1092	1312	899	337	855	1557	1563	1717	680	1279	996	1632
JZ	1540	476	2227	1969	2570	2576	2488	2908	3016	559	753	585	195	336	1245	1137	1357	944	290	848	1601	1584	1762	624	1323	1041	1677
LY	1748	666	2417	2160	2760	2766	2678	3098	3206	749	946	780	385	525	1435	1327	1547	1134	151	7706	1792	1769	1952	538	1513	1231	1867

(3)   Variables symbols

$$x_{ij} = \begin{cases} 1 & \text{if the demand point } i \text{ is assigned to reserve site;} \\ 0 & \text{otherwise} \end{cases}$$

$$y_{ij} = \begin{cases} 1 & \text{if the reserve site } j \text{ is selected;} \\ 0 & \text{otherwise;} \end{cases}$$

L is the maximum distance between demand point i and its covered reserve site j.

3.2.4 Objectives

An SPR storage site location model is a decision-making tool for identifying reserve locations in a landscape to achieve two objectives: minimum response time and minimum construction cost. Based on the current construction situation and the experience of other countries, the location should be near convenient traffic connections and refineries, which can decrease the response time and save costs. The first objective is the shortest response time, which means the distance between the emergency demand point and its nearest emergency reserve point is minimized. The formula can be described by the p-center model as follows (Current et al. 1990):

$$\min V_1 = L \tag{1}$$

The second objective is minimum response cost, which aims to ensure that the total weighted distance between each demand node and its closest reserve site is minimized. The p-median model is applied to minimize the response cost, which can be expressed as follows (Mladenović et al. 2007):

$$\min V_2 = \sum_{i=1}^{m} \sum_{j=1}^{n} w_i d_{ij} x_{ij}. \tag{2}$$

3.2.5 Constraints

A modeling constraint is a requirement for a candidate solution based on the objective function. In the SPR site selection model, the candidate base should be subject to the following constraints. Constraint (3) indicates that the distance between the SPR site and SPR demand should be less than the maximum L; Constraint (4) ensures the existing relation until the candidate site is selected. Constraint (5) means the frequency of the SPR site will be covered. Constraint (6) limits the total number of selected SPR sites. Constraints (7) and (8) present the volume limitation for SPR demand and supply, respectively.

$$\sum_{j=1}^{n} d_{ij} x_{ij} \leq L \tag{3}$$

$$x_{ij} \leq y_{ij} \tag{4}$$

$$\sum_{j=1}^{n} x_{ij} = g_i \tag{5}$$

$$\sum_{j=1}^{n} y_i = P \tag{6}$$

$$y_j C_j^- \leq \sum_{i=1}^{m} w_i x_{ij} \leq y_j C_j^+ \tag{7}$$

$$\sum_{i=1}^{m} \sum_{j=1}^{n} w_i x_{ij} \leq C \tag{8}$$

$$\max_i \left(\sum_{i=1}^{m} \sum_{j=1}^{n} d_i x_{ij} \right) \leq \lambda \tag{9}$$

$$x_{ij} \in \{0, 1\}, y_{ij} \in \{0, 1\}.$$

3.2.6 Multi-objective model for SPR site selection

Combining the objectives and constraints, a time–cost multi-objective model for SPR site selection is elaborated. Furthermore, some constraints should be clarified in considering the current construction situation. Specifically, Constraint 4 should be modified so the value is less than P, and Constraint 6 indicates the volume of SPR demand, which is equal to C.

3.2.7 Algorithms

As a discrete objective decision-making model, the algorithm of goal programming is presented to solve the multi-objective problem (Aouni and Kettani 2001). Then, the analytic hierarchy process is applied to determine the weights of each objective. Finally, Lingo software is used to obtain the results.

4 Results and discussion

4.1 Results

Based on the SPR site location model, the optimal solution for the first and second objective function is calculated by Lingo, $V_1^* = 812$ and $V_2^* = 999,194.1$. Under the conditions of the emergency delivery distance of 900 km, the total number of reserve sites (20 or 30), and the storage capacity (800 or 1000), Figs. 4 and 5 present the results of optimal SPR site locations selected from candidate cities under two scenarios. Considering the equal significance of fairness and efficiency, the weight value of the two objectives equals to 0.5 applied in the multi-objective programming model, which consults Zhang et al. for a reference (Zhang et al. 2008).

4.2 Discussion

4.2.1 Four-scenario discussion

Table 7 shows the selected results under four scenarios. For scenario 1, the assumption is that the upper limits of the reserve numbers and storage capacity are 20 and 1000 ($P \le 20$, $C \le 1000$), respectively. Nineteen cities are selected as the final reserve sites. Among them, eight coastal ports and seven cities belong to the flow nodes; inland ports and intersection make up only two. To some extent, these locations are consistent with China's SPR reality by excluding Shanshan and Lanzhou. For scenario 2 ($P \le 20$, $C \le 800$), the results contain eight coastal ports and eight flow node cities; the number of both inland port and transit is two. However, compared with the current locations, the model does not select Shanshan, Lanzhou, Qingdao, Jinzhou, or Huizhou. For scenario 3 ($P \le 25$, $C \le 1000$), 21 cities are selected. Among them, five types of cities (coastal harbors, source, nodes, inland port, and transit) are nine, seven, seven, two, and three, respectively. The selection performs without Shanshan and Lanzhou and the scale. In scenario 4 ($P \le 25$, $C \le 800$), 23 cities are selected, including eight coastal harbors, nine source regions, nine source nodes, two inland ports, and

four transits. However, Shanshan, Lanzhou, Qingdao, Jinzhou, and Huizhou do not feature in these results.

The results of scenarios 1 and 3 are superior to those of scenarios 2 and 4. Regarding the objective value, the maximum 10 million tonnes of storage capacity performs better than the maximum 8 million tonnes capacity. Therefore, for layout scheme matching, the scenario of the number of reserve sites being less than or equal to 25 and, meanwhile, the maximum capacity of reserve sites being 10 million tonnes provides the best the result.

Furthermore, 13 cities are selected in four scenarios, including Fuzhou, Dalian, Shanghai, Tianjin, Ningbo, Daqing, Xining, Xi'an, Dushanzi, Yueyang, Nanjing, Cangzhou, and Chengdu. Accordingly, the future construction work should mainly focus on these cities. In addition, concerning storage scales, Shanghai, Dalian, Tianjin, Ningbo, Xi'an, and Nanjing play an important role in reserving responsibility.

4.2.2 Model reliability and limitations

Model reliability can be verified because six cities in scenario 3 are in accordance with reality. Therefore, the results of scenario 3 should be adopted, taking full consideration of Tianjin, Qingdao, Dalian, Daqing, and Zhanjiang as future reserve sites. Meanwhile, Chengdu, Xi'an and Yueyang are expected to take reserve responsibility in China's SPR Phase III.

In fact, the SPR storage site location model is limited by several assumptions and constraints. For example, each parameter may change (storage capacity, distance) due to

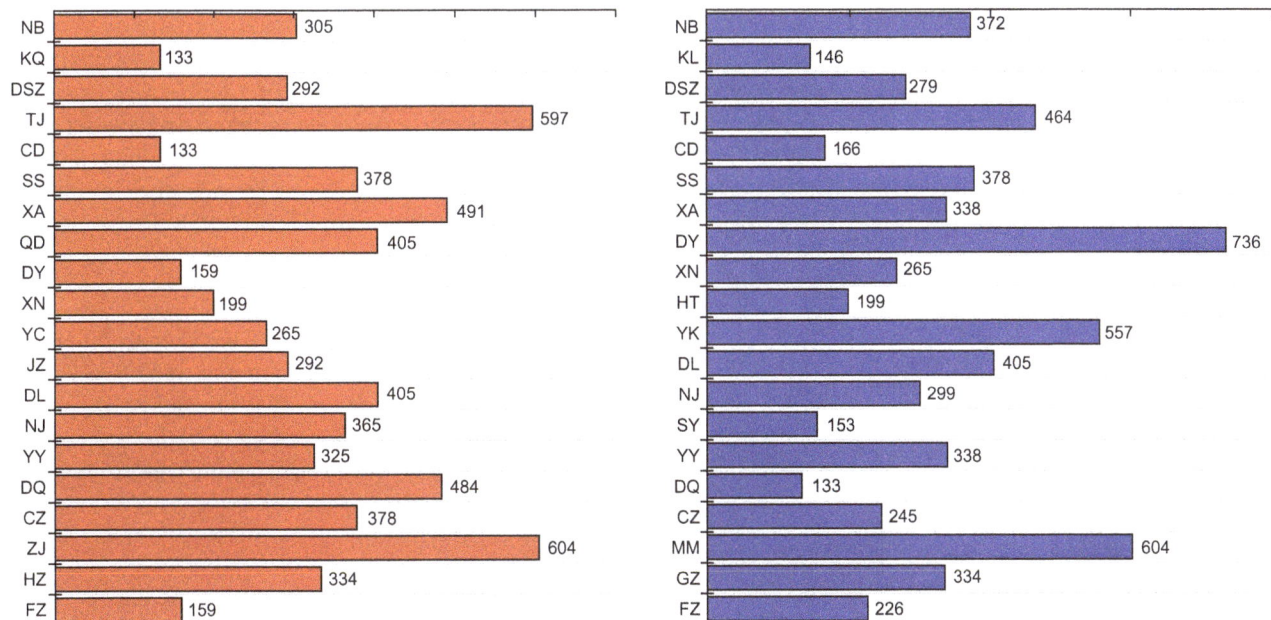

Fig. 4 Selection results in scenarios 1 and 2. *Red bar chart* represents the scenario 1 (reserve sites ≤20, storage capacity ≤1000 ten thousand tonnes). *Blue bar chart* represents the scenario 2 (reserve sites ≤20, storage capacity ≤800 ten thousand tonnes)

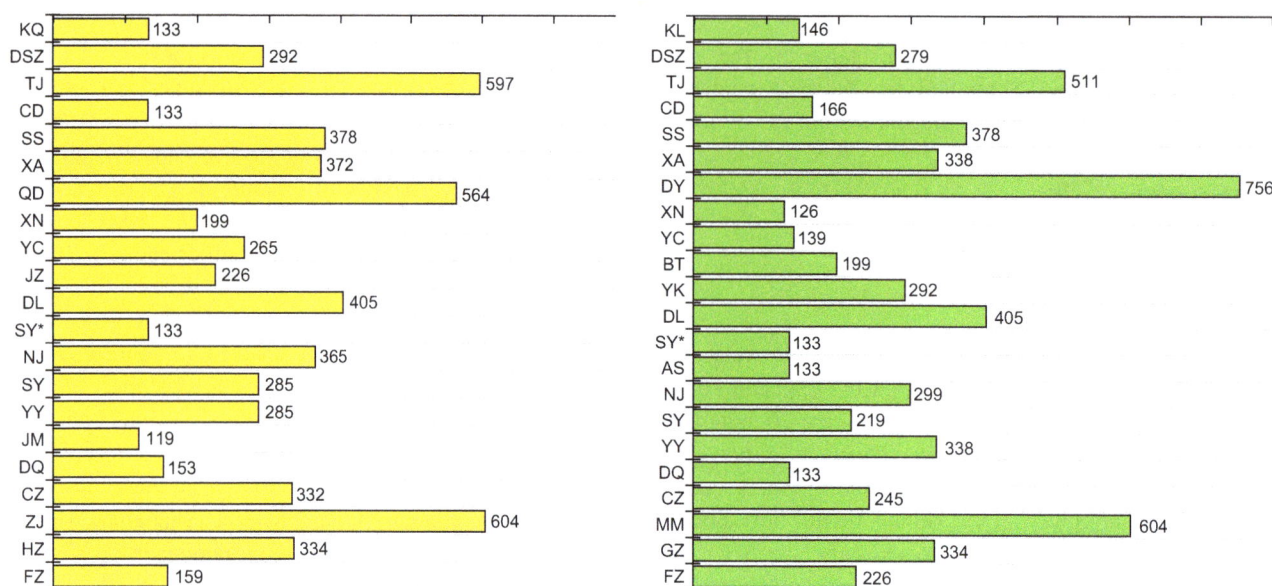

Fig. 5 Selection results in scenarios 3 and 4. *Yellow bar chart* represents the scenario 3 (reserve sites ≤25, storage capacity ≤1000 ten thousand tonnes). *Green bar chart* represents the scenario 4 (reserve sites ≤25, storage capacity ≤800 ten thousand tonnes)

Table 7 Four scenarios selected results

	Total amount ≤ 20		Total amount ≤ 25		Common
	Upper limit = 1000	Upper limit = 800	Upper limit = 1000	Upper limit = 800	
Coastal ports	FZ, HZ, DL, JZ, QD, SH, TJ, NB	FZ, GZ, MM, DL, YK, SH, TJ, NB	FZ, HZ, ZJ, DL, JZ, QD, SH, TJ, NB	FZ, GZ, MM, DL, YK, SH, TJ, NB	FZ, DL, SH, TJ, NB
Source	DQ, YC, XN, DY, XA, DSZ, KQ	DQ, SY, HH, XN, DY, XA, DSZ, KL	DQ, SY, YC, XN, XA, DSZ, KQ	DQ, SY, BT, YC, XN, DY, XA, DSZ, KL	DQ, XN, XA, DSZ
Inland ports	YY, NJ	YY, NJ	JM, YY, NJ	JM, YY, NJ	YY, NJ
Transit	CZ, CD	CZ, CD	CZ, CD, SY	CZ, AS, SY, CD	CZ, CD

economic, policy, and environmental influences, which can affect the accuracy location selection. However, considering the accessibility of domestic data, the model is only suited to a certainty situation. Generally, the emergency facility location model with uncertainty may estimate the locations more accurately. Although the SPR storage site location model takes into account the main influencing factors, the details of each reserve site, such as reserve type, reserve capacity, and future planning, are not discussed. Thus, policy makers should apply this model when more data are available to guide the SPR construction.

5 Conclusions

Determining optimal SPR storage site locations is an important and complicated problem. This study focuses on the problem of locating SPR storage sites in China. The location goals and influencing factors must be considered for resource accessibility, emergency efficiency, and a

plausible spatial pattern. We have made assumptions which might overlook many features of China's SPR construction. However, our application of the reserve site selection model generates many reasonable sites for future consideration. We believe that it can provide decision makers with a useful reference for Chinese SPR Phase III construction. The main results obtained for the SPR storage site location model are as follows:

(1)   The choice of SPR location is based on the minimum response time and minimum construction cost. The selection is based on the principles of justice, transparency, and efficiency. To achieve the goals, information about demand points and reserve sites and other related resources is explored for reserve site location determination.

(2)   Combined with the basic facilities location models, a multiple-objective programming model for SPR storage site location that satisfies the limitations of a set of constraints, such as reserve scale, storage

capacity, and emergency periods, is introduced. Then, we design an algorithm based on Lingo to further improve the model.

(3) According to the principle of proximity and the distribution of petroleum resources, the information on 52 demand points and 55 candidate reserve sites was collected. The optimal reserve site locations for China's SPR vary under different scenarios. Specifically, the performance of 10 million tonnes of storage capacity is better than that of 8 million tonnes with the same reserve numbers; 13 cities are selected including Fuzhou, Dalian, Shanghai, Tianjin, Ningbo, Daqing, Xining, Xi'an, Dushanzi, Yueyang, Nanjing, Cangzhou, and Chengdu under each scenario. The scenario of number of reserve sites being less than or equal to 25 and, meanwhile, the maximum capacity of the reserve site being 10 million tonnes offers the best solution. Thus, policy makers should consider adopting the results of scenario 3 selecting Tianjin, Qingdao, Dalian, Daqing and Zhanjiang, Chengdu, Xi'an, and Yueyang to take the reserve responsibility in China's SPR Phase III.

(4) Although the built model has demonstrated its effectiveness for China's SPR storage site determination, the influence from other uncertain factors should not be ignored. Researchers can further demonstrate the model's feasibility by testing various uncertain factors.

Acknowledgments We gratefully acknowledge that this work was supported by the National Natural Science Foundation of China (Nos. 71273277/71373285/71303258) and the Philosophy and Social Sciences Major Research Project of the Ministry of Education (No. 11JZD048). Helpful comments by anonymous reviewers are most appreciated.

Appendix 1

See Table 8.

Table 8 Cities abbreviations

No.	Abbr.	City	No.	Abbr.	City	No.	Abbr.	City
1	AQ	Anqing	20	DQ	Daqing	39	GM	Gelmud
2	BJ	Beijing	21	JM	Jinmen	40	HTG	Huatugou
3	FZ	Fuzhou	22	WH	Wuhan	41	DY	Dongying
4	LZ	Lanzhou	23	YY	Yueyang	42	QD	Qingdao
5	YM	Yumen	24	SY	Songyuan	43	YT	Yantai
6	GZ	Guangzhou	25	NJ	Nanjing	44	XA	Xi'an
7	HZ	Huizhou	26	LYG	Lianyungang	45	SH	Shanghai
8	MM	Maoming	27	JJ	Jiujiang	46	CD	Chengdu
9	ZJ	Zhanjiang	28	AS	Anshan	47	TJ	Tianjin
10	CZ	Cangzhou	29	SY*	Shenyang	48	BKQ	Baikouquan
11	SJZ	Shijiazhuang	30	PJ	Panjin	49	DSZ	Dushanzi
12	RQ	Renqiu	31	DL	Dalian	50	KMY	Karamay
13	QHD	Qinhuangdao	32	DD	Dongdan	51	KC	Kuqa
14	TS	Tangshan	33	JZ	Jinzhou	52	KL	Korla
15	LY	Luoyang	34	YK	Yingkou	53	SS	Shanshan
16	ZZ	Zhengzhou	35	BT	Baotou	54	KM	Kunming
17	NY	Nanyang	36	HT	Hohht	55	NB	Ningbo
18	PY	Puyang	37	YC	Yinchuan			
19	TL	Tieling	38	XN	Xining			

Asterisk is used to separate the two cities. SY is short for Songyuan, SY* is short for Shenyang

Appendix 2: Lingo programming

```
sets:
a/1..52/:w;!i;
b/1..55/:y;!j;
link1(a,b):d,x;!;
endsets
data:
d=@ole('d.xls');! Shortest path matrix;
w=@ole('w.xls');! Amount of each demand
    point;
enddata
min=L
@for(a(i):@sum(link1(i,j):d(i,j)*x(i,j))<=L);
@for(a(i):@for(b(j):x(i,j)<=y(j)));
@for(a(i):@sum(b(j):x(i,j))=1);
@sum(b(j):y(j))<=25;
@for(b(j):@sum(a(i):w(i)*x(i,j))<=y(j)*1000);
@for(b(j):@sum(a(i):w(i)*x(i,j))>=y(j)*100);
@for(a(i):@sum(link1(i,j):d(i,j)*x(i,j))<=900);
@for(link1(i,j):@bin(x(i,j)));
@for(b(j):@bin(y(j)));
End
```

References

Aouni B, Kettani O. Goal programming model: a glorious history and a promising future. Eur J Oper Res. 2001;133(2):225–31. doi:10.1016/S0377-2217(00)00294-0.

Bai Y, Zhou P, Tian L, et al. Desirable strategic petroleum reserves policies in response to supply uncertainty: a stochastic analysis. Appl Energy. 2016a;162:1523–9. doi:10.1016/j.apenergy.2015.04.025.

Bai Y, Zhou P, Zhou DQ, et al. Desirable policies of a strategic petroleum reserve in coping with disruption risk: A Markov decision process approach. Comput Oper Res. 2016b;66:58–66. doi:10.1016/j.cor.2015.07.017.

Chen HT. Location for strategic petroleum reserve base in China based on fuzzy comprehensive evaluation and maximal covering models. Sci Technol Manag Res. 2010;30(20):222–7 (in Chinese).

Chen H, Liao H, Tang BJ, et al. Impacts of OPEC's political risk on the international crude oil prices: an empirical analysis based on the SVAR models. Energy Econ. 2016;57:42–9. doi:10.1016/j.eneco.2016.04.018.

CNPC. Chinese refining industry under crude oil import right liberalization in 2015. https://eneken.ieej.or.jp/data/6405.pdf. Accessed 3 Nov 2015.

Current J, Min H, Schilling D. Multiobjective analysis of facility location decisions. Eur J Oper Res. 1990;49(3):295–307. doi:10.1016/0377-2217(90)90401-V.

Davis RM. National strategic petroleum reserve. Science. 1981;213(4508):618–22. doi:10.1126/science.213.4508.618.

De Vos NJ, Rientjes THM. Multi-objective training of artificial neural networks for rainfall-runoff modeling. Water Resour Res. 2008. doi:10.1029/2007WR006734.

Dong X, Zhou Z, Li H. Improve the government strategic petroleum reserves. Adv Chem Eng Sci. 2013. doi:10.4236/aces.2013.34A1001.

Dong KY, Sun RJ, Li H, et al. A review of China's energy consumption structure and outlook based on a long-range energy alternatives modeling tool. Pet Sci. 2017;14:214–27. doi:10.1007/s12182-016-0136-z.

Fan Y, Zhang XB. Modelling the strategic petroleum reserves of China and India by a stochastic dynamic game. J Policy Model. 2010;32(4):505–19. doi:10.1016/j.jpolmod.2010.05.008.

Farahani RZ, Asgari N, Heidari N, et al. Covering problems in facility location: a review. Comput Ind Eng. 2012;62(1):368–407. doi:10.1016/j.cie.2011.08.020.

Helm D. Energy policy: security of supply, sustainability and competition. Energy Policy. 2002;30(3):173–84. doi:10.1016/S0301-4215(01)00141-0.

Hubbard RG, Weiner RJ. Managing the strategic petroleum reserve: energy policy in a market setting. Annu Rev Energy. 1985;10(1):515–56. doi:10.1146/annurev.eg.10.110185.002503.

Jiao JL, Han KY, Wu G, et al. The effect of an SPR on the oil price in China: a system dynamics approach. Appl Energy. 2014;133:363–73. doi:10.1016/j.apenergy.2014.07.103.

Karl TL. Oil-led development: social, political, and economic consequences. Encycl Energy. 2007;4:661–72. doi:10.1016/B0-12-176480-X/00550-7.

Li WL, Tan JH. Study of the storage of strategic petroleum reserves in China. China Offshore Oil Gas Eng. 2002;14(3):9–14 (in Chinese).

Li X, Zhao Z, Zhu X, et al. Covering models and optimization techniques for emergency response facility location and planning: a review. Math Methods Oper Res. 2011;74(3):281–310. doi:10.1007/s00186-011-0363-4.

Liu MZ, Qu CZ, Feng YF. Development and application of an optimal layout model for national coal emergency reserve. China Saf Sci J. 2014;8:024 (in Chinese).

Majid NDA, Shariff AM, Loqman SM. Ensuring emergency planning & response meet the minimum process safety management (PSM) standards requirements. J Loss Prev Process Ind. 2016;40:248–58. doi:10.1016/j.jlp.2015.12.018.

Martínez-Palou R, de Lourdes Mosqueira M, Zapata-Rendón B, et al. Transportation of heavy and extra-heavy crude oil by pipeline: a review. J Pet Sci Eng. 2011;75(3):274–82. doi:10.1016/j.petrol.2010.11.020.

Mladenović N, Brimberg J, Hansen P, et al. The p-median problem: a survey of metaheuristic approaches. Eur J Oper Res. 2007;179(3):927–39. doi:10.1016/j.ejor.2005.05.034.

Niu YJ, Guo JK, Shao HY. Study of the location evaluation of national strategic depots for refined oil products. Logist Technol. 2010;29(16):149–50 (in Chinese).

Niu YJ, Mu X, Zhao CJ, et al. Evaluation of site selection for national strategic reserve depots of refined petroleum products. Adv Mater Res. 2013;779:1607–12. doi:10.4028/www.scientific.net/AMR.779-780.1607.

Oregon Department of Transportation Research Unit. Selection criterial for using nighttime construction and maintenance operations. http://www.oregon.gov/ODOT/TD/TP_RES/ResearchReports/SelCritNighttimeCon.pdf. Accessed May 2003.

Park YS. China's energy security strategy: implications for the future Sino-US relations. Int J Soc Sci Stud. 2015;3(2):30–40. doi:10.11114/ijsss.v3i2.670.

Roth R. Computer solutions to minimum-cover problems. Oper Res. 1969;17(3):455–65. doi:10.1287/opre.17.3.455.

The Oxford Institute for Energy Studies. The structure of China's oil industry: past trends and future prospects. 2016. https://www.oxfordenergy.org/wpcms/wp-content/uploads/2016/05/The-structure-of-Chinas-oil-industry-past-trends-and-future-pro spects-WPM-66.pdf. Accessed May 2016.

Tillerson JR. Geomechanics investigations of SPR crude oil storage caverns. Albuquerque: Sandia Labs; 1979.

Toregas C, Swain R, ReVelle C, et al. The location of emergency service facilities. Oper Res. 1971;19(6):1363–73. doi:10.1287/opre.19.6.1363.

Williams JC. Optimal reserve site selection with distance requirements. Comput Oper Res. 2008;35(2):488–98. doi:10.1016/j.cor.2006.03.012.

Wilson DT, Hawe GI, Coates G, et al. A multi-objective combinatorial model of casualty processing in major incident response.

Eur J Oper Res. 2013;230(3):643–55. doi:10.1016/j.ejor.2013.04.040.

Wu K. China's energy security: oil and gas. Energy Policy. 2014;73:4–11. doi:10.1016/j.enpol.2014.05.040.

Wu P, Wang XZ. Description of China's strategic oil reserves. Translated from China Oil Weekly. http://hexun.com/2011-01=17/126848881.html. Accessed on 2 May 2012.

Zhang T, Lv XD, Zhang YF. Location-allocation problems of strategic petroleum storage based on a hybrid intelligent algorithm. Oil Gas Storage Transp. 2008;27(10):12 **(in Chinese)**.

Zhu YC. Research on the position for strategic petroleum reserve. Future Dev. 2007;10:8–11 **(in Chinese)**.

11

Hydrophobic silica nanoparticle-stabilized invert emulsion as drilling fluid for deep drilling

Maliheh Dargahi-Zaboli[1] · Eghbal Sahraei[1] · Behzad Pourabbas[2]

Abstract An oil-based drilling fluid should be stable and tolerant to high temperatures for use in deep drilling. An invert emulsion of water in oil is a good choice as an oil-based drilling fluid which is a mixture of a solid phase and two immiscible liquid phases stabilized by a polymeric surfactant. In deep drilling, due to high temperatures, the polymeric surfactant degrades and a phase separation occurs. Here, octadecyltrimethoxysilane-modified silica nanoparticles were used to form a stable invert emulsion of water in oil for the drilling fluid model which resulted in a milky fluid with the formation of 60 μm water droplets. In addition, rheological study showed that using hydrophobic silica nanoparticles resulted in a stable water in oil invert emulsion with desired properties for a drilling fluid that can be modified by adjusting the nanoparticle nature and content. Aging experiments at 120 °C indicated that they also have good stability at high temperatures for challenging drilling operations.

Keywords Deep drilling · Drilling fluid model · High-temperature aging · Rheology · Silica nanoparticles · Stable invert emulsion

✉ Eghbal Sahraei
  sahraei@sut.ac.ir

[1] Department of Chemical Engineering, Petroleum Research Center, Sahand University of Technology, Tabriz, Iran

[2] Department of Polymer Engineering, Nanostructured Materials Research Center, Sahand University of Technology, Tabriz, Iran

Edited by Yan-Hua Sun

1 Introduction

Drilling technology has been widely used in the applied sciences and engineering, such as manufacturing industries, petroleum industries, pharmaceutical industries, aerospace, research laboratories, and from small-scale laboratories to heavy industry (Hossain and Al-Majed 2015). Modern drilling fluid (also called drilling mud) is an essential part of the rotary drilling system. The successful completion of a hydrocarbon well and its cost depend on the properties of drilling fluids to some extent (Bourgoyne et al. 1986; Hossain and Al-Majed 2015). Therefore, the selection of a suitable drilling fluid and routine control of its properties are the concern of the drilling operations (Hossain and Al-Majed 2015).

During oil and gas well drilling, the drilling fluid is used to (1) clean the borehole by carrying drilling cuts to the surface, (2) create sufficient hydrostatic pressure against the subsurface formation pressure, (3) keep the drilled borehole open for cementing casings in the hole, and (4) cool and lubricate the rotating bit (Bourgoyne et al. 1986). Therefore, to suspend the drilling cuts and keep the suspension when the circulation is stopped, the base fluid of the drilling fluid such as water or oil should be thickened to obtain a high viscosity. Furthermore, a drilling fluid must be nontoxic and stable during drilling as well as being low cost. To have all desired properties, a drilling fluid as a complex fluid includes different types of additives (Agarwal et al. 2013; Bourgoyne et al. 1986; Coussot et al. 2004; Shah et al. 2010). The type and amount of additives depend on the drilling requirements and the type of reservoir to be drilled and bring particular properties and rheological behavior for the drilling fluid. Drilling fluids are often described as thixotropic shear-thinning fluids with a yield stress which could be modified according to the flow and shear conditions (Agarwal et al. 2013; Coussot et al. 2004; Shah et al. 2010).

Oil-based drilling fluids such as an invert emulsion of water in an oil phase with various additives and water-based drilling fluids such as aqueous mixtures of clays and polymers are the two main categories of drilling fluid. Since the water-based fluids are relatively less expensive, they are more common, but their applications are limited to relatively low-temperature and low-pressure drilling operations. In the case of high-temperature and high-pressure drilling operations, diesel, fuel oil, mineral oil, or a linear paraffin are generally used as the base fluid in the oil-based fluids (Coussot et al. 2004; Shah et al. 2010; Agarwal et al. 2013).

In high-temperature drilling conditions (120 and 225 °C in some cases), the oil-based drilling fluids are favored because of better thermal stability and being less affected by contaminants which results in faster drilling. Oil-based drilling fluids are composed of oil as the continuous phase and water as the dispersed phase in conjunction with emulsifiers, viscosity modifiers, weighting materials, and wetting agents. Organophilic clays and barite are common viscosity modifiers and weighting material, respectively. In deep drilling, due to high temperatures, polymeric surfactants degrade and phase separation occur; therefore, an emulsifier tolerant of high temperatures is needed (Coussot et al. 2004; Shah et al. 2010; Agarwal et al. 2013). Also, the current trend in the drilling fluid development is to come up with novel environmentally friendly drilling fluids that will rival oil-based drilling fluids in terms of low toxicity level, performance, efficiency, and cost (Hossain and Al-Majed 2015). Therefore, nanosilica, nanographene, and other nano-based materials have been proposed for use as alternative drilling fluid additives. Nanomaterials in drilling fluid systems are expected to reduce the total solids and/or chemical content of such drilling fluid systems and hence to reduce the overall cost of drilling fluid system development (Hossain and Al-Majed 2015).

Silica nanoparticles (NPs) as an inorganic oxide are environmentally friendly nanoparticles, with a wide variety of industrial applications, in food and pharmaceuticals, catalysis, ceramics, and also as a stabilizer of emulsions. They are a good option for stabilizing the invert emulsion (Vignati and Piazza 2003; Dickinson 2010). Modified silica NPs are used as emulsifiers besides the commonly used surfactant molecules to stabilize an emulsion system by preventing the coalescence of droplets. This depends on the hydrophobicity of nanoparticles. These particle-stabilized emulsions are known as Pickering emulsions (Aveyard et al. 2003; Agarwal et al. 2013; Binks and Rodrigues 2003; Binks and Whitby 2005). In a Pickering emulsion, particles absorb on the liquid–liquid interface and the contact angle that particles make with the oil–water

interface (measured into the aqueous phase) determines the interface energy of attachment, tending to stabilize emulsions (Aveyard et al. 2003; Agarwal et al. 2013; Binks and Rodrigues 2003, 2005; Binks and Whitby 2005).

Since microparticles and nanoparticles of various shapes, sizes, and surface characteristics are commercially available; in early work Agarwal et al. (2011) used commercial hydrophobic nanosilica and organically modified bentonite clay for imparting stability to invert emulsions used as drilling fluids. Also, in the next work of Agarwal et al. (2013), formation of a stable invert emulsion using a combination of commercial hydrophobic nanosilica and organically modified nanoclay was documented while the rheology and morphology were examined (Agarwal et al. 2013).

The main objective of this study is to maintain morphology and rheological properties of oil-based drilling fluid while meeting the requirements of high-temperature operation. For first time, octadecyltrimethoxysilane (OTMOS)-modified core-shell silica NPs, synthesized by the authors, were used as the only emulsifier to stabilize the invert emulsion of water in oil (poly 1-decene) as a drilling fluid model. In addition, the rheological properties and morphology of the prepared stable invert emulsion were investigated, and the best model for prediction of flow behavior, yield stress, and plastic viscosity was determined.

2 Theoretical concepts

Emulsions are a common form of material in which emulsifiers, low molar mass surfactants, and surface-active polymers are used as stabilizers to help disperse one phase in the other. Less well-known stabilizers are solid particles, without forming micelles, and hence, solubilization phenomena are not present (Aveyard et al. 2003; Binks et al. 2005; Binks and Lumsdon 2000; Ding et al. 2005; Melle et al. 2005; Pickering 1907).

A proper formulation of an emulsion should be stable against droplet coalescence and macroscopic phase separation. In 1907, Pickering observed that colloidal particles situated at the oil–water interface can also stabilize emulsions of oil and water which are called either Pickering emulsions or solid-stabilized emulsions (Pickering 1907). Stabilization is achieved when nanometer to micrometer-sized particles diffuse to the interface between the dispersed and continuous phases and remain there in a stable mechanical equilibrium, forming rigid structures and minimizing coalescence. In a Pickering emulsion, the contact angle θ_{ow} in which the particle makes with the interface and shows the relative position of the particles at the oil–water interface (Fig. 1) is a critical parameter and specifies the type of emulsion, either water in oil (W/O) or

(a)

(b)

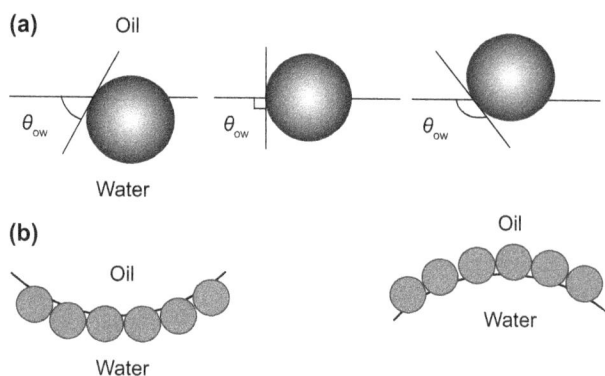

Fig. 1 a Position of a small spherical particle at a planar oil–water interface for a contact angle (measured through the aqueous phase) less than 90° (*left*), equal to 90° (*center*), and greater than 90° (*right*). b Corresponding probable positioning of particles at a curved interface. For $\theta_{ow} < 90°$, solid-stabilized O/W emulsions may form (*left*). For $\theta_{ow} > 90°$, solid-stabilized W/O emulsions may form (*right*)

oil in water (O/W) (Aveyard et al. 2003; Binks and Lumsdon 2000; Binks et al. 2005; Ding et al. 2005; Melle et al. 2005; Pickering 1907).

For hydrophilic particles, e.g., metal oxides, the contact angle θ_{ow} measured through the aqueous phase is normally less than 90° while a larger fraction of the particle surface is located in water than in oil. For hydrophobic particles, e.g., suitably modified silica, the contact angle θ_{ow} is generally greater than 90°, and the particles reside more in oil than in water. Particles which are either too hydrophilic or too hydrophobic are likely to be dispersed in either the aqueous or oil phase, respectively, resulting very unstable emulsions (Aveyard et al. 2003; Binks and Lumsdon 2000; Binks et al. 2005; Ding et al. 2005; Effati and Pourabbas 2012; Melle et al. 2005; Xue et al. 2009; Pickering 1907). Another way to determine the hydrophobic nature of particles is to measure the contact angle of a water drop on a surface covered with the nanoparticles, θ, and is exact enough to be applied in present work (Binks and Lumsdon 2000; Effati and Pourabbas 2012; Xue et al. 2009). During the formation of an emulsion, hydrophobic particles fabricate a shell around a water droplet. This layer in the interface acts as a capsule to separate water droplets from each other and minimizes the total interfacial energy (Binks 2002; Hsu et al. 2005).

From a practical point of view as the drilling fluid, once the flow of the drilling fluid is stopped for any reason, a desirable drilling fluid should possess a yield stress large enough to prevent the settling of suspended solids, such as drilling cuts and barite particles. However, when the flow is started again, the gel structure should break down quickly in order to minimize the pumping costs during an actual

drilling operation (Bourgoyne et al. 1986; Shah et al. 2010).

3 Experimental

3.1 Materials

OTMOS-modified silica NPs with variety of hydrophobicities (synthesized by authors as summarized in the next section) as concentrated colloid in ethanol (EtOH, 99.9 %, Fisher Scientific), methanol (MeOH, 99 %, Fisher Scientific), ammonium hydroxide (NH$_4$OH, 28 %–29 %, Fisher Scientific), tetraethoxysilane (TEOS, $\geq$98 %, Sigma-Aldrich), octadecyltrimethoxysilane (OTMOS, $\geq$97 %, Sigma-Aldrich), and poly(1-decene) (kinematic viscosity ν of 50 cSt at 40 °C, Sigma-Aldrich) were used as received without further purification. All other solvents were laboratory grade obtained mainly from Sigma-Aldrich and used as received without further treatments.

3.2 Synthesis of OTMOS-modified core-shell silica NPs

Silica NPs were prepared by hydrolysis of TEOS in methanol/ethanol mixtures, similar to the Stöber procedure (Stöber et al. 1968), but with an addition of OTMOS as the surface modifier. All reactions were carried out at 65–70 °C for 60 min with a TEOS concentration of 0.25 mol/L and a water/TEOS molar ratio of 38 (Effati and Pourabbas 2012). First, an alcohol solution with a volume of 23 mL was made from a mixture of EtOH and MeOH following by adding water and NH$_4$OH under vigorous stirring at room temperature. Next, the reaction mixture was heated to appropriate temperature (65–70 °C) under a N$_2$ atmosphere. Then, 1.8-mL TEOS was added to the mixture dropwise. Stirring was continued while adding OTMOS dropwise for 4 min (not all at once) until the desired OTMOS/TEOS molar ratio was reached. At the end of the reaction, precipitated silica NPs were separated by centrifugation at 10,000 rpm and 10 °C for 15 min. After discarding the supernatant, the particles were dispersed in water and EtOH by sonication. This precipitation was repeated 2 times; finally, the particles were re-dispersed in EtOH up to 10 mg/mL. A typical reaction yielded 500 mg of functionally modified silica NPs with an average diameter below 100 nm. Contact angle (θ) measurements showed that hydrophilic bare silica NPs ($\theta \sim 0°$) changed to hydrophobic silica NPs ($\theta = 92°$ and $\theta = 115°$) just by adding 0.01 and 0.02 molar ratio of OTMOS to TEOS, respectively.

3.3 Stabilization of an invert emulsion employing OTMOS-modified silica NPs

The following assumptions were considered: (1) The continuous phase in invert emulsions is the oil phase of poly 1-decene, (2) the dispersed phase is 30 volume percent deionized water (DI water), (3) to impart emulsion stability, synthesized OTMOS-modified silica NPs are employed.

Invert emulsions were fabricated in several experiments using specific procedures. First, OTMOS-modified silica NPs with different hydrophobicities as concentrated colloids in ethanol were diluted with amounts of water. Next, colloids were added to a specific amount of poly 1-decene to reach 30 vol% water. Simultaneously, the emulsion was homogenized with a PowerGen homogenizer at 14,000 rpm for 5 min following 20,000 rpm for 2 min. Colloids were added dropwise; otherwise, the dispersion of the water phase in the oil phase is weak and phase separation occurs. A creamy emulsion was formed which showed a stable invert emulsion and was used for morphology and rheology analysis. Before rheology measurements, all samples were kept under pre-shear of 850 s^{-1} for 1 h to create a shear history following by a 10-min rest. After that, in terms of mathematically representing the flow behavior, yield stress, and plastic viscosity, mathematical models including the Bingham plastic, Power law, Casson, and Herschel–Bulkley were considered to find the best fit (Agarwal et al. 2011; Hossain and Al-Majed 2015; Gupta 2000; Shah et al. 2010).

Finally, due to importance of high-temperature aging of the invert emulsion, an aging experiment was done at 120 °C. This was carried out using a 500-mL Fann aging cell pressurized to 4 atmospheres by nitrogen gas to prevent the evaporation of water contained in the invert emulsion. For aging under static conditions, the pressurized cell was placed in a preheated oven for 12 h. Next, the aged drilling fluids were depressurized and cooled to room temperature following by homogenizing at 20,000 rpm for 2 min. To compare results with Agarwal et al. (2011, 2013) study, some experiments were done at 225 °C and at 3.5 MPa (Agarwal et al. 2011, 2013).

3.4 Characterization

Measurements of contact angle, θ, were done using a laboratory-made contact angle instrument equipped with a camera. The films with OTMOS-modified silica NPs were prepared by a doctor blade coating process at 78 °C (around boiling point of EtOH) using 10 mg mL^{-1} OTMOS-modified silica NPs/ethanol solution on a glass slide. The average contact angle of 2 μL of DI water droplets, placed on the coated glass, was calculated by measuring the same sample at ten different positions at room temperature in a clean room (Xue et al. 2009; Effati and Pourabbas 2012).

Droplet sizes of the dispersed water phase in emulsions were determined from images from an optical microscope equipped with a camera. Since the emulsions were quite concentrated to characterize morphology, they were diluted by a factor of 20 with additional poly 1-decene as the oil phase. Then, a droplet of the diluted emulsion was placed on a slide for microscopy work.

The rheological properties of the invert emulsion were determined with the help of an AR 2000 EX Rheometer (TA Instruments, New Castle, DE) fitted with parallel plate fixtures. The diameter of each plate was 4 cm, and the gap was set at 1 mm for all the experiments. All measurements were taken at room temperature.

4 Results and discussion

4.1 Morphology of the emulsions

It has been found that hydrophobic silica NPs can be used to stabilize W/O emulsions (Aveyard et al. 2003; Binks and Whitby 2005; Binks et al. 2008). To judge the stability, the time taken for the emulsions to separate out into two phases was considered.

Figure 2 shows the effect of hydrophobicity of silica NPs on emulsion properties for an overall nanoparticle concentration of 2 wt%. In the case of the emulsions containing extremely hydrophilic or hydrophobic silica NPs (Fig. 2a, d–f), phase separation occurred after only a few minutes. By contrast, the emulsions containing the modified silica NPs with contact angles around 92° (Fig. 2b) and 115° (Fig. 2c) were stable for months and days, respectively.

Number-average droplet sizes of the dispersed phase for the prepared emulsions shown in Table 1 were determined from optical microscopy images in Fig. 3. From Table 1, it can be seen that the finest dispersed phase size is 60 μm

Contact angle:

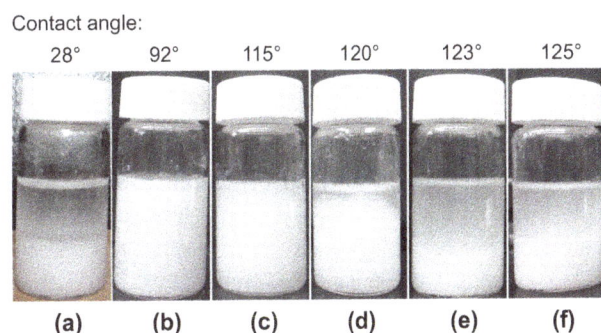

| 28° | 92° | 115° | 120° | 123° | 125° |
| (a) | (b) | (c) | (d) | (e) | (f) |

Fig. 2 Photographs of invert emulsions stabilized by silica NPs at an overall particle concentration of 2 wt%, 5 min after preparation at different OTMOS to TEOS molar ratios. a Bare silica NPs; b 0.01; c 0.02; d 0.03; e 0.04; f 0.05 OTMOS to TEOS molar ratio

Table 1 Average water droplet sizes in various invert emulsions stabilized by silica NPs

Emulsions (30 vol% water)[a]	Diameter, μm	Emulsion status
Stabilized by 1 wt% of 0.01[b] OTMOS-modified silica NPs, $\theta = 92°$	100–200	Stable for days
Stabilized by 2 wt% of 0.01[b] OTMOS-modified silica NPs, $\theta = 92°$	~60	Stable for weeks and months
Stabilized by 1 wt% of 0.02[b] OTMOS-modified silica NPs, $\theta = 115°$	~200	Stable for days
Stabilized by 2 wt% of 0.02[b] OTMOS-modified silica NPs, $\theta = 115°$	~100	Stable for days

[a] The nanoparticle content reported here is on the basis of oil

[b] OTMOS to TEOS molar ratio

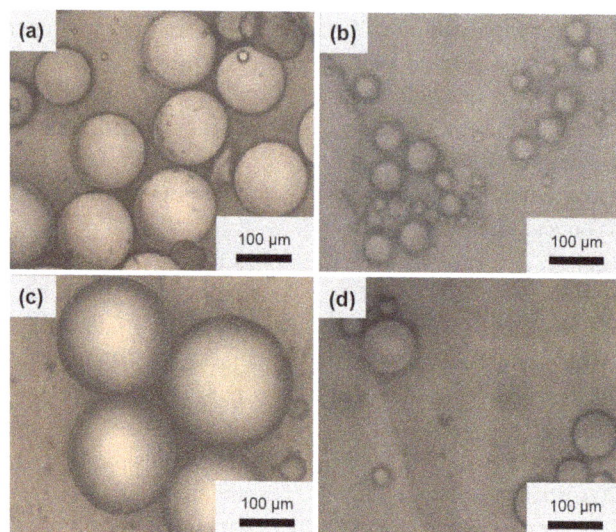

Fig. 3 Morphology of invert emulsions containing 30 vol% water stabilized by 1 wt% (**a**) and 2 wt% (**b**) of modified silica NPs with an OTMOS/TEOS molar ratio of 0.01, respectively; and 1 wt% (**c**) and 2 wt% (**d**) of modified silica NPs with an OTMOS/TEOS molar ratio of 0.02, respectively

which resulted in the emulsion stabilized by OTMOS-modified silica NPs, which was synthesized with an OTMOS to TEOS molar ratio of 0.01 and contact angle 92°.

4.2 Flow behavior

The flow behavior of the emulsions described in Table 1 is shown in Figs. 4 and 5. For all cases, the shear-thinning behavior observed was appropriate for a drilling fluid. It was found that the addition of OTMOS-modified silica NPs to the invert emulsion resulted in at least a doubling of viscosity at high shear rates and as much as one order of magnitude increase in viscosity at low shear rates in the case of using 2 wt% silica NPs (Binks et al. 2005). This desirable behavior of the invert emulsion is a consequence of a network structure which formed by NPs. In the case of the silica NPs synthesized with 0.01 OTMOS to TEOS molar ratio, there was a modest enhancement in viscosity when the addition amount of OTMOS-modified silica NPs

increased from 1 wt% to 2 wt% (black line and red line in Fig. 4 ($\theta = 92°$ in Table 1), and using more than 2 wt% of NPs did not change the viscosity significantly. In all cases, except perhaps when using 1 wt% silica NPs synthesized with 0.02 OTMOS to TEOS molar ratio which resulted in a contact angle of 115°, very high viscosity at low shear stress can be achieved. Figure 5 also showed that the point of transition from high viscosity to low viscosity was sharp, and the shear stress level could be altered over a modest range by changing the hydrophobicity and the amount of the added nanoparticles. Also considering stability of emulsions in Table 1, the invert emulsion stabilized by 2 wt% exhibited the best results among all prepared invert emulsions.

4.3 Mathematical modeling

By considering the shear-thinning behavior of the invert emulsions, the Bingham plastic, power law, Casson model, and Herschel–Bulkley model shown in Eqs. (1)–(4) were

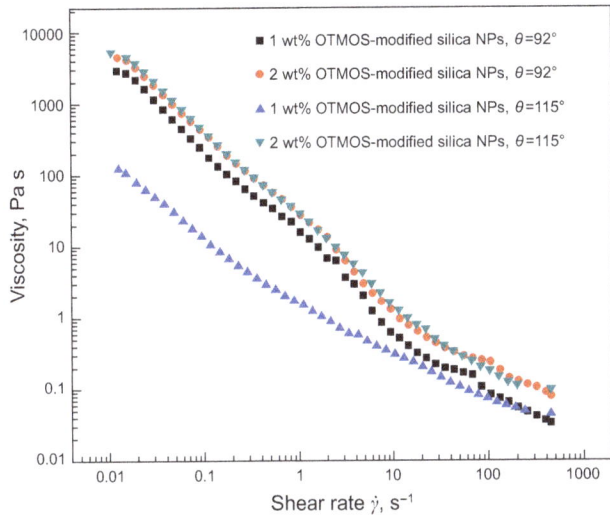

Fig. 4 Measured viscosity versus shear rate of invert emulsions containing 30 vol% water stabilized by OTMOS-modified silica NPs

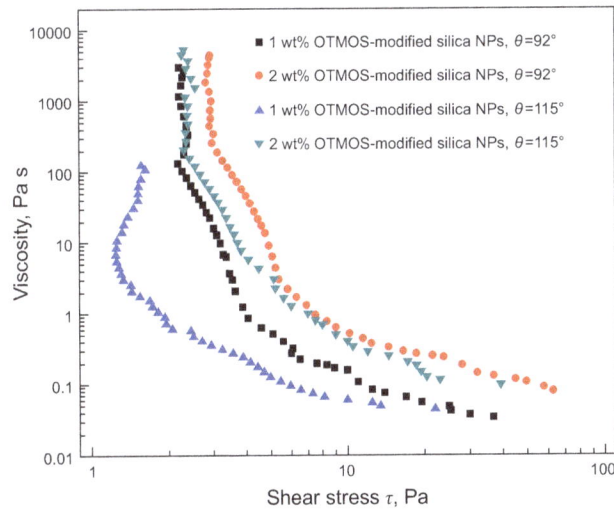

Fig. 5 Measured viscosity versus shear stress of invert emulsions containing 30 vol% water stabilized by OTMOS-modified silica NPs

considered to predict the flow behavior (Agarwal et al. 2011, 2013; Livescu 2012; Hossain and Al-Majed 2015).

$$\tau = \tau_0 + \eta_\infty \dot{\gamma} \tag{1}$$

$$\tau = \eta_\infty \dot{\gamma}^n \tag{2}$$

$$\sqrt{\tau} = \sqrt{\tau_0} + \sqrt{\eta_\infty \dot{\gamma}} \tag{3}$$

$$\tau = \tau_0 + \eta_\infty \dot{\gamma}^n \tag{4}$$

where τ is the shear stress; $\dot{\gamma}$ is the shear rate; τ_0 and η_∞ are the model constants representing the yield stress and the plastic viscosity, respectively; and n is the power-law index.

According to the patterns between shear stress and shear rate which are shown in Eqs. (1)–(4), flow curves (shear

stress vs shear rate) were drawn on both linear and logarithmic scales, and also, square root of shear stress versus shear rate was drawn on linear scale in Figs. 6, 7, and 8.

As shown in Fig. 6, the Bingham model shown in Eq. (1) is not a good model because the obtained figures are in the form of curves instead of straight lines. Similarly in Fig. 7, on the log scale, it can be said that the power-law model shown in Eq. (2) is not suitable, too. Since the data plotted in Fig. 6 follow the Herschel–Bulkley model shown in Eq. (3), this model could be considered as a desirable model for predicting shear-thinning behavior of the invert emulsion presented here. Also, according to the Casson model pattern shown in Eq. (4), the measured data are plotted in Fig. 8. Data plot nearly linearly in Fig. 8, which shows that the Casson model can be used to fit the behavior of the invert emulsions as well as the Herschel–Bulkley model. Since the Herschel–Bulkley model and Casson model were selected to predict the flow behavior, the validation of the models seemed to consist of nothing more than quoting the R^2 statistic from the fit. A linear regression was used to obtain the parameters of models using Figs. 9 and 10. The measured parameters of the Herschel–Bulkley model and the Casson model are shown in Table 2.

It is found that in the Casson model, the yield stress and the plastic viscosity vary over a modest range depending on the nature and amount of nanoparticles used, and the plastic viscosity has a negligible amount in comparison with the Herschel–Bulkley model. In the Herschel–Bulkley model, the changes of the shear stress are more than the Casson model, and the plastic viscosity is almost constant in the case of OTMOS-modified silica NPs with a contact angle of 115° (Agarwal et al. 2011, 2013).

Neither the Casson model nor the Herschel–Bulkley model predicts the experimental data perfectly, and the Herschel–Bulkley model is a complicated three-parameter model. The Casson model was found to provide the best fit for the data.

It is worth mentioning that depending on the requirements of a drilling operation, the viscosity can change from water-like values of around 0.001 Pa s to several folds higher. Similarly, the yield stress may range from being negligible to more than 10 Pa. The viscosity, yield stress, and other properties of the drilling fluid depend on the composition and could be tuned as required (Agarwal et al. 2011, 2013).

4.4 Barite effect

Barite as the weighting agent is used to impart density to all types of drilling fluids. Weighting agents are used to control formation pressures and prevent formation damage or blowup (Bourgoyne et al. 1986; Hossain and Al-Majed

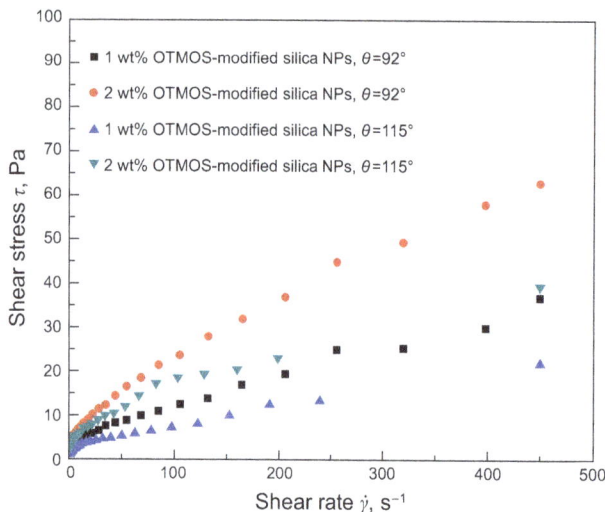

Fig. 6 Measured shear stress versus shear rate of invert emulsions containing 30 vol% water stabilized by OTMOS-modified silica NPs, linear scale

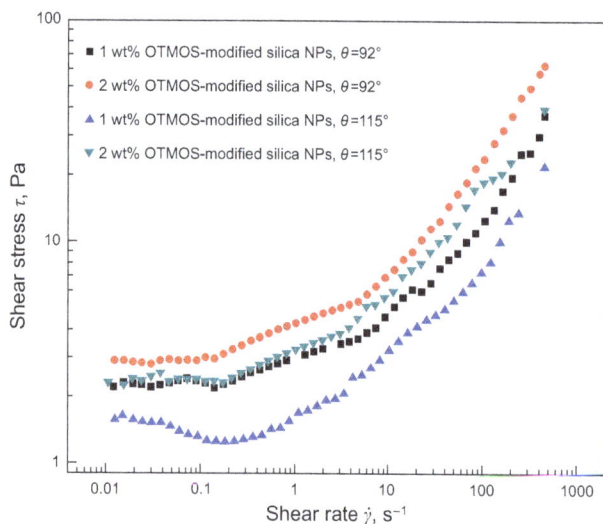

Fig. 7 Measured shear stress versus shear rate of invert emulsions containing 30 vol% water stabilized by OTMOS-modified silica NPs, logarithmic scale

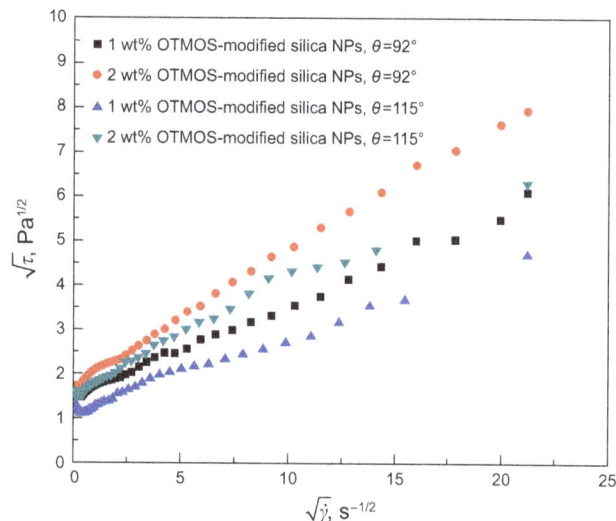

Fig. 8 Measured square root of shear stress versus square root of shear rate of invert emulsions containing 30 vol% water stabilized by OTMOS-modified silica NPs, linear scale

3 wt%, the yield stress also increased, which shows that the flow properties would be adjusted by changing the amount of nanoparticles to achieve the desired properties. As in the case without barite, the flow behavior can be predicted by the Casson model (Fig. 11b). Table 3 presents the parameters of the Casson model.

4.5 High-temperature aging effect

The invert emulsion prepared here which was stabilized by the OTMOS-modified silica NPs can be used as drilling fluid when it has good wellbore stability and high-temperature tolerance. In ultra-deep drilling, surfactants degrade due to high temperature, causing a phase separation (Agarwal et al. 2011, 2013). Figure 12 shows the optical microscopy image of the invert emulsion stabilized by OTMOS-modified silica NPs with a contact angle of 92° and aged at 120 °C for 12 h. As can be seen, the water phase was still emulsified as small droplets in the oil phase and less than 80 μm. Also, some experiments were done at 225 °C to compare results with work done by Agarwal et al. (2011, 2013).

Flow curves in Figs. 13a and 14a show similar behavior for the fresh invert emulsion. The parameters of the Casson model for the aged emulsion (in Table 4) show that the yield stress and the plastic viscosity reduced. As in other cases, the Casson model properly predicted the aged invert emulsion behavior (Figs. 13b, 14b). The reduced yield stress value indicates that the aged invert emulsion could not form a gel structure effectively. This behavior of yield stress is similar to the invert emulsion prepared by Agarwal et al. (2011, 2013) in which nanosilica and nanoclay were used as solid additives (Agarwal et al. 2011, 2013). Similar

2015). In this work, API-grade barite was used to prepare drilling fluids with a density of 1.40 g mL^{-1} (equal to 90 lb ft^3) from the unweighted fluid of 0.89 g mL^{-1}. The required amount of barite was added to the prepared OTMOS-modified silica NPs-stabilized emulsion, and the mixture was then homogenized for 2 min. Figure 11a shows flow curves of drilling fluids with and without barite. It can be seen that in the case of using barite, the yield stress decreased significantly while the plastic viscosity was mostly unaffected (Table 3). This occurs because that the gel structure formed by NPs is disrupted by the micro-sized particles of barite. However, if the content of the OTMOS-modified silica NPs increased from 2 wt% to

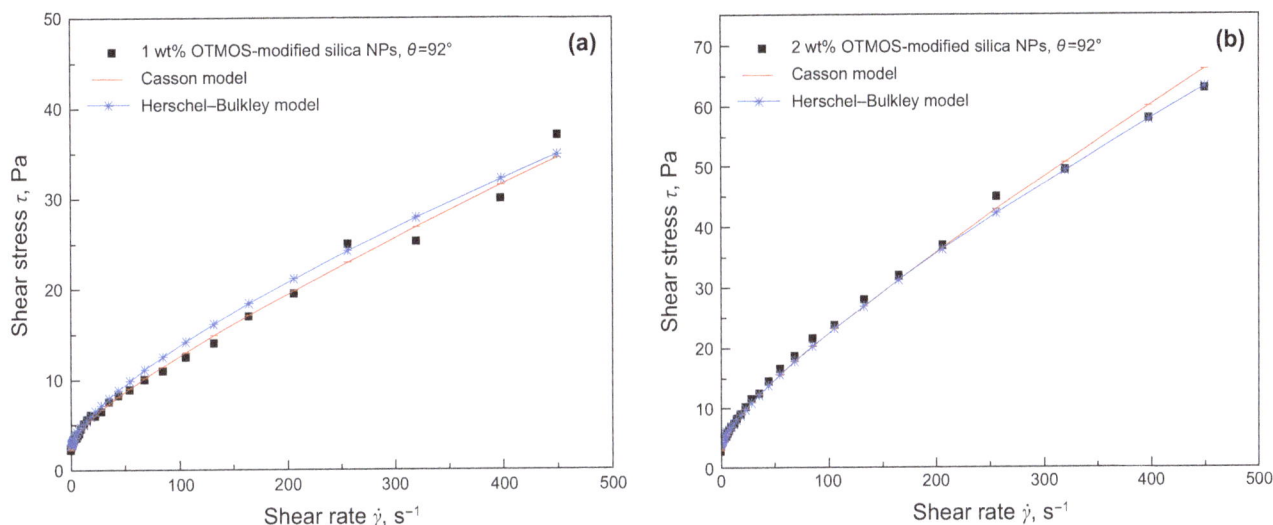

Fig. 9 Shear stress versus shear rate of invert emulsions containing 30 vol% water stabilized by 1 wt% OTMOS-modified silica NPs ($\theta = 92°$) (**a**) and 2 wt% OTMOS-modified silica NPs ($\theta = 92°$) (**b**). The *solid lines* without and with *stars* represent the best fit of the Casson model and the Herschel–Bulkley model, respectively. Parameters of models are given in Table 2

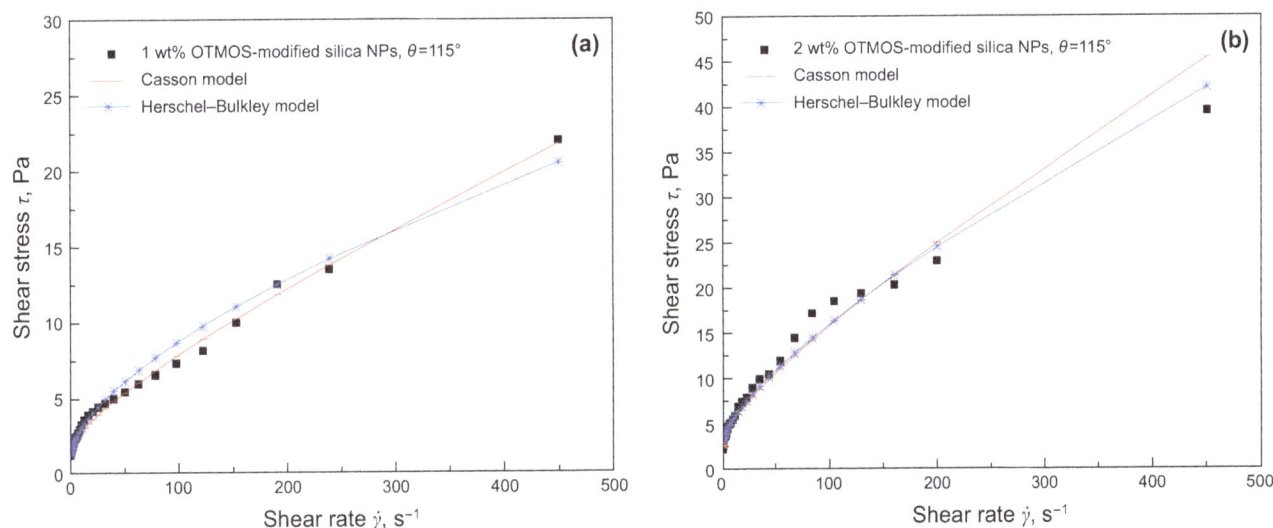

Fig. 10 Shear stress versus shear rate of invert emulsions containing 30 vol% water stabilized by 1 wt% OTMOS-modified silica NPs ($\theta = 115°$) (**a**) and 2 wt% OTMOS-modified silica NPs ($\theta = 115°$) (**b**). The *solid lines* without and with *stars* represent the best fit of the Casson model and the Herschel–Bulkley model, respectively. Parameters of models are given in Table 2

Table 2 Parameters of models for emulsions containing OTMOS-modified silica NPs

Invert emulsion (30 vol% water)	Casson model			Herschel–Bulkley model			
	τ_0, Pa	η_∞, Pa s	R^2	τ_0, Pa	η_∞, Pa s	n	R^2
Stabilized by 1 wt% of 0.01 OTMOS-modified silica NPs, $\theta = 92°$	2.2	0.043	0.9942	2.7	0.42	0.71	0.9870
Stabilized by 2 wt% of 0.01 OTMOS-modified silica NPs, $\theta = 92°$	2.9	0.092	0.9974	4.0	0.52	0.78	0.9968
Stabilized by 1 wt% of 0.02 OTMOS-modified silica NPs, $\theta = 115°$	1.3	0.028	0.9946	1.3	0.41	0.63	0.9836
Stabilized by 2 wt% of 0.02 OTMOS-modified silica NPs, $\theta = 115°$	2.4	0.060	0.9739	3.5	0.42	0.74	0.9779

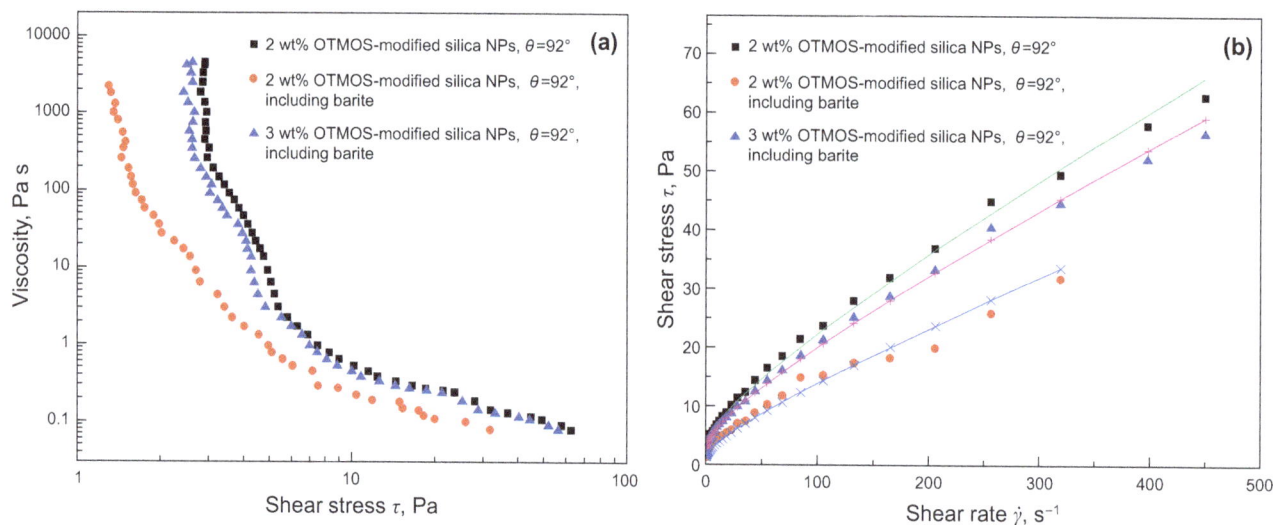

Fig. 11 Effect of barite on the flow behavior of invert emulsions containing 30 vol% water stabilized by OTMOS-modified silica NPs ($\theta = 92°$). **a** Viscosity versus shear stress. **b** Shear stress versus shear rate. The *solid lines* represent the best fit of the Casson model. Model parameters are given in Table 3

Table 3 Parameters of the Casson model for emulsions containing OTMOS-modified silica NPs and barite

Invert emulsion (30 vol% water)	Casson model		
	τ_0, Pa	η_∞, Pa s	R^2
Stabilized by 2 wt% of 0.01 OTMOS-modified silica NPs, $\theta = 92°$	2.90	0.092	0.9974
Stabilized by 2 wt% of 0.01 OTMOS-modified silica NPs, $\theta = 92°$, containing barite	1.25	0.069	0.9814
Stabilized by 3 wt% of 0.01 OTMOS-modified silica NPs, $\theta = 92°$, containing barite	2.60	0.082	0.9978

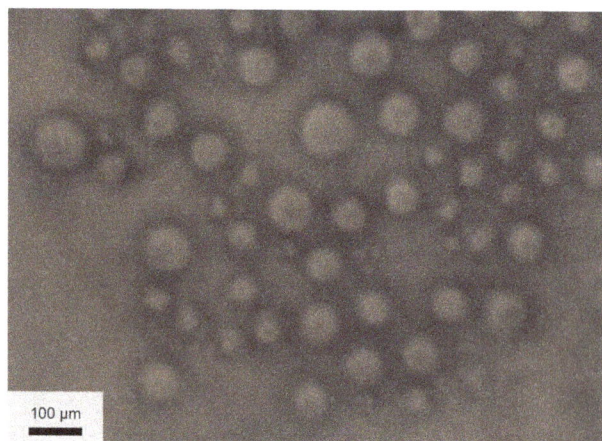

Fig. 12 Morphology of the invert emulsion stabilized by 2 wt% OTMOS-modified silica NPs ($\theta = 92°$) after aging at 120 °C

to the yield stress, the plastic viscosity reduced after aging and did not follow the invert emulsion prepared by Agarwal et al. (2011, 2013). This adverse behavior can be explained by the nanoclay platelets available in Agarwal study which resulted in better exfoliation and dispersion of

the nanoclay platelets by the heat treatment (Agarwal et al. 2011, 2013).

5 Conclusions

Synthesized functionally modified silica NPs as the only emulsifier were used to emulsify and stabilize the invert emulsion employing water and poly 1-decene oil phase. It was shown that the resulting morphology and flow behavior of the prepared invert emulsion meet drilling fluid requirements.

The flow behavior of the invert emulsions was investigated as a function of amount and nature of silica NPs, and it was found that the shear-thinning behavior could be represented by the Casson and Herschel–Bulkely models and the best results belong to 2 wt% OTMOS-modified silica NPs which make a 92° contact angle. On adding a weighting material of barite, there was a loss of the yield stress, but this could be recovered by increasing the content of OTMOS-modified silica NPs.

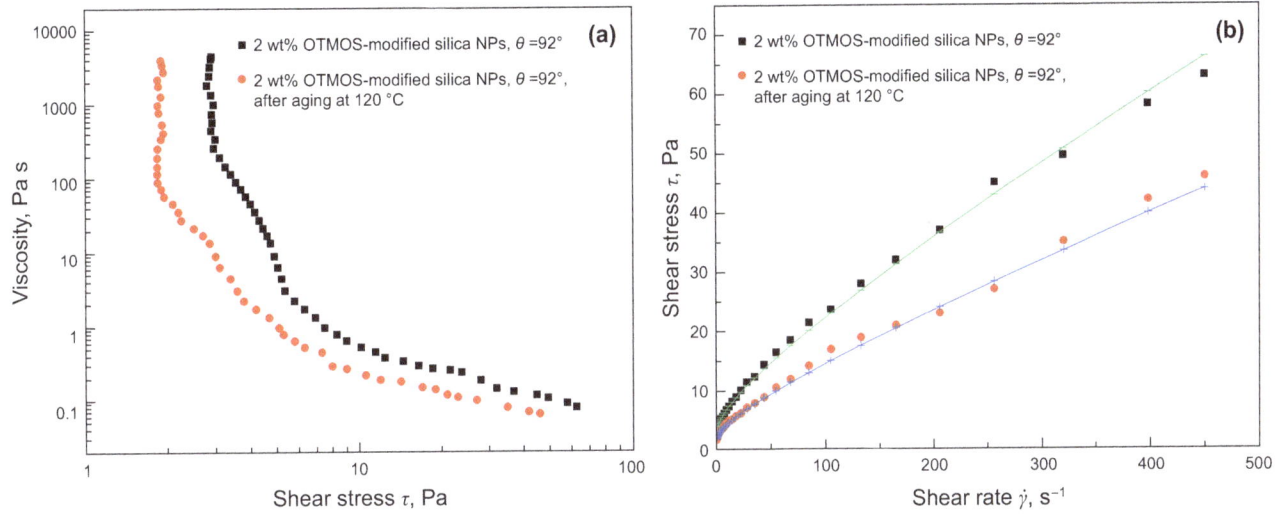

Fig. 13 Flow behavior of invert emulsions containing 30 vol% water stabilized by OTMOS-modified silica NPs ($\theta = 92°$) before and after aging at 120 °C. **a** Viscosity versus shear stress. **b** Shear stress versus shear rate. The *solid lines* represent the best fit of the Casson model. Model parameters are given in Table 4

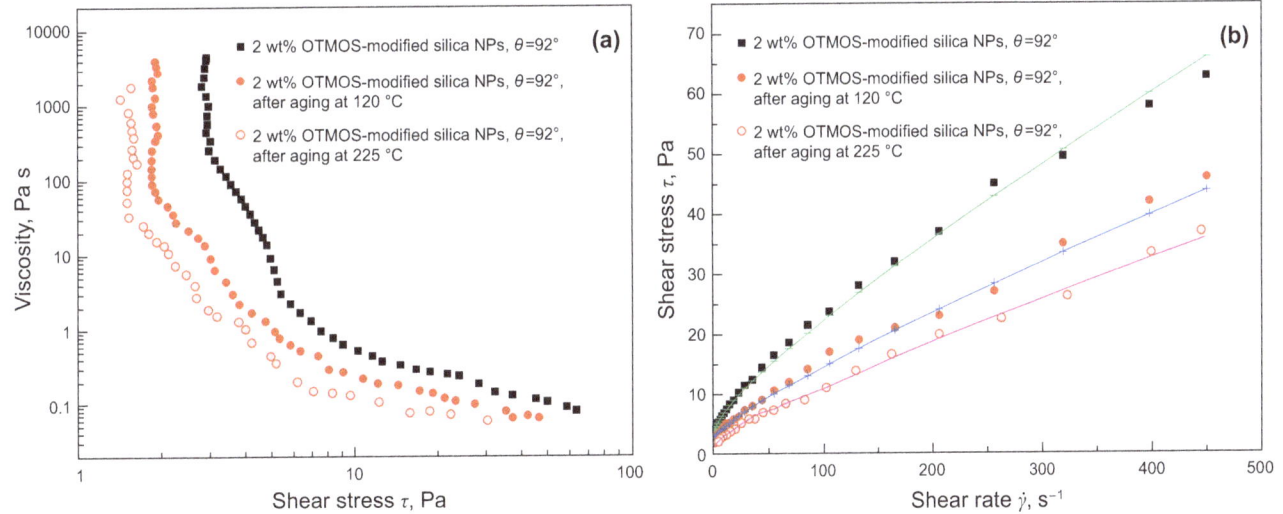

Fig. 14 Flow behavior of invert emulsions containing 30 vol% water stabilized by OTMOS-modified silica NPs ($\theta = 92°$) before and after aging at 225 °C. **a** Viscosity versus shear stress. **b** Shear stress versus shear rate. The *solid lines* represent the best fit of the Casson model. Model parameters are given in Table 4

Table 4 Parameters of the Casson model for fresh/aged emulsions containing OTMOS-modified silica NPs

Invert emulsion (30 vol% water)	Casson model		
	τ_0, Pa	η_∞, Pa s	R^2
Stabilized by 2 wt% of 0.01 OTMOS-modified silica NPs, $\theta = 92°$, fresh	2.9	0.092	0.9974
Stabilized by 2 wt% of 0.01 OTMOS- modified silica NPs, $\theta = 92°$, after aging at 120 °C	1.7	0.063	0.9957
Stabilized by 2 wt% of 0.01 OTMOS- modified silica NPs, $\theta = 92°$, after aging at 225 °C	1.3	0.051	0.9963

In addition, after aging at 120 and 225 °C for 12 h, the stabilized invert emulsion maintained its stability and flow properties. Thus, OTMOS-modified silica NPs could be a good candidate as emulsifier instead of other additives such as polymeric ones, especially for formulating drilling fluids at 120 °C and higher temperatures of drilling operations.

Acknowledgements The authors would like to extend lots of thanks to Mr. Nasser Maragheh, vice president of Sarve Energy Arya Sahand Engineering Company (SEAS Co.) and Dr. Mahmoud Dargahi for their invaluable discussion and advice.

References

Agarwal S, Phuoc TX, Soong Y, et al. Nanoparticle-stabilized invert emulsion drilling fluids for deep-hole drilling of oil and gas. Can J Chem Eng. 2013;9(1):1641–9. doi:10.1002/cjce.21768.

Agarwal S, Tran P, Soong Y, et al. Flow behavior of nanoparticle stabilized drilling fluids and effect of high temperature aging. American Association of Drilling Engineers (AADE). 2011; AADE-11-NTCE-3.

Aveyard R, Binks BP, Clint J. Emulsions stabilized solely by colloidal particles. Adv Colloid Interface Sci. 2003;100–102:503–46. doi:10.1016/S0001-8686(02)00069-6.

Binks BP. Particles as surfactants - similarities and differences. Curr Opin Colloid Interface Sci. 2002;7:21–41. doi:10.1016/S1359-0294(02)00008-0.

Binks BP, Liu W, Rodrigues JA. Novel stabilization of emulsions via the heteroaggregation of nanoparticles. Langmuir. 2008;24:4443–6. doi:10.1021/la800084d.

Binks BP, Lumsdon SO. Influence of particle wettability on the type and stability of surfactant-free emulsions. Langmuir. 2000;16(1):8622–31. doi:10.1021/la000189s.

Binks BP, Philip J, Rodrigues JA. Inversion of silica-stabilized emulsions induced by particle concentration. Langmuir. 2005;21(1):3296–302. doi:10.1021/la046915z.

Binks BP, Rodrigues JA. Types of phase inversion of silica particle stabilized emulsions containing triglyceride oil. Langmuir. 2003;19:4905–12. doi:10.1021/la020960u.

Binks BP, Rodrigues JA. Inversion of emulsions stabilized solely by ionizable nanoparticles. Angew Chem Int Ed. 2005;44:441–4. doi:10.1002/anie.200461846.

Binks BP, Whitby CP. Nanoparticle silica-stabilised oil-in-water emulsions: improving emulsion stability. Colloids Surf A Physicochem Eng Asp. 2005;253(1):105–15. doi:10.1016/j.colsurfa.2004.10.116.

Bourgoyne AT, Millheim KK, Chenevert ME, et al. Applied drilling engineering. 1st ed. SPE textbook series; 1986.

Coussot P, Bertrand F, Herzhaft B. Rheological behavior of drilling muds, characterization using MRI visualization. Oil Gas Sci Technol - Rev IFP. 2004;59(1):23–9. doi:10.2516/ogst:2004003.

Dickinson E. Food emulsions and foams: stabilization by particles. Eur Curr Opin Colloid Interface Sci. 2010;15:40–9. doi:10.1016/j.cocis.2009.11.001.

Ding A, Binks BP, Goedel WA. Influence of particle hydrophobicity on particle-assisted wetting. Langmuir. 2005;21:1371–6. doi:10.1021/la047858c.

Effati E, Pourabbas B. One-pot synthesis of sub-50 nm vinyl- and acrylate-modified silica nanoparticles. Powder Technol. 2012;219:276–83. doi:10.1016/j.powtec.2011.12.062.

Gupta RK. Polymer and composite rheology. 2nd ed. New York: CRC Press; 2000.

Hossain ME, Al-Majed AA. Fundamentals of sustainable drilling engineering. 1st ed. Hoboken: Wiley; 2015.

Hsu MF, Nikolaides MG, Dinsmore AD, et al. Self-assembled shells composed of colloidal particles: fabrication and characterization. Langmuir. 2005;21:2963–70. doi:10.1021/la0472394.

Livescu S. Mathematical modeling of thixotropic drilling mud and crude oil flow in well sand pipelines—a review. J Pet Sci Eng. 2012;98:174–84. doi:10.1016/j.petrol.2012.04.026.

Melle S, Lask M, Fuller GG. Pickering emulsions with controllable stability. Langmuir. 2005;21:2158–62. doi:10.1021/la047691n.

Pickering SU. Emulsions. J Chem Soc Trans. 1907;91:2001–21. doi:10.1039/CT9079102001.

Shah SN, Shanker NH, Ogugbue CC. Future challenges of drilling fluids and their rheological measurements. American Association of Drilling Engineers (AADE). 2010; AADE-10-DF-HO-41.

Stöber W, Fink A, Bohn E. Controlled growth of monodisperse silica spheres in the micron size range. J Colloid Interface Sci. 1968;26:62–9. doi:10.1016/0021-9797(68)90272-5.

Vignati E, Piazza R. Pickering emulsions: interfacial tension, colloidal layer morphology, and trapped-particle motion. Langmuir. 2003;19:6650–6. doi:10.1021/la034264l.

Xuc L, Li J, Fu J, et al. Super-hydrophobicity of silica nanoparticles modified with vinyl groups. Colloids Surfaces A Physicochem Eng Asp. 2009;338:15–9. doi:10.1016/j.colsurfa.2008.12.016.

12

Stress analysis on large-diameter buried gas pipelines under catastrophic landslides

Sheng-Zhu Zhang[1] · Song-Yang Li[2] · Si-Ning Chen[1] · Zong-Zhi Wu[3] ·
Ru-Jun Wang[1] · Ying-Quan Duo[1]

Handling editor: Jian Shuai

Abstract This paper presents a method for analysis of stress and strain of gas pipelines under the effect of horizontal catastrophic landslides. A soil spring model was used to analyze the nonlinear characteristics concerning the mutual effects between the pipeline and the soil. The Ramberg–Osgood model was used to describe the constitutive relations of pipeline materials. This paper also constructed a finite element analysis model using ABAQUS finite element software and studied the distribution of the maximum stress and strain of the pipeline and the axial stress and strain along the pipeline by referencing some typical accident cases. The calculation results indicated that the maximum stress and strain increased gradually with the displacement of landslide. The limit values of pipeline axial stress strain appeared at the junction of the landslide area and non-landslide area. The stress failure criterion was relatively more conservative than the strain failure criterion. The research results of this paper may be used as a technical reference concerning the design and safety management of large-diameter gas pipelines under the effects of catastrophic landslides.

Keywords Buried gas pipeline · Catastrophic landslide · Finite element analysis · Stress · Strain

✉ Sheng-Zhu Zhang
zhangshengzh5168@163.com

[1] China Academy of Safety Science and Technology, Beijing 100012, China

[2] College of Mechanical and Transportation Engineering, China University of Petroleum, Beijing 102249, China

[3] State Administration of Work Safety, Beijing 100713, China

Edited by Yan-Hua Sun

1 Introduction

Landslides are a common geological disaster frequently occurring in mountainous areas of the southwest provinces, the Loess Plateau in the northwest, as well as mountainous and hilly areas of the mid-south and southeast provinces in China. Newly built pipelines have to pass through small landslide areas, and if it is not possible to be avoided, then during the operation period of pipelines, soil piling, excavation and other third-party activities around the pipeline may damage the stability of rock and earth mass, resulting in a landslide (Yu 1989; Challamel and Debuhan 2003). The large-diameter gas pipeline connects the gas source to the consumer over a long route, and it often crosses over areas with complex conditions (Hucka et al. 1986; Hall et al. 2003). Therefore, we often suffered from landslide disasters. Before the year 2000, more than ten pipeline accidents were caused by landslides and other geological disasters occurred to the natural gas pipeline in Sichuan Province (Wang 2014). As reported by European Gas Pipeline Incident Data Group, the accidents caused by geological disasters accounted for 13% of the total amount of gas pipeline accidents which happened in Europe from 2004 to 2013, of which landslides accounted for 85.2% of all geological disasters, forming the leading geological disaster type (EGIG 2015). Especially under the effects of catastrophic landslides, the pipeline and even the surrounding people will be severely influenced (Honegger et al. 2010; Liu et al. 2010; Zheng et al. 2012). With the continuous construction of large-diameter gas pipelines, an increasing number of geological disasters such as landslides caused by various factors happen along the pipeline, which brings a great challenge to the safe operation of pipelines (Wang et al. 2008). Therefore, it is necessary to make a stress analysis on large-diameter buried gas pipelines under the effects of catastrophic landslides.

For the stress condition of pipelines under the effect of landslides, Rajani et al. (1995) studied the stress response of pipelines under the effect of a horizontal landslide using an analytical method. O'Rourke et al. (1995) studied the stress features of deviated pipeline landslide areas using the Ramberg–Osgood power hardening formula. Deng et al. (1998) studied the calculation methods of internal force and deformation under other soil mass loads, during horizontal landslides. Shuai et al. (2008) studied the failure characteristics of pipelines under the effect of landslides. Zhang et al. (2010) investigated the interaction of landslides and pipelines. Hao et al. (2012) studied the calculation of horizontal thrust imposed on pipelines by landslides. Lin et al. (2011) established a soil quality landslide model under fully buried conditions and carried out a test on the pipeline stresses under the effects of a landslide. However, studies of the stress condition of buried pipelines under the effects of catastrophic landslides are rarely made. Aiming at the working conditions of buried gas pipelines with large diameters, the authors created a soil spring analysis model for the pipeline under the effects of catastrophic landslides using ABAQUS finite component analysis software, as well as making a finite element analysis on the pipeline stresses and deformation, which can be used as a reference for safety design and operations in landslide areas.

2 The interactive process of landslide mass and pipeline under the effects of catastrophic landslides

Pipelines may pass through landslide masses through two methods, namely horizontal pass-through and vertical pass-through. This paper only analyzes the stress condition of pipelines under horizontal pass-through. When the pipeline passes through the landslide mass horizontally, the slide direction of the landslide mass is at right angles to the pipeline axis (Ma et al. 2006; Han et al. 2012). The evenly distributed load endured by the pipeline is in the horizontal and vertical directions. Under the effect of the landslide mass, the buried pipeline may displace for a certain degree. The accident process of the buried pipeline before and after a landslide is shown in Fig. 1. This finite component analysis considered the buried pipeline and the surrounding soil mass as a system.

3 Load calculation of pipelines across landslide mass

Considering the material nonlinearity and geometric non-linearity of the buried pipeline and soil mass, a nonlinear soil spring element was used to simulate the interaction of

soil and pipes (Wang and Yeh 1985; Zheng et al. 2015). The soil spring model disperses the effects of soil on the pipeline to be springs with different stiffness in the axial, vertical and horizontal directions (Kennedy et al. 1977; Iimura 2004; Yan et al. 2009). The axial soil spring parameters are decided by the backfill in the pipe ditch, while the horizontal and vertical soil spring is determined by the nature of soil mass around the pipeline burying site. The soil joint was set up connected to each joint outside of all the nodes on the pipeline, and the JOINTC unit was used to connect the pipeline to the soil joint to simulate the effects of a soil spring. A JOINTC unit can describe the interactive force generated by the change of relative displacement between two joints. The mechanical properties of the soil spring were described by ultimate soil resistance and yield displacement. T_u, Q_u, Q_d and P_u represent the ultimate resistance in the three directions, and Δt, Δq_d, Δq_u and Δp represent the yield displacement in the three directions, respectively, as shown in Fig. 2. The specific value can be calculated based on the *Guideline for the Design of Buried Steel Pipeline* published by American Lifelines Alliance (ALA 2001).

When the size of the structure in the direction of thickness is less than the sizes in the other two directions, and when the stress in the direction of thickness can be neglected, the shell element can be used for stimulation. Concerning a thin-walled pipeline with a large diameter, when the wall thickness is less than 1/10 of the overall structure size, the pipeline is dispersed as several rectangular shell elements to accurately describe the deformation of the pipeline under the effects of force or displacement loads, and the distribution of stress strain (Klar and Marshall 2008; Liu et al. 2014; Zhao and Zhao 2014). This paper established a long-distance pipeline model with four joints and reduced integration shell element S4R. The S4R element has six degrees of freedom on each joint, including 3 transitional degrees of freedom and 3 rotational degrees of freedom.

Considering the nonlinear characteristics of pipes, the Ramberg–Osgood (R–O) model was used to describe the stress–strain relationship of pipes, as shown in Eq. (1):

$$\varepsilon = \frac{\sigma_s}{E}\left[\frac{\sigma}{\sigma_s} + \frac{n}{1+r}\left(\frac{\sigma}{\sigma_s}\right)^r\right] \tag{1}$$

where E is the initial elastic modulus of pipe, Pa; ε is strain; σ is stress, Pa; σ_s is the yield stress of pipe, Pa; r and n are R–O model parameters.

During the calculation, fixed constraints were imposed on circular joints on both ends of the pipeline to simulate the effects of infinite extension of the pipeline on the studied pipe section. The fixed constraints were imposed on the area outside the landslide area, in order to limit its movement and simulate the constraints and supporting

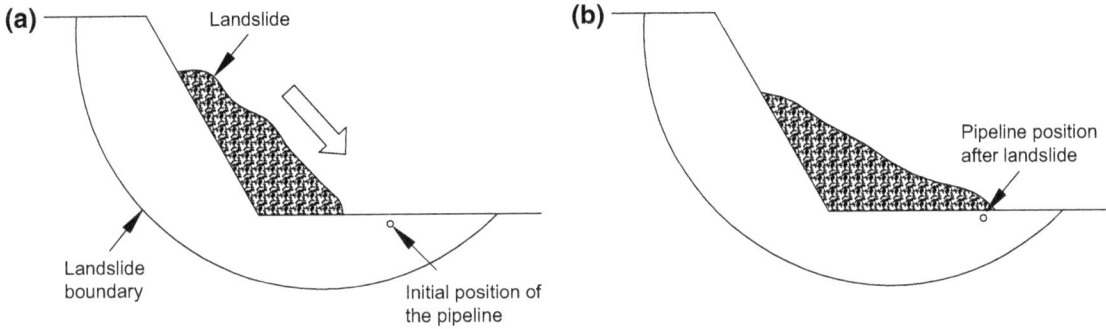

Fig. 1 Accident process of the buried pipeline before and after a landslide

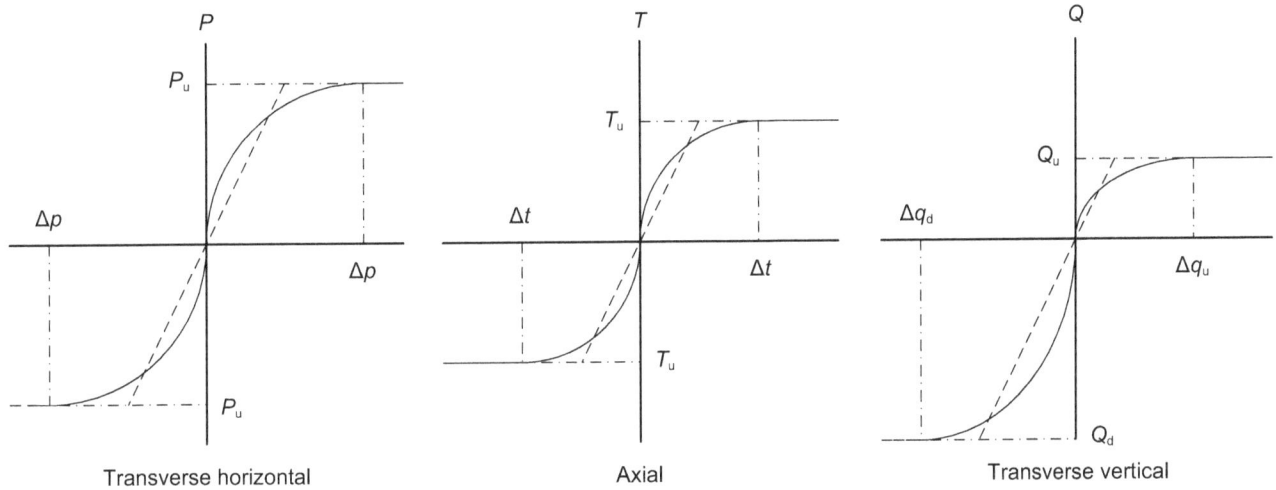

Fig. 2 Schematic diagram of soil spring effect

effects on the soil mass around the pipeline (Zhang and Huang 2012). When loading the load, the internal force is imposed on the entire management element, and then, a relative displacement is imposed based on the soil spring joint in the landslide area against the landslide direction, in order to simulate the effects of soil movement on the pipeline. Since the volume of landslide mass is larger than the pipeline, the integral movement of landslide is not influenced by the pipeline when the landslide is pushing and squeezing the pipeline. It is assumed that the movement speeds of the horizontal parts of the landslide are consistent, the displacement load of soil mass sliding downwards is evenly distributed, and the vertical friction of the soil mass against the pipeline is neglected.

4 Stress analysis of a pipeline in the landslide mass in a landslide area

4.1 Model overview

The corresponding value calculation model was established based on the design data of a large-diameter gas pipeline in

a landslide area. The entire section of the model was laid out by burying it underground. To build in the pipeline model, the pipeline parameters were assigned with values shown in Table 1. The values of the soil spring parameters were calculated according to the site survey results of a catastrophic landslide accident, and the formula specified in the ALA *Guideline for the Design of Buried Steel Pipeline*, as shown in Table 2. The pipeline was made of X70 steel, and the R–O model parameters r and n were assigned with 16.6 and 5.5, respectively, as described in the literature (Liu and Sun 2005).

4.2 Modeling

To reduce the effect of fixed end boundary conditions on the calculation results, circular joints at the end of the shell element pipeline and six degrees of freedom of the soil joints were all constrained. As described in the literature (Liu 2008), when the length of the non-slide area was four times the length of the slide area, the requirements for calculation accuracy can be assured. In this case, while the length of the landslide area is 370 m, the analysis length of the shell element model is 1850 m.

Table 1 Pipeline parameters

Pipe	Diameter, mm	Wall thickness, mm	Elastic modulus, GPa	Poisson's ratio	Minimum tensile strength, MPa	Minimum yield strength, MPa	Density of natural gas, kg m^{-3}	Transportation temperature, °C	Transportation pressure, MPa
X70	1016	21	207	0.3	570	485	95	20	3.8

Table 2 Soil spring parameters

Parameters	Pipe axis direction	Horizontal direction	Vertical upward	Vertical downward
Yield stress, kN m^{-1}	135	821	150	10,059
Yield displacement, m	0.008	0.0886	0.1708	0.2032

The pipeline is divided into 24 elements in the circumferential direction by mesh generation against the model. To ensure the simulation accuracy of pipeline stress strain in the landslide area, it is necessary to densify the mesh near the landslide area, and to set the axial length of the three sections of pipeline elements (Sects. 2, 3, 4) at 0.25 m, and the length of Sects. 1 and 5 at 1 m, as shown in Fig. 3. Based on the results of the site survey and analyzing the influence of different values of landslide displacements, the displacement is imposed on the pipeline slide area. The vertical displacement is set at 0.2 m, the side displacement is set at 0.1–0.8 m and the interval is assigned at 0.1 m.

4.3 Analysis of results

Under the effects of different landslide displacements, the maximum stress and the maximum strain when the pipeline passes through a catastrophic landslide mass horizontally are shown in Figs. 4 and 5, respectively.

When the landslide area through which the pipeline passes reaches 740–1110 m, the pipeline around it is often severely influenced by the landslide. To analyze the stress and strain conditions near the pipeline in this area in detail, it is required to define a route in the axial direction of the pipeline, and its stress and strain conditions are shown in Figs. 6 and 7.

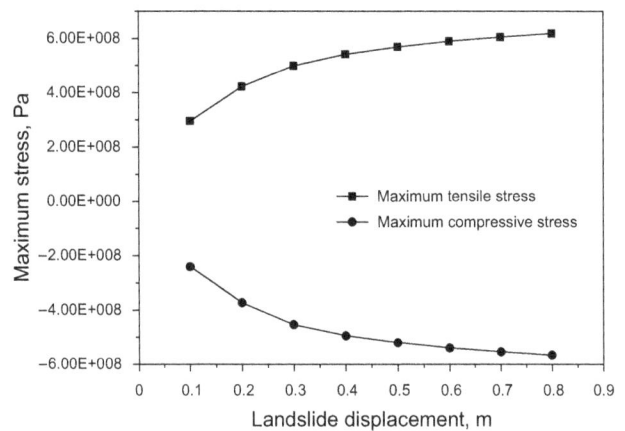

Fig. 4 Maximum stress of a pipeline under the effects of different landslide displacements

It can be observed from Figs. 4 and 5 that when the landslide displacement increases gradually, the maximum stress and strain grows accordingly. When the landslide displacement increases from 0.1 m to 0.8 m, the tensile stress increases from 312 MPa to 663 MPa, the tensile strain increases from 0.14% to 1.63%, the compressive stress increases from 240 MPa to 566 MPa and the compressive strain increases from 0.13% to 1.27%.

Under different landslide displacements, the limit values of the axial stress and strain of the pipeline appear at 740

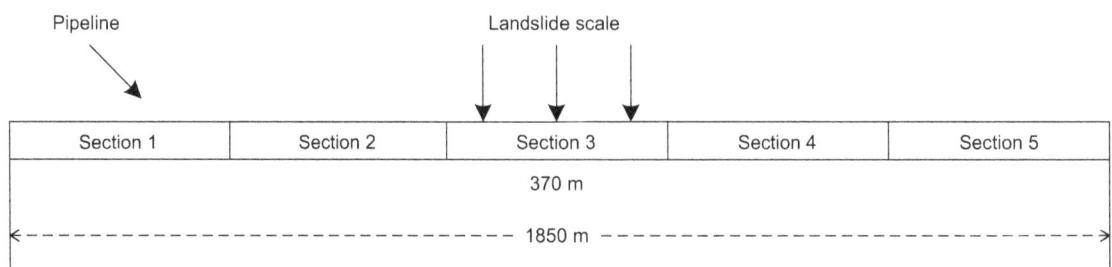

Fig. 3 Schematic diagram of the pipeline model passing through the landslide mass

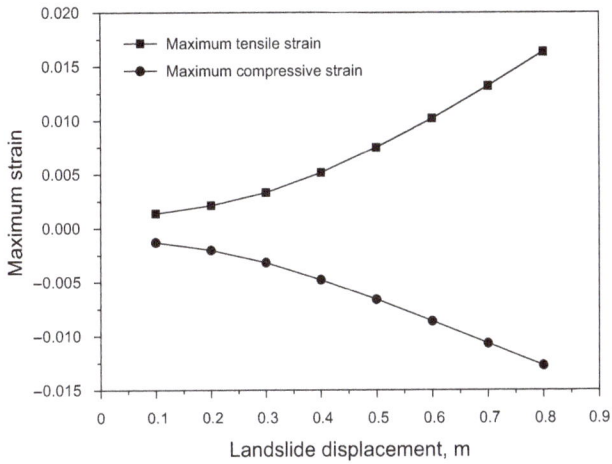

Fig. 5 Maximum strain of a pipeline under the effects of different landslide displacements

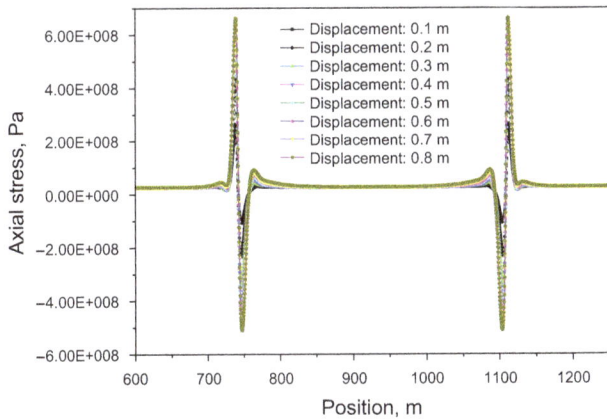

Fig. 6 Axial stress distribution of the pipeline

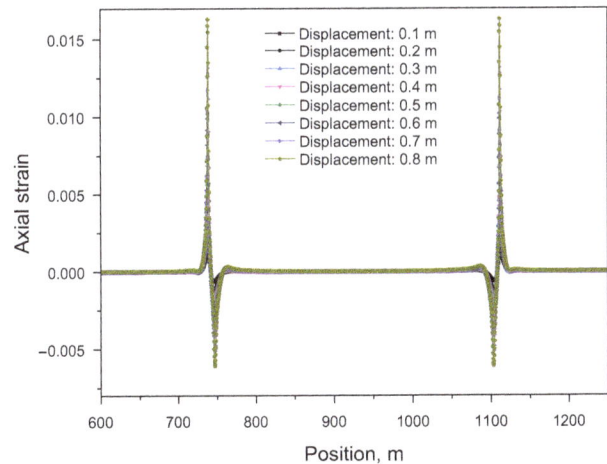

Fig. 7 Axial strain distribution of the pipeline

and 1100 m, which is at the junction of the landslide area and the non-landslide area. In addition, the limit values of stress and strain increase with an increase in the landslide displacement. On both sides of the landslide area, one side

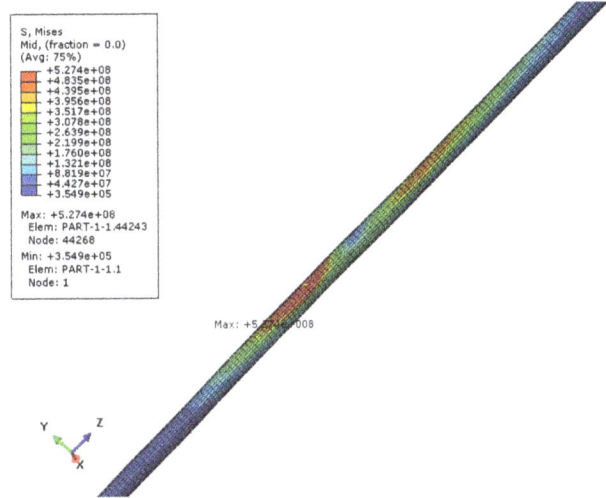

Fig. 8 Stress cloud of a pipeline near the landslide area

of the pipeline is pulled and the other side is stressed. At increasingly longer distances from the interval, the stress values reduce.

When the landslide displacement is 0.3 m, the maximum axial stress on the pipeline is 540 MPa. When the landslide displacement is 0.4 m, the maximum axial stress on the pipeline is 584 MPa. If the minimum yield strength of the pipe is exceeded, the pipeline is deemed having failed. Upon calculation, the critical value of the landslide displacement causing pipeline fracture is 0.36 m, as shown in Fig. 8.

When the landslide displacement is 0.5 m, the axial tensile strain is 0.75%, and the axial compressive strain is 0.663%. If the axial compressive strain exceeds the allowable compressive strain (allowable compressive strain of X70 pipeline is 0.66%), it is deemed that the pipeline suffers from yield damage, as shown in Fig. 9.

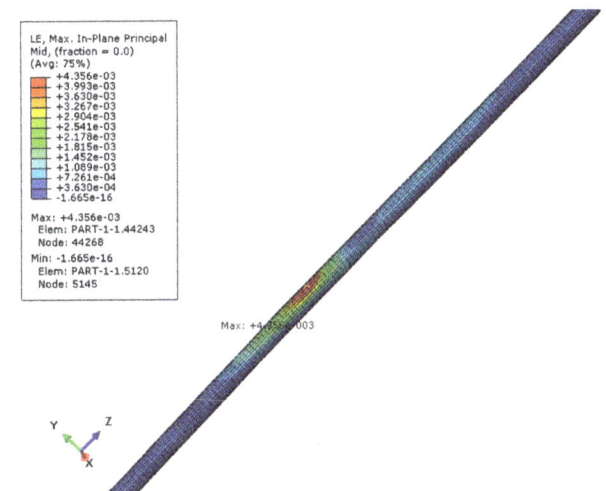

Fig. 9 Strain cloud of a pipeline near the landslide area

5 Conclusions

It can be found from the distributions of axial stress and strain along the pipeline that the limit values of the axial stress and strain appear near the junction of the slide and non-slide areas, which is the dangerous cross section of the pipeline.

The maximum stress and strain increase with an increase in the landslide displacement, and the limit value points are located at the top and bottom of the pipeline. Calibration and verification for strength at this place should be strengthened.

On the basis of the stress failure law, the pipeline fails when the landslide displacement reaches 0.36 m. In accordance with the strain failure law, the maximum strain does not reach the failure threshold in this case, indicating that the stress failure law is more conservative than the strain failure law.

Acknowledgements This research was funded by the National Science and Technology Support Program (2015BAK16B02 and 2015BAK16B01) and the Fundamental Research Funds of China Academy of Safety Science and Technology.

References

American Lifelines Alliance (ALA). Guideline for the design of buried steel pipeline. 2001.

Challamel N, Debuhan P. Mixed modelling applied to soil-pipe interaction. Comput Geotech. 2003;30(3):205–16. doi:10.1016/S0266-352X(03)00011-9.

Deng DM, Zhou XH, Shen YP. Calculation of pipeline inner force and distortion during transverse landslide body. Oil Gas Storage Transp. 1998;17(7):18–22 **(in Chinese)**.

European Gas Pipeline Incident Data Group (EGIG). In: 9th report of the european gas pipeline incident data group (period 1970–2013). 2015.

Hall WJ, Nyman DJ, Johnson ER, et al. Performance of the Trans-Alaska pipeline in the November 3, 2002 Denali fault earthquake. Long Beach, California. 2003. doi:10.1061/40687(2003)54.

Han B, Wang ZY, Zhao HL, et al. Strain-based design for buried pipelines subjected to landslides. Pet Sci. 2012;9(2):236–41. doi:10.1007/s12182-012-0204-y.

Hao JB, Liu JP, Jing HY, et al. A calculation of landslide thrust force to transverse pipeline. Acta Petrolei Sin. 2012;33(6):1093–7 **(in Chinese)**.

Honegger DG, Hart JD, Phillips R, et al. Recent PRCI guidelines for pipelines exposed to landslide and ground subsidence hazards. In: The 8th international pipeline conference, calgary: ASME, 2010. pp. 1–10. doi:10.1115/IPC2010-31311.

Hucka VJ, Blair CK, Kimball EP. Mine subsidence effects on a pressurized natural gas pipeline. Min Eng. 1986;38(10):980–4.

Iimura S. Simplified mechanical model for evaluating stress in pipeline subject to settlement. Constr Build Mater. 2004;18(6):469–79. doi:10.1016/j.conbuildmat.2004.01.002.

Kennedy R, Chow A, Williamson R. Fault movement effects on buried oil pipeline. J the Transp Eng Division ASCE. 1977;103(5):617–33.

Klar A, Marshall AM. Shell versus beam representation of pipes in the evaluation of tunneling effects on pipelines. Tunnell Undergr Space Technol. 2008;23(4):431–7. doi:10.1016/j.tust.2007.07.003.

Lin D, Lei Y, Xu KF, et al. An experiment on the effect of a transverse landslide on pipelines. Acta Pet Sin. 2011;32(4):728–32 **(in Chinese)**.

Liu H. Response analysis of buried pipeline subject to landslide. Master thesis. Dalian: Dalian university of technology; 2008 **(in Chinese)**.

Liu PF, Zheng JY, Zhang BJ, et al. Failure analysis of natural gas buried X65 steel pipeline under deflection load using finite element method. Mater Des. 2010;31(3):1384–91. doi:10.1016/j.matdes.2009.08.045.

Liu XB, Chen YF, Zhang H, et al. Prediction on the design strain of the X80 steel pipelines across active faults under stress. Natural Gas Ind. 2014;34(12):123–30 **(in Chinese)**.

Liu XJ, Sun SP. Strain design method for underground pipeline crossing fault. Spec Struct. 2005;2:81–5 **(in Chinese)**.

Ma QW, Wang CH, Kong JM. Dynamical mechanisms of effects of landslides on long distance oil and gas pipelines, Wuhan University. J Natural Sci. 2006;11(4):820–4. doi:10.1007/BF02830170.

O'Rourke MJ, Liu XJ, Flores-Berrones R. Steel pipe wrinkling due to longitudinal permanent ground deformation. J Transp Eng. 1995;121(5):443–51. doi:10.1061/(ASCE)0733-947X.

Rajani BB, Robertson PK, Morgenstern NR. Simplified design methods for pipelines subject to transverse and longitudinal soil movements. Can Geotech J. 1995;32(2):309–23. doi:10.1139/t95-032.

Shuai J, Wang XL, Zuo SZ. Breakage action and defend measures to pipeline under geological disaster. Weld Pipe Tube. 2008;31(5):9–15 **(in Chinese)**.

Wang LRL, Yeh YH. A refined seismic analysis and design of buried pipeline for fault movement. Earthq Eng Struct Dyn. 1985;13(1):75–96. doi:10.1002/eqe.4290130109.

Wang LW. Study on several typical geological disaster damage assessment methods for in service pipelines. Ph.D. dissertation. Beijing: China University of Science and Technology; 2014.

Wang X, Shuai J, Ye Y, et al. Investigating the effects of mining subsidence on buried pipeline using finite element modeling. In: The 7th international pipeline conference, Calgary: American Society of Mechanical Engineers, 2008. pp. 601–06. doi:10.1115/IPC2008-64250.

Yan X, Wang T, Yang X, et al. Study on stresses and deformations of suspended pipeline in collapsible loess areas based on elastic-plastic foundation model. In: ICPTT 2009, Shanghai: ASCE, 2009:1970–79. doi:10.1061/41073(361)208.

Yu LQ. Pipeline construction in landslide area. Oil Gas Storage Transp. 1989;8(6):60–4 **(in Chinese)**.

Zhang KY, Wang Y, Ai YB. Analytical solution to interaction between pipelines and soils under arbitrary loads. Chin J Geotech Eng. 2010;32(8):1189–93.

Zhang ZG, Huang MS. Boundary element model for analysis of the mechanical behavior of existing pipelines subjected to tunneling-induced deformations. Comput Geotech. 2012;46:93–103. doi:10.1016/j.compgeo.2012.06.001.

Zhao XY, Zhao Y. Strain response analysis of oil and gas pipelines subject to lateral landslide. J Natural Disasters. 2014;23(4):250–6 **(in Chinese)**.

Zheng JY, Zhang BJ, Liu PF, et al. Failure analysis and safety evaluation of buried pipeline due to deflection of landslide process. Eng Fail Anal. 2012;25:156–68. doi:10.1016/j.engfailanal.2012.05.011.

13

A review of China's energy consumption structure and outlook based on a long-range energy alternatives modeling tool

Kang-Yin Dong[1,2] · Ren-Jin Sun[1] · Hui Li[1,3] · Hong-Dian Jiang[1]

Abstract China's energy consumption experienced rapid growth over the past three decades, raising great concerns for the future adjustment of China's energy consumption structure. This paper first presents the historical evidence on China's energy consumption by the fuel types and sectors. Then, by establishing a bottom-up accounting framework and using long-range energy alternatives planning energy modeling tool, the future of China's energy consumption structure under three scenarios is forecast. According to the estimates, China's total energy consumption will increase from 3014 million tonnes oil equivalent (Mtoe) in 2015 to 4470 Mtoe in 2040 under the current policies scenario, 4040 Mtoe in 2040 under the moderate policies scenario and 3320 Mtoe in 2040 under the strong policies scenario, respectively, lower than those of the IEA's estimations. In addition, the clean fuels (gas, nuclear and renewables) could be an effective alternative to the conventional fossil fuels (coal and oil) and offer much more potential. Furthermore, the industry sector has much strong reduction potentials than the other sectors. Finally, this paper suggests that the Chinese government should incorporate consideration of adjustment of the energy consumption structure into existing energy policies and measures in the future.

Keywords Energy consumption structure · China-LEAP model · Scenario analysis · Clean fuels · Industrial sector

1 Introduction

Energy is essential for economic and social development and the improvement of life in all the countries (Bilgen 2014). Energy consumption is a key lever to achieve more rapid development (Rennings et al. 2012). Most scholars claimed that there is a strong relationship between China's energy consumption and economic growth (Li et al. 2014; Liao and Wei 2010; Zhang et al. 2011). China's energy consumption has increased dramatically since 2000 and is forecast to keep rising in the next several decades due to continuous economic growth. In the statistics of International Energy Agency (IEA), BP and the National Bureau of Statistics of China (NBS) (IEA 2015; BP 2016; NBS 2015), China's energy consumption increased from 131 million tonnes oil equivalent (Mtoe) (in 1965) to 3014 Mtoe (in 2015), with the GDP increasing from 172 billion yuan (in 1965) to 67,670 billion yuan (in 2015) (Fig. 1).

In China, the primary energy consumption includes five types, i.e., coal, oil, gas, nuclear and renewables, which are mainly used in the four sectors, i.e., transport, industry, building and others (Bilgen 2014). The China's 12th Five Year Plan set an ambitious goal, for which the adjustment of energy consumption structure should make significant progress during the 2011–2015 period. Thus, a number of studies have focused on China's energy consumption structure such as new energy development, energy

✉ Ren-Jin Sun
sunrenjin@cup.edu.cn

[1] School of Business Administration, China University of Petroleum-Beijing, 102249 Beijing, China

[2] Department of Agricultural, Food and Resource Economics, Rutgers, State University of New Jersey, New Brunswick, NJ 08901, USA

[3] Energy Systems Research Center, University of Texas at Arlington, Arlington, TX 76019, USA

Edited by Xiu-Qin Zhu

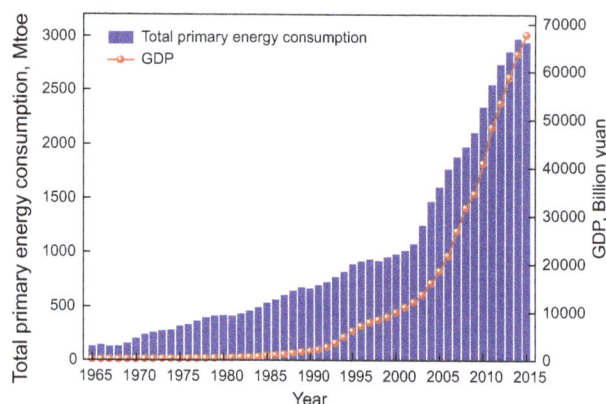

Fig. 1 Total energy consumption and GDP in China. *Data source* BP Statistical Review of World Energy (2016) and NBS China Statistical Yearbook (2015)

conservation and the improvement of energy efficiency (Lin and Wang 2015; Peng et al. 2015; Weidou and Johansson 2004). We need to consider China's energy consumption structure in the past, present and future. What energy plan should be worked out to guarantee the goal of the adjustment of energy consumption structure, as set by the State (China) for the 13th Five Year Plan, is achieved? These issues must be carefully solved before the energy strategies and policies are formulated. Hence, a review of China's energy consumption structure and outlook is valuable and may provide a guideline for policy-making.

The major aims of this paper are (1) to present a comprehensive and systematic investigation of China's energy consumption structure from the point of view of fuel types and sectors, (2) to analyze China's energy consumption structure in the future under three scenarios, by using the bottom-up accounting framework and LEAP (Stockholm Environment Institute, SEI 2014) energy modeling tool and (3) to describe the results and also propose policy suggestions.

2 A review of China's energy consumption structure

2.1 China's energy consumption by fuel types

2.1.1 Coal

Since the foundation of the People's Republic of China, China's energy has primarily come from coal (Govindaraju and Tang 2013). In comparison with oil and natural gas, coal is overwhelmingly abundant and more widely distributed in China. Therefore, coal is the principal energy source in China and it is given a strategic role in the economic growth of the country (Li and Leung 2012). Although the share of coal in the total energy consumption has fallen from 87.1% (in 1965) to 63.7% (in 2015)

(Fig. 2), coal has long been the dominant fuel type in China, soaring from 114 Mtoe (in 1965) to 1920 Mtoe (in 2015) (Fig. 3). Specifically, despite the increased economic growth and a continuous increase in China's coal consumption in 1965–1978, the share of coal in the total energy consumption gradually decreased from 87.1% to 71.3%. After the introduction of the reform and opening-up policy, the consumption of coal in China has increased rapidly from 283 Mtoe (in 1978) to 664 Mtoe (in 1995). Following that, with the adjustment of energy consumption structure in 1995–2001 and the supply of coal being tightly limited in China, the share of coal in the total energy consumption has fallen gradually. However, China's actual coal consumption has increased dramatically since the turn of the millennium, considered as a result of rapid economic development, urbanization, energy shortages, etc. The consumption of coal increased from 679 Mtoe (in 2000) to 1920 Mtoe (in 2015), with an average annual growth rate of 7.2% (Fig. 3).

2.1.2 Oil

In China, oil is one of the important primary energy sources and has a strategic role in promoting economic growth. Since it initiated an economic reform program in 1978, China has witnessed rapid economic growth and an improved living standard (Zheng and Luo 2013). Meanwhile, oil consumption in China increased rapidly from 91 Mtoe (in 1978) to 560 Mtoe (in 2015) (Fig. 3), with an average annual growth of 5.0%. However, China in not rich in oil resources, and its oil reserves account for only 2% of the world oil reserves. Hence, China is highly reliant on the oil imports, and about 61% of oil consumption was imported in 2015 (NBS 2015). Moreover, China became a net importer of crude oil in 1993, and the world's second-largest oil consumer in 2002. Since the early 1990s, with

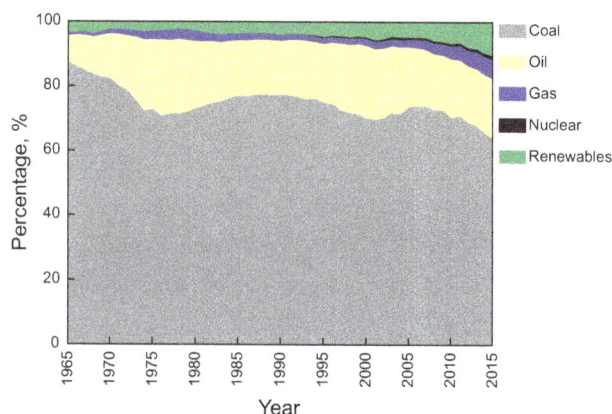

Fig. 2 Percentage of China's energy consumption structure by fuel types, 1965–2015. *Data source* BP Statistical Review of World Energy (2016) and NBS China Statistical Yearbook (2015)

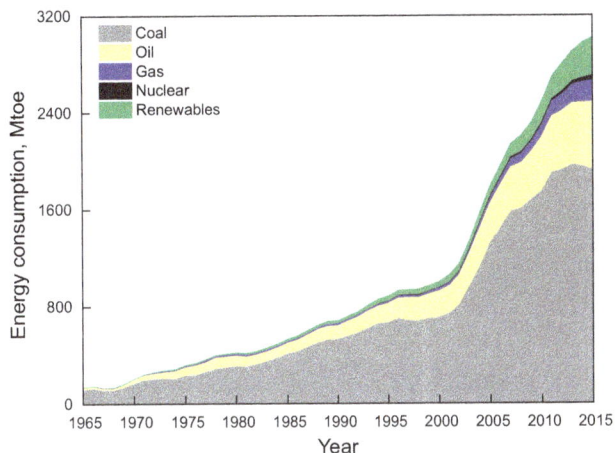

Fig. 3 History of China's energy consumption by different fuel types, 1965–2015. *Data source* BP Statistical Review of World Energy (2016) and NBS China Statistical Yearbook (2015)

increasing economic growth, improved quality of life and the booming development of automobile and aviation industry, total oil consumption in China has increased rapidly. The share of oil in total energy consumed has gradually increased, from 8.3% (in 1965) to 18.6% (in 2015) (Fig. 2).

2.1.3 Gas

Natural gas is an important energy source for power generation, chemical feedstock, residential usage, etc. (Mohr and Evans 2011). China possesses rich natural gas resources, and Chinese authorities have estimated that the TRR (technically recoverable resource) and URR (ultimately recoverable resources) of natural gas are 6.1 trillion cubic meters (tcm) and 37 tcm, respectively (Hou et al. 2015; Zou et al. 2015). Although China is rich in gas resources, the domestic natural gas industry in China developed slowly during its industrialized period. In 2015, natural gas consumption in China was 177.6 Mtoe, accounting for only 5.9% of the domestic energy needs (Fig. 2).

Recently, the government has begun to develop natural gas as a partial substitute for coal, due to the problems from high reliance on coal, such as air pollution, water contamination and greenhouse gas emissions. Natural gas consumption in China has soared since the dawn of the twenty-first century, from 24.7 Mtoe (in 2001) to 177.6 Mtoe (in 2015) (Fig. 3), with an average annual growth rate of 15.1%.

2.1.4 Nuclear

Considering the rising cost of oil and natural gas and the enormous environmental pressure resulting from coal

consumption, nuclear energy is an inevitable strategic option for China (Zhou and Zhang 2010). By the end of 2013, there were 17 nuclear power units in commercial operation in China (CNEA 2014; NRDC 2014), with production of 38.6 Mtoe (in 2015). Nuclear energy still only accounts for 1.3% of China's national energy needs (Fig. 2).

As one of the largest developing countries, the Chinese government began to develop nuclear energy in the 1980s. At the end of 1991, QNNP (Qinshan Nuclear Power Plant) was put into operation, and nuclear energy consumption in China increased slowly during the twentieth century. However, nuclear energy consumption in China has soared and continued to rise fast this century, from 4.0 Mtoe (in 2001) to 38.6 Mtoe (in 2015) (Fig. 3), with an average annual growth rate of 17.6%.

2.1.5 Renewables

The first account of renewable electricity consumption in China dated back to the 1950s, and nationwide development of renewable electricity started at the end of the 1970s and especially after the reform and opening-up in 1978 (Fang 2011). Generally speaking, renewable electricity in China includes hydroelectric, wind, bioenergy, geothermal, solar and other renewables.

The development of renewable electricity in China can be divided into four stages since 1973 (Hao 2013), i.e., starting stage (1973–1992), the preliminary stage of industrialization (1993–2004), fast developing stage (2005–2009) and industrial-scale stage (2010–). Firstly, in the 1970s, Chinese government began to develop renewable electricity in response to energy shortages, which were caused by the World Energy Crisis. Since then, renewable electricity consumption in China increased gradually from 8.3 Mtoe (in 1973) to 29.6 Mtoe (in 1992), with an average annual growth rate of 6.6%. Secondly, to accelerate the development of renewable electricity, some initiatives and laws were made by the government, e.g. China's Agenda 21 (enacted in 1992), Developing Program of New Energy and Renewable electricity during 1996–2010 (in 1995) as well as the Laws of Saving on Energy Resources in China (promulgated in 1998) (NBS 2015). In particular, China's Agenda 21 became effective on March 25, 1994, signaling China's new energy and renewable electricity industry stepped into the preliminary stage of industrialization. By the end of 2004, the renewable electricity consumption in China was 80.9 Mtoe, accounting for 5.5% of the national energy need. Thirdly, with the encouragement and support of the government, China's new energy and renewable electricity industry has made significant breakthroughs in technology during the period of 2005–2009. Furthermore, renewable electricity consumption in China increased rapidly from 90.9 Mtoe (in 2005) to 146.2 Mtoe (in 2009) (Fig. 3), with an average annual growth rate of 10.0%. Lastly, in the industrial-scale stage, renewable electricity consumption in China has continued to rise fast, with the consumption of 319.5 Mtoe, accounting for 10.6% of the domestic energy needs by the end of 2015 (Fig. 2).

In summary, the adjustment of China's energy consumption structure is closely related to the stage of social development. The adjustment of China's energy consumption by fuel type for the study period of 1965–2015 can be summarized as follows: The share of coal in total energy consumed has gradually declined, the share of oil has gradually increased, the share of natural gas has rapidly increased, and the utilization of nuclear and renewables has rapidly increased. Hence, China's energy consumption structure has displayed a diversified trend, and the share of clean energy has gradually increased.

2.2 China's energy consumption by sectors

According to the data of IEA, China's energy consumption can be commonly divided into four energy-consuming end-use sectors, namely transport sector (TS), industry sector (IS), building sector (BS) and other sectors (OS). Notably, although the electricity sector (ES) is not included in the energy-consuming end-use sectors, it is essential to the industrialization and urbanization process, and indeed is an essential element to the transport, industry, building and other sectors of society (Dincer et al. 2012; Marton and Eddy 2012).

2.2.1 Transport sector

China is currently in the development stage of rapid urbanization (Lin and Du 2015), so the transportation sector accounts for a major share of energy consumption in China, with about 8.2% of the total energy consumption in 2013. Furthermore, in the period of 1990–2013, the energy demand of China's transportation sector increased gradually from 34 Mtoe (in 1990) to 249 Mtoe (in 2013), with its share in the total energy consumed rising from 4.0% (in 1990) to 8.2% (in 2013) (Fig. 4).

In addition, the most important component of energy used by the transportation sector is oil, and the oil demand of the transportation sector increased from 71% (in 1990) to 91% (in 2013). In contrast, the share of coal in total transportation energy consumed dramatically decreased from 29% (in 1990) to 1% (in 2013), simply because the change from coal powered (steam) locomotives to diesel and electric trains. Additionally, against the background of China's national energy conservation and emission reduction aims, especially in light of how to reach the emission reduction targets as put forward in the "12th Five Year

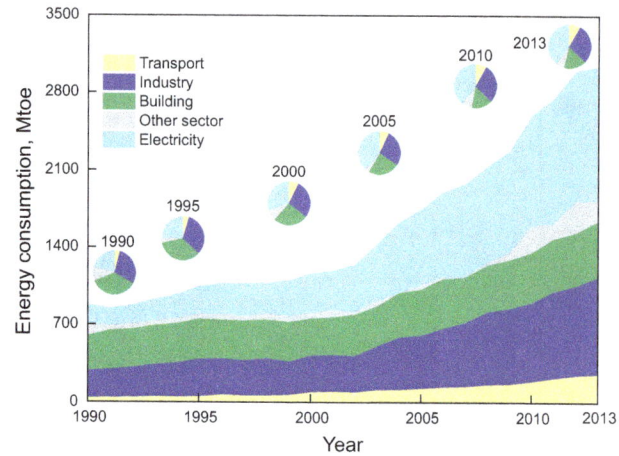

Fig. 4 Historical data of China's energy consumption by different sectors, 1990–2013. *Data source* BP Statistical Review of World Energy (2016), IEA World Energy Outlook (WEO) (2015) and NBS China Statistical Yearbook (2015)

Plan" (Lin and Du 2015), much more clean fuels (e.g. gas and renewables) are used in China's transportation sector. These three fuels (gas, electricity and renewables) accounted for 5%, 2% and 1% of transportation energy need, respectively, in 2013; however, these shares are still low (Table 1).

2.2.2 Industry sector

In the industrialization stage of China, economic growth is dominated by the industry sector (Ouyang and Lin 2015). The importance of China's industry sector is highlighted by its role in providing the massive employment opportunities and raw materials during the industrialization and urbanization process. With the rapid development of China's industry sector, energy consumption by the industry sector has increased rapidly over the past two decades, rising from 245 Mtoe (in 1990) to 881 Mtoe (in 2013), and the share in China total energy consumed rose from 27.9% (in 1990) to 29.0% (in 2013) (Fig. 4). Notably, a decrease in the industrial final energy use occurred during the period of 1995–2000, primarily because ownership restructuring in China's state industry was introduced (CEIC 2014).

Coal has dominated the energy consumption structure in China's industry sector for a long time (Ouyang and Lin 2015); however, the proportion of industrial final energy use decreased continuously from 74% (in 1990) to 54% (in 2013). In the meantime, the energy demand of China's industry sector for oil and gas is not massive, with the shares being 7% and 3% in 2013, respectively. In addition, the proportion of energy consumed in electrical power generation increased dramatically from 17% (in 1990) to 36% (in 2013) (Table 2).

Table 1 Energy consumption by transport sector in China during 1990–2013. *Data source* IEA WEO (2015)

	Energy consumption, Mtoe						Shares, %	
	1990	1995	2000	2005	2010	2013	1990	2013
Total	34	46	88	135	208	249	100	100
Coal	10	7	4	4	3	3	29	1
Oil	24	39	83	128	193	226	71	91
Gas	0	0	0	1	7	13	0	5
Electricity	0	0	1	2	4	5	0	2
Renewables	0	0	0	0	1	2	0	1

Table 2 Energy consumption by industry sector in China during 1990–2013. *Data source* IEA WEO (2015)

	Energy consumption, Mtoe						Shares, %	
	1990	1995	2000	2005	2010	2013	1990	2013
Total	244	339	329	531	725	881	100	100
Coal	180	249	215	331	403	475	74	54
Oil	20	24	30	44	60	60	8	7
Gas	3	4	5	10	17	33	1	4
Electricity	41	62	78	146	245	313	17	36

2.2.3 Building sector

As the second-largest energy consumer after the USA, China is also the second-largest building energy consumer worldwide (IEA 2015). Undoubtedly, energy consumption associated with China's building sector inevitably has displayed an upward trend along with the industrialization and urbanization process, and increased gradually from 314 Mtoe (in 1990) to 506 Mtoe (in 2013); however, its share in the total energy consumed dropped from 35.7% (in 1990) to 16.7% (in 2013) (Fig. 4). In the building sector, energy is used for equipment, providing heating, cooling, lighting and other household needs (Zhang et al. 2015). The energy types in BS have displayed a diversified trend. In the period of 1990–2013, the shares of oil, gas and electricity grew from 2% to 9%, 1% to 7% and 2% to 26%, respectively. In contrast, the shares of coal and renewables decreased from 29% to 15% and 66% to 43%, respectively (Table 3).

2.2.4 Other sectors

Losses occur when the efficiency of a device or process deviates from the efficiency that would occur if the device or process were ideal; and the low efficient heating system leads to enormous energy loss (Bilgen 2014). Generally, the energy consumption by the other sectors mainly refers to the energy losses in all the sectors (Dincer et al. 2012). Undoubtedly, with the energy consumption increasing, energy losses in all the sectors also increased rapidly over the past two decades, growing from 104 Mtoe (in 1990) to 184 Mtoe (in 2013), with the percentage in the total energy consumed of around 6.1% (Fig. 4).

Specifically, oil plays a key role in the energy losses in the sectors, and its share grew gradually from 41% (in 1990) to 50% (in 2013). In the meantime, the energy losses of electricity and renewables have inevitably displayed an upward trend, with their shares growing from 6% to 15% and 0% to 4%, respectively. Notably, a decrease in the share of coal occurred during the period of 1990–2013, primarily because of the accelerating cleaning process of coal (Table 4).

2.2.5 Electricity sector

Due to the large population, the rapid economic development, as well as the process of urbanization and industrialization, China has a much higher demand for electricity than before (Shiu and Lam 2004; Yuan et al. 2007; Zhou et al. 2015). Energy consumption by China's power sector has soared and continued to rise fast since 1990, from 181 Mtoe (in 1990) to 1218 Mtoe (in 2013), with its share in domestic total energy consumed rising from 20.6% (in 1990) to 40.1% (in 2013) (Fig. 4).

Coal is overwhelmingly abundant and thermal power accounts for a large proportion in different power generation methods in China, so it has always been the main source of power generation. In 2013, it accounted for 86% of the power generation energy. Additionally, with the development of micro grids, smart grids and smart energy-related concepts, techniques and systems, as well as the penetration of other energies, some unconventional power

Table 3 Energy consumption by the building sector in China during 1990–2013. *Data source* IEA WEO (2015)

	Energy consumption, Mtoe						Shares, %	
	1990	1995	2000	2005	2010	2013	1990	2013
Total	303	315	305	353	400	506	100	100
Coal	88	76	54	64	61	77	29	15
Oil	7	17	19	28	35	43	2	9
Gas	2	2	4	10	24	36	1	7
Electricity	6	16	25	48	78	132	2	26
Renewables	200	204	203	203	202	218	66	43

Table 4 Energy consumption by the other sectors in China during 1990–2013. *Data source* IEA WEO (2015)

	Energy consumption, Mtoe						Shares, %	
	1990	1995	2000	2005	2010	2013	1990	2013
Total	83	91	94	145	195	184	100	100
Coal	40	34	31	45	51	47	48	26
Oil	34	45	48	71	94	93	41	50
Gas	4	3	4	7	10	9	5	5
Electricity	5	8	10	19	30	28	6	15
Renewables	0	1	1	3	10	7	0	4

Table 5 Energy consumption of power generation in China during 1990–2013. *Data source* IEA WEO (2015)

	Energy consumption, Mtoe						Shares, %	
	1990	1995	2000	2005	2010	2013	1990	2013
Total	181	275	360	682	1004	1218	100	100
Coal	153	241	314	605	884	1047	85	86
Oil	16	13	16	18	9	5	8	0
Gas	1	1	5	7	20	27	1	2
Nuclear	0	3	4	14	19	29	0	2
Renewables	11	17	21	38	72	110	6	9

generation methods have been implemented in China, such as the renewable electricity power (wind power, solar photovoltaic and hydroelectric power, etc.) (Chang et al. 2015; Yang and Shen 2013). Therefore, the share of renewables in total power generation energy consumed gradually increased from 6% (1990) to 9% (2013) (Table 5).

3 Methodology

3.1 China-LEAP model framework

In this methodology, the LEAP model is used as an energy accounting modeling tool to calculate China's energy consumption. LEAP was developed by the Stockholm Environmental Institute (SEI-US) (Schnaars 1987; Heaps 2002, 2012). Specifically, LEAP is a scenario-based energy environment modeling tool for energy policy analysis and

can be used to track energy consumption, production and resource extraction in all the sectors (Chontanawat et al. 2014; Kemausuor et al. 2015). The aim of the model is to analyze the effects of multiple factors on energy consumption under different scenarios in an objective, quantitative and comprehensive way, to provide a reference for the policy makers and investors (Huang et al. 2011). As this paper is to study China's energy consumption structure by the fuel types and sectors, the LEAP model was chosen because (1) it allows users to build energy forecast systems based on existing energy demand and supply data, to prepare different long-run scenarios and to compare results with different scenarios (Ates 2015), (2) it has low initial data requirement and (3) it is free to use for developing country researchers and government agencies.

As Fig. 5 shows, a bottom-up accounting framework is established for China's-LEAP model to estimate China's energy consumption structure. From the perspective of the key drivers of energy use, seven key factors are considered

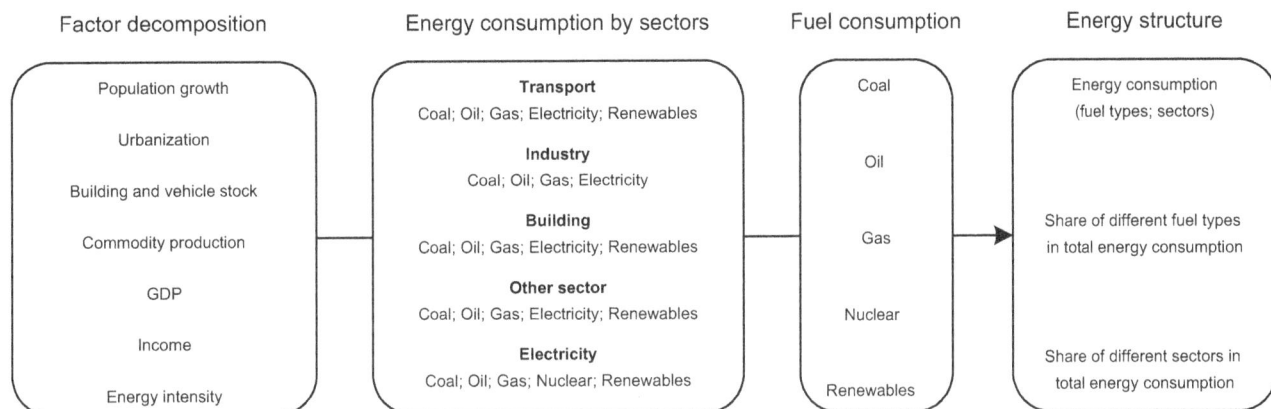

Fig. 5 Research framework for China-LEAP model

in the analysis, namely population growth, urbanization, building and vehicle stock, commodity production, GDP, income and energy intensity. In terms of the energy consumption sectors, they are mainly distributed in TS, IS, BS, OS and ES. Accordingly, fuel consumption considered in this study includes five types, namely coal, oil, gas, nuclear and renewables. Considering the accessibility and applicability of data information, Eqs. (1), (2) and (3) describe the calculation process employed in this study.

$$EC_r = \sum_i EC_{i,j} = \sum_j EC_{j,r} \qquad (1)$$

$$EC_{i,j} = \sum_j EC_{i,j,r} \qquad (2)$$

$$EC_{j,r} = \sum_i EC_{i,j,r} \qquad (3)$$

$$\alpha_r = \frac{EC_{i,r}}{EC_R} \qquad (4)$$

$$\beta_r = \frac{EC_{j,r}}{EC_r} \qquad (5)$$

where, EC_r is the total primary energy consumption in year r (Mtoe); $EC_{i,r}$ is the total energy consumption of type i fuel in year r (Mtoe); $EC_{j,r}$ is the total energy consumption of the sector j in year r (Mtoe); $EC_{i,j,r}$ is the energy consumption of type i fuel in the sector j in year r (Mtoe); α_r is the share of type i fuel in the total energy consumption in year r (%); β_r is the share of the sector j in the total energy consumption in year r (%).

The LEAP model consists of three blocks of programs, i.e., database, aggregation and scenarios (Shin et al. 2005). The LEAP model is based on exogenous input of the main parameters and factors (Perwez et al. 2015). In the LEAP model, the data set consists of various factors such as population growth, urbanization, GDP and energy intensity. To ensure the reliability of data, we have compiled data from (1) the available published literature, (2) the national official reports released by China's authorities such as NBS and (3) the international institutes' reports

such as the WEO from IEA, the Statistical Review of World Energy from BP and the Annual Energy Outlook (AEO) from the US Energy Information Administration (EIA), while the real GDP data employed in our analysis are obtained from the NBS and the energy consumption data are from IEA, BP, EIA, etc. In addition, to ensure comparability after the data collection, data expressed in the various units are converted to the same unit in this paper; for example, energy consumption and GDP are expressed in Mtoe and billion yuan, respectively. Notably, in order to calculate the future energy consumption, the LEAP model uses the 2013 energy consumption as the baseline, and the energy consumption projection is done from 2015 to 2040.

3.2 Scenario examination

3.2.1 Current policies scenario (CPS)

The CPS follows those policies and implemented measures which have been formally adopted; however, the policy adjustment effect on the CPS is limited. Since the CPS reflects what is expected to happen in terms of the policies and implemented measures, it is used as the reference scenario for evaluating China's energy consumption structure. Additionally, the assumptions of the key factors adopted for the CPS specified by the different sectors are shown in Table 6.

3.2.2 Moderate policies scenario (MPS)

The MPS assumes that policies and implemented measures have begun to affect China's energy markets, together with the successful improvement of energy efficiency. However, the relevant policies and specific measures need to be put into effect in this scenario. The assumptions of the key factors for the MPS specified by the different sectors are given in Table 7.

Table 6 Key assumptions of the CPS for all the sectors

Sectors	Key assumptions
TS	Efficiency improvements in fuel economy are limited; Subsidies for hybrid and electric vehicles; Promotion of fuel-efficient cars; Cap on passenger light-duty vehicles (PLDV) sales in some cities
IS	Small plant closures and phasing out of outdated production capacity; Mandatory adoption of coke dry quenching and top-pressure turbines
BS	Application for construction conservation design standards
OS	Compared with 2013, the growth rate of energy losses will be below 20% in 2020
ES	40 GW of new nuclear plants by 2050; Reaching 290 GW of installed hydro-capacity, 100 GW to wind, 35 GW to solar by 2015

Table 7 Key assumptions of the MPS for all the sectors

Sectors	Key assumptions
TS	Fuel economy target for PLDVs: 6.9 L/100 km by 2015, 5.0 L/100 km by 2020
IS	Contain the expansion of energy-intensive industries; Implementation of CO_2 pricing since 2020; Reduction in industrial energy intensity by 21% during 2011–2015
BS	Share of energy efficient building is 30% by 2020; Implementation of energy price policy, such as reform heating price; Introduction of energy standards; All fossil-fuel subsidies are phased out by 2020
OS	Compared with 2013, the growth rate of energy losses will be below 10% in 2020
ES	58 GW of nuclear capacity, 200 GW of wind, 100 GW of solar PV and 30 GW of bioenergy by 2020; Implementation of CO_2 pricing after 2020

Table 8 Key assumptions of the SPS for all the sectors

Sectors	Key assumptions
TS	Compared with 2010, 55% efficiency improvements by 2040 and support for the use of biofuels; Enhanced support to alternative fuels
IS	Introducing the CO_2 pricing by 2020; Enhanced energy efficiency standards; Support the introduction of CCS
BS	95% of new building achieve saving of 55%–65% in space heating compared to 1980
OS	Compared with 2013, the growth rate of energy losses will be below 5% in 2020
ES	Higher CO_2 pricing; Enhanced support for renewables; Continued support to nuclear capacity additions post-2020; Deployment of CCS from around 2020

3.2.3 Strong policies scenario (SPS)

Compared with the other scenarios, the SPS sets out a much more aggressive energy pathway, which is consistent with stronger energy efficiency policies. Furthermore, this scenario assumes more vigorous policy actions are to be implemented after 2020. The assumptions of the key factors for the SPS specified by the different sectors are shown in Table 8.

4 Results and discussion

4.1 Results

Based on the three scenarios introduced in Sect. 3.2, the future energy consumption structure of China is established, as presented in the energy consumption by the

sectors, energy consumption by the fuel types and total energy consumption.

4.1.1 Energy consumption by sectors

The results of China's energy consumption by different sectors under the three scenarios are shown in Figs. 6, 7 and 8, respectively. Figure 6 presents the energy consumption of different sectors under the CPS in 2010–2040. The energy consumption of the different sectors (TS, IS, BS, OS and ES) under the CPS will maintain the rising trend, reaching 600, 1253, 625, 393 and 2606 Mtoe, respectively, in 2040. However, as shown in Table 9, the growth rates of energy consumption by the different sectors under CPS will drop during the 2000–2040 period. For example, the growth of energy consumption by the TS under the CPS will be 10.0%, 6.2%, 4.6% and 1.3% during the ten-year periods from 2000 to 2040.

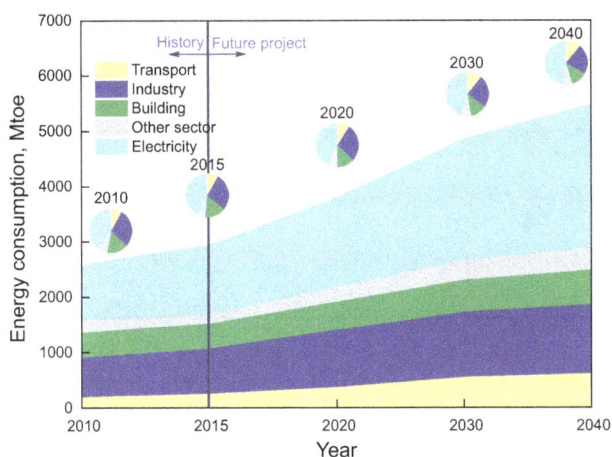

Fig. 6 Energy consumption of different sectors under the CPS in 2010–2040

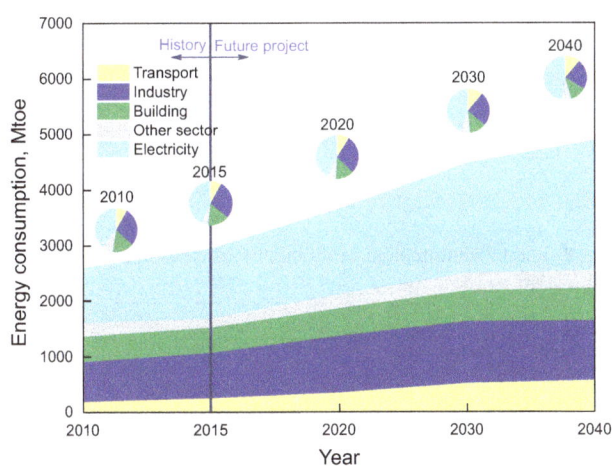

Fig. 7 Energy consumption of different sectors under the MPS in 2010–2040

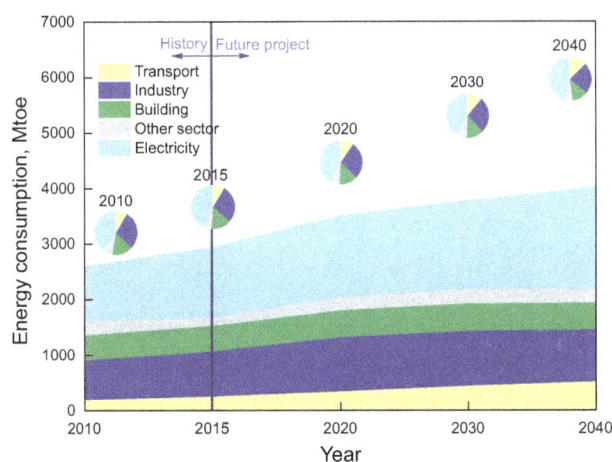

Fig. 8 Energy consumption of different sectors under the SPS in 2010–2040

As shown in Fig. 6, IS takes the dominating share of the total primary energy consumption under the CPS all the way through 2040, although the share will decrease from 27.5% in 2010 to 22.9% in 2040. Meanwhile, BS and OS have inevitably displayed a downward trend, with their shares decreasing from 17.4% (in 2010) to 11.4% (in 2040) and 9.3% (in 2010) to 7.2% (in 2040), respectively. Notably, although the energy consumption of the TS is not high, its share will increase from 7.1% in 2010 to 11.0% in 2040.

Figure 7 presents the energy consumption of different sectors under the MPS in 2010–2040. Compared with the CPS, energy consumption in the different sectors (TS, IS, BS, OS and ES) under the MPS will decrease, reaching 559, 1072, 580, 333 and 2337 Mtoe, respectively, in 2040. Moreover, as shown in Table 9, the growth of energy consumption under MPS will be lower than that under CPS. Notably, the growth of the energy consumption in the IS under the MPS will be −0.4% during the 2030–2040 period. The main reason is that policies and implemented measures under the MPS will have begun to affect China's energy markets, together with the successful improvement of energy efficiency.

In terms of the shares of the total primary energy consumption by different sectors under the MPS, IS still comprises the dominant share of the total primary energy consumption, maintaining its level at roughly 43.6% of the total primary energy consumption in 2040. BS is the second-largest energy-consuming end-use sector, with its share maintaining at around 22.8% of the total primary energy consumption in 2040. The share of the total primary energy consumption by the TS will increase from 14.0% in 2012 to 22.0% in 2040. Although the share of the total primary energy consumption by the OS under the MPS will keep rising, it will account for about 13.1% of the total primary energy consumption in 2040, lower than that under the CPS.

Figure 8 presents the energy consumption of different sectors under the SPS in 2010–2040. Compared with the other scenarios, the policies and implemented measures under the SPS will be more aggressive and effective. Energy consumption by the different sectors (TS, IS, BS, OS and ES) under the SPS will be lower than that of the other two scenarios, reaching 5.3, 937, 490, 243 and 1850 Mtoe, respectively, in 2040. Furthermore, as shown in Table 9, the growth rates of energy consumption under the SPS will also be lower than that under the CPS and MPS. Specifically, the growth rates of energy consumption in the IS, BS and OS under the SPS will be negative during the 2030–2040 period.

Additionally, as shown in Fig. 8, compared with the MPS, the shares of the total primary energy consumption by the IS and BS under the SPS will change slightly,

Table 9 Growth rates of energy consumption in the different sectors under the three scenarios in 2000–2040 (%)

	CPS				MPS				SPS			
	2000–10	2010–20	2020–30	2030–40	2000–10	2010–20	2020–30	2030–40	2000–10	2010–20	2020–30	2030–40
TS	10.0	6.2	4.6	1.3	10.0	5.8	4.4	1	10.0	5.2	2.9	1.9
IS	9.2	4.1	1.4	0.7	9.2	3.9	1.0	−0.4	9.2	3.4	0.1	−0.6
BS	3.1	2.6	1.5	0.9	3.1	2.4	1.2	0.6	3.1	2.1	0.4	−0.2
OS	8.4	3.6	3.2	1.1	8.4	2.7	2.6	0.8	8.4	2.1	0.9	−0.5
ES	12.1	5.5	3.5	1.9	12.1	5.0	2.8	1.8	12.1	4.4	1.0	1.5

reaching 43.1% and 22.5% in 2040. Meanwhile, with more effective measures implemented, the share of the total primary energy consumption by the TS under the SPS will increase dramatically from 14.0% in 2012 to 22.0% in 2040. Moreover, the share of the total primary energy consumption by the OS under the MPS will decrease significantly, lower than that under the other scenarios, accounting for about 11.2% of the total primary energy consumption in 2040.

4.1.2 Energy consumption by fuel types

China's energy consumption by fuel type under the three scenarios is shown in Figs. 9, 10 and 11. Figure 9 shows the energy consumption of different fuel types (coal, oil, gas, nuclear and renewables) under the CPS in 2010–2040. The forecast result in Fig. 9 indicates that the energy consumption of the different fuel types under the CPS will keep rising, reaching 2362, 785, 487, 284 and 549 Mtoe, respectively, in 2040. Generally, the growth rates of the energy consumption by the different fuel types under the CPS will be decreasing during the every ten-year period from 2000 to 2040 (Table 10). Specifically, the growth rate of the coal consumption under the CPS will decrease dramatically during the same period, namely 10.1%, 1.8%, 1.7% and 0.8%.

Figure 9 also shows that coal makes up the dominant share of the total energy consumption under the CPS all the way to 2040, although the share will decrease from 68.8% in 2010 to 52.9% in 2040. The share of oil consumed in the total energy consumption will hover around 17.8%. Meanwhile, the shares of the total consumption of gas, nuclear and renewables under CPS will keep rising, reaching 10.9%, 6.4% and 12.3%, respectively, in 2040.

Figure 10 shows the energy consumption of different fuel types under the MPS in 2010–2040. Compared with the CPS, energy consumption in the terms of coal and oil under the MPS will be lower, reaching 1867 and 687 Mtoe, respectively, in 2040. Meanwhile, as shown in Table 10, the growth rates of the coal and oil consumption under the MPS will be lower than that of the CPS. Furthermore, the consumption of gas, nuclear and renewables under the

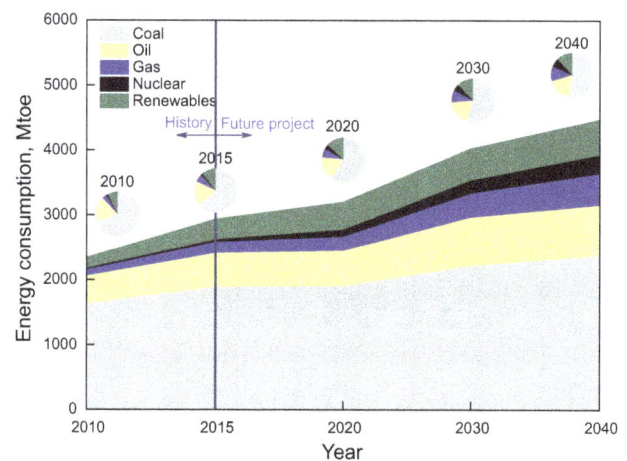

Fig. 9 Energy consumption of different fuel types under the CPS in 2010–2040

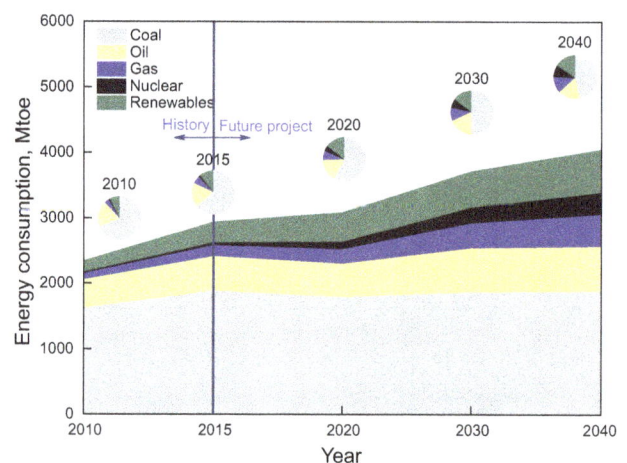

Fig. 10 Energy consumption of different fuel types under the MPS in 2010–2040

MPS will be higher than that of the CPS, reaching 502, 335 and 654 Mtoe, respectively, in 2040 (Fig. 10). In addition, the growth rate of the gas, nuclear and renewables consumption under the MPS will also be higher than that of the CPS, reaching 3.1%, 4.0% and 1.8%, respectively, in 2040, offering more potential than the conventional fossil fuels (coal and oil).

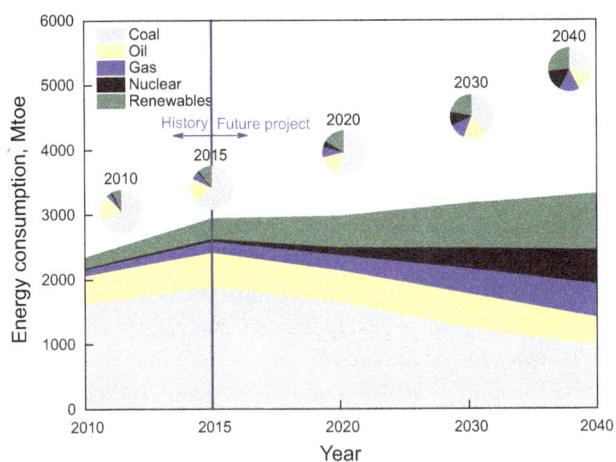

Fig. 11 Energy consumption of different fuel types under the SPS in 2010–2040

Figure 10 also shows that although coal and oil account for 87.6% and 63.1% of the total energy consumption in 2010 and 2040, respectively, their shares under the MPS will keep a decreasing trend. In contrast, gas, nuclear and renewables would be a major alternative to coal and oil, with their shares increasing to 12.4%, 8.3% and 16.2%, respectively, in 2040.

Figure 11 shows the energy consumption of different fuel types under the SPS in 2010–2040. Unlike the other two scenarios, Fig. 11 shows a downward trend both in the coal and in oil consumption under the SPS over the period from 2015 to 2040, reaching 958 and 445 Mtoe, respectively, in 2040. Furthermore, the growth rates of the coal and oil consumption under the SPS will be lower than that of the CPS and MPS. The growth rate of coal consumption under the SPS will be 10.1%, 0.2%, −3.1% and −2.7% during the ten-year periods from 2000 to 2040 (Table 10). Figure 11 shows an upward trend in the gas, nuclear and renewables consumption under the SPS over the period from 2015 to 2040, reaching 510, 529 and 879 Mtoe, respectively, in 2040. Table 10 shows that the growth rate of this section of energy consumption under the SPS will be lower than that of the other scenarios.

Figure 11 also shows that coal and oil no longer comprise the dominant share of the total energy consumption under the SPS in 2040, with their total shares standing at 45.0%. Moreover, with more aggressive energy efficiency policies and more vigorous policy action to be implemented, clean fuels (gas, nuclear and renewables) will keep an increasing trend, with their shares rising to 13.4%, 15.4% and 26.5%, respectively, in 2040.

4.1.3 Total energy consumption

Figure 12 presents the total energy consumption in China under the different scenarios in 2010–2040. It can be seen

Table 10 Growth rates of the energy consumption by different fuel types under the three scenarios in 2000–2040 (%)

	CPS				MPS				SPS			
	2000–10	2010–20	2020–30	2030–40	2000–10	2010–20	2020–30	2030–40	2000–10	2010–20	2020–30	2030–40
Coal	10.1	1.8	1.7	0.8	10.1	1.1	0.5	0.1	10.1	0.2	−3.1	−2.7
Oil	7.8	2.5	2.1	0.5	7.8	1.8	1.6	0.1	7.8	1.1	1.2	−2.1
Gas	17.8	9.7	6.0	3.0	17.8	9.9	6.1	3.1	17.8	10.6	6.3	3.2
Nuclear	17.4	22.8	7.2	3.9	17.4	23.3	8.9	4.0	17.4	24.6	11.1	5.8
Renewables	14.8	10.3	1.5	1.2	14.8	11.0	2.3	1.8	14.8	11.9	4.0	2.7

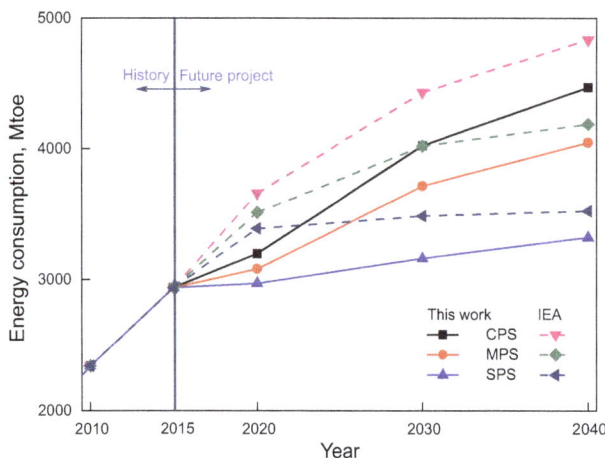

Fig. 12 Total energy consumption under different scenarios in 2010–2040

that by 2040, the total energy consumption in China will still be increasing under most of the scenarios. Total energy consumption will increase from 3016 Mtoe in 2015 to 4467 Mtoe in 2040 under the CPS. Compared with the CPS, it will decrease by 422 Mtoe to 4045 Mtoe in 2040 under the MPS, while it will decrease by 1146 Mtoe to 3321 Mtoe in 2040 under the SPS. The average annual growth rates of the total energy consumption in China under the different scenarios (CPS, MPS and SPS) will be 1.6%, 1.2% and 0.4%, respectively, over the period from 2015 to 2040. However, our estimates in this work are all lower than the IEA's ones.

4.2 Discussion

Prediction results of China's energy consumption by the transport sector (TS) in the present study and previous studies are shown in Table 11 for comparison purposes. As Table 11 shows, Ou et al. (2010) estimated Chinese TS energy consumption in 2020, 2030 and 2040 to be 375, 479 and 531 Mtoe, respectively, under the business as usual (BAU) scenario; and to be 350, 427 and 426 Mtoe, respectively, under the promoting electric vehicles (PEV) scenario. Compared with the values reported by Ou et al.

Table 11 Chinese TS energy consumption estimated in the previous studies (Mtoe) (Ou et al. 2010; IEA 2015) and in this work

	Ou et al. (2010)[a]		IEA (2015)			This study		
	BAU	PEV	CPS	NPS	450	CPS	MPS	SPS
2020	375	350	348	346	336	357	345	329
2030	479	427	482	465	414	535	509	425
2040	531	426	564	520	428	600	559	503

[a] The figures in BAU and PEV are estimated through the results calculated by Ou et al. (2010)

(2010), our results in this work are initially conservative for the first decade under the scenario and then become more optimistic. In addition, even though the trends of the IEA's predictions for the various scenarios were in line with those of our predictions; however, our predicted values were lower than those of the IEA for all three scenarios. The discrepancies between the IEA's predictions and ours could be attributed to the following three reasons. Firstly, the IEA's predictions are weighted heavily toward the 12th Five year Plan set in 2011, which underestimated the rigorous measures undertaken by the Chinese government for improving energy efficiency and reducing environmental impacts achieved in 2015. Secondly, China is not one of the IEA's 28-member countries. Hence, the IEA's data sources were likely obtained through third parties and might not reflect China's current situation. Thirdly, there are different assumptions used in this work and the IEA's calculations for the three scenarios.

Regarding the reduction potential of the energy consumption, there will be a maximum reduction of 1146 Mtoe of energy consumption under the SPS in 2040 compared with the CPS, and the SPS has much more reduction potentials. However, compared with much more aggressive energy efficiency policies under the SPS, the ongoing energy policies and measures in China show lower availabilities, efficiency and potentials. In this regard, some much more effective and advisable policies and measures should be put forward by the Chinese government.

When the results shown in Figs. 6, 7, 8, 9, 10, 11 and 12 are studied together, it can be concluded that the clean fuels (gas, nuclear and renewables) could be major alternatives to the conventional fossil fuels (coal and oil) and offer much more potential, accounting for 57.8% of the total energy consumption in 2040 under SPS, reaching 1918 Mtoe. In terms of the energy consumption by the sectors, IS has much more reduction potential than the other sectors. Compared with the CPS, the energy consumption will be reduced by 316–937 Mtoe in 2040 under the SPS. Hence, policy makers should pay attention to the development of Chinese clean fuels and to the energy reduction in the IS.

5 Conclusions and policy implications

5.1 Policy implications

From the comprehensive analysis and discussion carried out above, we obtain the following policy implications:

First, it will be necessary to incorporate consideration of the adjustment of energy consumption structure into the industrialization and urbanization process. Currently, compared with energy consumption, the government gives high priority to energy production and ignores the importance of adjustment of energy consumption structure. Hence, it is essential that in future the government should consider energy consumption structure from a strategic height and long-term perspective, adjusting the energy development strategy according to the adjustment of energy consumption structure.

Second, it is critical to establish and perfect the related policies and measures. This is the most important measure, chiefly because the China's energy market has experienced a lack of more aggressive energy efficiency policies and more vigorous policy actions for decades. For example, the resource tax in China's energy market has been comprehensively introduced. However, the stringency level of the resource tax has generally lagged behind that of developed countries. Furthermore, it is difficult to implement some energy market policies and measures in China.

Third, more focus should be laid on the development of clean fuels and energy reduction in the IS. In order to promote the development of clean fuels, the policy makers should improve existing pricing and subsidy policies. Besides, according to the adjustment of the energy consumption structure, the government should reasonably adjust the shares between the clean fuels and conventional fossil fuels. Furthermore, the government and enterprises should pay attention to the structural energy saving especially of the IS.

5.2 Conclusions

In this study, to forecast China's energy consumption structure in the future, we first present a comprehensive and systematic review of the development status of China's energy consumption structure by the fuel types from 1990 to 2015 and the sectors from 1990 to 2013. Then, under the CPS, MPS and SPS, a bottom-up accounting framework was developed and the LEAP model was used to forecast the China's energy consumption structure from 2015 to 2040. At last, the suggestions in the four aspects are proposed to further promote the adjustment of the China's energy consumption structure. The main conclusions drawn from this study are summarized as follows:

(1) From the perspective of energy consumption by the sectors, IS takes the dominant share of the total primary energy consumption under the three scenarios all the way to 2040 and the share will display an upward trend. In addition, compared with other energy-consuming end-use sectors, IS will offer more energy reduction potential. The energy consumption of the IS in 2040 under the SPS will be lower by 316 Mtoe than that under the CPS.

(2) From the perspective of energy consumption by the fuel types, coal and oil take the dominant share of the total energy consumption under the three scenarios all the way to 2040; however, the share will maintain a decreasing trend. In contrast, clean fuels will reach 1918 Mtoe, accounting for 57.8% of the total energy consumption in 2040 under the SPS. Therefore, the clean fuels will offer more development potential than conventional fossil fuels in the future.

(3) From the perspective of the total energy consumption, China's total energy consumption will increase continuously in all scenarios from 2015 to 2040. Specifically, the total energy consumption will increase from 3016 Mtoe in 2015 to 4467 Mtoe in 2040 under the CPS, 4045 Mtoe in 2040 under the CPS and 3321 Mtoe in 2040 under the CPS, respectively. It is notable that our estimates are all lower than the IEA's.

(4) From the perspective of the existing policies and measures, by analyzing the results under the three scenarios, we can find that the effect of the policies and measures under the CPS is poorer than those under the MPS and SPS, which means that existing policies and measures show lower availabilities, efficiency and potentials.

In summary, compared with that of the developed countries, China's energy consumption structure still needs the further improvement and adjustment. As a result, the Chinese government should incorporate consideration of the adjustment of energy consumption structure into existing energy policies and measures.

Acknowledgements This study is supported by National Natural Science Foundation (No. 71273277) and National Social Science Foundation (No. 13&ZD159). The authors appreciate the helpful reviews and comments by the anonymous reviewers.

References

Ates SA. Energy efficiency and CO_2 mitigation potential of the Turkish iron and steel industry using the LEAP (long-range energy alternatives planning) system. Energy. 2015;90:417–28. doi:10.1016/j.energy.2015.07.059.

Bilgen S. Structure and environmental impact of global energy consumption. Renew Sustain Energy Rev. 2014;38:890–902. doi:10.1016/j.rser.2014.07.004.

BP Statistical Review of World Energy 2016. http://www.bp.com/en/global/corporate/energy-economics/statistical-review-of-world-energy.html. Accessed 20 Jun 2016.

CEIC. China economic and industry data database, 2014. http://www.ceicdata.com/en/countries/china (**in Chinese**).

Chang K, Xue F, Yang W. Review of the basic characteristics and technical progress of smart grids in China. Autom Electric Power

Syst. 2015;33:10–5. doi:10.3321/j.issn:1000-1026.2009.17.003 (**in Chinese**).

Chontanawat J, Wiboonchutikula P, Buddhivanich A. Decomposition analysis of the change of energy intensity of manufacturing industries in Thailand. Energy. 2014;77:171–82. doi:10.1016/j.energy.2014.05.111.

CNEA. National nuclear security and operation in 2013. 2014. http://www.china-nea.cn/html/2014-02/28741.html. Accessed 11 Feb 2014 (**in Chinese**).

Dincer I, Rosen MA. Exergy: energy, environment and sustainable development. Newnes: Elsevier; 2012.

Fang Y. Economic welfare impacts from renewable energy consumption: the China experience. Renew Sustain Energy Rev. 2011;15:5120–8. doi:10.1016/j.rser.2011.07.044.

Govindaraju VC, Tang CF. The dynamic links between CO_2 emissions, economic growth and coal consumption in China and India. Appl Energy. 2013;104:310–8. doi:10.1016/j.apenergy.2012.10.042.

Hao XD. A study of the Sino-US energy consumption structures. Wuhan: Wuhan University; 2013 (**in Chinese**).

Heaps C. Long-range energy alternatives planning (Leap) system. [Software Version 2011.0043]. Somerville: Stockholm Environment Institute; 2012.

Heaps C. Integrated energy-environment modelling and LEAP. SEI, 2002. http://www.energycommunity.org/default.asp.

Hou Z, Xie H, Zhou H, et al. Unconventional gas resources in China. Environ Earth Sci. 2015;73:5785–9. doi:10.1007/s12665-015-4393-8.

Huang Y, Bor YJ, Peng CY. The long-term forecast of Taiwan's energy supply and demand: LEAP model application. Energy Policy. 2011;39:6790–803. doi:10.1016/j.enpol.2010.10.023.

IEA. World Energy Outlook 2015. http://www.worldenergyoutlook.org/ (2015). Accessed 20 Jan 2016.

Kemausuor F, Nygaard I, Mackenzie G. Prospects for bioenergy use in Ghana using long-range energy alternatives planning model. Energy. 2015;93:672–82. doi:10.1016/j.energy.2015.08.104.

Liao H, Wei YM. China's energy consumption: a perspective from Divisia aggregation approach. Energy. 2010;35:28–34. doi:10.1016/j.energy.2009.08.023.

Li F, Song Z, Liu W. China's energy consumption under the global economic crisis: decomposition and sectoral analysis. Energy Policy. 2014;64:193–202. doi:10.1016/j.enpol.2013.09.014.

Lin B, Wang A. Estimating energy conservation potential in China's commercial sector. Energy. 2015;82:147–56. doi:10.1016/j.energy.2015.01.021.

Li R, Leung GC. Coal consumption and economic growth in China. Energy Policy. 2012;40:438–43. doi:10.1016/j.enpol.2011.10.034.

Lin B, Du Z. How China's urbanization impacts transport energy consumption in the face of income disparity. Renew Sustain Energy Rev. 2015;52:1693–701. doi:10.1016/j.rser.2015.08.006.

Marton K, Eddy WF. Effective tracking of building energy use: improving the commercial buildings and residential energy consumption surveys. Washington: National Academies Press; 2012.

Mohr S, Evans G. Long term forecasting of natural gas production. Energy Policy. 2011;39:5550–60. doi:10.1016/j.enpol.2011.04.066.

NBS. China statistical yearbook 2015. http://www.stats.gov.cn/tjsj/ndsj/2015/indexeh.htm.

NRDC. Effective regulation of nuclear energy development, avoid repeating the mistakes of Fukushima, 2014. (**in Chinese**).

Ouyang X, Lin B. An analysis of the driving forces of energy-related carbon dioxide emissions in China's industrial sector. Renew Sustain Energy Rev. 2015;45:838–49. doi:10.1016/j.rser.2015.02.030.

Ou XM, Zhang XL, Chang SY. Scenario analysis on alternative fuel/vehicle for China's future road transport: life-cycle energy demand and GHG emissions. Energy Policy. 2010;38:3943–56. doi:10.1016/j.enpol.2010.03.018.

Peng L, Zeng X, Wang Y, et al. Analysis of energy efficiency and carbon dioxide reduction in the Chinese pulp and paper industry. Energy Policy. 2015;80:65–75. doi:10.1016/j.enpol.2015.01.028.

Perwez U, Sohail A, Hassan SF, et al. The long-term forecast of Pakistan's electricity supply and demand: an application of long range energy alternatives planning. Energy. 2015;93:2423–35. doi:10.1016/j.energy.2015.10.103.

Rennings K, Brohmann B, Nentwich J, et al. Sustainable energy consumption in residential buildings. New York: Springer; 2012.

Schnaars SP. How to develop and use scenarios. Long Range Plan. 1987;20:105–14. doi:10.1016/0024-6301(87)90038-0.

Shin HC, Park JW, Kim HS, et al. Environmental and economic assessment of landfill gas electricity generation in Korea using LEAP model. Energy Policy. 2005;33:1261–70. doi:10.1016/j.enpol.2003.12.002.

Shiu A, Lam PL. Electricity consumption and economic growth in China. Energy Policy. 2004;32:47–54. doi:10.1016/s0301-4215(02)00250-1.

Stockholm Environment Institute (SEI). Long range energy alternatives planning system 2014. Joint IEA–IEF–OPEC report. http://www.opec.org/opec_web/en/publications. Accessed 20 Oct 2015.

Weidou N, Johansson TB. Energy for sustainable development in China. Energy Policy. 2004;32:1225–9. doi:10.1016/s0301-4215(03)00086-7.

Yang S, Shen C. A review of electric load classification in smart grid environment. Renew Sustain Energy Rev. 2013;24:103–10. doi:10.1016/j.rser.2013.03.023.

Yuan J, Zhao C, Yu S, et al. Electricity consumption and economic growth in China: cointegration and co-feature analysis. Energy Econ. 2007;29:1179–91. doi:10.1016/j.eneco.2006.09.005.

Zhang M, Song Y, Yao L. Exploring commercial sector building energy consumption in China. Nat Hazards. 2015;75:2673–82. doi:10.1007/s11069-014-1452-5.

Zhang J, Deng S, Shen F, et al. Modeling the relationship between energy consumption and economy development in China. Energy. 2011;36:4227–34. doi:10.1016/j.energy.2011.04.021.

Zheng Y, Luo D. Industrial structure and oil consumption growth path of China: empirical evidence. Energy. 2013;57:336–43. doi:10.1016/j.energy.2013.05.004.

Zou C, Yang Z, Zhu R, et al. Progress in China's unconventional oil & gas exploration and development and theoretical technologies. Acta Geol Sin (English Edition). 2015;89:938–71. doi:10.1111/1755-6724.12491.

Zhou S, Zhang X. Nuclear energy development in China: a study of opportunities and challenges. Energy. 2010;35:4282–8. doi:10.1016/j.energy.2009.04.020.

Zhou K, Yang S, Shen C, et al. Energy conservation and emission reduction of China's electric power industry. Renew Sustain Energy Rev. 2015;45:10–9. doi:10.1016/j.rser.2015.01.056.

14

Efficient ozonation of reverse osmosis concentrates from petroleum refinery wastewater using composite metal oxide-loaded alumina

Yu Chen[1] · Chun-Mao Chen[1] · Brandon A. Yoza[2] · Qing X. Li[3] · Shao-Hui Guo[1] ·
Ping Wang[1] · Shi-Jie Dong[1] · Qing-Hong Wang[1]

Abstract Novel Mn–Fe–Mg- and Mn–Fe–Ce-loaded alumina (Mn–Fe–Mg/Al$_2$O$_3$ and Mn–Fe–Ce/Al$_2$O$_3$) were developed to catalytically ozonate reverse osmosis concentrates generated from petroleum refinery wastewaters (PRW-ROC). Highly dispersed 100–300-nm deposits of composite multivalent metal oxides of Mn (Mn^{2+}, Mn^{3+}, and Mn^{4+}), Fe (Fe^{2+} and Fe^{3+}) and Mg (Mg^{2+}), or Ce (Ce^{4+}) were achieved on Al$_2$O$_3$ supports. The developed Mn–Fe–Mg/Al$_2$O$_3$ and Mn–Fe–Ce/Al$_2$O$_3$ exhibited higher catalytic activity during the ozonation of PRW-ROC than Mn–Fe/Al$_2$O$_3$, Mn/Al$_2$O$_3$, Fe/Al$_2$O$_3$, and Al$_2$O$_3$. Chemical oxygen demand removal by Mn–Fe–Mg/Al$_2$O$_3$- or Mn–Fe–Ce/Al$_2$O$_3$-catalyzed ozonation increased by 23.9% and 23.2%, respectively, in comparison with single ozonation. Mn–Fe–Mg/Al$_2$O$_3$ and Mn–Fe–Ce/Al$_2$O$_3$ notably promoted ·OH generation and ·OH-mediated oxidation. This study demonstrated the potential use of composite metal oxide-loaded Al$_2$O$_3$ in advanced treatment of bio-recalcitrant wastewaters.

Keywords Petroleum refinery wastewater · Reverse osmosis concentrate · Catalytic ozonation · Composite metal oxide

1 Introduction

The need for freshwater and its conservation are motivating factors for treatment of wastewaters generated by petroleum refining industries. Reverse osmosis (RO) systems are widely used during the treatment and reclamation processes for the effluent from petroleum refinery wastewater (PRW) plants (Pérez-González et al. 2012). In China, 70 wt%–80 wt% of the effluent is reclaimed using RO systems and used as the high-quality feed water for steam production. The remaining 20 wt%–30 wt% of RO concentrate (ROC) contains petroleum-derived chemicals (Chen et al. 2016). Direct discharge of the ROC threatens the ecological environment and human health. The organics in ROC generated from PRW reclamation (PRW-ROC) need to be reduced to eliminate these negative impacts and to meet increasingly stringent discharge standards (Moreira et al. 2017). Previous work has already investigated the use of physicochemical and biological treatments; however, low concentrations and biologically recalcitrant organic matter suggest these methods are unsuitable (Bagastyo et al. 2013).

Advanced oxidation processes (AOPs) are the preferred advanced treatment method during reclamation of various municipal and industrial wastewater ROC products (Joo and Tansel 2015; Ren et al. 2016). These methods, including ozonation (Dialynas et al. 2008), Fenton oxidation (Zhou et al. 2012), photocatalysis (Joo and Tansel 2015), photooxidation (Umar et al. 2016), sonolysis (Pérez-González et al. 2012), or electrochemical oxidation

Yu Chen and Chun-Mao Chen have contributed equally to this work.

✉ Qing-Hong Wang
wangqhqh@163.com

1 State Key Laboratory of Petroleum Pollution Control, Beijing Key Laboratory of Oil and Gas Pollution Control, China University of Petroleum, Beijing 102249, China

2 Hawaii Natural Energy Institute, University of Hawaii at Manoa, Honolulu, HI 96822, USA

3 Department of Molecular Biosciences and Bioengineering, University of Hawaii at Manoa, Honolulu, HI 96822, USA

Edited by Xiu-Qin Zhu

(Bagastyo et al. 2011; Van Hege et al. 2002), can provide efficient removal of low concentration and recalcitrant organics. Among these, the heterogeneous catalytic ozonation processes (COPs) are the most promising, as they are economical, highly efficient, and simple in their application. The predominant role of catalysts for heterogeneous COP treatment is decomposing ozone into more active species such as hydroxyl radicals ($\cdot OH$), and/or for the adsorption of specific organics that can react with dissolved ozone (Chen et al. 2015). A wide variety of catalysts have been developed for COPs; however, most synthesized or prepared catalysts are costly, limiting their industrial application.

Alumina (Al_2O_3) and metal oxide-loaded Al_2O_3 have been widely applied in COPs. These materials have highly active and large surface area, good mechanical properties and are stable (Einaga and Futamura 2005; Pocostales et al. 2011; Keykavoos et al. 2013; Vittenet et al. 2015). Industrial grade γ-Al_2O_3 particles have been used for enhanced ozonation of petrochemical effluents (Vittenet et al. 2015). Ti/Al_2O_3 (Bing et al. 2017), Ru/Al_2O_3 (Zhou et al. 2007), and V/Al_2O_3 (Qi et al. 2009) are efficient catalysts for the treatment of recalcitrant organics such as dimethyl phthalate and 1, 2-dichlorobenzene. Mn/Al_2O_3 is another powerful catalyst for removal of bio-recalcitrant organics. Mn/Al_2O_3 can significantly produce $\cdot OH$ when it reacts with ozone, resulting in enhanced catalytic degradation of atrazine (Rosal et al. 2010a), and fenofibric acid (Rosal et al. 2010b). Fe/Al_2O_3 exhibited both significant inhibition

of BrO_3^- formation and high total organic carbon removal for a Br-containing raw water (Nie et al. 2014). Numerous studies have revealed that composite metal oxides loaded on Al_2O_3 supports exhibited high catalytic activity compared with single metal oxide loading (Tong et al. 2010). Mn–Fe–Cu/Al_2O_3 enhanced catalytic ozonation of PRW compared with single or double metal oxide-loaded Al_2O_3, a result from interactions between the composite metal oxides on the Al_2O_3 surface (Chen et al. 2015). Magnesium oxides, that include MgO nanocrystals and MgO/granular activated carbon (GAC), have potential for ozonation of bio-recalcitrant organics including phenols (Moussavi et al. 2014), benzene homologues (Rezaei et al. 2016), and dye pollutants (Moussavi and Mahmoudi 2009). During the ozonation of catechol that is catalyzed by MgO/GAC, $\cdot OH$ was responsible for its degradation and mineralization, and this reaction rate constant was six times greater than single ozonation (Moussavi et al. 2014). Catalysts containing Ce have also been studied for the ozonation of p-chlorobenzoic acid (Bing et al. 2013), bezafibrate (Xu et al. 2016), dimethyl phthalate (Yan et al. 2013), and tonalide (Santiago-Morales et al. 2012). However, their applicability for wastewater treatment has not yet been reported.

In this study, novel composite metal oxide-loaded Al_2O_3 catalysts, including Mn–Fe–Mg/Al_2O_3 and Mn–Fe–Ce/Al_2O_3, were prepared and characterized. The potential use of these catalysts for the advanced treatment of PRW-ROC using COP was investigated. Insights into these catalytic mechanisms are also provided.

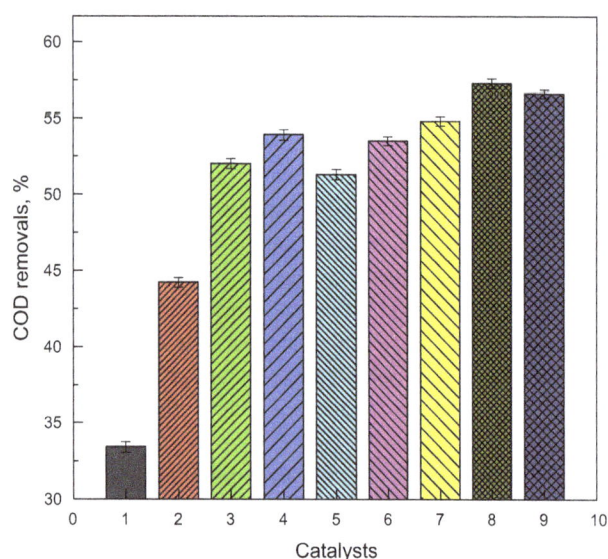

Fig. 1 COD removals for PRW-ROC using single ozonation and various COPs (0.5 g catalyst, 5 mg/min ozone, 30 °C, and 40 min). *1* single ozonation, *2* Al_2O_3-COP, *3* Fe/Al_2O_3-COP, *4* Fe/Al_2O_3(s)-COP, *5* Mn/Al_2O_3-COP, *6* Mn/Al_2O_3(s)-COP, *7* Mn–Fe/Al_2O_3-COP, *8* Mn–Fe–Mg/Al_2O_3-COP, and *9* Mn–Fe–Ce/Al_2O_3-COP

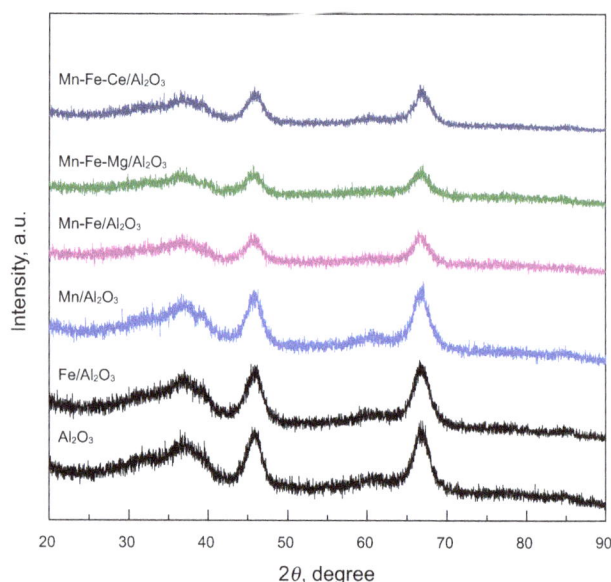

Fig. 2 XRD patterns of Al_2O_3 and metal oxide-loaded Al_2O_3 catalysts

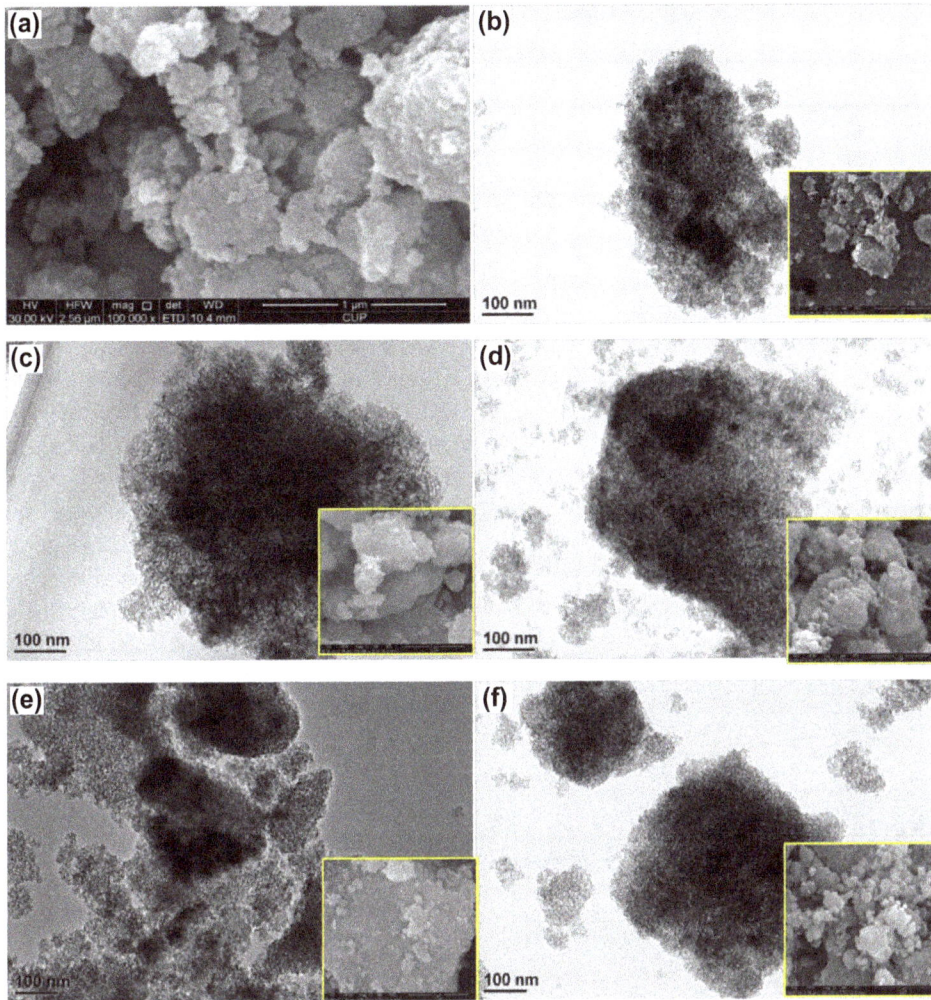

Fig. 3 SEM micrographs of Al$_2$O$_3$ (**a**), and SEM–TEM micrographs of Fe/Al$_2$O$_3$ (**b**), Mn/Al$_2$O$_3$ (**c**), Mn–Fe/Al$_2$O$_3$ (**d**), Mn–Fe–Mg/Al$_2$O$_3$ (**e**), and Mn–Fe–Ce/Al$_2$O$_3$ (**f**) catalysts

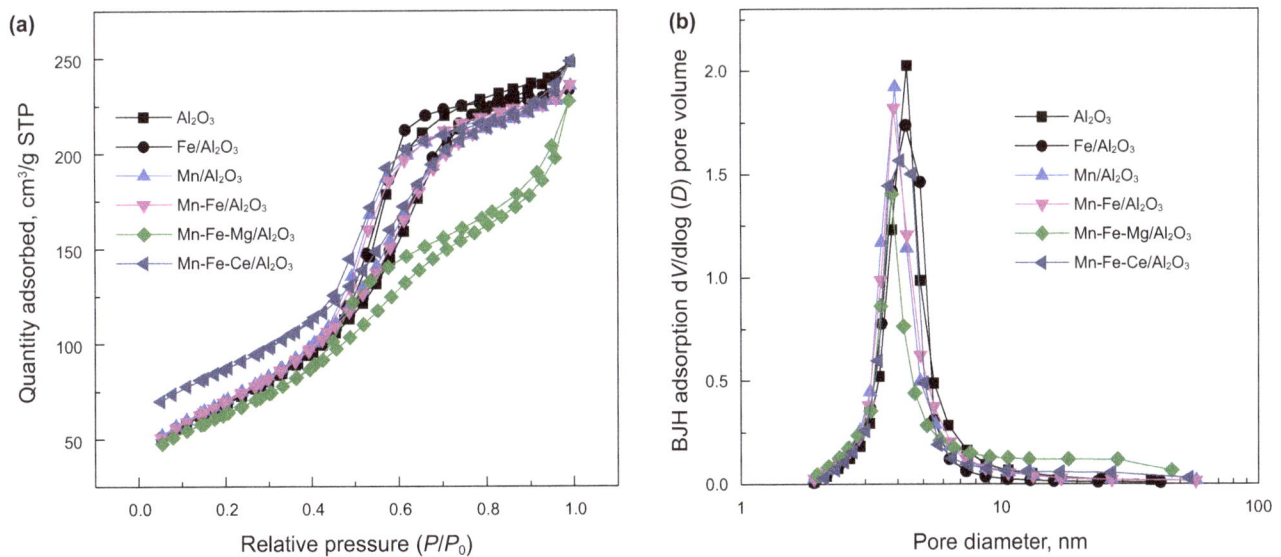

Fig. 4 Isotherms (**a**) and pore distribution curves (**b**) of Al$_2$O$_3$ and metal oxide-loaded Al$_2$O$_3$ catalysts by N$_2$ adsorption–desorption

Table 1 Surface areas and pore structures of Al_2O_3 and metal oxide-loaded Al_2O_3 catalysts

Catalysts	S_{BET}, m^2/g	V_P, cm^3/g	D_a, nm
Al_2O_3	250	0.38	6.1
Fe/Al_2O_3	252	0.36	5.7
Mn/Al_2O_3	258	0.36	5.6
$Mn–Fe/Al_2O_3$	255	0.37	5.7
$Mn–Fe–Mg/Al_2O_3$	230	0.35	6.1
$Mn–Fe–Ce/Al_2O_3$	242	0.35	5.8

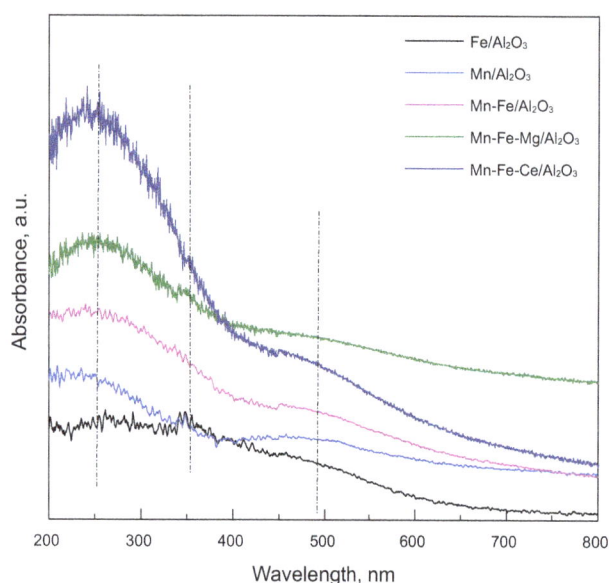

Fig. 5 UV-Vis patterns of metal oxide-loaded catalysts

2 Experimental

2.1 Preparation of catalysts

Commercial pseudoboehemite (65.6 wt% of Al_2O_3) was purchased from Chalco Shandong Co., Ltd. (China). $Fe(NO_3)_3\cdot9H_2O$ (≥98.5 wt%), $Mn(NO_3)_2\cdot4H_2O$ solution (50 wt%), $Mg(NO_3)_2\cdot6H_2O$ (≥99.0 wt%), and $Ce(NO_3)_3\cdot6H_2O$ (≥99.0 wt%) were obtained from Beijing Chemical Reagents Co., China. The catalysts were prepared according to the incipient wetness impregnation method. 60.0 g boehmite was impregnated with the mixture solution of 4.57 g $Mn(NO_3)_2\cdot4H_2O$, 4.57 g $Fe(NO_3)_3\cdot9H_2O$ and 0.79 g $Mg(NO_3)_2\cdot6H_2O$, or 0.35 g $Ce(NO_3)_3\cdot6H_2O$ (≥99.0 wt%) to yield Mn–Fe–Mg/ Al_2O_3 or Mn–Fe–Ce/Al_2O_3 catalysts.

Impregnation of 60.0 g boehmite with a mixture solution of 4.55 g $Mn(NO_3)_2\cdot4H_2O$ and 4.56 g $Fe(NO_3)_3\cdot9H_2O$ yielded Mn–Fe/Al_2O_3 catalyst. Impregnation of 60.0 g boehmite with 4.45 or 9.00 g $Mn(NO_3)_2\cdot4H_2O$ yielded Mn/ Al_2O_3 or Mn/Al_2O_3 (s) catalysts, respectively.

Impregnation of 60.0 g boehmite with 4.46 or 9.00 g $Fe(NO_3)_3\cdot9H_2O$ yielded Fe/Al_2O_3 or Fe/Al_2O_3 (s) catalysts, respectively. The impregnated samples were calcined at 550 °C for 4 h in air after drying at 120 °C for 12 h. Al_2O_3 was prepared from pseudoboehemite by calcination at 550 °C for 4 h in air.

2.2 Characterization of catalysts

The crystal forms were observed by X-ray powder diffraction (XRD) using a D8 advance X-ray powder diffractometer (Bruker, Germany) with 40.0 kV working voltage and 40.0 mA current and a copper target X-ray tube. The specific surface area and pore size distribution were determined using a Tristar II 3020 surface area and porosity analyzer (Micromeritics, USA) with liquid nitrogen cooling at −196 °C. The total surface areas (S_{BET}) and total pore volume (V_p) were calculated according to Brunauer–Emmett–Teller (BET) and Barrett–Joyner–Halenda (BJH) methods, respectively. The bulk chemical composition was determined by X-ray fluorescence (XRF) analysis with an AX XRF analyzer (Axiosm, Netherlands). The elemental surface distribution was determined by X-ray photoelectron spectroscopy (XPS) analysis with a PHI Quantera SXM X-ray photoelectron spectrometer (ULVAC, USA), where all measured values of the binding energy (BE) were referred to the C_{1S} line at 284.8 eV. The diffuse reflectance spectra were recorded on a U-4100 UV–Vis spectrophotometer (Hitachi, Japan). The surface morphology was observed with a Tecnai G2 F20 transmission electron microscope (TEM) and a Quanta 200F scanning electron microscope (SEM) (FEI, USA). The point of zero charge (pH_{pzc}) was determined according to the pH drift method (Altenor et al. 2009).

2.3 Ozonation of PRW-ROC

The PRW-ROC used was collected directly from the RO unit of a wastewater treatment plant in Liaohe Petrochemical Co., China National Petroleum Corp. The ranges of pH values, 5-day biochemical oxygen demand (BOD_5), chemical oxygen demand (COD), and electric conductivity (25 °C) were determined. These were 8.0 to 8.5, 9.2 to 16.3, 105.6 to 125.3 mg/L, and 4438 to 5130 μS/cm, respectively. The COD concentration of PRW-ROC failed to meet the current Emission Standard of Pollutants for Petroleum Refining Industry of China (GB 31570-2015) in which the allowable COD concentration is lower than 60 mg/L. The BOD_5/COD ratios of PRW-ROC were ranged from 0.09 to 0.13. Due to low biodegradability, COP was determined as an efficient advanced treatment method for PRW-ROC.

The experimental system was constructed with an oxygen tank, RQ-02 ozone generator (Ruiqing, China), 200-mL quartz column reactor, flow meter, and an exhaust

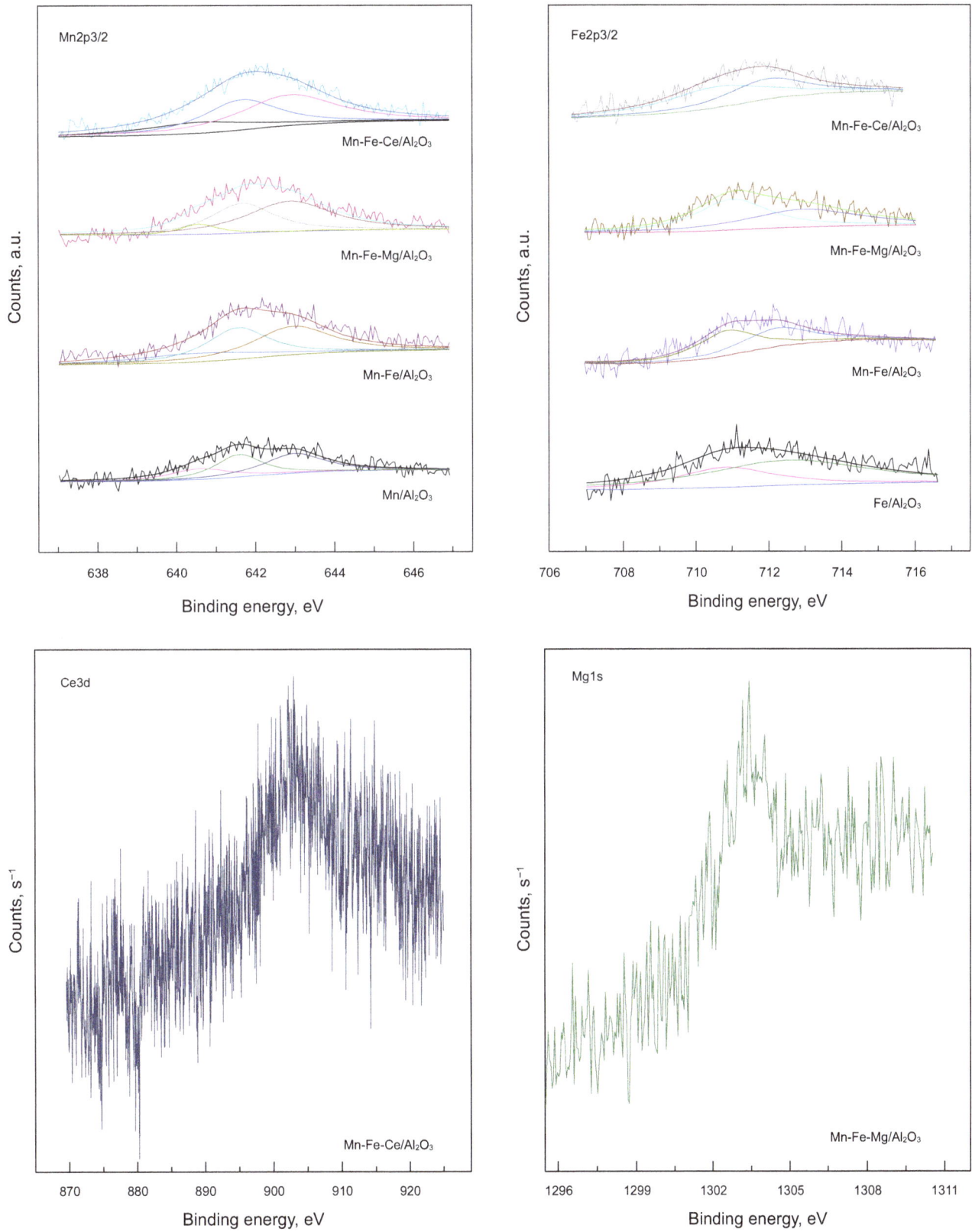

Fig. 6 XPS spectra of Mn2p, Fe2p, Ce3d, and Mg1s of metal oxide-loaded catalysts

Table 2 Binding energies and surface atomic ratios of Mn, Fe, Mg, and Ce elements on catalysts

Items	Fe/Al$_2$O$_3$	Mn/Al$_2$O$_3$	Mn–Fe/Al$_2$O$_3$	Mn–Fe–Mg/Al$_2$O$_3$	Mn–Fe–Ce/Al$_2$O$_3$
Binding energies					
Mn2p	–	641.8	642.2	642.2	642.2
Fe2p	711.1	–	711.9	711.5	711.7
Ce3d	–	–	–	–	903.0
Mg1s	–	–	–	1303.2	
Metal oxides value state					
Mn^{2+}:Mn^{3+}:Mn^{4+}	–	0.22:0.37:0.41	0.22:0.36:0.42	0.10:0.38:0.53	012:0.39:0.51
Fe^{2+}:Fe^{3+}	0.31:0.69	–	0.52:0.48	0.50:0.50	0.57:0.43
Surface atomic ratio					
Mn2p/Al2p	–	0.015	0.016	0.014	0.015
Fe2p/Al2p	0.016	–	0.018	0.011	0.013
Mg1s/Al2p	–	–	–	0.010	
Ce3d/Al2p	–	–	–		0.004
(Mn2p + Fe2p + Mg1s + Ce3d)/Al2p	0.016	0.015	0.034	0.035	0.032

gas collector. An aliquot of 100 mL of PRW-ROC and 0.5 g of catalyst were added in the reactor at 30 °C. The gaseous ozone was then introduced through a porous diffuser at the bottom of the reactor with a flow rate of 5 mg/min. The experiments were carried out under varying the initial pH values (adjusted with 1 N NaOH or HCl) and reaction times. After treatment, dried oxygen was blown into the PRW-ROC at a rate of 3.0 L/min to quench the reaction and eliminate the residual ozone. The resulting suspension was filtered (Whatman Qualitative No. 5) to separate catalyst particles prior to further analysis at various intervals. The ·OH quenching experiments were performed to determine the oxidation mechanism. The ·OH scavengers, *tert*-butanol (*t*BA), and sodium bicarbonate (NaHCO$_3$) were added into PRW-ROC (0.5 and 1.0 g/L, respectively) prior to experiments. All the experiments were performed in triplicate.

The pH and conductivity were measured with a MP 220 pH meter (Mettler Toledo, Switzerland) and a CD400 conductivity meter (Alalis, China), respectively. The leaching of Ce and Mg elements was measured with an AAnalyst atomic absorption spectrometer (PerkinElmer, USA) using a nitrous oxide/oxygen–acetylene flame. The BOD$_5$ was tested on a BODTrak II BOD meter (HACH, USA). The COD was measured with a CTL-12 COD meter (HATO, China). The COD removal was calculated using the following equation:

$$\text{COD removal} = \left([COD]_0 - [COD]_1\right)/[COD]_0 \quad (1)$$

3 Results and discussion

3.1 Catalytic performances of catalysts

The COD removal from PRW-ROC using COPs increased using Al$_2$O$_3$ and metal oxide-loaded Al$_2$O$_3$. Ozonation catalyzed using Mn–Fe–Mg/Al$_2$O$_3$ (Mn–Fe–Mg/Al$_2$O$_3$-COP) resulted in an increased COD removal (57.3%) compared with the other catalysts, Mn–Fe–Ce/Al$_2$O$_3$-COP (56.6%), Mn–Fe/Al$_2$O$_3$-COP (55.8%), Fe/Al$_2$O$_3$-COP (52.0%), Fe/Al$_2$O$_3$ (s)-COP (53.8%), Mn/Al$_2$O$_3$-COP (51.3%), Mn/Al$_2$O$_3$ (s)-COP (53.5%), and Al$_2$O$_3$-COP (45.2%). It was especially significant when compared with single oxide ozonation (33.4%) using a 40-min treatment (Fig. 1). The surface content of loaded Fe$_2$O$_3$ or MnO was about 4.3 wt% for Fe/Al$_2$O$_3$ (s) and Mn/Al$_2$O$_3$ (s), almost double compared with Fe/Al$_2$O$_3$ and Mn/Al$_2$O$_3$ (2.1 wt%). The increased Fe and Mn oxide contents resulted only in a limited performance improvement for COD removal. In comparison, composite metal oxide-loaded Al$_2$O$_3$ was more effective at equivalent loadings (about 4.2 wt%), suggesting synergistic effects. For Mn–Fe–Mg/Al$_2$O$_3$, Mn–Fe–Ce/Al$_2$O$_3$, Mn–Fe/Al$_2$O$_3$, Fe/Al$_2$O$_3$, and Mn/Al$_2$O$_3$ composites, the content of Fe$_2$O$_3$ and MnO was 2.1 wt%, and MgO and CeO$_2$ were above 0.3 wt% based on XRF analysis. Further investigation focused on utilization of Mn–Fe–Ce/Al$_2$O$_3$, Mn–Fe–Mg/Al$_2$O$_3$, Mn–Fe/Al$_2$O$_3$, Mn/Al$_2$O$_3$, Fe/Al$_2$O$_3$, and Al$_2$O$_3$ catalysts.

3.2 Characteristics of catalysts

The metal oxide-loaded Al$_2$O$_3$ showed typical γ-Al$_2$O$_3$ diffraction peaks (Fig. 2). Obvious XRD diffraction peaks from the metal oxides were not observed, due to the low loading or amorphous status. Using TEM, it was determined that the deposited metal oxides formed micro-agglomerates in irregular shapes and sizes on the surface of Al$_2$O$_3$ (Fig. 3), and that based on SEM, the surface morphology of Al$_2$O$_3$ itself was little changed.

Adsorption–desorption isotherms and pore distributions varied among Al_2O_3 and metal oxide-loaded Al_2O_3 catalysts (Fig. 4a, b). According to IUPAC classification, the isotherms of these catalysts suggest a typical type IV mesopore structure (Xu and Pang 2004). A hysteresis loop attributed to type H_1 was observed for Al_2O_3, Fe/Al_2O_3, Mn/Al_2O_3, $Fe–Mn/Al_2O_3$, and $Mn–Fe–Ce/Al_2O_3$, suggesting uniform shape and pore size. The hysteresis loop determined for $Mn–Fe–Mg/Al_2O_3$, however, resulted in a combination of H_1 and H_3 types. This suggests the potential presence of silt pores that are a result of metal oxide particle accumulation. Different hysteresis types are potentially a result of interactions between the oxides and the Al_2O_3 support. All deposited oxide catalysts have a pronounced pore distribution peak at 5–6 nm. The surface areas (S_{BET}), pore volumes (V_P), and average pore sizes (D_a) were 230–250 m^2/g, 0.35–0.38 cm^3/g, and 5.7–6.1 nm, respectively (Table 1). $Mn–Fe–Mg/Al_2O_3$ showed the lowest S_{BET} and V_P values among these catalysts.

Figure 5 shows the UV–Vis spectra of metal oxide-loaded catalysts. Fe/Al_2O_3 had a wide absorbance peak with a lower intensity centered at 200–600 nm, attributed to isolated Fe^{3+}, oligomeric FeO_x clusters, and large Fe_2O_3 particles (Santhosh Kumar et al. 2004). Mn/Al_2O_3 displayed an absorption peak centered at 250 nm and a wide peak with lower intensity centered at 400–600 nm, associated with a charge transfer (CT) $O^{2-} \rightarrow Mn^{2+}$ and poorly resolved absorbance bands (d-d transitions) from Mn^{3+}- and Mn^{4+}-oxo species, respectively (Wu et al. 2015). $Mn–Fe/Al_2O_3$ exhibited peaks at 250, 350, and 480 nm, due to composite Fe and/or Mn oxide presence. $Mn–Fe–Mg/Al_2O_3$ and $Mn–Fe–Ce/Al_2O_3$ showed significantly increased absorbance intensity at 250 nm, suggesting homogenous dispersion of Ce and Mg oxides.

Figure 6 shows the XPS spectra of Mn2p, Fe2p, Mg1s, and Ce3d for metal oxide-loaded catalysts. Table 2 shows the binding energies and surface atomic ratios for these metallic elements. The Mn2p3/2 peaks of Mn/Al_2O_3, $Mn–Fe/Al_2O_3$, $Mn–Fe–Mg/Al_2O_3$, and $Mn–Fe–Mg/Al_2O_3$ are attributed to Mn^{2+} oxides (MnO or $Mn(OH)_2$), Mn^{3+} oxides (Mn_2O_3 or MnOOH), and Mn^{4+} oxide (MnO_2) (Zhang et al. 2015). The Fe2p3/2 peaks of Fe/Al_2O_3, $Mn–Fe/Al_2O_3$, $Mn–Fe–Mg/Al_2O_3$, and $Mn–Fe–Mg/Al_2O_3$ are related to Fe^{2+} oxides (FeO or $Fe(OH)_2$) and Fe^{3+} oxides (Fe_2O_3 or FeOOH) (Shwana et al. 2015). The surface atomic ratios of Fe2p to Al2p (0.011 ~ 0.018) and Mn2p to Al2p (0.014 ~ 0.016) for catalysts changed little and were close to the bulk molar ratio of Fe to Al (0.0129) and Mn to Al (0.01547). The asymmetrical distribution of the Mg1s peak is likely a result of interactions between Mg, Mn, and Fe. The surface atomic ratio (0.010) of Mg1s to

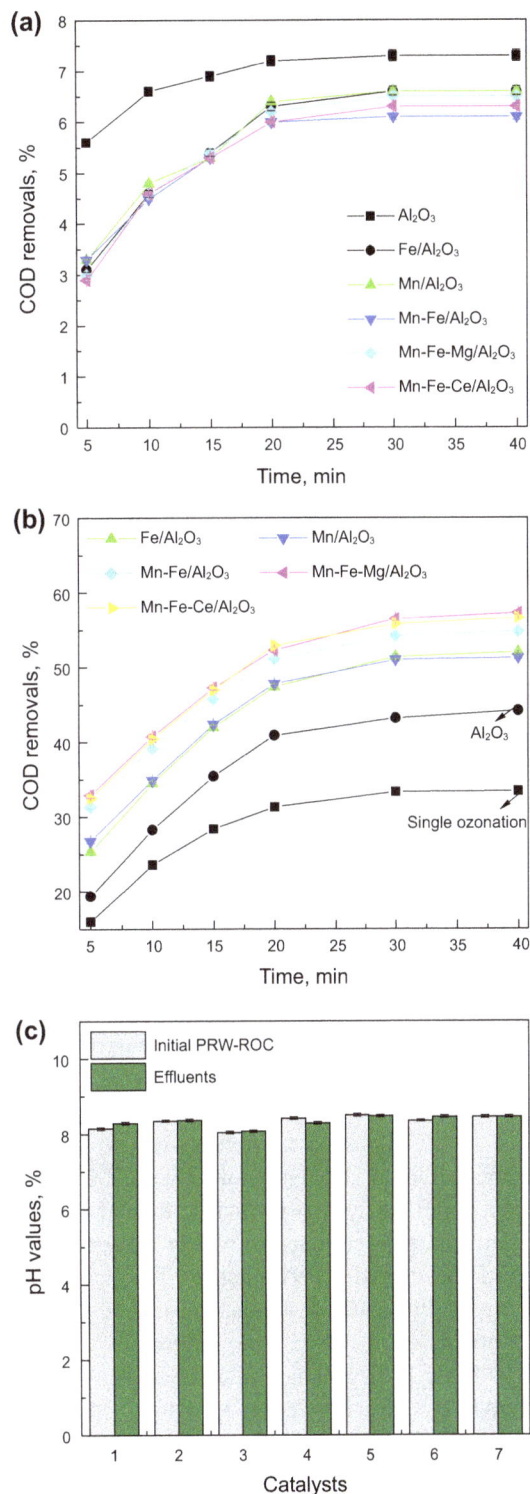

Fig. 7 COD removal from PRW-ROC by adsorption (**a**) and over single ozonation and various COPs (**b**); pH value changes of PRW-ROC over single ozonation and various COPs (**c**): *1* single ozonation, *2* Al_2O_3-COP, *3* Fe/Al_2O_3-COP, *4* Mn/Al_2O_3-COP, *5* $Mn–Fe/Al_2O_3$-COP, *6* $Mn–Fe–Mg/Al_2O_3$-COP, and *7* $Mn–Fe–Ce/Al_2O_3$-COP

Table 3 Influences of ·OH scavengers on COD removals of PRW-ROC using Mn–Fe/Al₂O₃-COP, Mn–Fe–Mg/Al₂O₃-COP, and Mn–Fe–Ce/Al₂O₃-COP

Systems	COD removals (%)				
	No ·OH scavenger	*t*BA		NaHCO₃	
		0.5, g/L	1.0, g/L	0.5, g/L	1.0, g/L
Mn–Fe/Al₂O₃-COP	55.8	24.9	22.7	30.5	28.7
Mn–Fe–Mg/Al₂O₃-COP	57.3	37.6	14.9	34.3	31.3
Mn–Fe–Ce/Al₂O₃-COP	56.6	34.9	16.9	31.7	27.8

0.5 g catalyst, 5 mg/min ozone, 30 °C, and 40 min

Al2p for Mn–Fe–Mg/Al₂O₃ is far greater than that in bulk (0.0038), suggesting high surface area distribution of the Mg oxide. Similarly the surface atomic ratio (0.004) of Ce3d to Al2p for Mn–Fe–Ce/Al₂O₃ is higher than that in bulk (0.0009), again suggesting high surface area dispersion of Ce^{4+} oxide (Ding et al. 2016). The surface atomic ratios of the sum of Mn2p + Fe2p + Mg1s + Ce3d to Al2p for Mn–Fe–Mg/Al₂O₃, Mn–Fe–Ce/Al₂O₃, and Mn–Fe/Al₂O₃ catalysts were similar to each other and doubled compared to Mn/Al₂O₃ and Fe/Al₂O₃.

3.3 Mechanisms of catalytic ozonation

The adsorption onto the catalysts reached saturation by 40 min. Adsorption on Al₂O₃, Fe/Al₂O₃, Mn/Al₂O₃, Mn–Fe/Al₂O₃, Mn–Fe–Mg/Al₂O₃, and Mn–Fe–Ce/Al₂O₃ contributed to COD removals by 7.3%, 6.6%, 6.6%, 6.1%, 6.5%, and 6.3%, respectively (Fig. 7a). No significant differences for adsorption capacity among catalysts were observed, likely due to similar surface areas (Table 1). COD removals using various COPs increased by 23.9%–11.8% compared with single ozonation (Fig. 7b). These results are significantly greater than simple adsorption (7.3%–6.1%). The observed difference can be attributed to the application of catalytic ozonation. Mn–Fe/Al₂O₃ exhibited better catalytic performance than Mn/Al₂O₃ and Fe/Al₂O₃, and the introduction of Mg and/or Ce further improved the catalytic performance. The active surface areas for all the material comparisons were similar and would not have an impact on differences in catalytic activity (Table 1). The enhanced catalytic activity of Mn–Fe–Mg/Al₂O₃, Mn–Fe–Ce/Al₂O₃, and Mn–Fe/Al₂O₃ resulted from the metal oxide components themselves, interactions between the metal oxides, interactions between the metal oxides and Al₂O₃ support, as well as its environment (Table 2). Similar relationships between reaction times and COD removals using different COPs were obtained (Fig. 7b), suggesting the similarity of catalytic mechanisms for the various catalysts. Small pH changes in the effluents from single ozonation and COPs were found compared to that of the initial PRW-ROC (Fig. 7c).

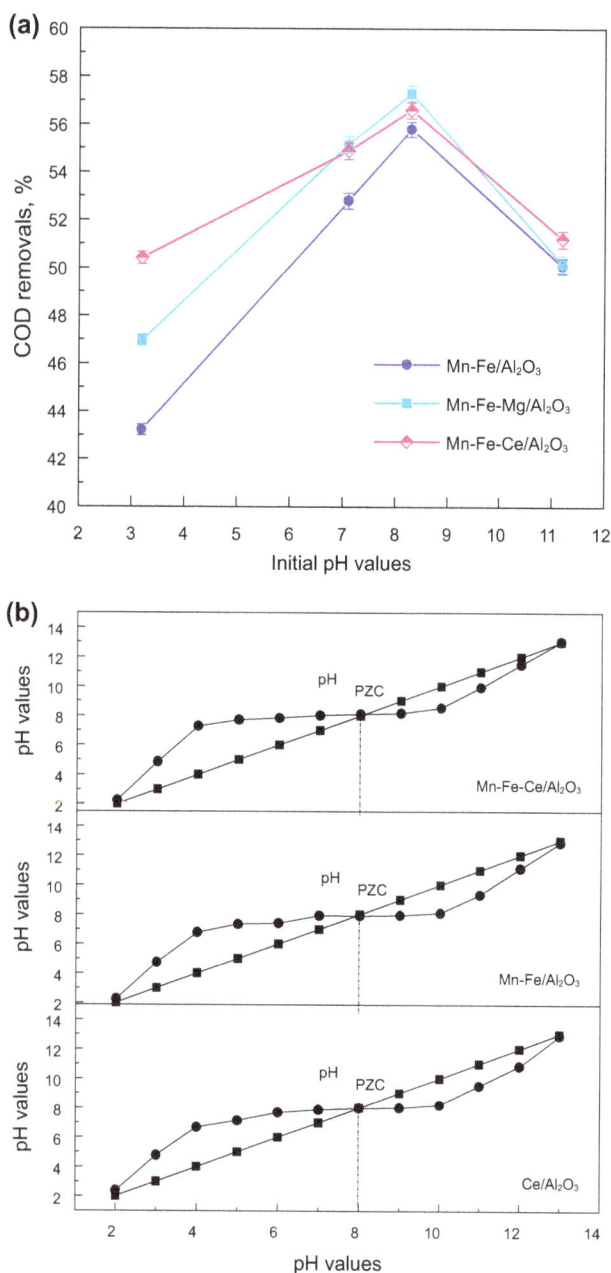

Fig. 8 **a** Influences of the initial pH values on COD removals using Mn–Fe/Al₂O₃-COP, Mn–Fe–Mg/Al₂O₃-COP, and Mn–Fe–Ce/Al₂O₃-COP; and **b** pH_pzc values of three catalysts

Several reports have suggested that ·OH generation induced by catalysts increased removal of pollutants during COP treatment (Qi et al. 2013). In order to identify whether these treatments result in the generation of ·OH, COD removals in the presence of tBA and NaHCO$_3$ were examined. From this it was determined that the COD removal of PRW-ROC was lessened due to the introduction of tBA and NaHCO$_3$ in bulk (Table 3). NaHCO$_3$ may impair catalytic decomposition of ozone into ·OH on the catalyst surface because of its high affinity to Lewis acid sites on the catalyst surface. In contrast, tBA can quench aqueous ozone decomposition by reacting with ·OH in bulk, generating inert intermediates. As such, the catalytic ozonation of PRW-ROC was dominated by ·OH both on the catalyst surface and in bulk in COPs over Mn–Fe–Mg/Al$_2$O$_3$, Mn–Fe–Ce/Al$_2$O$_3$, and Mn–Fe/Al$_2$O$_3$. The decreased extent of COD removals by both tBA and NaHCO$_3$ in Mn–Fe–Mg/Al$_2$O$_3$-COP and Mn–Fe–Ce/Al$_2$O$_3$-COP was greater than that in Mn–Fe/Al$_2$O$_3$-COP. It is reasonable to then expect that the introduction of small concentrations of Mg or Ce can be used to further promote ·OH generation. In addition, the inhibitory effect for ·OH generation using tBA was greater than that by NaHCO$_3$, suggesting more ·OH oxidation occurred in bulk rather than on the catalyst surface. COD was still, however, reduced in spite of the addition of ·OH scavengers and can be ascribed to direct ozonation.

Initial pH values that were either acidic, at the pH$_{pzc}$, or alkaline, significantly influenced the COD removal during COP treatment of PRW-ROC (Fig. 8a). An initial pH value of 3 resulted in reduced efficiency of COP and was probably a result of the metal species being leached from the

catalyst. Initial pH values of 8.3 (Fig. 8a) were close to the pH$_{pzc}$ (Fig. 8b) of the three composite catalysts and resulted in efficient COD removal. Alkaline pH values around 11 may result in the formation of carbonate and bicarbonate during organic mineralization, again decreasing the efficiency of COP (Xiong et al. 2003). The positive impact of surface hydroxyl groups (–OH), on active surfaces, has been determined for the removal of organics (Lu et al. 2014; Zhang et al. 2007, 2008). It is believed that pH values near the pH$_{pzc}$ of the catalyst can result in accelerated ·OH generation, due to a neutral –OH state (Qi et al. 2012). Based on ·OH quenching experiments and the impact of initial pH values, the surface –OH likely does have a significant role in COP treatment of PRW-ROC. The surface MnOOH and FeOOH of catalysts are the principle drivers for COP according to XPS results (Table 2). The enhanced COD removal compared with Mn–Fe–Mg/Al$_2$O$_3$ and Mn–Fe–Ce/Al$_2$O$_3$ may suggest greater ·OH-related activity due to interactions between the various metal oxides and/or changes of metallic states influenced by the environment. Ozone reacts with surface –OH groups during COP treatment and results in highly active ·OH generation in bulk and/or on the surface, resulting in organics oxidation (Fig. 9).

3.4 Reusability and stability of catalysts

Catalysts were reused ten times for COD removal from PRW-ROC with Al$_2$O$_3$-COP, Mn–Fe–Mg/Al$_2$O$_3$-COP, and Mn–Fe–Ce/Al$_2$O$_3$-COP. Conditions for experiments included, 0.5 g catalyst, 5 mg/min ozone, 30 °C, 40 min treatment, and initial pH values. COD removal from PRW-

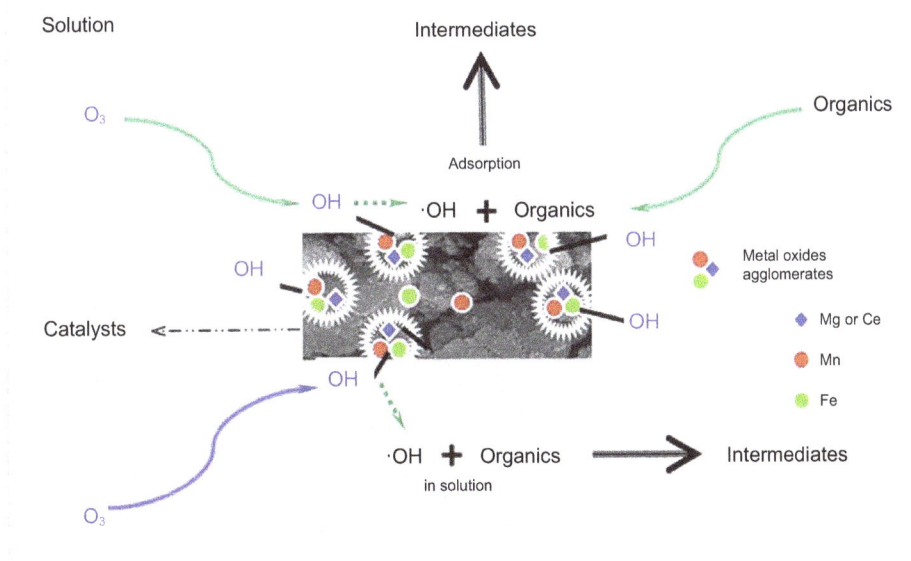

Fig. 9 Proposed ozonation mechanisms of organics in PRW-ROC upon Mn–Fe–Mg/Al$_2$O$_3$ and Mn–Fe–Ce/Al$_2$O$_3$

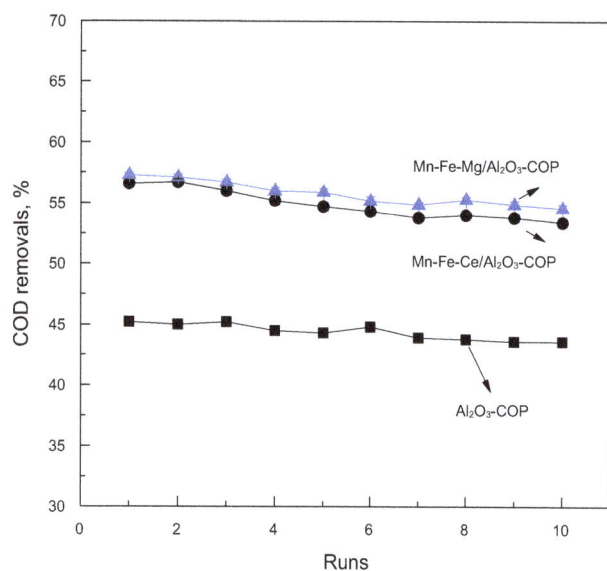

Fig. 10 COD removals in 10 COP runs using Al_2O_3, Mn–Fe–Mg/ Al_2O_3, and Mn–Fe–Ce/Al_2O_3

ROC was maintained within the range of 57.3%–54.6% for Mn–Fe–Mg/Al_2O_3-COP and 55.6%–53.4% for Mn–Fe–Ce/ Al_2O_3-COP, far greater than Al_2O_3-COP (45.2%–43.6%) (Fig. 10). Single ozonation and adsorption had relatively low efficiency for the removal of bio-recalcitrant organics from PRW-ROC. By using Mn–Fe–Mg/Al_2O_3-COP and Mn–Fe–Ce/Al_2O_3-COP, bio-recalcitrant organics in PRW-ROC were degraded by the greater generation of ·OH compared with single ozonation (Table 3). The PRW-ROC was successfully treated by utilizing the composite metal oxide-loaded Al_2O_3 during catalyzed ozonation to reduce the COD value to below 60 mg/L. This value met the Emission Standard of Pollutants for Petroleum Refining Industry of China (GB 31570-2015). No significant leaching of Ce or Mg elements was detected in Mn–Fe–Ce/ Al_2O_3-COP or Mn–Fe–Mg/Al_2O_3-COP. These results showed that Mn–Fe–Mg/Al_2O_3 and Mn–Fe–Ce/Al_2O_3 catalysts were stable and reuseable and could increase the efficiency of ozonation treatments of PRW-ROC.

4 Conclusions

The composite metal oxide-loaded Al_2O_3 catalysts, Mn–Fe– Mg/Al_2O_3 and Mn–Fe–Ce/Al_2O_3, were prepared and used for efficient ozonation of PRW-ROC. The 100–300 nm multivalent metal oxides of Mn (Mn^{2+}, Mn^{3+} and Mn^{4+}), Fe (Fe^{2+} and Fe^{3+}), and Mg (Mg^{2+})/Ce (Ce^{4+}) were highly dispersed on the Al_2O_3 support. COD removals by Mn–Fe– Mg/Al_2O_3-COP and Mn–Fe–Ce/Al_2O_3-COP increased by 23.9% and 23.2%, respectively, relative to single ozonation. Mn–Fe–Mg/Al_2O_3 and Mn–Fe–Ce/Al_2O_3 promote ·OH

generation and ·OH-mediated oxidation and are effective at degrading bio-recalcitrant organics in PRW-ROC. The composite metal oxide-loaded Al_2O_3-catalyzed ozonation exhibited great potential and industrial feasibility for advanced treatment of bio-recalcitrant PRW-ROC.

Acknowledgements This project was supported in part by the National Science and Technology Major Project of China (No. 2016ZX05040-003).

References

Altenor S, Carene B, Emmanuel E, et al. Adsorption studies of methylene blue and phenol onto vetiver roots activated carbon prepared by chemical activation. J Hazard Mater. 2009;165:1029–39. doi:10.1016/j.jhazmat.2008.10.133.

Bagastyo AY, Keller J, Poussade Y, et al. Characterization and removal of recalcitrants in reverse osmosis concentrates from water reclamation plants. Water Res. 2013;45:2415–27. doi:10. 1016/j.watres.2011.01.024.

Bagastyo AY, Radjenovic J, Mu Y, et al. Electrochemical oxidation of reverse osmosis concentrate on mixed metal oxide (MMO) titanium coated electrode. Water Res. 2011;45:4951–9. doi:10. 1016/j.watres.2011.01.024.

Bing J, Hu C, Zhang L. Enhanced mineralization of pharmaceuticals by surface oxidation over mesoporous γ-Ti–Al_2O_3 suspension with ozone. Appl Catal B Environ. 2017;202:118–26. doi:10. 1016/j.apcatb.2016.09.019.

Bing J, Wang X, Lan B, et al. Characterization and reactivity of cerium loaded MCM-41 for p-chlorobenzoic acid mineralization with ozone. Sep Purif Technol. 2013;118:479–86. doi:10.1016/j. seppur.2013.07.048.

Chen C, Yoza BA, Wang Y, et al. Catalytic ozonation of petroleum refinery wastewater utilizing Mn–Fe–Cu/Al_2O_3 catalyst. Environ Sci Pollut Res. 2015;22(7):5552–62. doi:10.1007/s11356-015- 4136-0.

Chen X, Zhang Z, Liu L, et al. RO applications in China: history, current status, and driving forces. Desalination. 2016;397:185–93. doi:10.1016/j.desal.2016.07.001.

Dialynas E, Mantzavinos D, Diamadopoulos E. Advanced treatment of the reverse osmosis concentrate produced during reclamation of municipal wastewater. Water Res. 2008;42:4603–8. doi:10. 1016/j.watres.2008.08.008.

Ding J, Lin J, Xiao J, ZhangY Zhong Q, Zhang S, Guo L, Fan M. Effect of fluoride doping for catalytic ozonation of low-temperature denitrification over cerium-titanium catalyst. J Alloy Compd. 2016;665:411–7. doi:10.1016/j.jallcom.2016.01.040.

Einaga H, Futamura S. Oxidation behavior of cyclohexane on alumina-supported manganese oxides with ozone. Appl Catal B: Environ. 2005;60:49–55. doi:10.1016/j.apcatb.2005.02.017.

Joo SH, Tansel B. Novel technologies for reverse osmosis concentrate treatment: a review. J Environ Manag. 2015;150:322–35. doi:10. 1016/j.jenvman.2014.10.027.

Keykavoos R, Mankidy R, Ma H, et al. Mineralization of bisphenol A by catalytic ozonation over alumina. Sep Purif Technol. 2013;107:310–7. doi:10.1016/j.seppur.2013.01.050.

Lu X, Huang X, Ma J. Removal of trace mercury (II) from aqueous solution by in situ formed Mn–Fe(hydr) oxides. J Hazard Mater. 2014;280:71–8. doi:10.1016/j.jhazmat.2014.07.056.

Moreira FC, Boaventura RAR, Brillas E, et al. Electrochemical advanced oxidation processes: a review on their application to synthetic and real wastewaters. Appl Catal B Environ. 2017;202:217–61. doi:10.1016/j.apcatb.2016.08.037.

Moussavi G, Aghapour AA, Yaghmaeian K. The degradation and mineralization of catechol using ozonation catalyzed with MgO/GAC composite in a fluidized bed reactor. Chem Eng J. 2014;249:302–10. doi:10.1016/j.cej.2014.03.059.

Moussavi G, Mahmoudi M. Degradation and biodegradability improvement of the reactive red 198 azo dye using catalytic ozonation with MgO nanocrystals. Chem Eng J. 2009;152:1–7. doi:10.1016/j.cej.2009.03.014.

Nie Y, Hu C, Li N, et al. Inhibition of bromate formation by surface reduction in catalytic ozonation of organic pollutants over β-FeOOH/Al₂O₃. Appl Catal B Environ. 2014;147:287–92. doi:10.1016/j.apcatb.2013.09.005.

Pocostales P, Álvarez P, Beltrán FJ. Catalytic ozonation promoted by alumina-based catalysts for the removal of some pharmaceutical compounds from water. Chem Eng J. 2011;168:1289–95. doi:10.1016/j.watres.2011.10.046.

Pérez-González A, Urtiaga AM, Ibáñez R, et al. State of the art and review on the treatment technologies of water reverse osmosis concentrates. Water Res. 2012;46:267–83. doi:10.1016/j.watres.2011.10.046.

Qi F, Xu B, Chen Z, et al. Catalytic ozonation of 2-isopropyl-3-methoxypyrazine in water by γ-AlOOH and γ-Al₂O₃: comparison of removal efficiency and mechanism. Chem Eng J. 2013;219:527–36. doi:10.1016/j.apcatb.2012.04.003.

Qi F, Xu B, Chen Z, et al. Influence of aluminum oxides surface properties on catalyzed ozonation of 2,4,6-trichloroanisole. Sep Purif Technol. 2009;2:405–10.

Qi F, Xu B, Zhao L, et al. Comparison of the efficiency and mechanism of catalytic ozonation of 2, 4, 6-trichloroanisole by iron and manganese modified bauxite. Appl Catal B Environ. 2012;121–122:171–81. doi:10.1016/j.apcatb.2012.04.003.

Ren Y, Yuan Y, Lai B, et al. Treatment of reverse osmosis (RO) concentrate by the combined Fe/Cu/air and Fenton process (1stFe/Cu/air-Fenton-2ndFe/Cu/air). J Hazard Mater. 2016;302:36–44. doi:10.1016/j.jhazmat.2015.09.025.

Rezaei F, Moussavi G, Bakhtiari AR, Yamini Y. Toluene removal from waste air stream by the catalytic ozonation process with MgO/GAC composite as catalyst. J Hazard Mater. 2016;306-8. doi:10.1016/j.jhazmat.2015.11.026.

Rosal R, Gonzalo MS, Rodríguez A, et al. Catalytic ozonation of fenofibric acid over alumina-supported manganese oxide. J Hazard Mater. 2010a;183:271–8. doi:10.1016/j.jhazmat.2010.07.021.

Rosal R, Gonzalo MS, Rodríguez A, et al. Catalytic ozonation of atrazine and linuron on MnOₓ/Al₂O₃ and MnOₓ/SBA-15 in a fixed bed reactor. Chem Eng J. 2010b;165:806–12.

Santhosh Kumar M, Schwidder M, Grünert W, Brückner A. On the nature of different iron sites and their catalytic role in Fe-ZSM-5 DeNOx catalysts: new insights by a combined EPR and UV/VIS spectroscopic approach. J Catal. 2004;227:384–97. doi:10.1016/j.jcat.2004.08.003.

Santiago-Morales J, Gómez MJ, Herrera S, et al. Oxidative and photochemical processes for the removal of galaxolide and tonalide from wastewater. Water Res. 2012;46:4435–47. doi:10.1016/j.watres.2012.05.051.

Shwana S, Jansson J, Olsson L, et al. Chemical deactivation of H-BEA and Fe-BEA as NH₃-SCR catalysts—effect of potas-

sium. Appl Catal B Environ. 2015;166–167:277–86. doi:10.1016/S1001-0742(09)60298-9.

Tong S, Shi R, Zhang H, et al. Catalytic performance of Fe₃O₄-CoO/Al₂O₃ catalyst in ozonation of 2-(2,4-dichlorophenoxy)propionic acid, nitrobenzene and oxalic acid in water. J Environ Sci China. 2010;22:1623–8. doi:10.1016/S1001-0742(09)60298-9.

Umar M, Roddick F, Fan L. Impact of coagulation as a pre-treatment for UVC/H₂O₂-biological activated carbon treatment of a municipal wastewater reverse osmosis concentrate. Water Res. 2016;88:12–9. doi:10.1016/j.watres.2015.09.047.

Van Hege K, Verhaege M, Verstraete W. Indirect electrochemical oxidation of reverse osmosis membrane concentrates at boron-doped diamond electrodes. Electrochem Commun. 2002;4:296–300. doi:10.1016/S1388-2481(02)00276-X.

Vittenet J, Aboussaoud W, Mendret J, et al. Catalytic ozonation with γ-Al₂O₃ to enhance the degradation of refractory organics in water. Appl Catal A Gen. 2015;504:519–32. doi:10.1016/j.apcata.2014.10.037.

Wu G, Gao Y, Ma F, et al. Catalytic oxidation of benzyl alcohol over manganese oxide supported on MCM-41 zeolite. Chem Eng J. 2015;271:14–22. doi:10.1016/j.cej.2015.01.119.

Xiong Y, He C, Karlsson HT, et al. Performance of three-phase three-dimensional electrode reactor for the reduction of COD in simulated wastewater-containing phenol. Chemosphere. 2003;50:131–6. doi:10.1016/S0045-6535(02)00609-4.

Xu B, Qi F, De Sun, et al. Cerium doped red mud catalytic ozonation for bezafibrate degradation in wastewater: efficiency, intermediates, and toxicity. Chemosphere. 2016;146:22–31. doi:10.1016/j.chemosphere.2015.12.016.

Xu RR, Pang WQ. Chemistry of zeolites and porous materials. In: Structural analysis and properties of porous materials characterization. Li XP, editor. Beijing: Science Press; 2004. p. 145–9 (**in Chinese**).

Yan H, Lu P, Pan Z, et al. Ce/SBA-15 as a heterogeneous ozonation catalyst for efficient mineralization of dimethyl phthalate. J Mol Catal A Chem. 2013;377:57–64. doi:10.1016/j.molcata.2013.04.032.

Zhang ZF, Liu BS, Wang F, et al. High-temperature desulfurization of hot coal gas on Mo modified Mn/KIT-1 sorbents. Chem Eng J. 2015;272:69–78. doi:10.1016/j.cej.2015.02.091.

Zhang G, Qu H, Liu R, et al. Preparation and evaluation of a novel Fe-Mn binary oxide adsorbent for effective arsenite removal. Water Res. 2007;41:1921–8. doi:10.1016/j.watres.2007.02.009.

Zhang L, Ma J, Yu M. The microtopography of manganese dioxide formed in situ and its adsorptive properties for organic micropollutants. Solid State Sci. 2008;10:148–53. doi:10.1016/j.solidstatesciences.2007.08.013.

Zhou M, Tan Q, Wang Q, et al. Degradation of organics in reverse osmosis concentrate by electro-Fenton process. J Hazard Mater. 2012;215–216:287–93. doi:10.1016/j.jhazmat.2012.02.070.

Zhou Y, Zhu W, Liu F, et al. Catalytic activity of Ru/Al₂O₃ for ozonation of dimethyl phthalate in aqueous solution. Chemosphere. 2007;66:145–50. doi:10.1016/j.chemosphere.2006.04.087.

15

Multiple-stacked Hybrid Plays of lacustrine source rock intervals: Case studies from lacustrine basins in China

Shu Jiang[1,2] · You-Liang Feng[3] · Lei Chen[4] · Yue Wu[5] · Zheng-Yu Xu[6] ·
Zheng-Long Jiang[7] · Dong-Sheng Zhou[7] · Dong-Sheng Cai[8] · Elinda McKenna[1]

Abstract Hydrocarbon-producing lacustrine basins are widely developed in the world, and China has a large number of lacustrine basins that have developed since the early Permian. The organic-rich shale-dominated heterogeneous source rock intervals in Chinese lacustrine basins generally contain frequent thin interbeds of stratigraphically associated sandstone, siltstone, marl, dolomite, and limestone. The concept of "Hybrid Plays" as put forth in this article recognizes this pattern of alternating organic-rich shale and organic-lean interbeds and existence of mixed unconventional and conventional plays. Hybrid Plays in lacustrine source rock intervals present a unique closed petroleum system hosting continuous hydrocarbons. The interbedded organic-lean siliciclastic and/or carbonate plays are efficiently charged with hydrocarbons via short migration pathways from the adjacent organic-rich shale that is often also a self-sourced play. We assert "Hybrid Plays" provide the most realistic exploration model for targeting multiple-stacked and genetically related very tight shale, tight and conventional plays together in the entire source rock interval rather than individual plays only. The Hybrid Play model has been proven and works for a wide variety of lacustrine rift, sag and foreland basins in China.

Keywords Lacustrine basin · Hybrid Plays · Shale · Interbed · Source rock · Petroleum system

✉ Shu Jiang
sjiang@cgi.utah.edu

[1] Energy and Geoscience Institute, University of Utah, Salt Lake City, UT 84108, USA

[2] Research Institute of Unconventional Oil and Gas and Renewable Energy, China University of Petroleum (East China), Qingdao 266580, Shandong, China

[3] Research Institute of Petroleum Exploration and Development, PetroChina, Beijing 100083, China

[4] School of Geoscience and Technology, Southwest Petroleum University, Chengdu 610500, Sichuan, China

[5] Sinopec Petroleum Exploration and Production Research Institute, Beijing 100083, China

[6] PetroChina Hangzhou Institute of Geology, Hangzhou 310023, Zhejiang, China

[7] School of Ocean Sciences, China University of Geosciences, Beijing 100083, China

[8] CNOOC, Beijing 100027, China

Edited by Jie Hao

1 Introduction

Lacustrine shales play significant roles in providing source rock in a number of producing lacustrine basins in China, SE Asia, Brazil and Africa (Hu 1982; Kelts 1988; Katz 1990; Carroll 1998; Bohacs et al. 2000; Doust and Sumner 2007; Katz and Lin 2014). In China, lacustrine shales were deposited across a wide variety of tectonic and depositional environments both onshore and offshore since the early Permian (Huang et al. 2015; Jiang et al. 2015, 2016). Due to the relatively smaller size of lacustrine water bodies, lakes experience higher rates of environmental change compared to marine systems, resulting in small reservoir bodies characterized by rapid lateral and vertical facies changes, sensitivity to climate, broad ranges of salinity, pH, etc. (Kelts 1988; Graham et al. 1990; Lambiase 1990; Carroll and Bohacs 1999; Bohacs et al. 2000; Cohen 2003; Gierlowski-Kordesch 2010; Jiang et al. 2013a).

Chinese companies have been aggressively exploring shale gas and shale oil resources in many basins in China

(Zhang et al. 2012a). Besides the successful development of Silurian marine Longmaxi shale gas in Sinopec's Fuling Shale Gas Field in the SE Sichuan Basin (Jiang 2014), Chinese companies have been exploring onshore unconventional hydrocarbon plays in the mature lacustrine basins of China in order to meet and maintain oil and gas production needs. Even with this dedicated effort, the lacustrine shale gas and shale oil exploration and production are moving slowly since lacustrine shales in China are clay-rich, heterogeneous with a high wax content and have poor hydrocarbon flow capacity compared to marine shale (Jiang et al. 2013b, 2014, 2016; Li et al. 2013a; Jiang 2014; Katz and Lin 2014). More studies on lacustrine shale are needed to help find more realistic exploration and production models.

Naturally fractured lacustrine shales have yielded some production in China since 1960s from the Songliao and Bohai Bay Basins in NE China, the Subei Basin in East China, the Jianghan and Nanxiang Basins in Central China, the Sichuan Basin in SW China, and the Qaidam, Junggar and Turpan-Hami Basins in NW China (Li et al. 2006; Zhang et al. 2014). Table 1 lists examples of production in the naturally fractured shale reservoirs of various lacustrine basins across China. The production is considered to be from the free state hydrocarbons stored in the naturally open fractures (Li et al. 2006). These shales, however, have only attracted exploration interest in the last 5 years following the shale gas success in the USA. Currently, the primary targeted tight oil/shale oil plays are listed below in age order (Wang et al. 2014):

1. Lower Permian Fengcheng tight dolomitic reservoir in the Junggar Basin and Middle Permian Lucaogou tight dolomite in the Santanghu Basin and the Jimsar depression in the Junggar Basin,
2. Upper Triassic Yanchang7 (Chang7) tight sandstone in the Ordos Basin,
3. Tight sandstone in Upper Cretaceous in the Songliao Basin,
4. Jurassic tight lacustrine carbonate in the Sichuan Basin,
5. Paleogene Shahejie Fm. in the Bohai Bay Basin,
6. Paleogene Hetaoyuan Fm. in the Biyang Basin and
7. Oligocene Shangganchaigou Fm. in the Zhahaquan area in the Qaidam Basin.

For example, the Zha7 well in the Qaidam Basin produced oil from tight sandstones in the Oligocene Shangganchaigou Fm. at an average rate of 42 tonnes (308 barrels)/day (1 tonne = 7.33 barrels); the Qiping1 well in the Songliao Basin produced oil from hydraulically fractured tight sandstone interbeds in organic-rich shale at a rate of 10 tonnes (73 barrels)/day; Lei88 well in the Liaohe Sub-basin in the Bohai Bay Basin produced oil from tight carbonates at rate of 47 tonnes (344 barrels)/day; Well in the Xi233 tight oil demonstration area in the Ordos Basin produced oil from Upper Triassic Yanchang7 (Chang7) tight sandstone at an average rate of 88+ tonnes (645+ barrels)/day (Du et al. 2014). Due to the overwhelming focus and priority of business decisions to target shale plays, E&P targeting thin sandstone, siltstone and

Table 1 Historical production rates from naturally fractured shales in lacustrine basins in China (compiled from Li et al. 2006; Jiu et al. 2013; Zhang et al. 2014)

Basin	Age and Fm.	Well	Year	Average production rate per day
Zhanhua Sag, Bohai Bay Basin	Paleogene Shahejie3 shale	He54	1973	Oil: 91.4 tonnes (670 barrels) Gas: 2740 m³ (95,900 ft³)
Zhanhua Sag, Bohai Bay Basin	Paleogene Shahejie3 shale	Luo42	1990	Oil: 79.7 tonne (584 barrels) and Gas: 7746 m³ (271,110 ft³)
Liaohe Sub-basin, Bohai Bay Basin	Paleogene Shahejie3 shale	Shugu165	2011	Oil: 21 tonnes (154 barrels)
Dongpu Sub-basin, Bohai Bay Basin	Paleogene Shahejie3 shale	Pushen-18-1	2011	Oil: 370 tonnes (2712 barrels)
Biyang Sub-Basin, Nanxiang Basin	Paleogene Hetaoyuan3 shale	Biye HF-1 well (Horizontal well)	2011	Oil: 20.8 tonnes (152 barrels)
Songliao Basin	Upper Cretaceous Qingshankou shale	Xin19-11 well and Ying12 well		Oil: 26.4-35.2 tonnes (194-258 barrels) for well Xin19-11 and 4 tonnes (29 barrels) for Ying12
Jianghan Basin	Paleogene Qian3 shale	Wang4-12-2	1989	Oil: 49.3 tonnes (361 barrels)
East Sichuan Basin	Jurassic Dongyuemiao shale	Jian111	2010	Gas: 3925 m³ (137,375 ft³)
NE Sichuan Basin	Jurassic Dongyuemiao shale	Yuanba9	2010	Gas: 11,500 m³ (402,500 ft³)

carbonate interbeds within lacustrine shales seems to have been ignored previously, just as the shale plays were historically ignored in the tight sandstone oil and gas play exploration focus areas. Our recent detailed investigation of shale gas and shale oil plays in China found that the source rock intervals in China's lacustrine basins are highly heterogeneous, having dominantly organic-rich shale source rock with thin interbeds of organic-lean siltstone, sandstone and carbonates (Jiang et al. 2015, 2016). For the most part, these thin zones within the source rock interval received only minor attention in the past since hydrocarbons do not flow without stimulation.

Sporadic reports of the production of shale oil, shale gas, tight oil, and tight gas have not closed the knowledge gap that continues to exist between naturally fractured shale reservoirs, self-sourced shale reservoirs and coarse-grained interbed intervals. Previously they were separately targeted and studied during different time frames and exploration schedules as either shale plays or tight plays. The lack of systematic studies linking the organic-rich and organic-lean reservoirs inhibits a full understanding of the petroleum systems of the source rock intervals. Based on the geologic characteristics of lacustrine source rock intervals and successful oil production from the self-sourced organic-rich shale, e.g., Woodford shale and organic-lean low permeability carbonates, e.g., Niobrara and Bakken, we put forth the "Hybrid Plays" concept for multiple-stacked shale related plays (stratigraphically associated siliciclastic shale, siltstone, sandstone, and carbonate plays) within the lacustrine source rock intervals. The main objective of this study is to summarize the characteristics, distribution and production potential of "Hybrid Plays" of lacustrine source rocks using typical lacustrine basins in China as case studies. The studies are based on regional geologic overviews, data observations and interpretation of lithofacies, geochemistry, mineralogy, log responses and well tests, evaluation of petroleum systems for lacustrine source rock intervals, and production analysis of both organic-rich shales, and organic-lean siliciclastic and carbonate interbeds. This integrated approach will help to better characterize and understand the types of shale-related mixed resource plays and their potentials in the China lacustrine basins and stimulate further exploration and study of lacustrine basins all over the world.

2 Concepts and methodology

The concept of 'Hybrid Plays' presented in this paper grew out of our comprehensive review of many terms, descriptions and geological interpretations related to unconventional reservoirs in the last two decades. Historically, there have been many terms used to describe the shale-related

unconventional plays, e.g., "continuous accumulation," "unconventional reservoir/play," "shale play or shale reservoir," "hybrid shale-oil resource system," "tight (oil) play," and "resource play" (see Table 2). "Continuous accumulation" was mainly used to characterize the reservoirs charged with continuous hydrocarbons, e.g., basin-centered tight gas (Schmoker 1995); it does not define the types of unconventional reservoirs. "Unconventional reservoir/play" is a broadly generic term that has been used by industry for a wide range of play types including "coal bed methane (CBM)" "tight gas," "oil shale," "shale gas," "shale oil," "tight oil," "halo oil," and others (Passey et al. 2010; Clarkson and Pedersen 2011). This usage has created a certain amount of ambiguity in academia, researchers, and industry regarding the meaning of these terminologies since they are either too broad or too narrow. The definition of "shale play or shale reservoir" is a continuous fine-grained tight plays associated with organic-rich shale source rock that is also a reservoir (Curtis 2002; Bustin 2006; Passey et al. 2010). The lithologic definition of shale is a rock consisting of extremely fine-grained particles to variable amounts of silt-size particles with a wide range in mineral composition (clay, quartz, feldspar, heavy minerals, etc.) (Passey et al. 2010). "Hybrid shale-oil resource system" defines all fine-grained organic-rich and organic-lean plays as a system where hydrocarbons are produced from both organic-rich shales and juxtaposed organic-lean lithofacies (Jarvie 2012a, b). "Tight (oil) play" was used to define tight or shale reservoirs producing light oil (Clarkson and Pedersen 2011; EIA 2013); it is not clear that it refers to tight sandstone or carbonate reservoirs or include both organic-rich shale and organic-lean tight sandstone or carbonate reservoirs. "Resource play" was used describe an accumulation of hydrocarbons existing over a large area in tight reservoirs producing unconventional oil and gas via hydraulic fracturing (e.g., Sonnenberg 2014); it is too general and not clear in play types.

In recent decades, most organic-rich fine-grained tight plays, e.g., Barnett, Niobrara, Eagle Ford, and Bakken have been labeled "shale plays" and the terms of "shale oil," "tight oil," and "resource play" are often used interchangeably in public discourse (Curtis 2002; Bustin 2006; Passey et al. 2010; Jarvie 2012a; EIA 2013). In reality, the shale interval is only a subset of all low permeability fine-grained tight rocks in the source rock interval, including sandstone, siltstone, carbonates and shale. For example, among the so-called "Shale Plays" in North America large variability exists for vertical and lateral lithologies, organic matter content, petrophysical properties, etc. Only the Barnett and Tuscaloosa meet the previous definition of "Shale Plays." So-called shale plays like the Bakken/Exshaw play (USA and Canada) actually consist of key

Table 2 Comparison between historical terms for unconventional plays

Term	Source	Advantage	Disadvantage
Continuous accumulation	Schmoker (1995)	Emphasize the continuous distribution of unconventional hydrocarbons	Did not define the unconventional reservoir types
Unconventional reservoir/play	Passey et al. (2010), Clarkson and Pedersen (2011)	Define reservoirs with tight nature comparing to conventional reservoirs	Too broad or too narrow
Shale play or shale reservoir	Curtis (2002), Bustin (2006), Passey et al. (2010), Jarvie (2012a, b), EIA (2013)	Define reservoirs associated with organic-rich fine-grained tight plays	Confusion of self-sourced very fine-grained organic-rich shale reservoir and fine-grained organic-lean facies sourced from neighboring organic-rich shale
Hybrid shale-oil resource system	Jarvie (2012a, b)	Define all fine-grained organic-rich and organic-lean plays as a system	Still focused on shale and did not capture the reservoir types in the source rock interval
Tight (oil) play	Clarkson and Petersen (2011), EIA (2013)	Tight/shale reservoirs producing light oil	Not clear, either was used to refer to tight sandstone or carbonate reservoirs or include both organic-rich shale and organic-lean tight sandstone or carbonate reservoirs
Resource play	Sonnenberg (2014)	Tight reservoirs producing unconventional oil and gas via hydraulic fracturing	Too general and not clear in play types

permeable tight dolomite, siltstone and sandstone interbeds, shales, and carbonate reservoirs (LeFever 1991; Zhang et al. 2016). The Niobrara play is a low permeability chalk and self-sourced marl reservoir with continuous hydrocarbon accumulation (Sonnenberg 2011). The organic-lean interbeds interbedded within organic-rich shale are more conductive than shale in accumulating oil and gas reserves and allowing oil and gas to travel to the wellbore due to their higher porosity and permeability (Chalmers et al. 2012; Jarvie 2014).

The organic-rich lacustrine source rock intervals usually have a mixed succession of frequently interbedded shale, sandstone and/or carbonate. This interval used to be called shale due to fine-grained nature in most cases. The reality is, neither "Shale Play" or "Tight Play" and other historical terms mentioned above are not appropriate geologic terms for lacustrine petroleum systems having stacked, diverse, and rapidly changing lithofacies since they either refer to only a single lithofacies or they lump together rocks that are lithologically diverse and have different petrophysical properties in the source rock interval. We propose "Hybrid Plays" as a term to characterize all plays associated with source rock deposits in lacustrine basins. The following four characteristics define Hybrid Plays and can be used as the criteria to identify different Hybrid Plays:

1. The source rock intervals contain a mixed succession of genetically and stratigraphically associated organic-rich shale and organic-lean sandstone, siltstone and/or carbonate occurring within distinct geographical areas.

The organic-lean sandstone, siltstone, and/or carbonate are usually sandwiched between or juxtaposed against mature shale source rock.

2. Reservoirs include three end members of conventional reservoirs with high inorganic porosity and Darcy range permeability, tight reservoirs with low microdarcy range permeability, and very tight fine-grained shale reservoirs with low permeability and low inorganic and organic porosity.

3. Unlike conventional reservoirs, the petroleum accumulations in "Hybrid Plays" are not formed primarily through buoyancy. Instead, oil and gas were either generated and trapped in the source rock as in situ hydrocarbon or have efficiently migrated into fine- to coarse-grained siltstone and sandstone directly adjacent to the source rocks (Raji et al. 2015). Such hybrid conventional-tight-very tight resource plays are basically a closed petroleum system containing the oil and gas originating from organic-rich shale and is now stored within the source rock stratigraphic interval. The considerable hydrocarbon retention hosted in "Hybrid Plays" is expected to mimic the overall characteristics of continuous accumulation as defined by Schmoker (1995).

4. Horizontal drilling and hydraulic fracturing are needed to release and produce the hydrocarbons from Hybrid Plays.

The methodology in this paper follows the approach from conceptual hypothesis, observation, data description, data integration, interpretation, and then to validation. (1) The concept of "Hybrid Plays" was proposed based on a paucity of scientific appreciation for the play types and experience and observations with lacustrine systems. (2) Extensive literature reviews, geologic evaluation, and analysis were conducted of unconventional petroleum systems in typical lacustrine basins in China. (3) The reservoirs were evaluated to identify all the proven and potential Hybrid Plays in the source rock intervals of lacustrine basins in China using the criteria of "Hybrid Plays" based on well log data, lithology, facies, geochemistry, porosity and permeability, mineralogy, petrology, hydrocarbon shows and well test data. (4) Properties and models of "Hybrid Plays" in different types of lacustrine basins are summarized.

3 Geology of source rocks of lacustrine basins in China

China has a series of widely distributed lacustrine basins ranging in age from early Permian to Quaternary. During the late Paleozoic (early Permian) after the Hercynian Orogeny, lacustrine foreland basins started to develop in northwest China, e.g., the Junggar and Turpan-Hami basins (Wang and Mo 1995; Wang and Li 2004; Zou et al. 2010; Figs. 1, 2). Beginning in the Mesozoic, the continent of Asia experienced a period of intracontinental tectonic events when Indosinian, Yanshanian and Himalayan orogenies were the predominant forces shaping the landmass in China (Wang et al. 2005). The sea retreated in most regions during this time and the dominant depositional systems were marked by a preponderance of fluvial-lacustrine environments creating a high number of lacustrine basins throughout China (e.g., Triassic foreland Ordos Basin in North China, Jurassic foreland Junggar and Qaidam Basins in Southwest China, Sichuan Basin in Southwest China, Cretaceous Songliao sag Basin in NE China, Paleogene rift Bohai Bay, Jianghan, Nanxiang and Subei basins in East China, etc. (Figs. 1, 2). The early Permian to Quaternary organic-rich shales in China occur in more than 50% of the lacustrine basins (Wang et al. 2014). These organic-rich lacustrine shales comprise the majority of China's source rocks for three-quarters of China's oil resources and approximately one-half of the gas resources and are considered potential shale plays (Jia and Chi 2004; Zhang et al. 2008a; Zou et al. 2010, 2014; Lin et al. 2013; Jiang 2014; Fig. 2).

The rift development can be divided into pre-rift, syn-rift and post-rift stages according to the initiation and termination of rifting (Schlische 2003). Lacustrine basins in China with similar age and tectonic settings share comparable geological history, e.g., Cenozoic rift lacustrine

basins can also be described as a succession of pre-rift, syn-rift, and post-rift stages, allowing them to be regionally correlated for tectonic development, sedimentary evolution, source rock and reservoir distribution (Jiang et al. 2013a, b). Source rocks have a fundamental role in controlling the petroleum systems in the lacustrine basins in China. High-quality source rocks were deposited during the stable syn-rift phase and post-rift high lake level phase (Jiang et al. 2013a, b). Characteristic rock types in source rock intervals are associated with different lacustrine basins in China due to different tectonic histories, paleoclimate and sediment supply conditions (Fig. 1):

1. Early Permian source rocks in Junggar and Santanghu basins in Northwest China are characterized by shale and carbonates developed in hypersaline settings;
2. Upper Triassic Yanchang7 (Chang7) Fm. source rock in the Ordos Basin is characterized by shale, sandstone, and siltstone deposited mainly in a freshwater lake (Zhu et al. 2004);
3. Eocene to Oligocene Shahejie shale source rock intervals are characterized by gypsum, limestone, dolomite, sandstone, marl, and shale deposited in brackish to fresh water.

In the following case studies, each primary lacustrine source rock in the major basins will be characterized in detail for their lithofacies association, reservoir and hydrocarbon accumulation, and "Hybrid Play" potential.

4 Case studies of Hybrid Plays within source rock intervals of major lacustrine basins in China

After nearly 70 years of exploration and production, the exploration of onshore lacustrine basins in China has become mature for conventional plays. The exploration focus in these lacustrine basins has been shifting toward individual, unconventional tight sandstones, limestones and dolomites, and recently to shale reservoirs to increase reserves and maintain production at current levels. As we mentioned in the Introduction section, the recent exploration results for unconventional petroleum systems demonstrate that many of these lacustrine basins are likely to be prospective for shale oil and shale gas in organic-rich shale and tight oil and gas from interbedded sandstone and carbonate rocks within source rock intervals. To date, little attention has been paid to advancing the study of the petroleum systems for the entire source rock interval.

This study targets typical lacustrine basins in China representing different basin types, different lithofacies associations, various salinity levels and distinct tectonic settings in order to characterize and compare potential reservoirs within source rocks and identify common

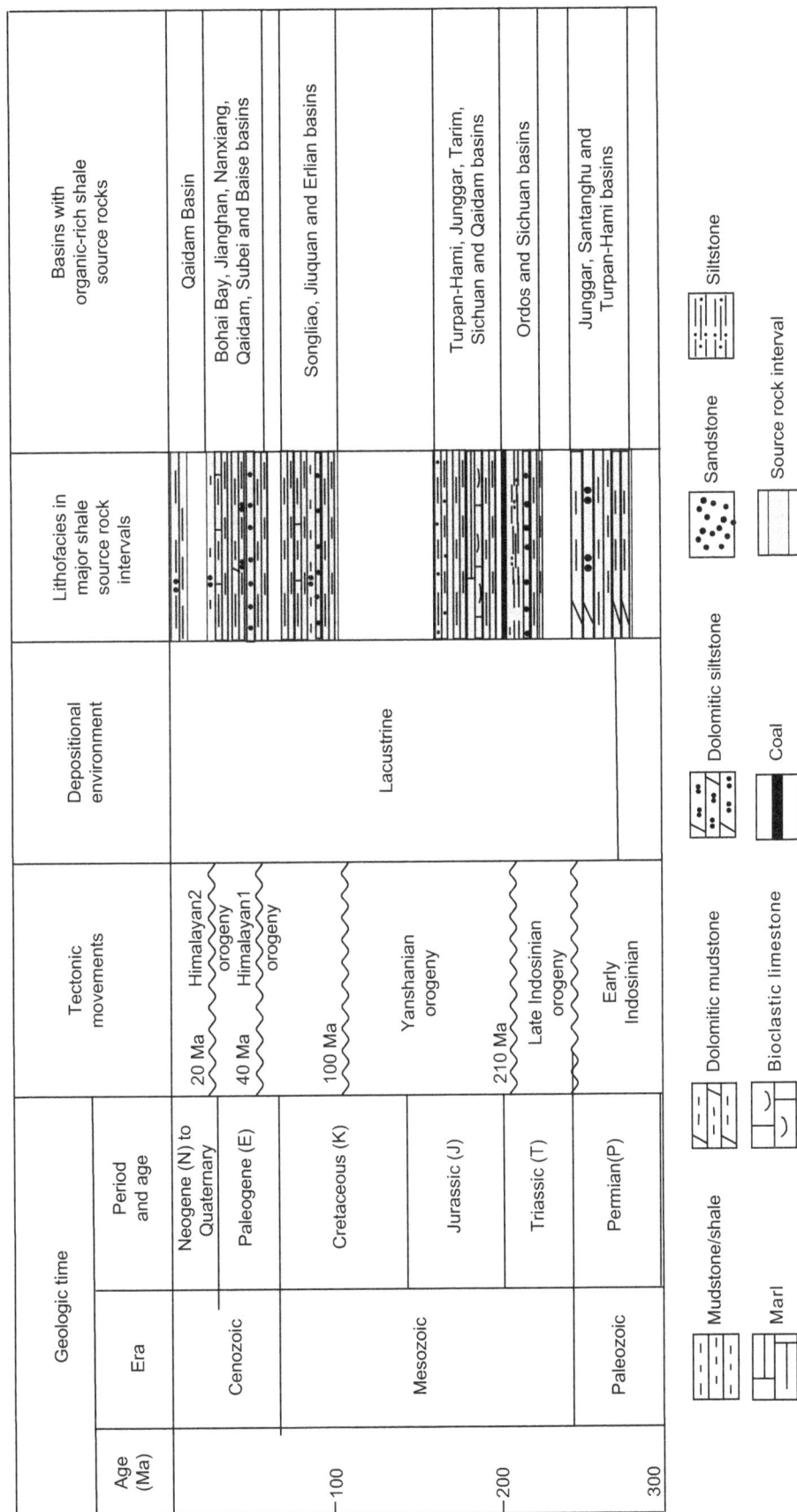

Fig. 1 Early Permian to Quaternary geologic history, development of shale source rocks, and lithofacies associated with organic-rich shales in lacustrine basins in China (compiled from Wang et al. 2005; Wang and Mo 1995; Wang and Li 2004; Zou et al. 2010)

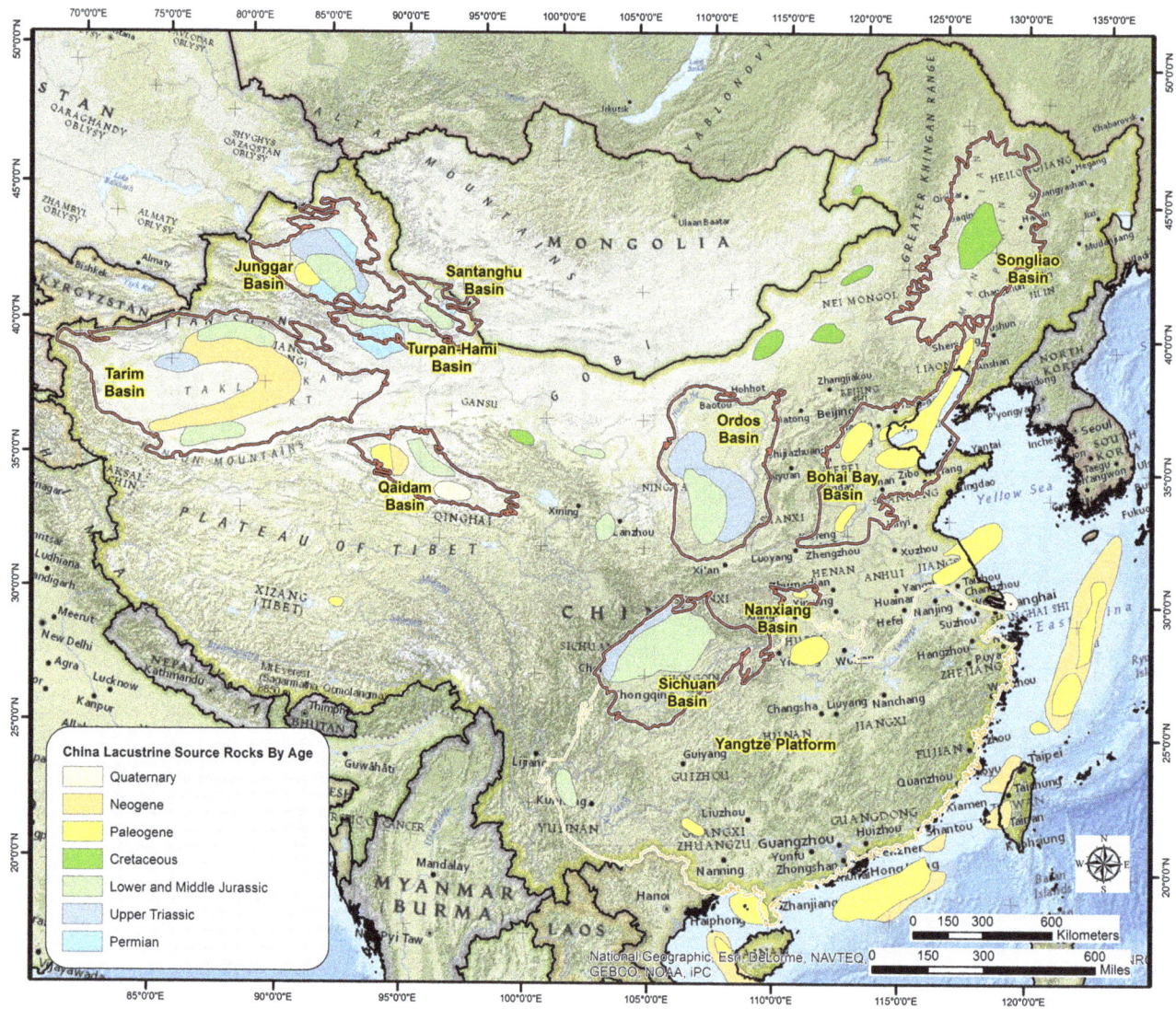

Fig. 2 Ages of the lacustrine basins in China and distribution of onshore lacustrine source rocks

"Hybrid Plays." The high-graded lacustrine basins ordered in age include:

1. Early to Middle Permian Junggar
2. Middle Permian Santanghu
3. Upper Triassic Ordos
4. Triassic to Jurassic Sichuan
5. Jurassic Turpan-Hami
6. Upper Cretaceous Songliao
7. Jurassic to Neogene Tarim
8. Paleogene Bohai Bay
9. Paleogene Nanxiang, and
10. Paleogene to Neogene Qaidam (Fig. 2).

4.1 Hybrid Plays cases within Middle Permian mixed siliciclastics and carbonate source rocks in the Junggar and Santanghu basins in Northwest China

The Junggar Basin and neighboring Santanghu and Turpan-Hami basins in NW China have similar geology. The Carboniferous, Permian, Triassic, and Jurassic sections all contain source rocks (Cao et al. 2005; Hao et al. 2011). Recognizing the importance of many source rock intervals including the Lower Permian Fengcheng shales (Hu et al. 2015; Xiang et al. 2015), the Middle Permian Lucaogou and equivalent Pingdiquan Fm. are considered the most important source rock intervals due to wide distribution, large thickness, high TOC, and good hydrocarbon potential (Carroll 1998). The Lucaogou Fm. in the Junggar and Santanghu Basins spans a wide variety of depositional environments from shallow to deep hypersaline lake. The Lucaogou Fm. consists of silty dolomite, marl, dolomitic siltstone, dolomitic shale, and shale, which are of mixed siliciclastics and carbonate. The dolomitic shale and silty dolomite were deposited in areas of low clastic supply, either in central basin areas or at basin margins starved of clastic input. We use the Jimsar Depression in the Junggar Basin and the Malang Depression in the Santanghu Basin as representatives to characterize the Hybrid Plays of Middle Permian mixed siliciclastics and carbonate source rock.

4.1.1 Hybrid Plays within Middle Permian mixed siliciclastics and carbonate source rock in the Jimsar Depression in the Junggar Basin

The Jimsar Depression with an area of 1500 km^2 and located in the eastern portion of the Junggar Basin is a key area for current tight oil exploration targeting the Middle Permian Lucaogou Fm. The area with source rock greater than 200 m thick extends more than 300 km^2 in the Jimsar Sag. The effective source rocks include black shale, calcareous shale, dolomitic shale, and silty shale with Type I to II$_1$ kerogen in the Lucaogou source rock interval. Each of the source rock intervals generally have an average TOC of >3% (up to 31%) and an average $S_1 + S_2$ of 21.0 mg g^{-1}, and R_o ranging between 0.5% and 1.2%. The storage types for these dolomitic reservoirs include large quantities of nanoscale to microscale interparticle pores, dissolution pores, and natural fractures. The average porosity and permeability of the tight dolomitic siltstone and silty dolomite reservoirs are about 8.8% and 0.05 mD, respectively (Kuang et al. 2012).

Exploration of structural highs of Permian dolomitic shale reservoirs in the Junggar Basin began in 1981. Since 2010, many wells have been drilled targeting dolomitic tight oil prospects (Kuang et al. 2012). Our study integrating facies mapping, geochemistry, and type well analysis indicates the "Hybrid Plays" in the Jimsar Depression include the low permeability dolomite, silty dolomite, shaly dolomite, dolomitic siltstone, organic-rich dolomitic shale, and shale in the Lucaogou source rock interval (Fig. 3a, b). Using the vertical Ji174 well as an example (Fig. 3b), the heterogeneous source rock interval consists of black to gray shale, silty shale, dolomitic shale, dolomitic siltstone, siltstone, dolomite, and limestone. Effective source rocks include black shale, calcareous shale, dolomitic shale, and silty shale with TOC ranging from 2% to

11% (average of 4.0%), 0.2% to 13% (average of 3.7%), 0.77% to 7% (average of 3.3%) and 0.4% to 10.2% (average of 3.6%), respectively. The presence of mobile oil within the entire source rock interval is easily seen from the abundant oil stains on naturally fractured fine-grained cores and is evidenced by T_{max} of 430–460 °C, R_o of 0.78%–0.95% and the abundant extract of chloroform bitumen "A" in all the lithologies. The high free oil content indicated by chloroform bitumen "A" in the organic-lean dolomitic siltstone reveals that the oil migrated from juxtaposed organic-rich source rock intervals (Fig. 3b). The well was fractured targeting four zones with a gross thickness of 55 m in the source rock interval and oil was produced at a rate of 15 tonnes (109 barrels)/day. The production is from a "Hybrid Play" system consisting of shale, dolomitic shale, dolomitic siltstone and siltstone reservoirs (Fig. 3b). The best quality reservoir is from the dolomitic siltstone with high oil saturation (>90%), high porosity (12%–16%) and high permeability (0.1–1 mD) compared to the organic-rich shale interval with relatively lower oil saturation (<70%), low porosity (<5%) and low permeability (<0.01 mD) (Jin et al. 2015). The presence of good-quality reservoir facies (brittle dolomitic rock), well-developed fracture networks, mature source rock (oil window), moderate structural deformation, and thick regional cap rock make the Lucaogou a "Hybrid Play" system with continuous oil accumulation. Many recent commercial wells in the Jimsar Depression confirm this designation as a Hybrid Play system, e.g., the Lucaogou source rock interval in Ji25, Huobei2 and Fengnan7 wells reportedly produced oil from hybrid dolomitic siltstone, dolomitic shale, siltstone, and silty dolomite plays. Initial Production (IP, based on 24-h test) rates were measured at 18.16 tonnes (134 barrels)/day, 14.39 tonnes (106 barrels)/day and 12.3 tonnes (91 barrels)/day, respectively, after moderate reservoir stimulation (Kuang et al. 2012). The variation in production for these wells located in different areas is attributed to the spatial heterogeneity in facies change, structural setting, source rock maturity, source rock type, reservoir type, pressure, etc.

4.1.2 Hybrid Plays within Middle Permian mixed siliciclastics and carbonate source rock in the Malang Sag in the Santanghu Basin

The Santanghu Basin is located to the southeast of the Junggar Basin in NW China. The key area—Malang Sag—is located in the SE Santanghu Basin. Due to little terrestrial input and hypersaline depositional conditions similar to the Jimsar Depression in the Junggar Basin, the Lucaogou Fm. is mainly composed of alternating mixed siliciclastic mudstone and carbonate rocks. A localized fan delta has a limited extent in the southwestern margin of the

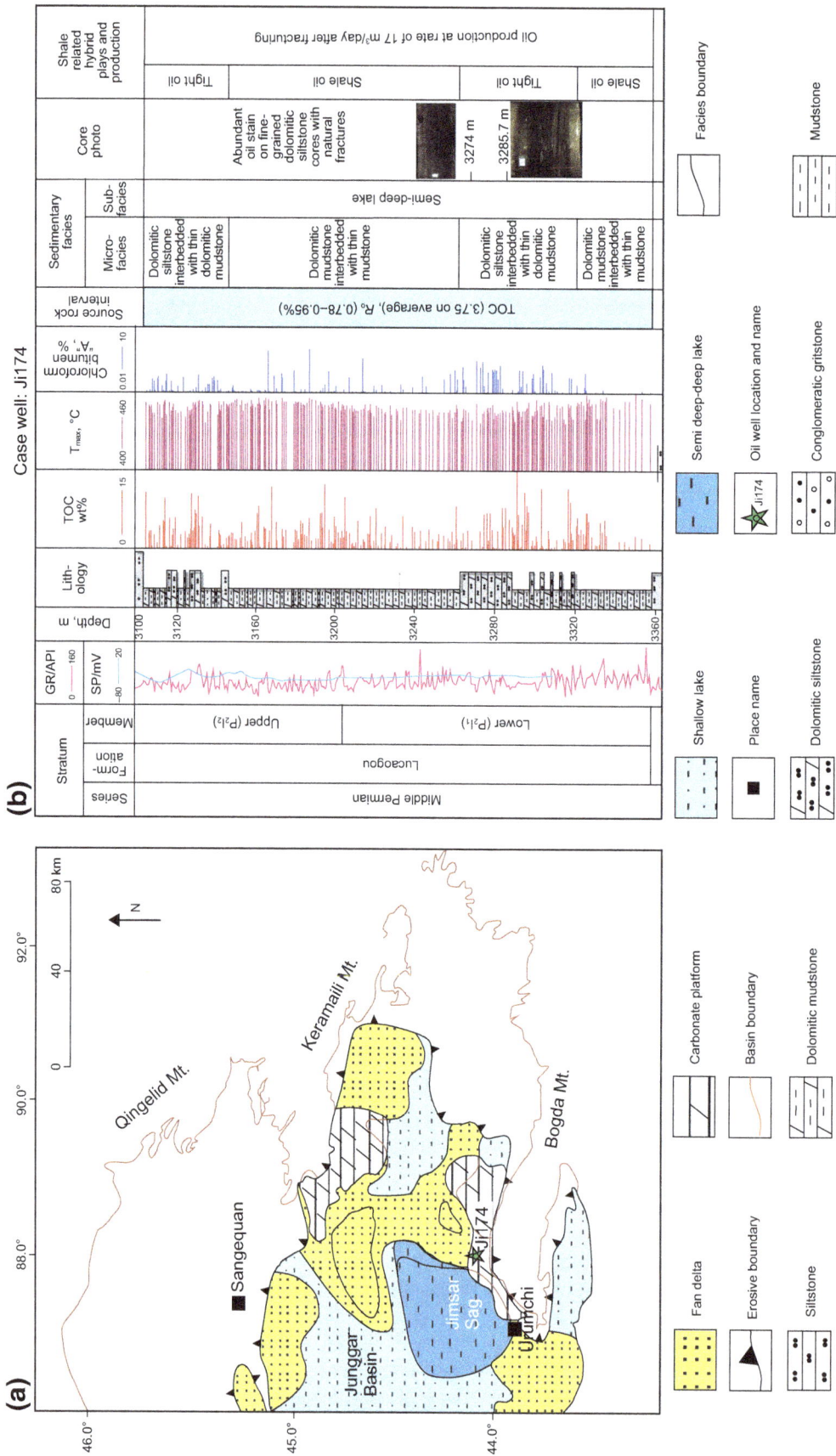

Fig. 3 **a** Sedimentary facies distribution for the Middle Permian Lucaogou Fm. in the eastern Junggar Basin and **b** Vertical well Ji174 demonstrating production from "Hybrid Plays" within the mixed siliciclastics and carbonate source rock interval of the Jimsar Sag, Santanghu Basin. Lithology modified from Kuang et al. (2012) and geochemical data modified from Kuang et al. (2014). The facies were mapped and identified based on seismic interpretation, lithologic description of core, completion report, petrologic analysis, and XRD mineralogy test of samples. The Hybrid Plays were identified based on the lithofacies, geochemistry, petrophysics, and production test results

Malang sag. Volcanic rocks are mainly distributed in the Mazhong structural belt (Fig. 4a). The primary source rock is the Middle Permian Lucaogou Fm. The oil reservoirs of the Permian interval mainly consist of mudstone, dolomitic shale or limestone-rich mudstone/marl, argillaceous limestone, and calcareous dolomite within the Lucaogou Formation (P_2l) source rock of the middle Permian (Liang et al. 2014), making it overall a self-sourced reservoir. The first well Ma1 was drilled in 1996 targeting interbedded intervals of fractured tight carbonates producing at a rate of 3.3 tonnes (24 barrels)/day. Since that discovery tight oil, exploration has been the focus in this basin. According to a variety of sources, the Malang Sag is estimated to have 1.9 billion tonnes (14 billion barrels) of tight oil resources.

Based on the vertical Ma7 well drilled and hydraulically fractured in 2004, the heterogeneous Lucaogou source rock interval is mainly composed of dolomitic mudstone, tuffaceous mudstone, and calcareous shale, making it a high-quality source rock with TOC content ranging from 1% to 7% and R_o values from 0.5% to 0.9% (Fig. 4b). This "Hybrid Play" produced oil from the source rock interval interbedded with dolomitic mudstone, tuffaceous mudstone, calcareous shale, and argillaceous limestone reservoirs at a rate of 22 tonnes (160 barrels)/day after hydraulic fracturing. For this Ma7 well, the lower member of Lucaogou Fm. consisting of dolomitic mudstone and tuffaceous mudstone is interpreted mainly as a shale oil play and the upper member of the Lucaogou Fm. consisting

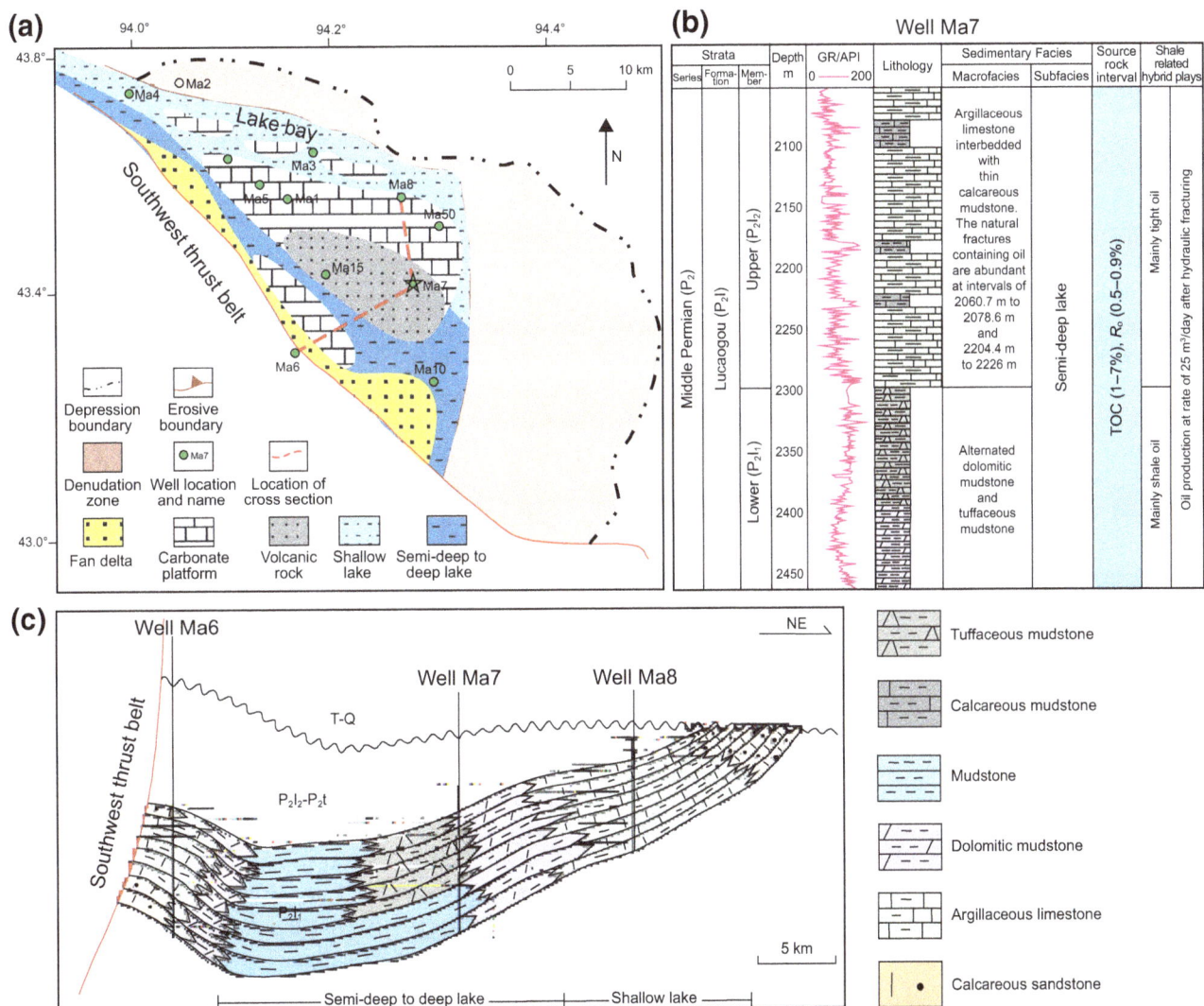

Fig. 4 a Sedimentary facies distribution for the Middle Permian Lucaogou Fm. in the Malang Sag of the Santanghu basin. **b** Stratigraphic column showing lithofacies, geochemistry, and drilling results of vertical Ma7 Well showing "Hybrid Plays" in the Middle Permian Lucaogou Fm. **c** Cross-well Ma6-Ma7-Ma8 profile illustrating spatial lithofacies variation in the Lower member of the Lucaogou Fm. from the margin to the central sag. The facies were mapped and identified based on seismic interpretation, lithologic description of core, completion reports, petrologic analysis, and XRD mineralogy test of samples. The Hybrid Plays were identified based on the lithofacies, geochemistry, petrophysics, and production test results

of naturally fractured argillaceous limestone interbedded with thin, organic-rich calcareous mudstone is regarded mainly as a tight oil reservoir (Fig. 4b). The regional well correlation shows that the lithofacies for the lower member of the Lucaogou Formation varies from calcareous sandstone, to argillaceous limestone, and dolomitic mudstone, and then to mudstone from the margin toward the central sag (Fig. 4c). The multiple lithofacies, continuous and regional distribution of the Lucaogou source rock interval clearly place it in the "Hybrid Play" model. Many wells (e.g., Ma58, Ma1, Ma6, Ma7, Ma8, Ma19, Ma21, Ma24) confirm the "Hybrid Play" characteristics of the Lucaogou source rock interval. The recent horizontal well and multistage hydraulic fracturing further improved the production rates and increased the economic viability. The horizontal Ma58H well produced oil at a rate of 115 tonnes (839 barrels)/day (based on an average 24-h test result) from a lateral of 804 m in a fractured section of the Lucaogou shale and argillaceous carbonate. Based on the drilling results, the oil is primarily produced from the tight argillaceous dolomite play with a second zone of production from the self-sourced Lucaogou dolomitic shale, calcareous shale, and limestone-rich shale/marl plays. The production is due in part to the brittle nature of the argillaceous dolomite allowing more efficient hydraulic fracturing. The marl, dolomitic shale, and calcareous shale are more ductile.

Based on the analysis of the Lucaogou source rock interval in the Junggar and Santanghu Basins, the major production of liquid hydrocarbons is from the Lucaogou dolomitic siltstone and argillaceous dolomite, which are analogous to the Bakken play in the Williston Basin in USA and Canada. The dolomite-rich mixed siliciclastics and carbonate formation is comparable to the prolific oil producing Middle Bakken mixed dolomite and siltstone.

4.2 Hybrid Plays within Upper Triassic siliciclastic source rock in Ordos Basin

The Ordos Basin is situated in the North China Block and is bounded by the Yin, Lüliang, Qinling, Liupan, and Helan mountains to the north, east, south, and west, respectively. The stratigraphy of the Ordos Basin records an evolution from a cratonic-passive margin basin through the Paleozoic to a Triassic intracratonic foreland basin. During late Triassic, the basin evolved into a continental setting due to the Indosinian tectonic movements (Yang et al. 2005; Hanson et al. 2007; Zou et al. 2012). No faults and folds developed in the basin and the gentle structure is a smooth monocline with an east to west dip of less than 1° (He 2002).

The stratigraphy of the Yanchang Formation reveals the temporal evolution of a freshwater siliciclastic lacustrine setting based upon the lack of evaporates and carbonates.

The source rock interval is the 7th member of the Upper Yanchang (or Chang7) shale and was deposited mainly in semi-deep to deep lake settings (Chen et al. 2007). The hot shale (TOC > 2%, up to 36% and oil bearing) thickness is between 20 and 40 m in the semi-deep to deep lake setting in most parts of the Ordos Basin (Fig. 5a). The heterogeneous organic-rich shale source rock is usually interbedded with fine-grained, massive sandy debrite, fine-grained siltstone to sandstone turbidites, and sandy slumps (Zou et al. 2012, 2015; Liu et al. 2017; Fig. 5b, c), indicating that tight oil and shale oil potentials coexist in the Chang7 source rock interval. From the L189 well (Fig. 5), we can clearly see that organic-rich shale is mainly developed in the transgressive systems tract (TST). The test in the TST interval consisting of organic-rich shale and sandstone yielded oil of 13.2 tonnes (98 barrels)/day. The production tests are influenced by both the shale oil and tight sandstone oil plays based on perforation locations, lithofacies association, and log interpretation. The presence of shale oil accumulation is identified based on high gamma ray, high resistivity, low density, and high acoustic travel time (Fig. 5b). In the high-stand systems tract (HST), the tight sandstone is well-developed supporting the interpretation of a tight (sandstone) oil play (Fig. 5b, c). A regional stratigraphic cross section for the proven tight sandstone oil producing play further confirms the shale play potential based on overpressure inferred by residue pressure (the difference between formation pressure and hydrostatic pressure) in this organic-rich shale interval (Fig. 5d). Slight overpressure also exists in the tight sandstone oil reservoir in this area even though the Chang7 Fm. in the Ordos Basin is generally underpressured. The higher residue pressure in the organic-rich shale compared to tight sandstone indicates abundant generated oil is still trapped in the shale. At the same time, the relatively high permeability with microdarcy to millidarcy range is associated with the sandstone interval (Fig. 5d), explaining why the sandstone is a target for tight oil exploration and development. A significant portion of the generated oil from the organic-rich shale migrated to the adjacent interbedded tight sandstone. The interbedded organic-rich oil-bearing shale and thin sandstone layers form "sandwich" style hydrocarbon generation, charge, and accumulation system, the hydrocarbons from the overpressured organic-rich shale migrated into and charged the sandstone interbeds very efficiently due to direct contact of shale source rock and sandstone reservoir (Ran et al. 2013). As organic-lean lithofacies are typically brittle and have the best production for shale resource plays in North America (Jarvie 2014), a similar occurrence is seen in the thin sandstones with porosity of 2%–12% and abundant high angle tectonic natural fractures in the lacustrine Ordos Basin. These thin sandstones are better candidates for successful hydraulic

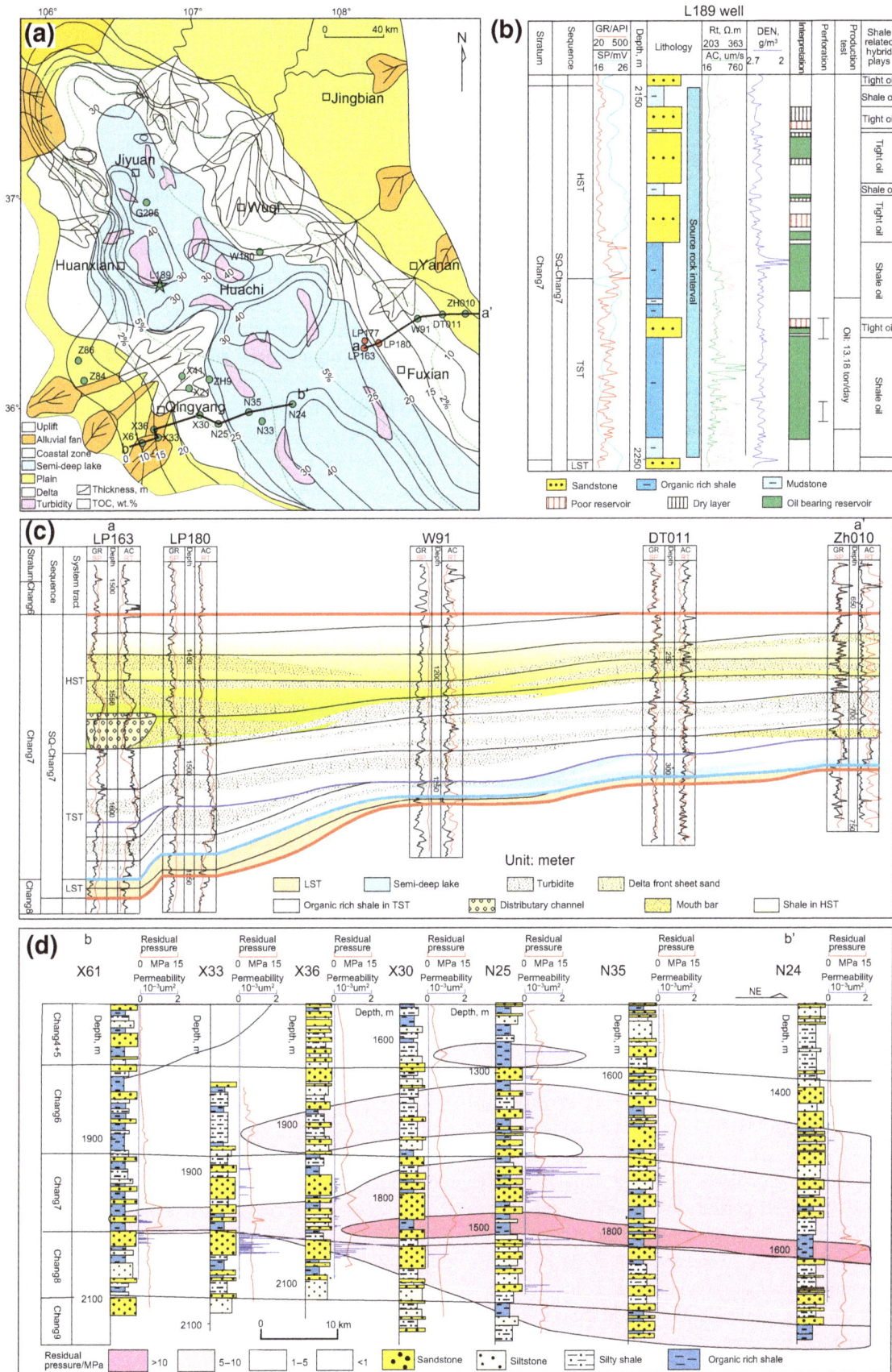

Figure (a) Map with locations including Jingbian, Jiyuan, Wuqi, Huanxian, Huachi, Yanan, Qingyang, Fuxian and well locations. Legend: Uplift, Alluvial fan, Coastal zone, Semi-deep lake, Plain, Delta, Turbidity, Thickness, m, TOC, wt.%

Figure (b) L189 well: Stratum, Sequence, GR/API 20–500, SP/mV 16–26, Depth, m, Lithology, Rt, Ω.m 203–363, AC, um/s 16–760, DEN, g/m³ 2.7–2, Interpretation, Perforation, Production test, Shale related hybrid plays. Legend: Sandstone, Organic rich shale, Mudstone, Poor reservoir, Dry layer, Oil bearing reservoir.

Figure (c) Cross-section a–a′ with wells LP163, LP180, W91, DT011, Zh010. Unit: meter. Legend: LST, Semi-deep lake, Turbidite, Delta front sheet sand, Organic rich shale in TST, Distributary channel, Mouth bar, Shale in HST.

Figure (d) Cross-section b–b′ with wells X61, X33, X36, X30, N25, N35, N24. Residual pressure 0 MPa 15, Permeability 10⁻³um². Legend: Residual pressure/MPa >10, 5–10, 1–5, <1, Sandstone, Siltstone, Silty shale, Organic rich shale.

◀**Fig. 5** **a** Sedimentary facies, TOC and thickness of the organic-rich 7th Member of the Upper Triassic Yanchang Formation (Chang7) in the Ordos Basin. Thickness maps where TOC > 2% (modified from Yang et al. 2013). The green and red dots denote oil and gas production, respectively; **b** Properties of Chang7 source rock interval in L189 well (After Yang et al. 2013); **c** West-to-east cross section showing the stratigraphic variation of the Chang 7 source rock interval across the Xifeng Oilfield (Section location in **a**). The facies were identified based on regional facies distribution, well log characteristics, lithologic association, and well correlation; **d** West-to-east cross section showing the residual pressure (difference between formation pressure and hydrostatic pressure) distribution and permeability of the Chang7 source rock interval (Section location in **a**) (After Yao et al. 2013). The "Hybrid Plays" were identified based on the lithofacies, geochemistry, petrophysics, pressure, and production test results

fracturing and are expected to deliver hydrocarbons better than the clay-rich Chang7 shale characterized by low porosity (<5%) and extremely low permeability (Zeng and Li 2009; Zhang et al. 2013). For the interbedded sandstone and siltstone in the Chang7 source rock, the predominantly tight sandstone and siltstone reservoirs exhibit overall low permeability (Yang et al. 2013; Yao et al. 2013). The interbedded sandstone and siltstone could be classified as conventional turbidite sandstones and siltstones, and sandy debrite reservoirs if the permeability value is higher than 0.1 mD (Zou et al. 2012). As an example, the 10–20 m sandstone interbeds having permeability at approximately 2 mD from the X33, X36 and N25 wells could be classified as a conventional play.

A "Hybrid Play" system for the Chang7 source rock interval has been proven by recent E&P. Production was recorded from both organic-rich shale, and sandstone and siltstone interbeds. In 2011, Yanchang Petroleum drilled China's first lacustrine shale gas well, the vertical Liuping177 in the organic-rich shale with $R_o > 1.1\%$ in the SE Ordos Basin (Xu and Bao 2009; Zou et al. 2010). This well confirmed the presence and potential of shale gas reservoirs in lacustrine basin sediments. The initial production (IP) of a 24-h test flowed at the rate of 2350 m^3 (82,250 ft^3)/day from the organic-rich shale interval, affirming the shale gas potential in the Ordos Basin. Several other wells drilled in this area all produced shale gas at IP up to 8000 m^3 (280,000 ft^3)/day from "Hybrid Plays" with adjacent beds of shale and tight sandstone. For the sand-dominant reservoirs within the organic-rich Chang7 shale, most have thicknesses of 5–25 m, porosity of 4%–10% and permeability less than 0.3 mD and are juxtaposed with the Chang7 shale source rock. The sand-rich zones have been targeted for continuous accumulation of tight oil hydrocarbons. For example, in the Zhidan area in the central west Ordos Basin, tight oil production has been tested in the sandstone interbeds within the Yanchang7 shale source rock interval with TOC content ranging from 3% to 8% in

127 wells. There are 29 wells with average oil production rates of 8.8+ tonnes (64+ barrels)/day. For the shale oil potential in the Ordos Basin, PetroChina Changqing Oilfield drilled and fractured the G295 well and the test result showed this well produced oil at 20.4 tonnes (150 barrels)/day from the organic-rich Yanchang7 (Chang7) shale with a thickness of 28 m and some natural fractures. Hybrid fracturing has been used to target the hybrid hydrocarbon-bearing organic-rich shale and organic-lean sandstone reservoirs within Yanchang7 (Chang7) source rock interval (Zhou et al. 2012).

4.3 Hybrid Plays within Jurassic carbonate-containing source rock in the Sichuan Basin

The Sichuan Basin is a large composite petroliferous basin located in the southwest part of China. It consists of thick Precambrian to Silurian marine shales and Triassic to Jurassic lacustrine shales. Recently, commercial shale gas production began from the Silurian Longmaxi marine shale in the Fuling Shale Gas Field located in SE Sichuan Basin (Jiang 2014; Jiang et al. 2015, 2016). Tight gas, shale gas, tight oil, and shale oil also have been sporadically produced from the Triassic Xujiahe and Jurassic Ziliujing Formations. This paper focuses on the Jurassic "Hybrid Plays" in the Sichuan Basin.

During the Early Jurassic, organic-rich shale and bioclastic shoals (mainly bioclastic or shelly limestone) were deposited in the Sichuan Basin (Fig. 6a). The Lower Jurassic Ziliujing lacustrine shale has a TOC content of 1%–2% (Fig. 6a). From the Yuanlu4 well in the northeast Sichuan Basin, it can be seen that gas is present in the entire Da'anzhai member source rock interval with lithologies ranging from organic-rich shale and marl, to organic-lean bioclastic limestone (Fig. 6b). This is a typical "Hybrid Play" system containing shale gas and tight carbonate gas reservoirs. A comparable trend is identified in the Jianye1HF horizontal well in the Jiannan Gas Field in the eastern Sichuan Basin. Good hydrocarbon shows (based on total hydrocarbon anomalies by mud logging) are reported in the Dongyuemiao member (Da'anzhai member equivalent) source rock interval. The lower Dongyuemiao has very good hydrocarbon shows from both organic-rich shale and organic-lean bioclastic limestone and marl. The Upper Dongyuemiao Member was hydraulically fractured and based on a 24-h performance test produced gas at rate of 12,000 m^3 (420,000 ft^3)/day from the shale and tight carbonate (marl) "Hybrid Plays" (Fig. 6c).

Currently, the lacustrine shale gas and shale oil exploration in Sichuan Basin is active in the northeast, east, and southeast regions of the basin. The vertical Yuanba21 well in the northeast produced gas at 50.7 × 10^4 m^3 (1774 × 10^4 ft^3)/day from the "Hybrid Plays" consisting

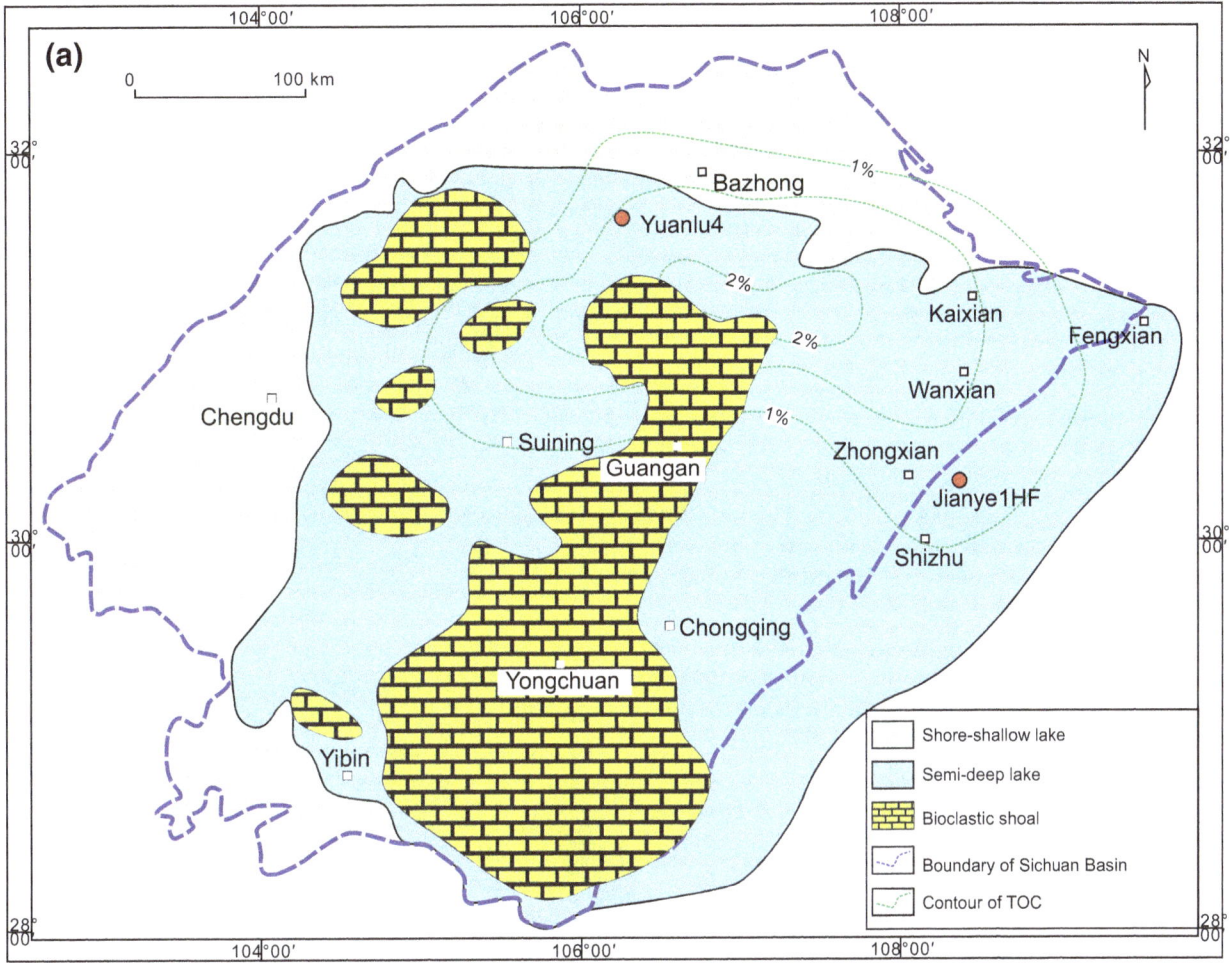

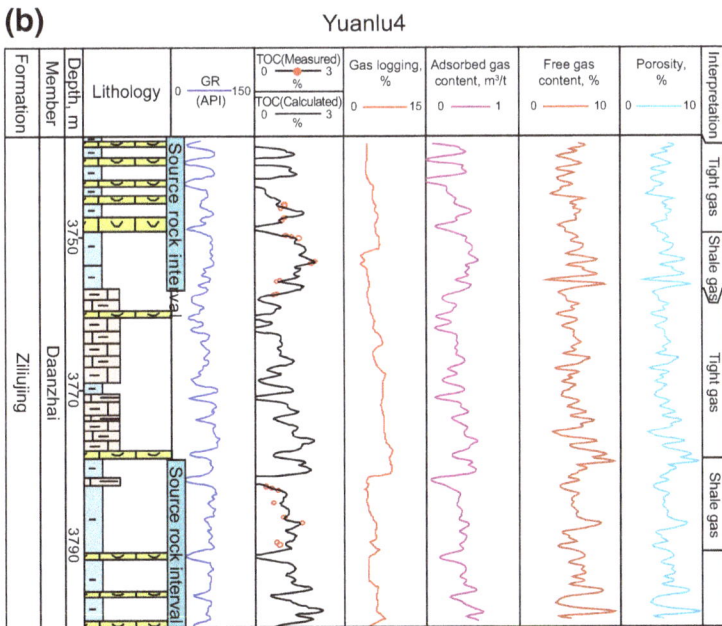

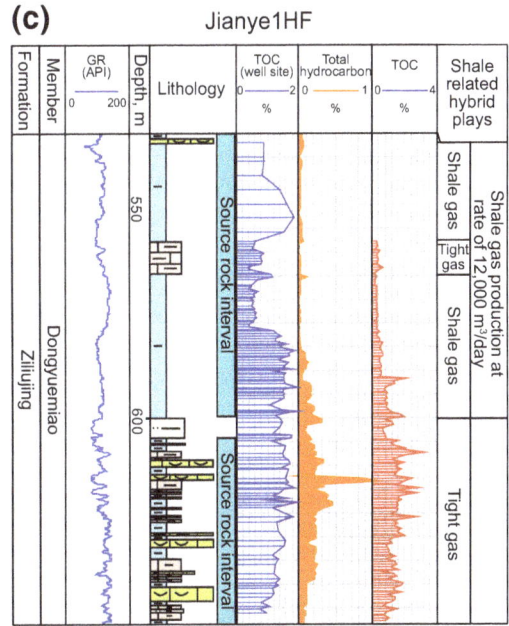

◀ **Fig. 6 a** The sedimentary facies distribution and TOC contour map of the Da'anzhai Member of the Lower Jurassic Ziliujing Formation in the Sichuan Basin (After Jin et al. 2013 and Li et al. 2013b); **b** "Hybrid Plays" in the Lower Jurassic Ziliujing Formation in the vertical Yuanlu4 well, original data courtesy of Sinopec; **c** "Hybrid Plays" in the Lower Jurassic Ziliujing Formation in the horizontal Jianye1HF well (modified and interpreted from Jin et al. 2013). This is a horizontal well and the data shown in **c** are from the drilling and logging results before the well was sidetracked to a horizontal lateral. The "Hybrid Plays" were identified based on lithofacies, geochemistry, petrophysics, and production test results

of shale and tight carbonate plays in the Da'anzhai Member of the Lower Jurassic Ziliujing Formation in northeastern Sichuan Basin. The vertical XL101 well in southeastern Sichuan Basin yielded shale gas flows of 11×10^4 m^3 $(385 \times 10^4$ ft^3)/day from similar "Hybrid Plays." The horizontal Shiping 2-2H well in the northeastern Sichuan Basin produced condensate oil from "Hybrid Play" shales and carbonate reservoirs at rates of 33.79 tonnes (250 barrels)/day based on 24-h performance test. Many oilfields in Central North Sichuan Basin (e.g., Gongshanmiao, Shilongchang, Zhongtaishan, Guihua, etc.) share similar characteristics of regionally continuous hydrocarbon accumulations. Historically, tight oil production from the Gongshanmiao Oilfield was primarily from the targeted bioclastic limestone interbeds (porosity of 2% and permeability of 0.057 mD) within the organic-rich Da'anzhai shale source rock interval (Jin et al. 2014). All these analyses and production results demonstrate the similarities between the Jurassic lacustrine Ziliujing play and the Bakken or Niobrara "Hybrid Plays." In all cases, the major oil production contribution is related to tight carbonates and minor production contributions are from shale with maturity levels ranging from 0.7% to 1.1%. The organic-rich shale plays a significant role in gas production from this "Hybrid Play" system within source rock intervals with maturity levels of 1.1%–1.4%.

4.4 Hybrid Plays within Upper Cretaceous siliciclastic source rock in the Songliao Basin

The Songliao Basin covers 260,000 km^2 and is a large-scale Mesozoic–Tertiary lacustrine basin in northeast China (Fig. 1). The largest oilfield—Daqing Oilfield—is located in the north-central area of this basin. The Songliao Basin originated with extensional faulting during the Triassic to the Middle Jurassic, followed by a rifting stage in the Late Jurassic. The post-rift stage with a large-scale thermal subsidence depression extended from Early to Late Cretaceous and uncomfortably overlays the syn-rift strata (Feng et al. 2010). The fault-bounded sediment fill in the syn-rift and widespread post-rift sediment fill generally thin toward the basin margins, resulting in steer's-head

geometry in cross section (Feng et al. 2010). The sedimentary filling of this basin is dominated by Upper Cretaceous strata, which serve as the main petroleum source and reservoir system. The main source rock in this basin is the thick, dark, and organic-rich Upper Cretaceous Qingshankou shale deposited in an anoxic deep lake setting during a period of high-stand lake levels (Liu et al. 2014).

The Qingshankou Formation (Fm.) was deposited in a semi-deep to deep lake that occupied a large area of the Songliao Basin. The Qingshankou shale source rocks cover an area of 40,000 km^2 and have TOC ranges >2% in the basin center, R_o values of 0.7%–1.5% and thicknesses of 50–400 m (Feng et al. 2010; Fig. 7a). Within the Qingshankou source rock interval, the organic-rich shale in semi-deep to deep lake settings are often associated with thin siliciclastic silty shale, turbidite sandstones, siltstone, and sandy mass transport deposits (MTDs), even occasionally associated with ostracod grainstone and marls (Pan et al. 2014; Fig. 7b–d). Altogether these regionally correlated, interbedded lithologies within the Qingshankou interval form "Hybrid Plays."

Another example of a "Hybrid Play" in the northern Songliao Basin is shown in the vertical Ha14 well yielding oil at rate of 1.36 tonnes (10 barrels)/day and gas at a rate of 1253 m^3 (43,855 ft^3)/day (based on 24-h performance test from the Qingshankou source rock interval consisting of siliciclastic shale and tight siltstone interbeds (Fig. 7c). Its neighbor well Ha18 also produced oil at a rate of 10 tonnes (73 barrels)/day from a "Hybrid Play" system consisting of interbedded shale and tight siltstone. The NGN1 well in the southern Songliao Basin clearly demonstrates that the siltstone, silty shale and sandstone interbeds, and shale have significant oil content (Fig. 7b), confirming the reservoir potential of shale, tight siltstone and sandstone reservoirs in this "Hybrid Play."

4.5 Hybrid Plays within Paleogene mixed siliciclastics and carbonate source rock in the onshore Bohai Bay Basin

The Bohai Bay Basin located in eastern China has been producing conventional oil and gas since 1961. The major source rock intervals include the Eocene Upper 4th member of the Shahejie Fm. (Es$_4$) deposited in a saline lake environment and the 3rd member of the Shahejie Fm. (Es$_3$) deposited in a brackish to fresh lake setting (Zhu et al. 2004). The high TOC associated with gypsum and carbonate is due to the good preservation of organic matter in the stratified saline to brackish lakes, where anoxic depositional settings and low detrital dilution allow the deposition of laminated organic-rich shale. Regionally, the Es$_3$ shale with thicknesses up to 500 m and TOC up to 14% is the most significant source rock for the conventional

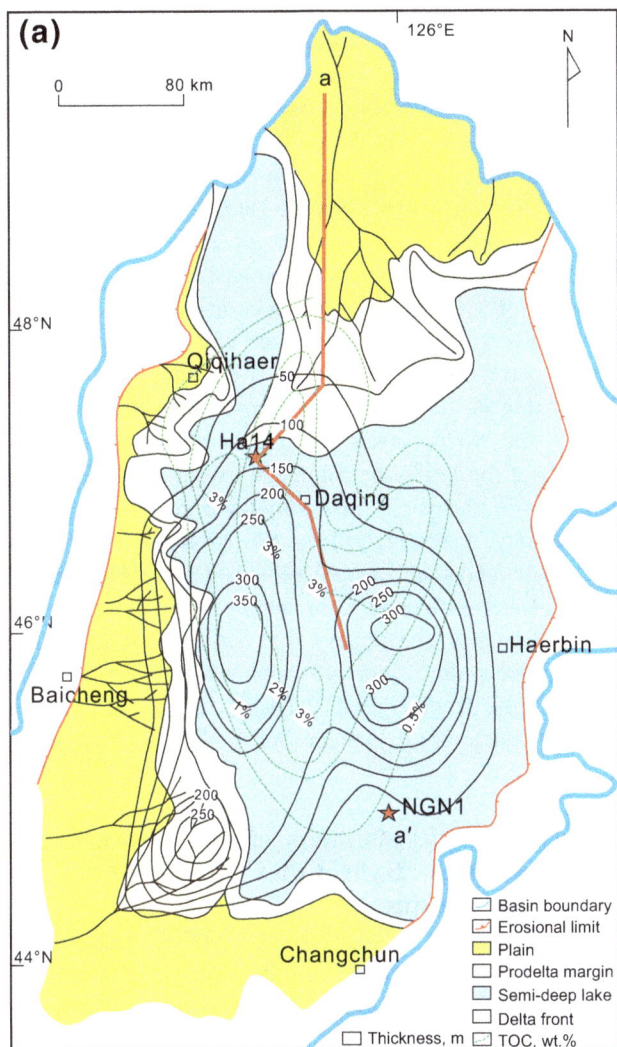

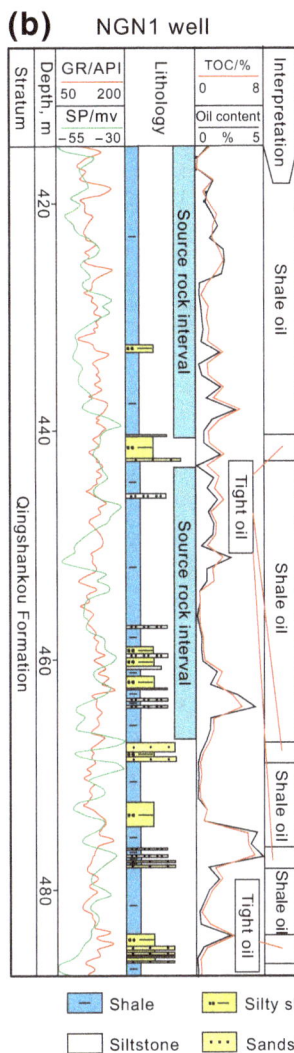

(b) NGN1 well

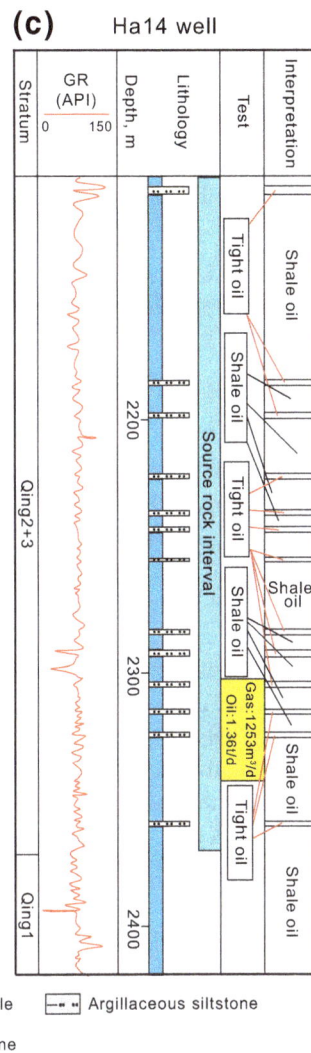

(c) Ha14 well

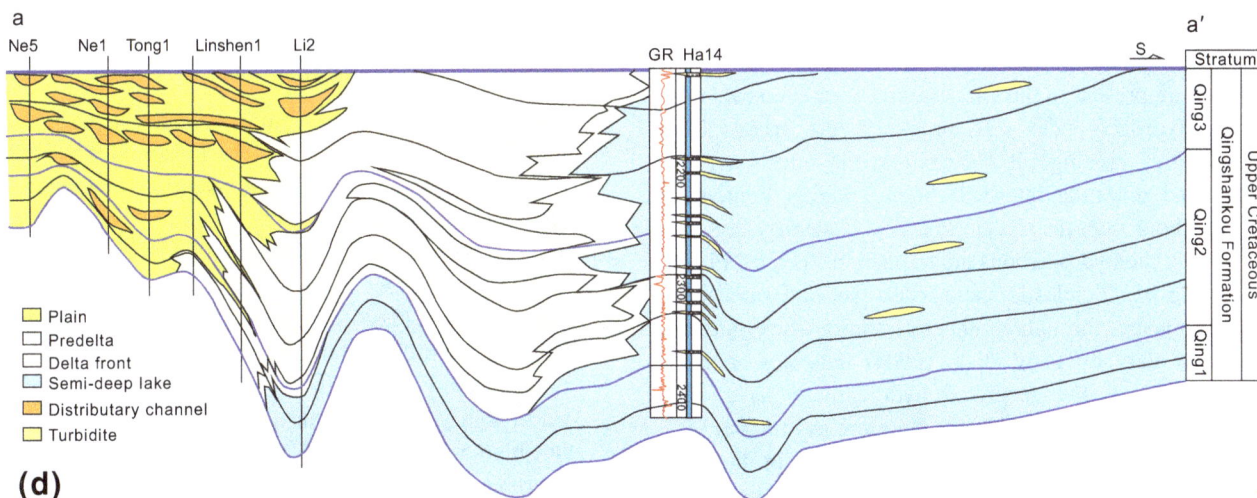

(d)

◄**Fig. 7 a** The sedimentary facies distribution and TOC contour of the Upper Cretaceous Qingshankou Fm. in the Songliao Basin (modified from Feng et al. 2010); **b** "Hybrid Plays" in the Upper Cretaceous Qingshankou source rock interval in the vertical NGN1 well (reinterpreted from Liu et al. (2011) based on lithofacies characteristics and production test results). The "Hybrid Plays" were identified based on the lithofacies, geochemistry, and production test results. See **a** for well location; **c** "Hybrid Plays" in the Upper Cretaceous Qingshankou source rock interval in vertical Ha14 well (modified and reinterpreted from Li et al. 2012). See **a** for well location; **d** North–south cross section showing the sedimentary facies distribution of the Upper Cretaceous Qingshankou Fm. in the Songliao Basin. See **a** for the location of the cross section

deltaic sandstone reservoirs in Bohai Bay Basin. From an exploration and interpretation point of view, the Bohai Bay Basin is a mature basin. Exploration has focused on the lenticular turbidite sandstones, thin carbonate, tight sandstones, and shale resources in Es_3 across many sub-basins that share similar depositional history in the Bohai Bay Basin (Zhang et al. 2012b; Jiu et al. 2013; Jiang et al. 2013a, b; Yang et al. 2016). For the purpose of characterizing "Hybrid Plays," the Jiyang Sub-basin is used as a representative petroleum system in the Bohai Bay Basin.

During the Paleogene Shahejie Fm. depositional period, the semi-deep to deep lake facies dominated the majority of the Jiyang Depression with deltas and alluvial fans developing near the uplifts (Fig. 8a, c). Due to turbidite deposition in semi-deep to deep lake settings (Fig. 8a, c), the source rock interval is commonly interbedded with sandstone and siltstone (Fig. 8b). From the Niu38, He160 and Yi170 wells, it was determined that the Es_3 source rock interval has a TOC of 1%–4% throughout the stratigraphic succession of organic-rich shale interbedded with sandstone, siltstone, gypsodolomite, and dolomite. The hydrocarbons are present in the whole source rock interval based on the hydrogen index (HI) or total hydrocarbons from mud logging readings. It appears that the sandstone, siltstone, and dolomite interbeds have a higher content of hydrocarbons than the shales (Fig. 8b). Generally, the data reveal that the hydrocarbons are trapped in "Hybrid Plays" consisting of siliciclastic shale, silestone, and sandstones and carbonates throughout the Es_3 source rock interval. Previous study by Pang et al. (2004) indicates approximately 40% of the generated hydrocarbons have been expelled into conventional traps, therefore as much as 60% of the generated hydrocarbons may be hosted in the "Hybrid Plays" of the Es_3 source rock interval in the Jiyang Basin. Besides the hybrid unconventional tight siliciclastic and carbonate plays, conventional turbidite reservoirs may also be present in the Es_3 source rock interval and form a combination of hybrid unconventional and conventional plays depending on stratigraphic levels and rock properties,

e.g., the conventional turbidite reservoir from the 2930–2950 m interval in the Niu38 well in Fig. 8b. Historical production data and recent drilling have confirmed the abundance of "Hybrid Plays" in the Jiyang Sub-basin and other sub-basins of the Bohai Bay Basin. For example, the shale oil and gas production from the naturally fractured Es_3 shale in the Zhanhua Depression of the Jiyang Sub-basin was first reported in the early 1960s. So far, about 130 wells have encountered shale oil and gas shows and over 30 wells have produced commercial quantities of shale oil with high production rates of 90+ tonnes (666+ barrels)/day. The He54 well in the Dongying Depression in the Jiyang Sub-basin produced at a rate of 91.3 tonnes (676 barrels)/day (Li et al. 2006; Jiu et al. 2013) and the Lei88 and Shugu165 wells produced oil at rates of 47 tonnes (345 barrels)/day and 21 tonnes (154 barrels)/day, respectively, from Es_3 hybrid tight carbonate and shale plays in Liaohe Sub-basin. The Pushen18-1, Wengu4 and horizontal Baimiaoping1 wells produced oil at initial rates of 420 tonnes (3108 barrels)/day, 120 tonnes (888 barrels)/day, and 40 tonnes (296 barrels)/day, respectively, from Hybrid Plays of tight sandstone and fractured shale in the Dongpu Sub-basin.

5 Characteristics summary of source rock intervals and "Hybrid Plays" in China's lacustrine basins

Based on extensive synthesis of the geology, geochemistry, mineralogy, petroleum systems, and production data for the key representative Permian to Cenozoic source rocks of lacustrine basins in China, the properties of the identified "Hybrid Plays" are summarized in Table 3. The common characteristics of these "Hybrid Plays" include:

1. They are all organic-rich (mostly >2%) shale related plays in the source rock interval in terms of genetic, stratigraphic and depositional association, hydrocarbon generation, and migration;
2. They have mixed plays with siliciclastic reservoirs and/or carbonates;
3. Most "Hybrid Play" reservoirs are very tight (permeability <0.1 mD) with the exception of a few conventional reservoirs with high porosity and high permeability e.g., turbidite sandstone;
4. In China, most "Hybrid Plays" in lacustrine basins are overpressured with pressure coefficients of 1.2–2 except for the regionally underpressured (pressure coefficients of 0.7–0.9) Upper Triassic Hybrid Plays in the Ordos Basin (Table 3). Overpressure is a production benefit and can enhance the flow of hydrocarbons to the wellbore.

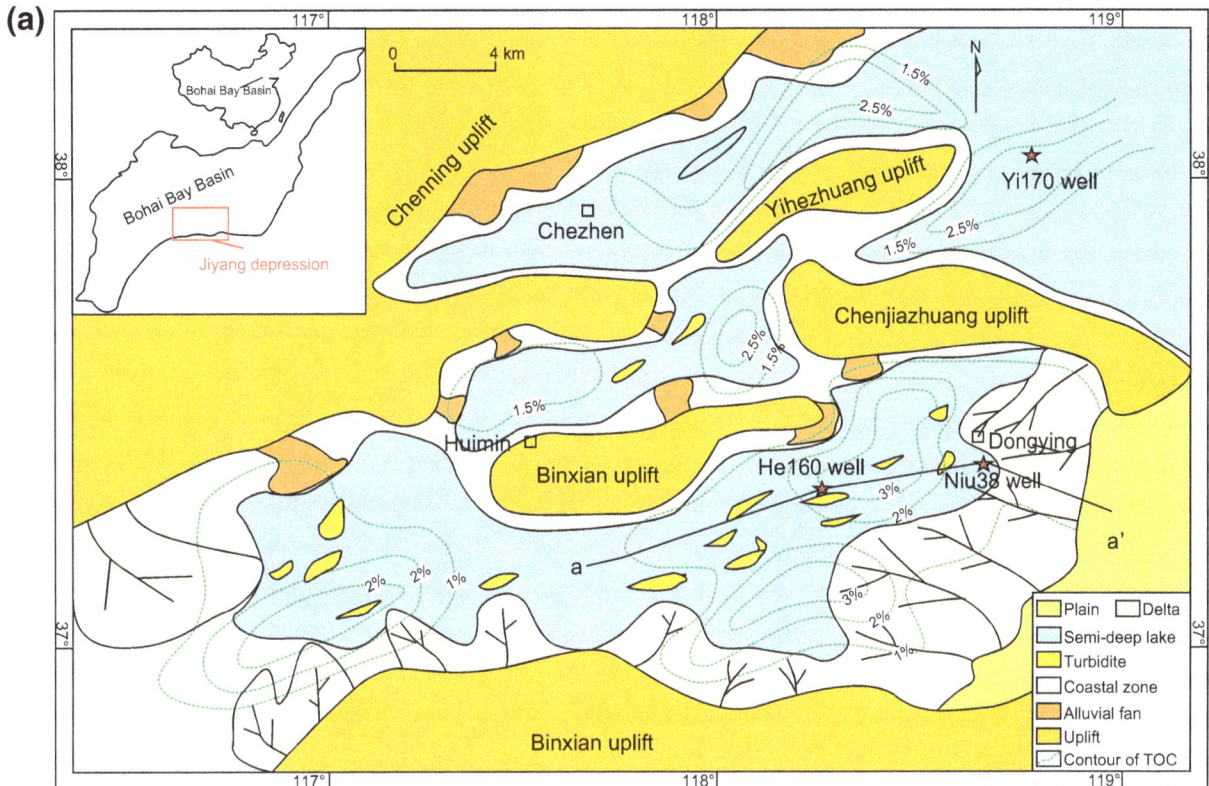

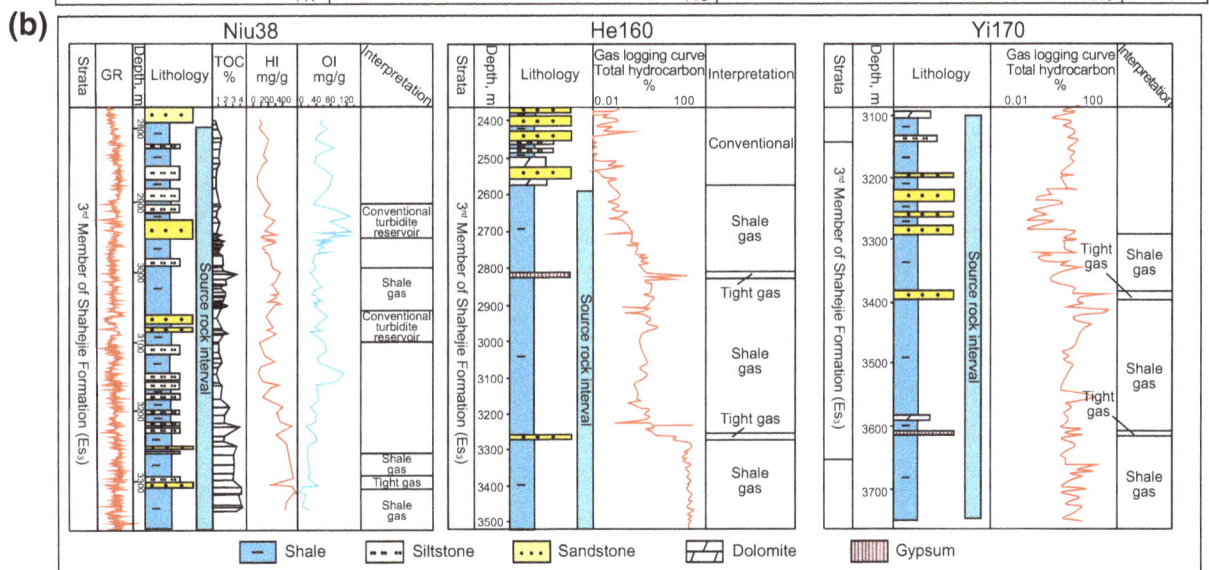

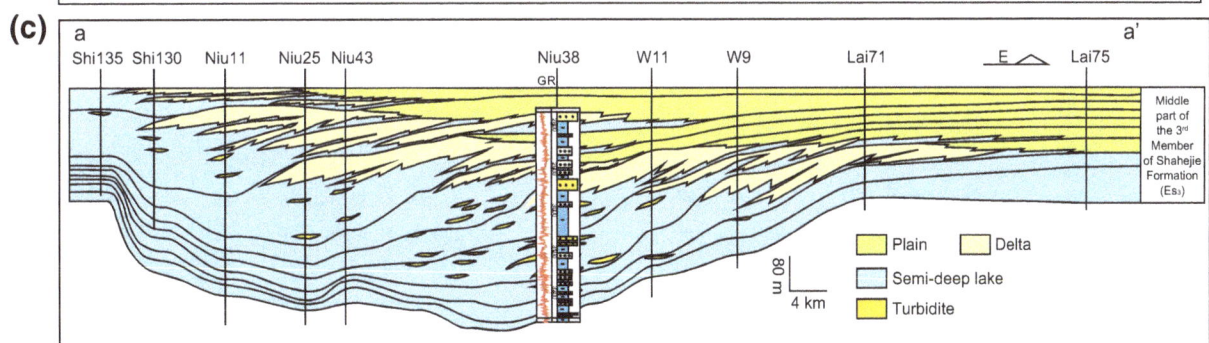

◄**Fig. 8** **a** The sedimentary facies distribution and TOC contour map of the 3rd Member of the Shahejie Formation (Es₃) in the Jiyang Depression, Bohai Bay Basin (Modified from Jiu et al. 2013); **b** Hybrid Plays in the 3rd Member of Shahejie Formation (Es₃) source rock interval in vertical Niu 38, He160 and Yi170 wells (original well data from Zhang et al. (2008b), see **a** for well locations). The Hybrid Plays were identified based on the lithofacies, geochemistry, and production test results; **c** West-east cross section showing the Es₃ sedimentary facies variation (Section location in **a**), compiled and reinterpreted from data published by Zhang et al. (2008b) and Jiu et al. (2013)

The Permian to Cenozoic lacustrine source rocks in China were deposited in shallow to deep lakes in hyper-saline to fresh water, in the basin center far away from clastic dilution. The source rock intervals have unique hybrid lithofacies associations based on age, tectonics, and depositional settings and are potential reservoirs for "Hybrid Plays" include organic-rich facies and organic-lean facies exampled below:

1. Organic-rich black shales interbedded with organic,-carbonate and silt rich shale (e.g., carbonate-rich shale with TOC > 3% in the Middle Permian Lucaogou source rock interval in the Junggar Basin in Fig. 3b);

2. Organic-lean carbonate (e.g., marl and bioclastic limestone interbedded in Jurassic Ziliujing source rock intervals in the Sichuan Basin in Fig. 6b, c);

3. Organic-lean sandstone and siltstone (e.g., in Upper Triassic Chang7 source rock in the Ordos Basin, Upper Cretaceous Qingshankou source rock interval in the Songliao and Cenozoic Es₃ source rock interval in the Bohai Bay Basin).

China has lacustrine rift, sag and foreland basins according to their tectonic settings (Fig. 9a–c), e.g., the Paleogene Bohai Bay Basin and Jianghan Basin are lacustrine rift basins, the late Cretaceous Songliao Basin is a sag basin, and the Middle Permian Santanghu Basin, late Triassic Sichuan Basin, Jurassic Junggar Basin, and Triassic Ordos Basin are lacustrine foreland basins. Taking into account the lacustrine basin forming regime, lithofacies, geochemistry, reservoir properties, and hydrocarbon generation and accumulation for the "Hybrid Plays" in the lacustrine source rock interval and conventional plays, petroleum system models have been established for China's lacustrine rift, sag and foreland basins (Fig. 9a–d), which clearly show the types of "Hybrid Plays" in the lacustrine source rock interval and their relationships with

Table 3 Properties of "Hybrid Plays" in source rock intervals of major lacustrine basins in China (most data based on 1000 + sample tests compiled from Zou et al. (2010, 2013a, b, 2014) and Jiang et al. (2015)

Basin	Age	Typical Hybrid Plays	TOC range/ average (%)	Maturity/ hydrocarbon type	Tight rock porosity (%)	Tight rock permeability, mD (10⁻³ μm²)	Pressure coefficient
Bohai Bay Basin	Paleogene (E)	Shahejie shales (Es₃ and Es₄), tight carbonate, tight sandstone and siltstone	0.5–14/ 3.5	0.5–1.7/oil and gas	0.5–12 for carbonate and sandstone < 6 for shale	0.001–1 for dolomite 0.001–0.05 for shale	1.3–1.9
Songliao Basin	Cretaceous (K)	Qingshankou shale oil, shale gas, tight sandstone oil	0.5–5/2.2	0.7–1.5/oil and gas	8–12 for siltstone 3–6 for shale	0.1–1 for siltstone <0.15 for shale	1.2–1.6
Sichuan Basin	Jurassic (J)	Ziliujing tight carbonate and shale oil	1–2.8/1.5	0.7–1.4/oil and gas	2–5 for shale and carbonate	0.03–0.65 for carbonate <0.001 for shale	1.2–1.7
Sichuan Basin	Late Triassic (T)	Xujiahe5 tight sandstone gas and shale gas	0.4–16/ 2.35	1.1–1.7/gas	3–10 for sandstone <5 for shale	<0.1 for sandstone <0.001 for shale	1.3–2
Ordos	Late Triassic (T)	Yanchang7 (Chang7) shale gas, tight sandstone oil, shale oil	0.3–36/ 8.3	0.5–1.2/ Mainly oil and some gas	Tight sand: 7–10 Shale: 2–5	0.01–0.1 for sandstone <0.01 for shale	0.7–0.9
Junggar, Santanghu, Turpan-Hami	Late Permian (P)	Lucaogou tight dolomite oil, shale oil	0.7–21/3	0.5–1.2/ Mainly oil and some gas	2–9 for dolomite <5 for shale	0.01–0.1 for dolomite <0.01 for shale	1–1.2

The porosity and permeability data for Ordos and Bohai Bay basins are based on 56 samples that were measured by Schlumberger-TerraTek in Salt Lake City

conventional plays. Even though the rift, sag and foreland lacustrine basins vary in basin configuration, sediments filling sequence, lithofacies, evolution of environment, structure, traps, and play types (Fig. 9a–c), these basins share the similar features of "Hybrid Plays" including the stratigraphically associated shale gas, turbidite conventional gas, shale oil, tight oil and turbidite conventional oil in the source rock interval (Fig. 9d). The shale source rocks in these three types of basins have high organic matter content and may contain Type I to Type III kerogen depending on the depositional environment and degree of terrestrial influence (Fig. 9). The type I and type II kerogen are usually present during the early rift stage and the type II and type III are present in the late rift stage with influence of terrestrial plants. The majority of generated hydrocarbons are still trapped in organic-rich shale forming self-sourced shale plays. A portion of the expelled hydrocarbons migrated to adjacent, coarser (compared to shale) siliciclastic and carbonate interbeds that are in direct contact with shale source rocks. These coarser interbeds form tight or conventional plays depending on reservoir porosity, permeability and hydrocarbon mobility. The juxtaposed tight/conventional carbonate and sandstone and shale oil accumulations occur where source intervals are in the oil-generating window in the shallow parts of the basins, whereas the juxtaposed tight/conventional and shale gas accumulations are usually located within the gas-generating window in deeper parts of the basins. These multiple-stacked siliciclastic and carbonate plays are associated with the organic-rich shale from which the oil and gas have been generated, expelled, and capped. The classic conventional hydrocarbons (except for turbidite conventional traps in source rock intervals) generally accumulate in conventional structural and stratigraphic traps above, below or around the source rock pod (Fig. 9a–d). Usually, the "Hybrid Play" systems in the source rock intervals are closed petroleum systems that are self-sourced or adjacently sourced. The reservoirs of the "Hybrid Plays" range from extremely tight shale to tight siltstone, sandstone, carbonate, and to conventional reservoirs. These hybrid plays with mixed siliciclastics and/or carbonate reservoirs are multiple-stacked in the source rock interval (Fig. 9a–d). The oil and gas accumulation in these "Hybrid Plays" is continuous.

6 Discussion and conclusions

Similar to the families of lacustrine basins in Southeast Asia, Rocky Mountains in Western US, Atlantic Margins, etc., China has a series of Mesozoic to Cenozoic hydrocarbon-producing lacustrine basins. Widespread and world-class source rock intervals in these lacustrine basins commonly consist of very thick and regionally distributed organic-rich shale source rocks (TOC abundance up to 36%) and thin, organic-lean siltstone, sandstone, and/or carbonate interbeds within it, making the organic-rich shale interval juxtaposed with organic-lean facies. The organic-rich shale is formed in anoxic semi-deep to deep lake settings. The siltstone and sandstone are usually turbidite or mass transport deposits (MTDs). The carbonate is generally formed in saline evaporite settings when the water level is low and less clastics are input into the calm lake. For China's various lacustrine basins with different tectonic and depositional settings, the source rock intervals are both laterally and vertically heterogeneous in terms of lithofacies. The Middle Permian lacustrine mixed siliciclastic siltstone and carbonates (mainly dolomitic rock) associated with shale source rocks are present in the Junggar and Santanghu foreland basins in NW China. The siliciclastic sandstone and siltstone in association with shale source rock deposits are typically present in Upper Triassic Chang7 source rocks in the Ordos foreland basin in northern China. The Jurassic carbonates (bioclastic limestone) are predominantly found in lacustrine source rocks in the Jurassic rift stage of the Sichuan Basin. The siliciclastic thick organic-rich shale and thin sandstone and siltstone are dominant in Upper Cretaceous Qingshankou source rock in the Songliao sag Basin. Paleogene Es$_3$ age equivalent source rocks consisting of mixed siliciclastic organic-rich shale, sandstone, siltstone and carbonate are pervasive in lacustrine rift basins in East China. The integrated evaluation of geology, geochemistry, mineralogy, petrophysics, and synthesis of scattered production tests indicate that thick organic-rich black shales serve as both source rock and reservoir and can be potential shale gas and shale oil plays depending on the maturity level of source rocks. The lenticular and blanket sandstone, carbonate and siltstone interbeds are directly in contact with shale source rocks creating an efficient charging system and are sometimes comparable to unconventional type plays that are mostly tight oil and gas plays similar to the Bakken dolomite and Niobrara chalk tight oil reservoirs. Even though the lacustrine interbeds with limited lateral extent are thinner than typical marine deposits, they could host significant amounts of hydrocarbons and be primary reservoirs since they are more porous and permeable than the adjacent shales. Industry data confirms that in some instances gas or oil was produced through more permeable and brittle sand or silt layers interbedded within the shale (Jarvie 2014). In the past, they were overlooked or ignored to some extent because the focus was directed toward unconventional exploration of regionally continuous shale and massive tight sand plays. We conclude both shale and related interbeds should be treated together as "Hybrid Plays" and expect they can host enormous continuous oil

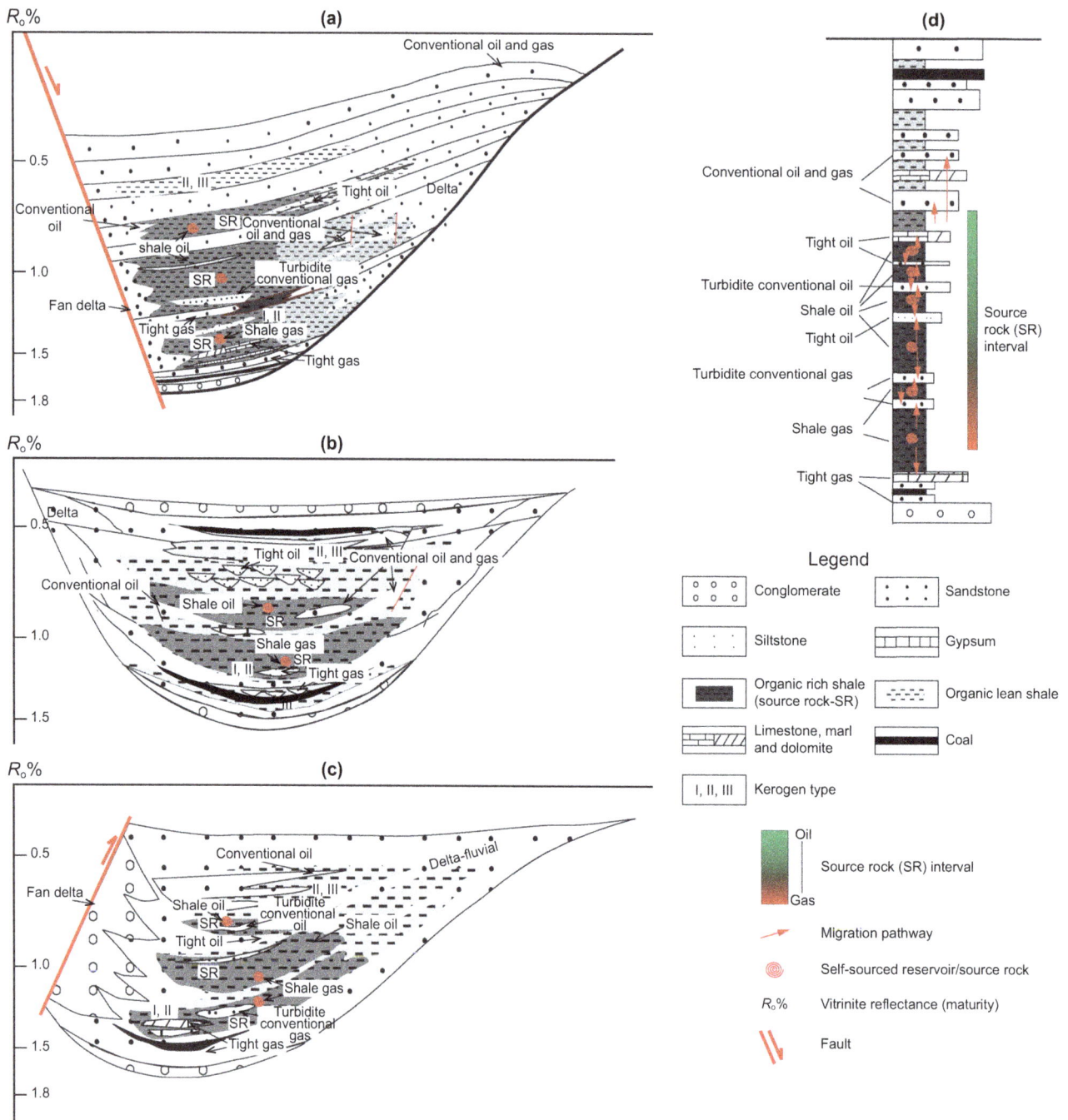

Fig. 9 Schematic diagrams showing the distribution of "Hybrid Plays" in the source rock intervals and conventional plays in the lacustrine rift basin (**a**), lacustrine sag basin (**b**), and lacustrine foreland basin (**c**). **d** shows the vertical association of various stacked "Hybrid Plays" including shale gas, turbidite conventional gas, shale oil, tight siliciclastic and carbonate oil, and turbidite conventional oil in the lacustrine source rock interval

and gas accumulations in all the producing lacustrine basins in China. The "Hybrid Plays" concept revamps our understanding of lacustrine source rock reservoirs and genetically synthesizes the source to accumulation of unconventional to conventional systems in the source rock interval. "Hybrid Plays" present unique petroleum systems hosting continuous hydrocarbons stored in thick, self-sourcing shale reservoirs and adjacent thin coarse-grained siliciclastic and carbonate interbeds, which form hybrid siliciclastic and carbonate unconventional plays/reservoirs in addition to conventional plays/reservoirs throughout the whole source rock interval. In undrilled areas and untested intervals of proven source rock, the potential plays can be identified based on regional geological and geophysical interpretation, geochemical, petrophysical, mineralogical analysis, log characteristics, and oil and gas shows in mud logs, etc. These "Hybrid Plays" represent a new exploration target in the multiple-stacked lithologies of organic-lean rocks juxtaposed against organic-rich shale. The potential for "Hybrid Plays" in source rocks will continue to open more prospects in previously uneconomic areas and what were considered to be uneconomic stratigraphic intervals.

The source rock interval has a much lower matrix permeability and much smaller pore throats than conventional reservoir intervals and unconventional tight siliciclastic and carbonate intervals outside the source rock. The remaining unexpelled and short distance migration hydrocarbons are available to be produced when sufficient fracture conductivity is induced by hydraulic fracturing. Many historical production cases and recent tight and shale play exploration endeavors proved the producibility of gas from both relatively coarse-grained interbedded intervals with lean organic matter and naturally fractured very fine-grained tight shale with rich organic matter. The natural fractures contribute both hydrocarbon storage and production. In essence, most "Hybrid Plays" in lacustrine basins in China are multiple-stacked unconventional and conventional prospects and the resources can be produced by placing horizontal wells in the brittle sandstone or carbonate interbeds and employing multistage hydraulic fracturing techniques targeting the mixed "Hybrid Plays" to ensure the fractures are extended from the sandstone or carbonate intervals into the adjacent organic-rich shale intervals. These clay-poor and organic-lean interbeds with coarser brittle minerals usually have better reservoir quality with higher porosity and permeability and are more conductive than shale allowing oil and gas to travel to the wellbore (Chalmers et al. 2012). Since the lacustrine rocks are very heterogeneous and shale deposited in lake settings is usually clay-rich, understanding the reservoir property distribution, and the regime of waxy crude oil flow in shale along with new hydraulic fracturing technology to unlock the hydrocarbons from the shale and interbeds together require a multidisciplinary approach involving teams of geologists, engineers and other industry professionals. With the advancement of hydraulic fracturing technology for clay-rich shale, the pervasive oil accumulation in the shale interval could become far more economic viable.

We assert "Hybrid Plays" in the heterogeneous lacustrine source rock provide the most realistic exploration and production model for multiple, horizontally, and/or vertically stacked mixed conventional and unconventional siliciclastic and/or carbonate plays in order to avoid relying solely on clay-rich shale targets. The "Hybrid Plays"

model has been proven and works for a wide variety of lacustrine rift, sag, and foreland basins in China. As they are exploited as future E&P targets, "Hybrid Plays" are expected to decrease production costs by increasing the use of multiple laterals from one pad/platform and multistage hydraulic fracturing to stimulate the "Hybrid Plays" together at one time.

Acknowledgements Authors want to acknowledge National Natural Science Foundation of China (Grant Numbers 40872077 and 41272122), China National Key Technology Research and Development Program (Grant Number 2001BA605A09-1), Open fund from Sinopec's Petroleum Exploration and Production Research Institute (Grant No. G5800-15-ZS-WX038), and EGI's China Shale Gas and Shale Oil Plays Consortia (100980) sponsored by 20 multi-national oil companies. Appreciation is expressed to PetroChina and Sinopec for the privilege of working on the industry data and many other Chinese institutes for collaborating. We are also grateful for Weatherford and Schlumberger-TerraTek for their sample testing. Ms. Shuang Chen is thanked for her drafting of revised figures.

References

Bohacs KM, Carroll AR, Neal JE, et al. Lake-basin type, source potential, and hydrocarbon character: an integrated sequence-stratigraphic-geochemical framework. Lake basins through space and time. AAPG Stud Geol. 2000;46:3–4. doi:10.1306/St46706C1.

Bustin MR. Geology report: where are the high-potential regions expected to be in Canada and the US? Capturing opportunities in Canadian shale gas. In: The Canadian Institute's 2nd Annual Shale Gas Conference, January 31–February 1, Calgary; 2006.

Cao J, Zhang Y, Hu W, et al. The Permian hybrid petroleum system in the northwest margin of the Junggar Basin, northwest China. Mar Pet Geol. 2005;22(3):331–49. doi:10.1016/j.marpetgeo.2005.01.005.

Carroll AR. Upper Permian lacustrine organic facies evolution, southern Junggar Basin, NW China. Org Geochem. 1998;28(11):649–67.

Carroll AR, Bohacs KM. Stratigraphic classification of ancient lakes: balancing tectonic and climatic controls. Geology. 1999;27(2):99–102. doi:10.1130/0091-7613.

Chalmers GRL, Bustin RM, Bustin AAM. Geological controls on matrix permeability of the Doig-Montney hybrid shale-gas–tight-gas reservoir, northeastern British Columbia (NTS 093P); in Geoscience BC Summary of Activities 2011, Geoscience BC, Report 2012-1; 2012. p. 87–96.

Chen Q, Li W, Gao Y, et al. The deep-lake deposit in the Upper Triassic Yanchang Formation in Ordos Basin, China and its significance for oil-gas accumulation. Sci China Ser D Earth Sci. 2007;50(2):47–58. doi:10.1007/s11430-007-6029-7.

Clarkson CR, Pedersen PK. Production analysis of Western Canadian unconventional light oil plays. In: Canadian Unconventional Resources Conference 2011 Jan 1. Society of Petroleum Engineers. Calgary, 15–17 November, Alberta, Canada.

Cohen AS. Paleolimnology: the history and evolution of lake systems. Oxford: Oxford University Press; 2003. p. 350.

Curtis JB. Fractured shale-gas systems. AAPG Bull. 2002;86(11):1921–38. doi:10.1306/61EEDDBE-173E-11D7-8645000102C1865D.

Doust H, Sumner HS. Petroleum systems in rift basins–a collective approach in Southeast Asian basins. Pet Geosci. 2007;13(2):127–44. doi:10.1144/1354-079307-746.

Du J, Liu H, Ma D, et al. Discussion on effective development techniques for continental tight oil in China. Pet Explor Dev. 2014;41(2):217–24. doi:10.1016/S1876-3804(14)60025-2.

EIA. EIA/ARI World Shale Gas and Shale Oil Resource Assessment Technically Recoverable Shale Gas and Shale Oil Resources: An Assessment of 137 Shale Formations in 41 Countries outside the United States. US Energy Information Administration. 2013. www.adv-res.com.

Feng ZQ, Jia CZ, Xie XN, et al. Tectonostratigraphic units and stratigraphic sequences of the nonmarine Songliao basin, northeast China. Basin Res. 2010;22(1):79–95. doi:10.1111/j.1365-2117.2009.00445.x.

Gierlowski-Kordesch EH. Lacustrine carbonates. Developments in sedimentology, vol. 61. Amsterdam: Elsevier; 2010. pp. 1–101.

Graham SA, Brassell S, Carroll AR, et al. Characteristics of selected petroleum source rocks, Xianjiang Uygur Autonomous Region, Northwest China. AAPG Bull. 1990;74(4):493–512.

Hao F, Zhang Z, Zou H, et al. Origin and mechanism of the formation of the low-oil-saturation Moxizhuang field, Junggar Basin, China: implication for petroleum exploration in basins having complex histories. AAPG Bull. 2011;95(6):983–1008. doi:10.1306/11191010114.

Hanson AD, Ritts BD, Moldowan JM. Organic Geochemistry of oil and source rock strata of the Ordos Basin, north-central China. AAPG Bull. 2007;91(9):1273–93. doi:10.1306/05040704131.

He ZX. Tectonic evolution and petroleum in the ordos basin. Beijing: Petroleum Industry Press; 2002. p. 390 (in Chinese).

Hu C. Source bed controls hydrocarbon habitat in continental basins, East China. Acta Pet Sin. 1982;2(2):9–13.

Hu T, Pang X, Yu S, et al. Hydrocarbon generation and expulsion characteristics of Lower Permian P1f source rocks in the Fengcheng area, northwest margin, Junggar Basin, NW China: implications for tight oil accumulation potential assessment. Geol J. 2015;51:880–900. doi:10.1002/gj.2705.

Huang C, Zhang J, Wang H, et al. Lacustrine shale deposition and variable tectonic accommodation in the Rift Basins of the Bohai Bay Basin in Eastern China. J Earth Sci. 2015;26(5):700–11. doi:10.1007/s12583-015-0602-3.

Jarvie DM. Shale resource systems for oil and gas: part 1—shale-gas resource systems. In: Breyer JA, editor. Shale reservoirs—Giant resources for the 21st century, vol. 97. Tulsa: AAPG Memoir; 2012a. p. 69–87. doi:10.1306/13321446M973489.

Jarvie DM. Shale resource systems for oil and gas: Part 2—shale-oil resource systems. In: Breyer JA, editor. Shale reservoirs—Giant resources for the 21st century, vol. 97. Tulsa: AAPG Memoir; 2012b. p. 89–119. doi:10.1306/13321447M973489.

Jarvie DM. Components and processes affecting producibility and commerciality of shale resource systems. Geol Acta. 2014;12(4):307–25. doi:10.1344/GeologicaActa2014.12.4.3.

Jia CZ, Chi YL. Resource potential and exploration techniques of stratigraphic and subtle reservoirs in China. Pet Sci. 2004;1(2):1–2.

Jiang S, Henriksen S, Wang H, et al. Sequence-stratigraphic architectures and sand-body distribution in Cenozoic rifted lacustrine basins, east China. AAPG Bull. 2013a;97(9):1447–75. doi:10.1306/030413I2026.

Jiang S, Dahdah N, Pahnke P, Zhang J. Comparison between marine shales and lacustrine shales in China, search and discovery article #30316. In: AAPG Annual Convention and Exhibition, May 19–22, Pittsburgh, Pennsylvania; 2013b.

Jiang S. Prospects for shale gas development in China. Issues Environ Sci Technol. 2014;39:181–97. doi:10.1039/9781782620556.

Jiang S, Zhang J, Jiang Z, et al. Geology, resource potentials, and properties of emerging and potential China shale gas and shale oil plays. Interpretation. 2015;3(2):SJ1–3. doi:10.1190/INT-2014-0142.1.

Jiang S, Xu Z, Feng Y, et al. Geologic characteristics of hydrocarbon-bearing marine, transitional and lacustrine shales in China. J Asian Earth Sci. 2016;115:404–18. doi:10.1016/j.jseaes.2015.10.016.

Jin Z, Bo G, Wu X. Characteristics and resource potential of lacustrine shale oil and gas in China, AAPG Search and Discovery Article #80359. In: AAPG ACE, May 19–22, Pittsburgh; 2013.

Jin T, Chen L, Guo X, et al. Geological investigation for the Da'anzhai tight oil in Gongshanmiao oilfield in Sichuan Basin. Nat Gas Explor Dev. 2014;37(3):19–23 (in Chinese).

Jin J, Xiang B, Yang Z, et al. Application of experimental analysis technology to research of tight reservoir in Jimsar Sag. Lithol Reserv. 2015;27(3):18–25. doi:10.3969/j.issn.1673-8926.2015.03.003 (in Chinese).

Jiu K, Ding W, Huang W, et al. Fractures of lacustrine shale reservoirs, the Zhanhua Depression in the Bohai Bay Basin, eastern China. Mar Pet Geol. 2013;48:113–23. doi:10.1016/j.marpetgeo.2013.08.009.

Katz BJ. Controls on distribution of lacustrine source rocks through time and space: Chapter 4. In: Katz BJ, editor. Lacustrine basin exploration-case studies and modern analogs, vol. 50. Tulsa: American Association of Petroleum Geologists Memoir; 1990. p. 61–76.

Katz B, Lin F. Lacustrine basin unconventional resource plays: key differences. Mar Pet Geol. 2014;56:255–65. doi:10.1016/j.marpetgeo.2014.02.013.

Kelts K. Environments of deposition of lacustrine petroleum source rocks: an introduction. Geol Soc Lond Spec Publ. 1988;40(1):3–26. doi:10.1144/GSL.SP.1988.040.01.02.

Kuang L, Tang Y, Lei D, et al. Formation conditions and exploration potential of tight oil in the Permian saline lacustrine dolomitic rock, Junggar Basin, NW China. Pet Explor Dev. 2012;39(6):700–11. doi:10.1016/S1876 3804(12)60095-0.

Kuang LC, Gao G, Xiang BL, et al. Lowest limit of organic carbon content in effective source rocks from Lucaogou Formation in Jimsar Sag. Pet Geol Exp. 2014;36(02):224–9. doi:10.11781/sysydz201402224 (in Chinese).

Lambiase JJ. A model for tectonic control of lacustrine stratigraphic sequences in continental Rift Basins: Chapter 16. In: Katz BJ, editor. Lacustrine exploration: case studies and modern analogues, vol. 50. Tulsa: AAPG Memoir; 1990. p. 265–76.

LeFever JA. History of oil production from the Bakken Formation, North Dakota. In: Hanson WB, editor. Guidebook to geology and horizontal drilling of the Bakken Formation. Billings: Montana Geological Society; 1991. p. 3–17.

Li Y, Zhong JH, Wen ZF, et al. Oil-gas accumulation types and distribution of mudstone rocks in the Jiyang Depression. Chin J Geol. 2006;41:586–600. doi:10.1111/1755-6724.12742 (in Chinese).

Li D, Li J, Wang S, et al. Tight oil reservoir characterization of Upper Cretaceous in Songliao Basin. In: International Symposium on Shale Oil Technologies, April 16–17, 2012, Wuxi, China; 2012.

Li M, Li Z, Jiang Q, et al. Reservoir Quality, Hydrocarbon Mobility and Implications for Lacustrine Shale Oil Productivity in the Paleogene Sequence, Bohai Bay Basin. In: Adapted from oral presentation presented at AAPG annual convention and exhibition, May 19–22; 2013a. Pittsburgh, Pennsylvania.

Li Y, Feng Y, Liu H, et al. Geological characteristics and resource potential of lacustrine shale gas in the Sichuan Basin, SW China. Pet Explor Dev. 2013b;40(4):454–60. doi:10.1016/S1876-3804(13)60057-9.

Liang H, Li X, Ma Q, et al. Geological features and exploration potential of Permian Tiaohu Formation tight oil, Santanghu Basin, NW China. Pet Explor Dev. 2014;41(5):616–27. doi:10.1016/S1876-3804(14)60073-2.

Lin L, Zhang J, Li Y, et al. The potential of China's lacustrine shale gas resources. Energy Explor Exploit. 2013;31(2):317–35. doi:10.1260/0144-5987.31.2.317.

Liu Z, Sun P, Jia J, et al. Distinguishing features and their genetic interpretation of stratigraphic sequences in continental deep water setting: a case from Qingshankou Formation in Songliao Basin. Earth Sci Front. 2011;18(4):171–80 (in Chinese).

Liu B, Lü Y, Ran Q, et al. Geological conditions and exploration potential of shale oil in Qingshankou Formation, Northern Songliao Basin. Oil Gas Geol. 2014;33:280–5 (in Chinese).

Liu XX, Ding XQ, Zhang SN, et al. Origin and depositional model of deep-water lacustrine sandstone deposits in the 7th and 6th members of the Yanchang Formation (Late Triassic), Binchang area, Ordos Basin, China. Pet Sci. 2017;. doi:10.1007/s12182-016-0146-x.

Pan S, Wei P, Wang T, et al. Mass-transport deposits of the upper cretaceous Qingshankou Formation, Songliao Terrestrial Basin, Northeast China: depositional characteristics, recognition criteria and external geometry. Acta Geol Sin (English Ed). 2014;88(1):62–77. doi:10.1111/1755-6724.12183.

Pang X, Li S, Jin Z, et al. Quantitative assessment of hydrocarbon expulsion of petroleum systems in the Niuzhuang sag, Bohai Bay Basin, East China. Acta Geol Sin (English Ed). 2004;78(3):615–25. doi:10.1111/j.1755-6724.2004.tb00174.x.

Passey QR, Bohacs K, Esch WL, et al. From oil-prone source rock to gas-producing shale reservoir-geologic and petrophysical characterization of unconventional shale gas reservoirs. In International oil and gas conference and exhibition in China 2010 Jan 1. Society of Petroleum Engineers. Beijing, China; 2010.

Raji M, Gröcke DR, Greenwell HC, et al. The effect of interbedding on shale reservoir properties. Mar Pet Geol. 2015;67:154–69. doi:10.1016/j.marpetgeo.2015.04.015.

Ran XQ, Zhu XM, Yang H, et al. Petroleum enrichment theory and practice for low permeability reservoir in large continental basin. Earth Sci Front. 2013;20:147–54.

Schlische RW. Progress in understanding the structural geology, basin evolution, and tectonic history of the eastern North American rift system. In: LeTourneau PM, Olsen PE, editors. Aspects of Triassic-Jurassic Rift Basin Geoscience. New York: Columbia University Press; 2003. p. 21–54.

Schmoker JW. Method for assessing continuous-type (unconventional) hydrocarbon accumulations. In: Gautier DL, Dolton GL, Takahashi KI, et al. editors. National assessment of United States oil and gas resources—Results, methodology, and supporting data. U.S. Geological Survey Digital Data Series. DDS–30; 1995.

Sonnenberg SA. The Niobrara petroleum system: a new resource play in the Rocky Mountain region. In: Estes-Jackson JE, Anderson DS, editors. Revisiting and Revitalizing the Niobrara in the Central Rockies, Rocky Mountain Association of Geologists, Denver, Colorado; 2011, pp. 12–32.

Sonnenberg SA. The upper Bakken Shale resource play, Williston Basin. URTeC:1918895, In: Unconventional resources technology conference, 25–27 August; 2014. Denver, Colorado.

Wang HZ, Mo XX. An outline of the tectonic evolution of China. Episodes. 1995;18:6–16.

Wang H, Li S. Tectonic evolution of China and its control over oil basins. J China Univ Geosci. 2004;15(1):1–8.

Wang H, Zhang SH, He GQ. China and mongolia. In: Selley RC, Cocks LRM, Plimer IR, editors. Encyclopedia of geology. Amsterdam: Elsevier; 2005. p. 345–58.

Wang H, Ma F, Wang Z, et al. Hydrocarbon plays and unconventional hydrocarbon distribution in lacustrine basin, China.URTeC:1921691. In: Unconventional resources technology conference (URTeC), 25–27 August, 2014. Denver, Colorado.

Xiang B, Zhou N, Ma W, et al. Multiple-stage migration and accumulation of Permian lacustrine mixed oils in the central Junggar Basin (NW China). Mar Pet Geol. 2015;59:187–201. doi:10.1016/j.marpetgeo.2014.08.014.

Xu SL, Bao SJ. Preliminary analysis of shale gas resource potential and favorable areas in Ordos Basin. Nat Gas Geosci. 2009;20(3):460–65 (in Chinese).

Yang Y, Li W, Ma L. Tectonic and stratigraphic controls of hydrocarbon systems in the Ordos basin: a multicycle cratonic basin in central China. AAPG Bull. 2005;89(2):255–69. doi:10.1306/10070404027.

Yang H, Li S, Liu X. Characteristics and resource prospects of tight oil and shale oil in Ordos Basin. Acta Pet Sin. 2013;34(1):1–11. doi:10.7623/syxb201301001.

Yang T, Cao YC, Wang YZ, et al. The coupling of dynamics and permeability in the hydrocarbon accumulation period controls the oil-bearing potential of low permeability reservoirs: a case study of the low permeability turbidite reservoirs in the middle part of the third member of Shahejie Formation in Dongying Sag. Pet Sci. 2016;13(2):204–24. doi:10.1007/s12182-016-0099-0.

Yao J, Deng X, Zhao Y, et al. Characteristics of tight oil in Triassic Yanchang Formation, Ordos Basin. Pet Explor Dev. 2013;40(2):161–9. doi:10.1016/S1876-3804(13)60019-1.

Zeng L, Li X. Fractures in sandstone reservoirs with ultra-low permeability: a case study of the Upper Triassic Yanchang Formation in the Ordos Basin, China. AAPG Bull. 2009;93(4):461–77. doi:10.1306/09240808047.

Zhang J, Xu B, Nie H, et al. Exploration potential of shale gas resources in China. Nat Gas Ind. 2008a;28(6):136–40.

Zhang LY, Li Z, Zhu RF, et al. Resource potential of shale gas in the Paleogene in the Jiyang depression. Nat Gas Ind. 2008b;12:009.

Zhang D, Li Y, Zhang J, et al. National survey and assessment of shale gas resource potential in China. Beijing: Geologic Publishing House; 2012a (in Chinese).

Zhang S, Zhang L, Li Z, et al. Formation conditions of Paleogene shale oil and gas in Jiyang depression. Pet Geol Recov Effic. 2012b;19(6):1–5 (in Chinese).

Zhang NN, Liu LF, Su TX, et al. Comparison of Chang 7 member of Yanchang Formation in Ordos Basin with Bakken Formation in Williston Basin and its significance. Geoscience. 2013;27(5):1120–30 (in Chinese).

Zhang L, Li J, Li Z, et al. Advancements in shale oil/gas research in North America and considerations on exploration for lacustrine shale oil/gas in China. Adv Earth Sci. 2014;29(6):700–11. doi:10.11867/j.issn.1001-8166.2014.06.0700 (in Chinese).

Zhang XS, Wang HJ, Ma F, et al. Classification and characteristics of tight oil plays. Pet Sci. 2016;13(1):18–33. doi:10.1007/s12182-015-0075-0.

Zhou Z, Mu L, Li X, et al. Hybrid fracturing treatments unleash tight oil reservoirs consisting of sand/shale sequences in the Changqing Oilfield. In: IADC/SPE Asia Pacific drilling technology conference and exhibition 2012 Jan 1. Society of Petroleum Engineers. IADC/SPE Asia Pacific Drilling Technology Conference and Exhibition. 9–11 July; 2012. Tianjin, China.

Zhu G, Jin Q, Zhang S, et al. Combination characteristics of lake facies source rock in the Shahejie Formation, Dongying Depression. Acta Geol Sin. 2004;78(3):416–27 (in Chinese).

Zou CN, Dong D, Wang S, et al. Geological characteristics and resource potential of shale gas in China. Pet Explor Dev. 2010;37(6):641–53. doi:10.1016/S1876-3804(11)60001-3.

Zou CN, Wang L, Li Y, et al. Deep-lacustrine transformation of sandy debrites into turbidites, Upper Triassic, Central China. Sedim Geol. 2012;265:143–55. doi:10.1016/j.sedgeo.2012.04.004.

Zou CN, Tao S, Hou L, et al. Unconventional petroleum geology. 2nd ed. Beijing: Geological Publishing; 2013a.

Zou CN, Yang Z, Cui J, et al. Formation mechanism, geological characteristics and development strategy of nonmarine shale oil in China. Pet Explor Dev. 2013b;40(1):15–27. doi:10.1016/S1876-3804(13)60002-6.

Zou CN, Yang Z, Zhang G, et al. Conventional and unconventional petroleum "orderly accumulation": concept and practical significance. Pet Explor Dev. 2014;41(1):14–30. doi:10.1016/S1876-3804(14)60002-1.

Zou CN, Yang Z, Hou LH, et al. Geological characteristics and "sweet area" evaluation for tight oil. Pet Sci. 2015;12(4):606–17. doi:10.1007/s12182-015-0058-1.

16

Architecture mode, sedimentary evolution and controlling factors of deepwater turbidity channels: A case study of the M Oilfield in West Africa

Wen-Biao Zhang[1] · Tai-Zhong Duan[1] · Zhi-Qiang Liu[1] · Yan-Feng Liu[1] · Lei Zhao[1] · Rui Xu[1]

Abstract Turbidity channels have been considered as one of the important types of deepwater reservoir, and the study of their architecture plays a key role in efficient development of an oil field. To better understand the reservoir architecture of the lower Congo Basin M oilfield, semi-quantitative–quantitative study on turbidity channel depositional architecture patterns in the middle to lower slopes was conducted with the aid of abundant high quality materials (core, outcrop, logging and seismic data), employing seismic stratigraphy, seismic sedimentology and sedimentary petrography methods. Then, its sedimentary evolution was analyzed accordingly. The results indicated that in the study area, grade 3 to grade 5 architecture units were single channel, complex channel and channel systems, respectively. Single channel sinuosity is negatively correlated with the slope, as internal grains became finer and thickness became thinner from bottom to top, axis to edge. The migration type of a single channel within one complex channel can be lateral migration and along paleocurrent migration horizontally, and lateral, indented and swing stacking in section view. Based on external morphological characteristics and boundaries, channel systems are comprised of a weakly confining type and a non-confining type. The O73 channel system can be divided into four complex channels named S1–S4, from bottom to top, with gradually less incision and more accretion. The study in this article will promote deeper understanding of turbidity channel theory, guide 3D geological modeling in reservoir development and contribute to efficient development of such reservoirs.

Keywords Reservoir architecture · Turbidity channel · Sedimentary evolution · Deep water · Shallow seismic · Controlling factors

1 Introduction

Deep water channels are considered one of the significant types of reservoir, sometimes containing rich oil and gas resources. There have been studies abroad on their depositional architecture, mainly focusing on architecture element recognition and description through core and other observations. Achievements have been made by independent oil companies (IOCs) and institutions, with the assistance of the rapid development of deepwater drilling, geophysics theory and sonar scanning, on sedimentary configuration of deepwater turbidite channels. Domestic researchers are doing similar studies using deepwater sediments in the South China Sea area. These studies and related achievements contribute a lot to reducing risks in early-stage deepwater exploration (Heiniö and Davies 2007; Posamentier and Kolla 2003; Slatt 2006; Menard 1955; Li et al. 2008; Wang et al. 2009; Deng et al. 2008; Lü et al. 2008). For continental slope areas in the Lower Congo Basin, West Africa, there have been studies of deepwater channel sedimentary patterns and the factors controlling them in terms of sedimentology and sequence stratigraphy.

However, with the gradual development of more deepwater oilfields, plenty of dynamic data reveal complex superimposition of single sand bodies inside channel systems and various high-heterogeneity rock types in single

✉ Wen-Biao Zhang
  zwb.syky@sinopec.com

[1] Petroleum Exploration and Production Research Institute, SINOPEC, Beijing 100083, China

Edited by Jie Hao

sand bodies. This impedes further and more effective development of deepwater oilfields. Therefore, it is necessary to analyze configurations of both complex and single sand bodies to clarify single sand body distribution patterns, quantitative scales, superimposition and lithology inside single sand bodies.

We took the deepwater M oilfield in West Africa as an example to study semi-quantitative–quantitative sedimentary configuration patterns and their spatial evolution in this article. Based on abundant drilling, core and high-quality seismic data, the study was carried out at three different levels, including channel system, complex channel and single channel levels, employing methods such as core description, log recognition and seismic attribute slices. The study in this article may promote turbidite channel theory understanding and benefit 3-D geomodeling, making it useful in developing this type oilfield more efficiently.

2 Geological background

The M oilfield is one of the most favorable exploration and development petroliferous basins located in the lower Congo Basin, with typical passive continental margin characteristics (Liu and Li 2009; Xiong et al. 2005; Kolla et al. 2001). The research zone lies in the middle–lower slope, between compressive and extensional zones, 186 km away from Luanda, Angola. The architecture is not severely impaired by Cretaceous gypsum activities, and the main target layer is Oligocene. Its sedimentary type is considered to be a deep water turbidite channel system under a regressive background with a water depth of 1400–1800 m (Fig. 1). There are 18 wells in the research zone of the Oligocene O73 reservoir, and they are characterized by a core depth of 168 m; average well distance more than 1000 m; dominant seismic frequency of 35 Hz; and sand recognition 15–20 m. The Pleistocene layer of the near-seabed area has turbidite channel sediments under a regressive background as well. The provenance is from the Eastern Congo River. The dominant seismic frequency reaches 65 Hz, and vertical sand recognition is 6 m. Therefore, it is reasonable to make an analogy between the target layer (deep zone) and the near-seabed layer (shallow zone), due to the similar turbidite channel sediments. The basic data used in this article include core data, logging data, high-density seismic acquisition, and shallow channel high-quality seismic data (single channel sand recognizable) and onsite outcrop measurement data. These types of information can be crosschecked and used in complementary manners in geological research models, e.g., the combination of shallow channel high-frequency seismic data and onsite outcrop data serves to complement the

limited deepwater seismic resolution, making full-scale analysis of channel architecture type and scale possible; core data provide complementary information for research into inner filling models. Comprehensive use of available information is a key method to study sedimentary architecture.

3 Turbidite channels hierarchical division

Several criteria were proposed for the hierarchical division of turbidite channel architecture. For example, Mutti and Normark proposed a five-turbidite-facies scheme in 1987, mainly based on the genetic type of sand bodies. Later, a seven-turbidite-facies scheme was put forward by Zhao et al. (2012a, b) and Lin et al. (2013). In this division, they gave more thought to hydrodynamic characteristics, sedimentation and contact relations of the formation of channel sand bodies. In this article, we adopted the seven-turbidite-facies scheme to study sedimentary architecture patterns on three levels, which are single channel, complex channel and channel system levels, to demonstrate the influence on reservoir distribution from macro perspectives (Table 1).

There are certain genetic connections between units of channels at different levels (Zhao et al. 2012a, b). The channel system is characterized by complex channels of different periods, while the formation of complex channels is subject to the migration patterns of single channels. There are differences in the scale of architectural units at different levels, thereby influencing the data type required for study. The architectural units of large-scale channel systems are recognized mainly through seismic data (comprising of seismic facies and seismic reflection structures), while that of the intermediate scale (complex channel) can be recognized also through seismic data (strata slicing). With regard to the small scale (single channels), drilling data (coring and well logging data) and high-resolution 3D seismic data near sea bottom are applied. The relations among various hierarchies are shown in Fig. 2, and higher level sedimentary characteristics are usually subjected to the perturbation and sedimentology of lower level architectural units.

4 Sedimentary architectural patterns of turbidite channel

It is the best to study the architectural patterns through detailed investigation of each hierarchy's characteristics and its origin. Nevertheless, considering the impact of reservoir distribution on real well development and production, we did the research from a 3-level perspective, and this was the channel system, complex channel and single

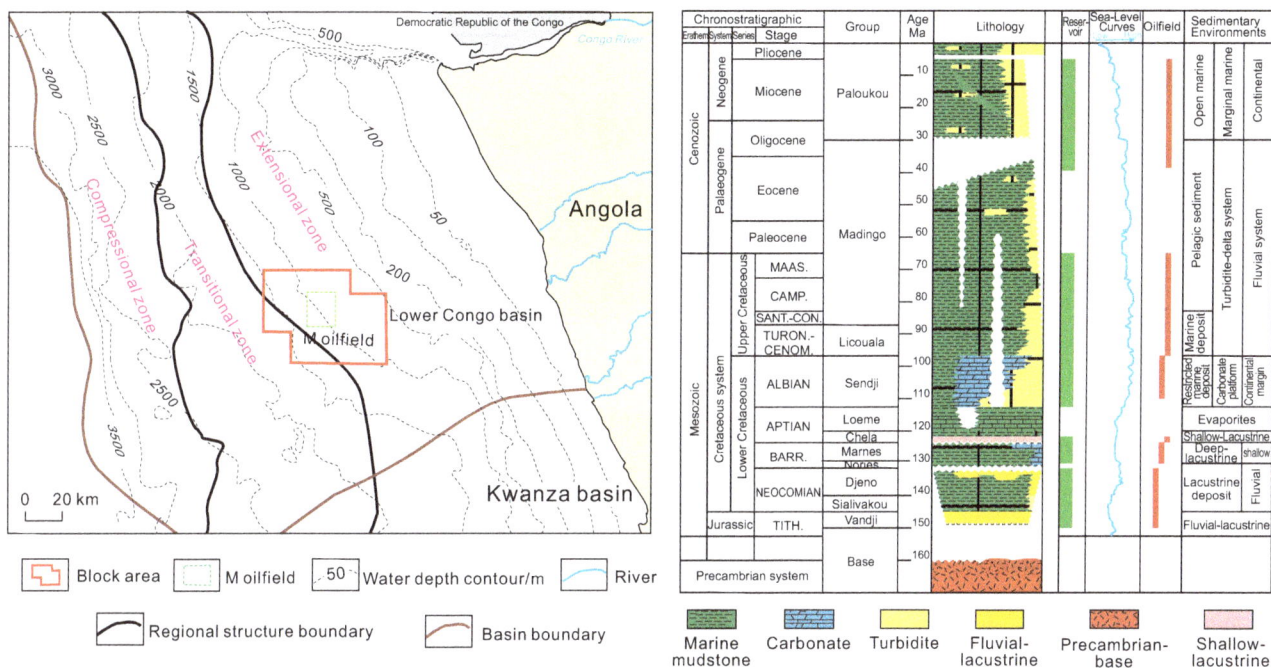

Fig. 1 Location and comprehensive stratigraphic column of study area

Table 1 Comparison of different schemes of hierarchical division of sedimentary configuration of turbidity channels (Lin et al. 2013, revised)

Mutti and Normark (1987)		Lamb (2003)		Lin et al. (2013)	
1	Basin filling, fan complex	6	Complete set of strata	7	Submarine fan complex
2	Single fan	5	Composition of several 4-level configurational units, which are distinguishable	6	Single submarine fan
3	Fan development stage	4	Sedimentation products of various sedimentary environments and patterns of flow	5	Channel system
				4	Complex channel
4	Natural levee microrelief of channel	3	Products under the same genetic mechanism	3	Single channel
5	Lithofacies, bedding microrelief	2	Single sedimentation unit	2	Sedimentation unit inside single channel (e.g., Bouma sequence)
		1	Further segmentation of a single sedimentation unit	1	Rhythmical layers inside the sedimentation unit

channel levels. We believed that the channel system influenced the vertical development layer selection, while the scale of the complex channel and the relations of its inner single channels played a key role in determining well spacing.

4.1 Hierarchical architectural patterns of single channels

Single channels are formed by repeated gravity-flow deposits along one channel over a period of time, a major origin unit in turbidite channels. So it is important to study

the architectural patterns and internal filling features in order to understand reservoir development.

4.1.1 Geometrical morphology characteristics

Elements of the geometrical channel morphology usually include the channel widths, depth, sinuosity, arc, length of curve, wave length (Wood and Mize-Spansky 2009). Based on studies of modern and ancient submarine fans, scholars found that due to the combined effects of ancient sedimentary environments, tectonic subsidence and eustatic sea level changes, there were telling differences in the

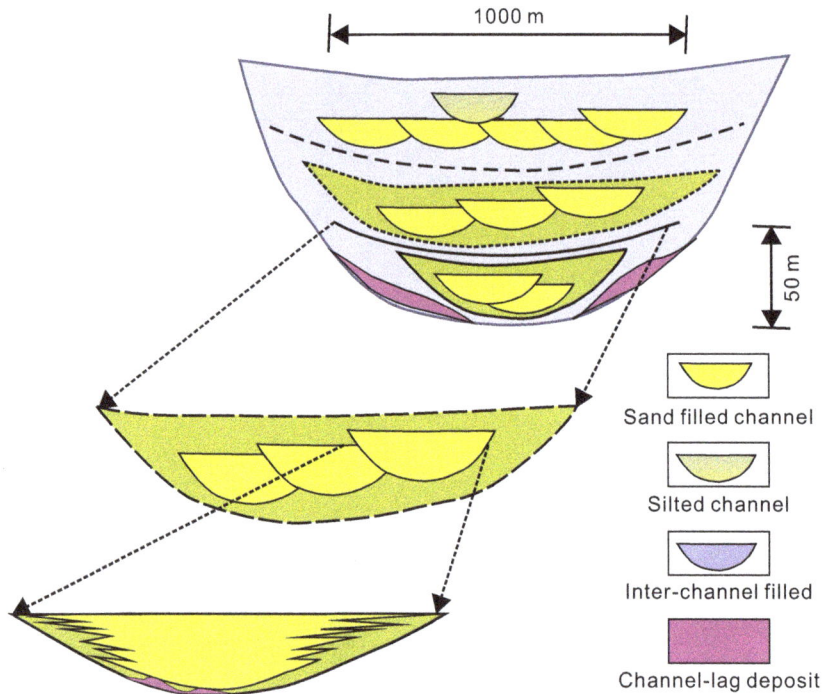

Fig. 2 Configuration unit sedimentary pattern of turbidity channels

geometrical morphology characteristics of turbidite channels. Because of the relative small scale of single channels (min. depth being 10 meters), while dominant frequency of M oilfield seismic data is 40 Hz, it is very difficult to recognize a single channel. However, the M oilfield near-seabed area high-frequency seismic data show a dominant frequency of 70 Hz and similar sedimentary background and type, which makes a critical complement to single channel studies. Since it is hard to extract ideal single channels from real oilfield seismic data, we analyze geometrical morphology and characteristics of single channels with the aid of shallow high-frequency seismic data. It has been found that in seismic profiles, single channels are U or V shaped, presenting medium-strong amplitude inner side, parallel or wave-like reflection, and good consistency as depicted by Fig. 3.

Sinuosity is one essential parameter in the study ($k = h_a/l_a$, k being sinuosity, while h_a being the winding length and l_a being the valley length, see Fig. 3a). It is measured through samples extracted from shallow seismic data. Statistical analysis revealed that the sinuosity distribution of single channels ranges between 1.0 and 5.4, averaging 1.87. As a result, single channels are classified as low-sinuosity channels and high-sinuosity channels. The average sinuosity of low-sinuosity channels was 1.2, while that of high-sinuosity channels was 1.8. Such difference is due to various geological factors (Wynn et al. 2007; Deptuck et al. 2003; Peakall et al. 2000) and may have much to do with the gradient of the ancient continental

slope. Against the backdrop of similar sedimentary origin and sedimentation background, we considered the channels located near the sea bottom due to their being less impacted by tectonic movement. Then, analysis of the relationship between single channel sinuosity and topographic slope ($\theta = \arctan(h/l)$, θ being slope, h being the width, l being the length, see Fig. 4a) was done, which shows a negative correlation between the two and a correlation coefficient of 0.8. See Fig. 4b for image. As the slope becomes steep, the downcutting enhances while lateral migration weakens. Whereas when the slope is gentle, provenance supply drops off and sedimentation becomes finer, enabling weaker downcutting and increased lateral migration. Thus, high-sinuosity single channels are formed. This analysis can also support estimating paleotopography slope based on the current single channel sinuosity. As for single channel width and depth (thickness), they are obtained from shallow high-frequency seismic data coupled with outcrop measurements, on account of sparse well distribution and difficulty to determine single sand body boundaries from such well spacing. The result shows that in the study area, the depth of single channels (d) ranged between 10 and 35 m, and width (w) generally between 150 and 450 m (Fig. 3b).

4.1.2 Lithofacies filling model

Lithofacies directly reflect the nature of the sedimentary environment. Different lithofacies have different genetic

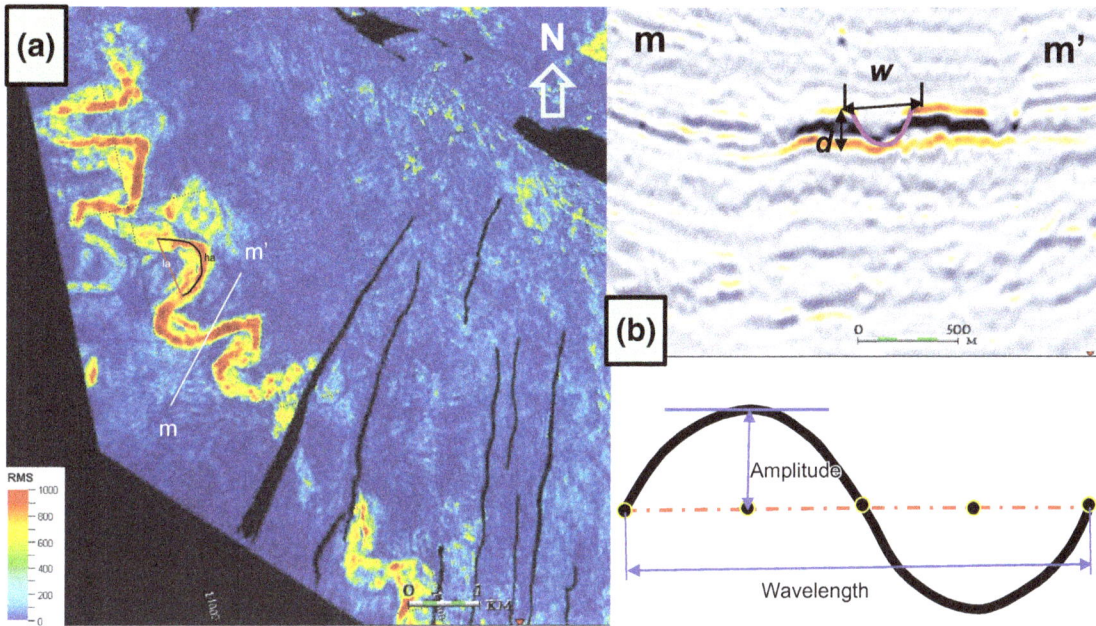

Fig. 3 Geometric elements of turbidity channels (the shallow seismic data in the study area)

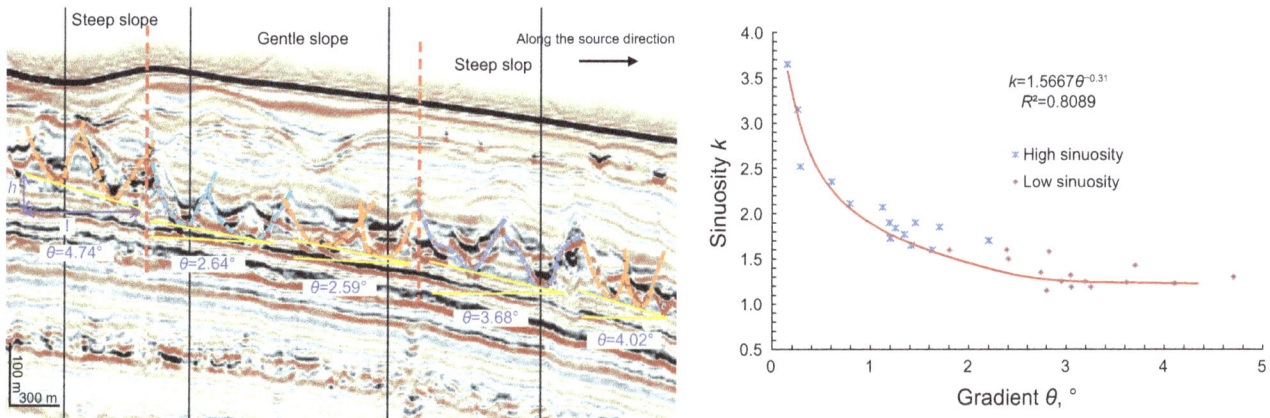

Fig. 4 Correlation of tortuosity and slope gradient (the shallow seismic data in the study area)

mechanisms, indicating different permeable capacity (Bouma 1985; Habgood et al. 2003). Therefore, exact identification of lithofacies types is required for studies on the genetic mechanism of turbidite channels and analysis on permeable discrepancy. The coring data show that obvious turbidite channel sedimentary characteristics can be found in the M oilfield. Lithologically, turbidite channels in the area are mainly composed of massive sandstones, mixed with a little fine-grained sediment. Based on sedimentary tectonics, they have Bouma sequence features and are mostly blocky structure. Cross and parallel beddings can be seen on the top. Erosive bases are generally developed, mixed with retention sediments (mudstone fragments) at the bottom of the channel. Multiple washings can be seen in the main body of the channel (secondary erosion). Mudstone interlayers and argillaceous slumps can

also be seen on some sites. Floating gravels are visible inside the massive sandstones (Fig. 5). The sand bodies are generally fining upward. That is, lithofacies inside the channel, from bottom upward, form a configuration pattern of retention sediment ∼ massive gravelly coarse sandstones (which may contain mud-sized grain) ∼ massive middle-fine sandstone ∼ interlaced bedded sandstone ∼ fine-grained sediment, with sand bodies becoming thinner from bottom to top.

The specific sedimentary sequences inside the channels produce corresponding logging responses. Single channel responses for individual wells are as follows. Natural gammas are mainly bell-wise and nearly box shaped, while the resistivity curve is slightly funnel shaped (Zhao et al. 2010). Finally, based on such information from field outcrops, cores of the target stratum, etc., the internal filling

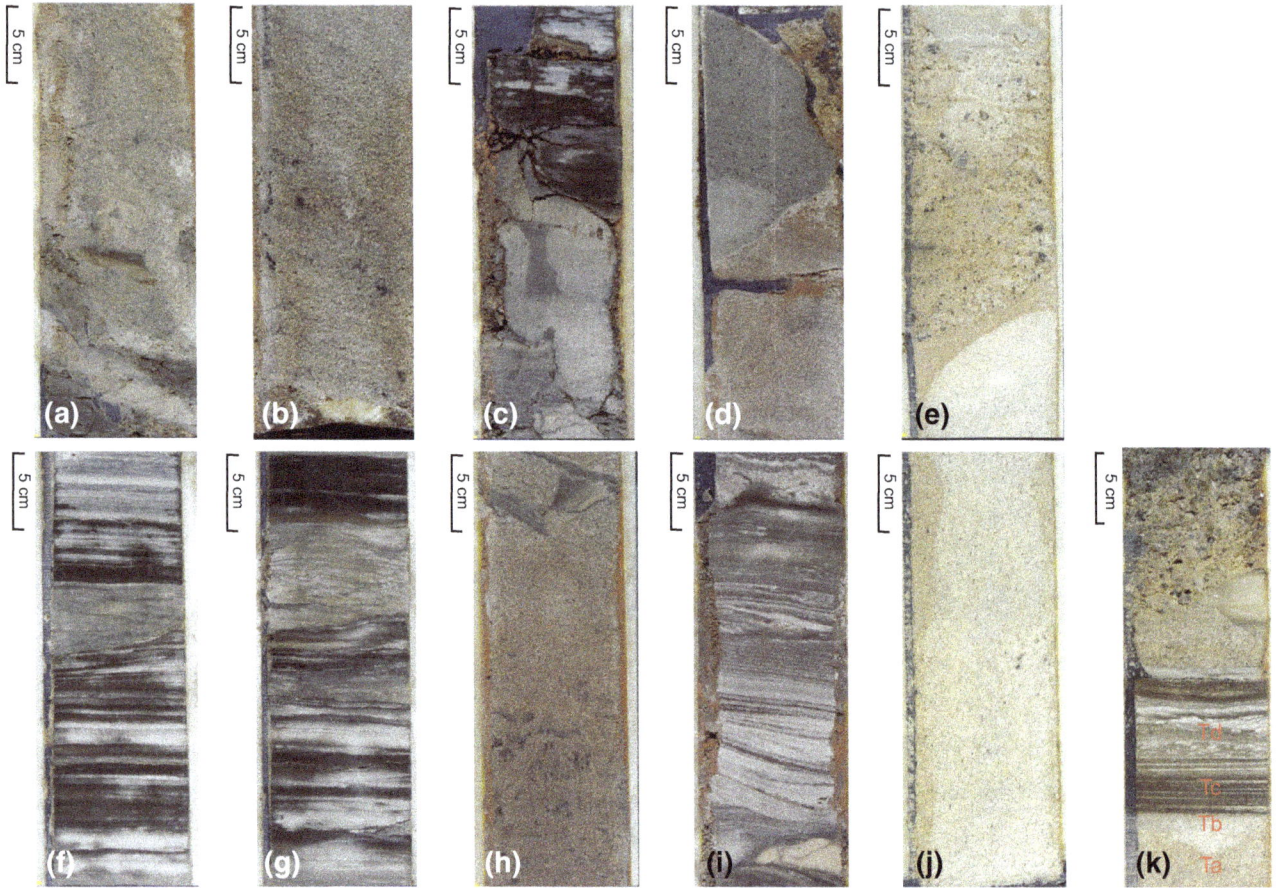

Fig. 5 Internal lithofacies characteristics of single channels in the M oilfield, West Africa (well position as shown in Fig. 9). From *top left* **a** massive gravelly coarse sandstone, well-1A, 3196.25 m; **b** massive mud-sized grain—coarse sandstone, well-1A, 3197.16 m; **c** retention sediments at the bottom, well-1A, 3199.06 m; **d** massive gravelly coarse sand facies, well-2B, 3202.29 m; **e** massive sandstone (see the erosion surface), well-2B, 3213.03 m; **f** thin-layer muddy silt facies, well-1A, 3205.0 m; **g** corrugated-bedding siltstone, well-1A, 3206.12 m; **h** massive gravelly medium-coarse sand facies (floating boulder clays on the top), well-2B, 3202.08 m; **i** corrugated-bedding siltstone, well-2G, 3209.26 m; **j** massive medium-fine-grained sand facies, well-2B, 3213.67 m; **k** Bouma sequence, well-2B, 3217 m

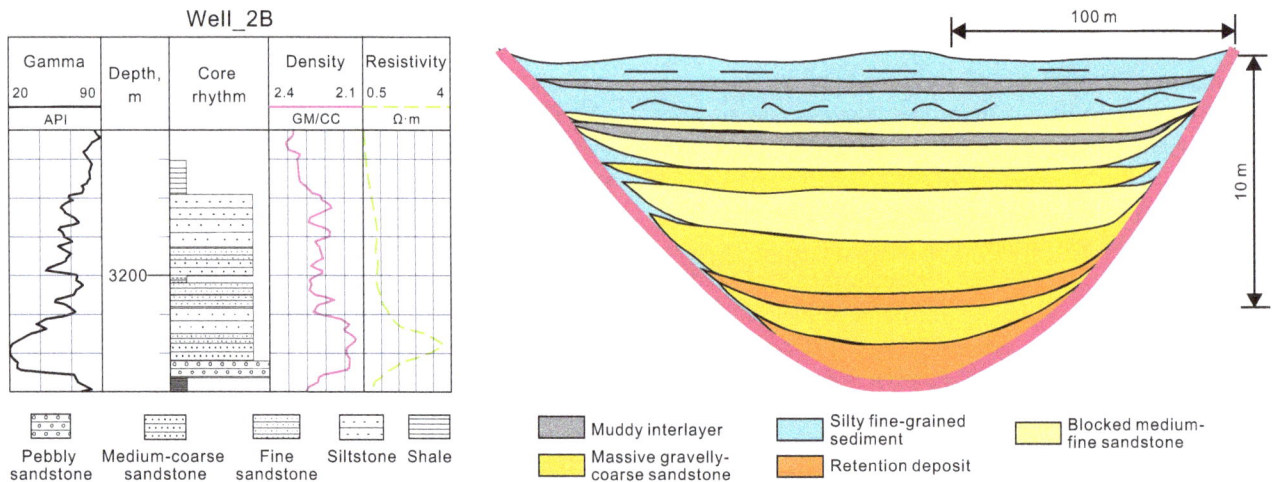

Fig. 6 Logging and lithofacies characteristics of single channel (well position as shown in Fig. 9)

patterns of single channels are summarized (Fig. 6). The inner part of single channels is filled in "bundles," with sand bodies becoming thinner and finer from axis to the edge. Vertically, retention gravels are usually deposited at the bottom of the channel. Sand bodies of the channel change progressively, mixed with thin muddy or clay layers. From the bottom upward, the pattern of sand tends to be thinner and finer. The sedimentary tectonics of thin sand-shale interbed at the top of the channel form parallel bedding-corrugated bedding.

4.2 Hierarchical architecture patterns of complex channels

A complex channel consists of single channels that are laterally or vertically superimposed, mainly controlled by an autogenetic cycle. It is a medium-sized architectural unit (Mutti and Normark 1987; Kane et al. 2007; Abreu et al. 2003; Anka et al. 2009). We studied the internal architectural patterns using both shallow high-frequency and deep well-to-seismic integration data.

4.2.1 Horizontal migration pattern

By drawing the RMS amplitude attributes slice of channel deposits in the M oilfield, we can see that there are two migration types in single channels: in lateral and paleocurrent direction (Zhao et al. 2012a, b) from the channel plan view. So we built the migration pattern of single channels inside the complex channel according to the seismic response features, as shown in Fig. 7. Lateral migration causes sand body distribution more continuous, which expands the connected range of sand bodies. With migration in the paleocurrent direction, since the single channels are vertically superposed, the sand bodies show great heterogeneity in the vertical direction. In plan view, channels migrate along the provenance direction, presenting wide stripes for sand bodies (Liu et al. 2013). The width of complex channels ranges between 300 and 1500 m in the study area. Due to different migration patterns of different channel segments, there are ambiguous relations between the width and depth for complex channels in our study. Another important factor to describe migration patterns in complex channels is sinuosity, which is due to single channel migration. It is measured to be 1.0–1.8, averaging at 1.3 in our study, using the central line in the boundary area of the complex channel as baseline. This is much smaller than that of single channels.

4.2.2 Profile migration pattern

The profile migration patterns of complex channels are also subjected to single channel migrations. As we know, the profile migration pattern of single channels can be classified into lateral and vertical migrations (Labourdette 2007; Hubbard et al. 2009). The lateral migration causes single channels joining together in the lateral direction horizontally, so the sand body thickness of complex channels is very close to that of a single channel (Fig. 8a). As such, the vertical heterogeneity is relatively weak. Also, the vertical migration can be categorized into two types: indented and swing (Fig. 8b, c), causing channels to superimpose in the vertical direction. The thicknesses of sand bodies are generally larger than the depth of single channels. Moreover, as retention slump-block deposits (inferior filter beds) usually occur at the bottom of the channel, the vertical heterogeneity of such compound sand bodies is relatively strong. Results show that the depths of complex channels in the study area lie between 20 and 185 m. As affected by profile migration features and the lithological changes inside, there are obvious well-seismic response characteristics in the boundary regions. The mudstone content rises at the boundary, making the sand body thinner, mostly manifested by weakened amplitude intensity. In addition, the lateral migration of channels can create imbricate responses on the seismic profile (Fig. 8).

4.3 Hierarchical architecture pattern of channel systems

The channel system (canyon or large incised waterway) is regarded as a large-scale unit, superimposed by various complex channels (Xie et al. 2012; Yu et al. 2012). Affected by erosive power, the reservoir architecture models of the same deepwater turbidity channel system differ greatly in different deposit locations. Spatial structure characteristics can be represented by the seismic information. Researchers categorize channel systems into restricted (with incised valley), weakly restricted (with incised valley) and non-restricted (without incised valley) according to geomorphic features (Hubbard et al. 2009; Zhao et al. 2012a, b; Lin et al. 2013; Clark and Pickering 1996; Deptuck et al. 2003; Sun et al. 2014; Chen et al. 2015).

Different superimposition patterns of complex channels lead to different architectures inside the channel system. Inside restricted channel systems, deeply incised indented and swing complex channels are the major part and barely develop large natural levees. With weakly restricted channel system, there are deeply incised indented and swing complex channels combined with weakly incised horizontal migration patterns, with deposit overflow along incised valleys, developing large natural levees on both sides. Non-restricted channel system features weakly incised indented and swing channels, along with horizontal migration patterns. There are occasional deeply incised channels at the bottom of non-restricted channel systems,

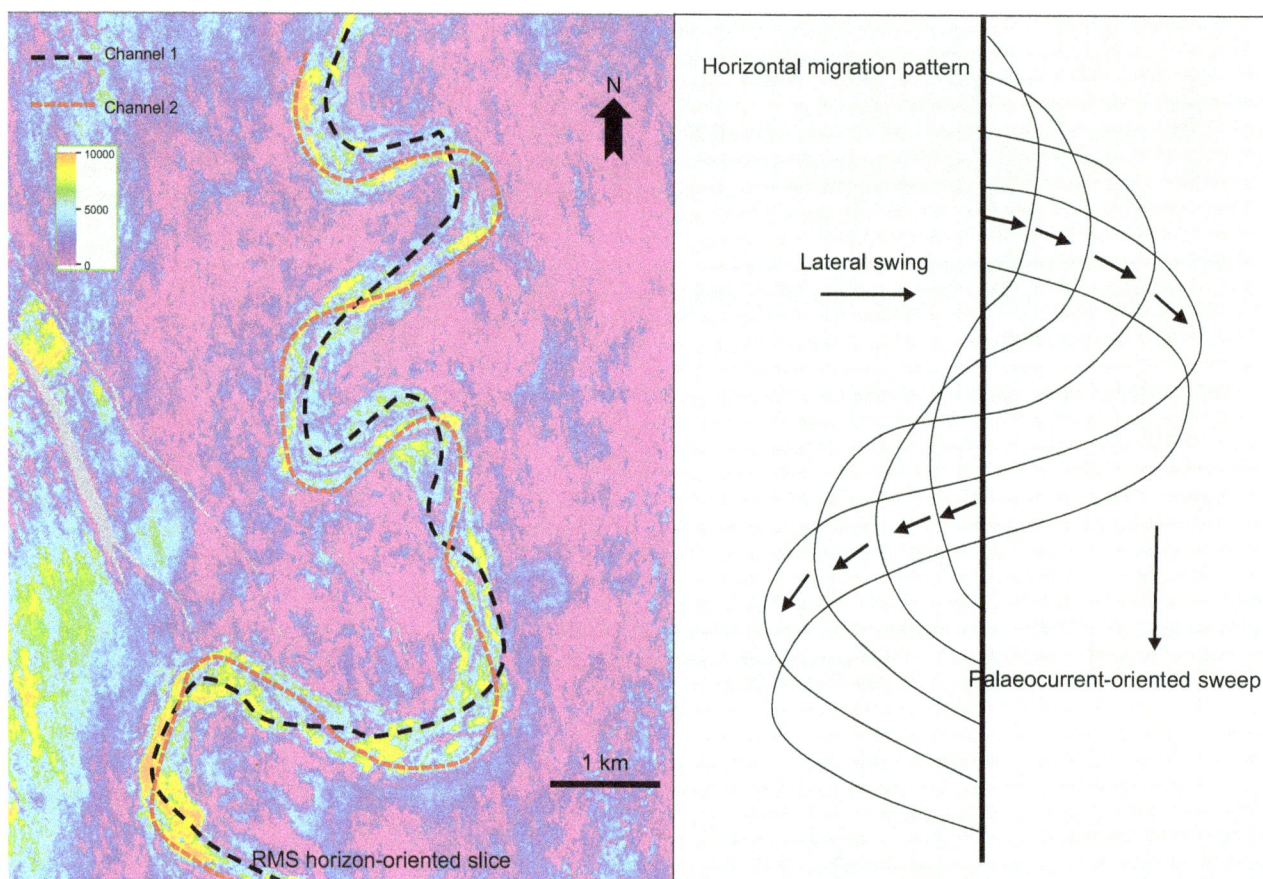

Fig. 7 Plane migration patterns of single channel based on layer slicing in M oilfield (slice position as shown in Fig. 1)

with no evident natural levee development. Especially in the late stage of channel system development, it is difficult to distinguish between fine-particle-filled channel deposits and natural levee deposits. To sum up, the spatial patterns of channel system vary significantly due to deposition patterns of complex channels.

Weakly restricted and non-restricted channel systems are considered as major categories developed in the O73 oilfield in the study area, and their seismic response characteristics are shown in Fig. 9. There show certain evolution trends in the plane. The closer it is to the provenance direction, the stronger the channels incision. Weakly restricted channel systems have distinct incised sections, and wedges (large natural levee deposit) are developed on both sides. The further it is from the provenance direction, the weaker the incision, and the stronger the aggradation and lateral migration. Non-restricted channel systems develop with no evident boundary features (without developing large incised valleys). Typical vertical evolution patterns can be found inside non-restricted channel systems, depicted in four stages. Great incision in the early-stage deposits and large incise valleys are developed. But in later stages, because of rising sea level, incision abates while aggradation and lateral migration enhances, resulting

in inconspicuous incised valleys. Results show widths of 1000–3000 m and depth of 80–280 m for the O73 channel system, a large-scale one.

5 Sedimentary evolution of turbidity channels

It is through analyzing the evolution of channel systems that we understand the deposition processes and architectural genesis. More importantly, it will strengthen the credibility of inter-well prediction for the architectural characterization based on well-log and seismic data under wide spacing.

5.1 Sedimentary evolution characteristics

Through well-to-seismic calibration, we make a comprehensive explanation of S1-S4 (Fig. 9) complex channel sediments in the O73 channel system of the M oilfield. Seismic attributes demonstrating sand body distributions, such as RMS amplitude, were extracted (Fig. 10). Combined with core data, we have managed to explain evolution characteristics of each stage (Fig. 11).

The O73 reservoir of the M oilfield shows noticeable sequence sedimentary characteristics. The early

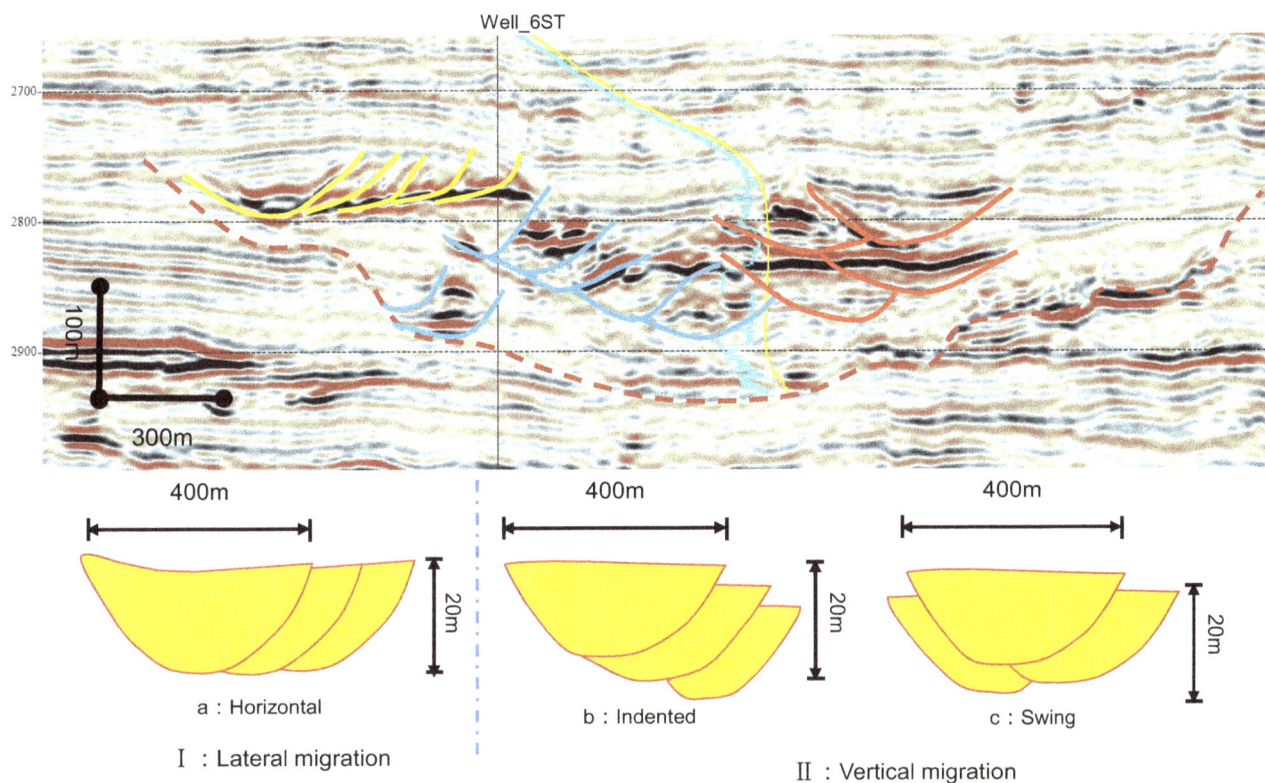

Fig. 8 Profile migration patterns and vertical evolution of single channels of zone O73 in the M oilfield

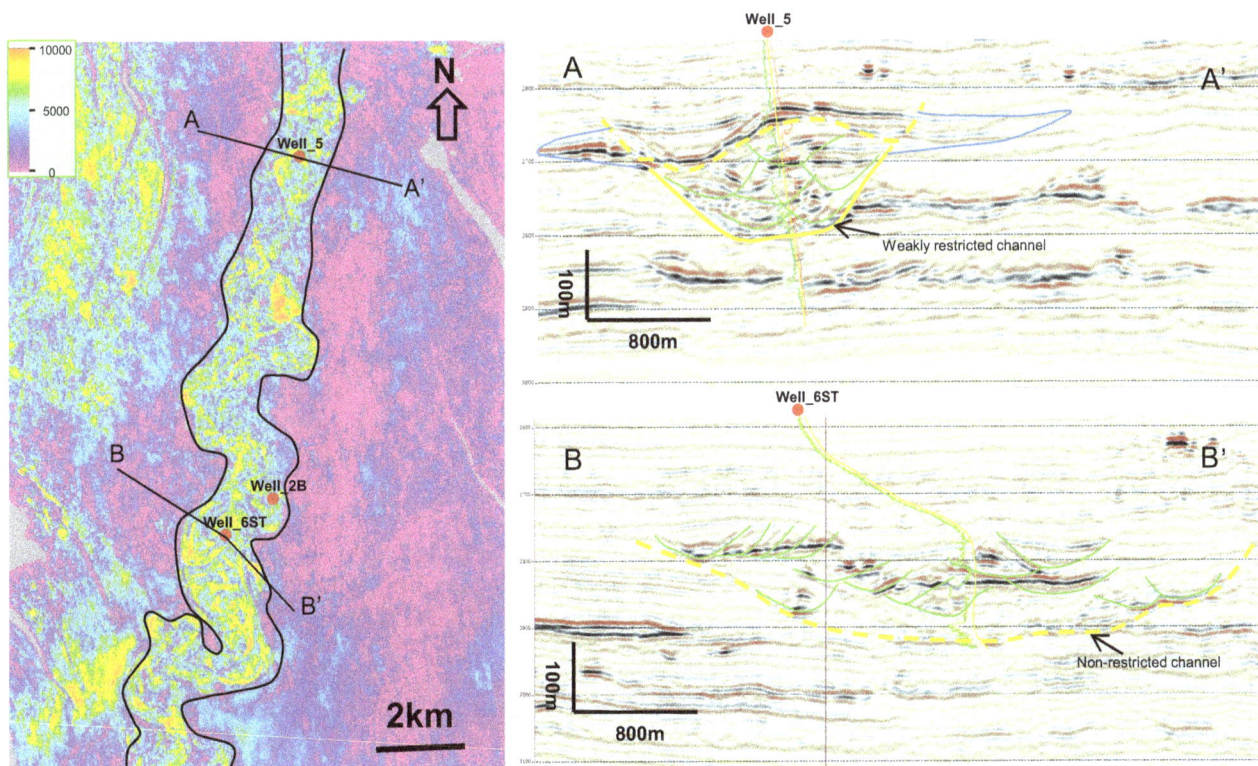

Fig. 9 Plane evolution relation of the O73 channel system (slice position as shown in Fig. 1)

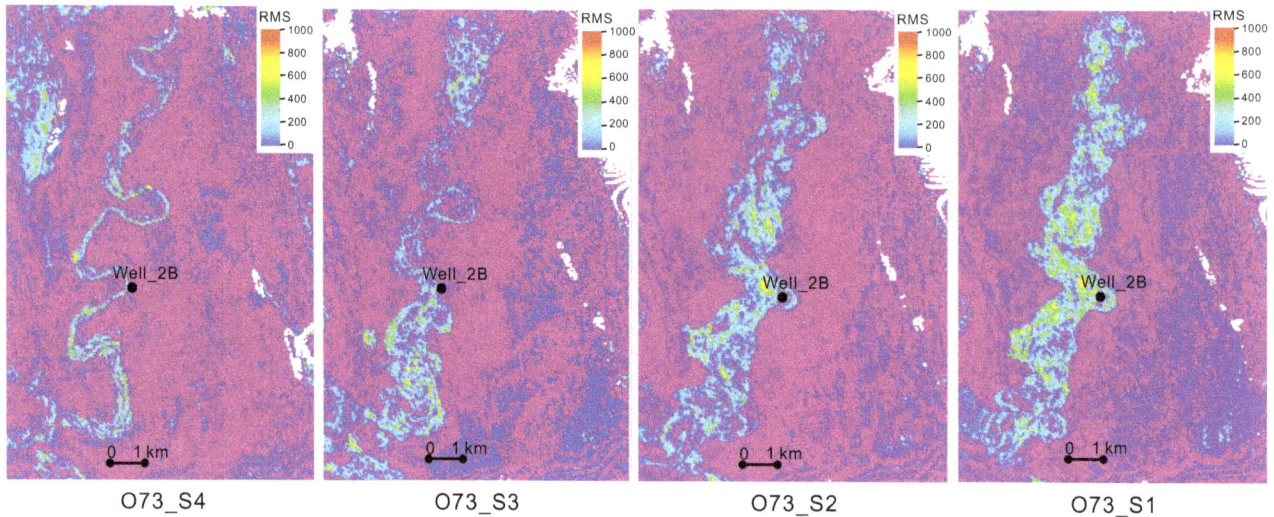

Fig. 10 Plane characteristics of O73 channel system in the M oilfield (slice position as shown in Fig. 9)

sedimentary stage (S1) belongs to deepwater stratified sediments. It shows distinct restricted characteristics with abundant supply and strong erosive power, widely distributed in the riverbed at the bottom of the large submarine canyon of the O73 reservoir. Well-2B coring shows that the lowermost part of the sequence mainly consists of a coarse sandstone stratigraphic unit, which often contains coarse to boulder-level conglomerates and mudstone fragments. It can be deduced that its high sedimentary energy enables it to downcut the older conglomerate layer and clastic layer. The middle part of the sequence is a mixed sedimentary unit of thicker coarse sand and medium sand, probably high-density turbidite sediment, and it progressively changed into a finer-grained Tb and Tc type low-density turbidite layer in an upward direction. The uppermost part is a mudstone layer, indicating gradually waning energy.

During the S2 stage, sediment supply is still sufficient, but the sea level began to rise forming a slightly restricted channel. Here the single channel shows lateral migration, as a weakly restricted channel, and thus widely distributed. As shown in Fig. 11, the S2 sequence channel sediment presented a transition trend from the main channel axis to the edge, with muddy interlayers gradually developing in the edge. Core analysis indicates that the lithology of S2 is mainly thick massive sandstone, partially interbedded with thin mudstone layers. The top layer gradually changed to siltstone and mudstone layers, with occasional ripple bedding, showing gradual abandonment characteristics.

During the S3 stage, the sea level continued to rise, and sediment supply started to decrease. Channel sediments show distinct non-restricted characteristics. Lateral migration and vertical downcutting are both strong for single channels, as well as high sinuosity. The channel on the planar graph (Fig. 10) is very clear. Superposition of multi-

period channel sand bodies results in expanded distribution of sand bodies. For well-2B, the meander section of the outer S3 sequence channel complex was drilled, so the core only represents a partial sedimentary filling sequence. There is a layer of gravelly sediments (about 4 m in thickness) at the bottom of the coring section, which is covered by a hard sand clastic layer (about 2 m in thickness) and then comes a 6-m-thick muddy siltstone (the top sediment of the S3 sequence is draped by the mudstone layer). For well-2G, drilling of the meander section of the interior S3 sequence channel revealed that S3 sequence channel intensively eroded S2 sequence sediments (Fig. 11e), which is possibly related to weak consolidation of the early channel sediment and the supply channel formed due to the negative topography on the edge of the channel.

The S4 sequence is the last layer of the O73 channel system, belonging to the sediment shrinkage stage when the sea level reached a peak. The channel is still highly sinuous, but the scale is much smaller than that of the S3 sequence. The downcutting depth decreases, with strengthened lateral restriction. Core analysis indicates that the bottom of the sequence is well-sorted medium-fine massive sandstone, with good consistency in seismic response within the whole oil field (Fig. 10). It is the final product of channel filling, and there will be the abandonment stage of the O73 channel system afterward.

5.2 Sedimentary controlling factors and their evolution

5.2.1 Sedimentary controlling factors

Deepwater detrital deposits are controlled by autogenetic cycles and allogenetic cycles. The controlling factors

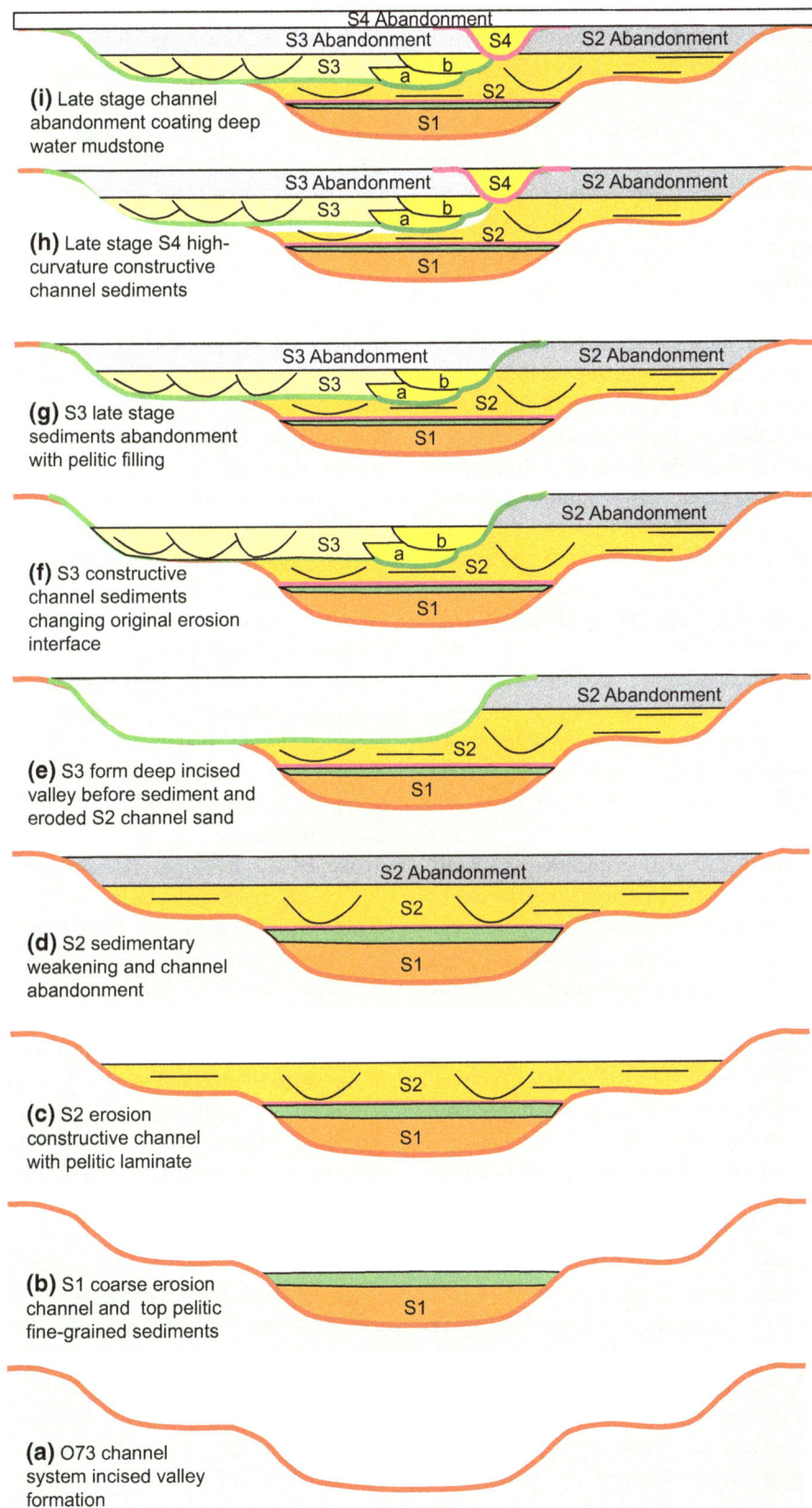

Fig. 11 Vertical evolutionary model of O73 channel system in the M oilfield

include eustacy, basin tectonic movement, sediment types and supply rates. Moreover, events such as earthquakes and tsunamis may also allow the clastic particles to reach the deep sea after traversing the continental shelf and slope valley, forming deepwater sediments (Stow et al. 1996; Shanmugam 2008). The combination of many controlling factors causes the difference in the erosive power of channels, resulting in the complex and diverse superposed relationship of sand bodies. These controlling factors include the provenance distance, provenance types, climate in the provenance area, sea level eustacy, topographic slope. Under normal circumstances, the closer to the provenance, the greater the topographic slope and the more sea level drops, the more abundant the sediment supply, the greater the load density, the higher the deposit velocity, and the stronger the erosive power (He et al. 2011; Zhuo et al. 2013; Li et al. 2011). These factors carry various weights in influencing channel systems, and they correspond to various types of channel systems. For instance, fast sedimentary flow and powerful erosion in steep slopes favor restricted or weakly restricted channel systems, whereas non-restricted channel systems are often observed in gentle slope areas. Likewise, when sea level falls, there is ample sediment supply and all kinds of channel systems can form. Otherwise, in times of rising sea level where sedimentary supply is scarce, turbidite channel systems are seldom developed. Furthermore, allogenetic cycles are more evident in high sea level periods while autogenetic cycles dominate in low sea level periods (Posamentier and Kolla 2003; Prather 2003). For the O73 channel system of the M oilfield, eustacy, tectonic movement and topographic slope play a key controlling role in reservoir architecture and distribution, and abundant sediment from the Congo River thanks to the moist climate in the Oligocene is also another vital factor in turbidite channel formation (Booth et al. 2003; Violet et al. 2005; Beydoun et al. 2002). Tectonic movements such as differential uplift keep modifying both the macro- and micro-topography, which alters the energy of gravitational flow, and then alters development location and distribution of deepwater sedimentary units. From a macro-perspective, the differential uplift of the Congo Basin in the Angola area causes the sedimentary center to move north. Meanwhile, from a micro-perspective, tectonics like salt diapir accompanied by partial salt rock movement also greatly affects deepwater channel systems (Anka et al. 2009; Broucke et al. 2004; Kolla 2007; Pirmez and Imran 2003). Eustacy influences the development of deepwater channel systems too. Deepwater sediments in West Africa developed in the Upper Cretaceous when global sea level fell. At that time, the scale of deepwater channel systems expanded as sea level fell and they advanced toward the sea. Furthermore, the planar features of deepwater channels turned from wide and thin to narrow

and thick. Because of the contemporaneous falling sea level and continent uplifting, it reduces the distance between provenance and deepwater sedimentary supply, which is beneficial to form sediments. Therefore, the above analysis indicates that sediment type of the O73 channel system is subjected to controlling factors including sediment supply, deepwater gravitational flow and density.

5.2.2 Evolution discussion

Although models of the development of the channel system are affected by multiple factors, they follow certain evolutionary trends (Posamentier and Kolla 2003; Prather 2003; Liu et al. 2008). Horizontally, the development is mainly manifested as the evolution of different channel system types; and vertically, the development is primarily represented by the evolution of its internal complex channels.

Horizontally, along the provenance direction, there are certain trends in changes on account of differences in erosion of sediments. As the root of the channel system is nearer to the sediment source, large size, high flow rate and strong erosive power, large incised valleys can be formed and restricted channel systems that focus on transporting sediments were mainly developed, leading to a large amount of fragmental flow, turbidite and slump sediments developed in it. In the middle of the channel system, sediments become finer with decreased flow rate, resulting in weakened downcutting and strengthened aggradation, so the weakly restricted channel system (e.g., O73 channel system as shown in Fig. 9) is mainly developed at this point, in which channels with some degree of bending are developed and filled with an amount of fragmental flow and slump substances. While at the distal end of the channel system, the supply energy wanes and sediments are of the smallest size and lowest flow rate. At this point, the sediments downcutting capacity is weak, but the lateral migration capacity is strong, developing non-restricted channel systems mainly in which the single channels are mostly moderately to highly bent.

Vertically, influenced by eustacy and delivery rate of sediments, the development of internal complex channels inside the channel system also follows evolutionary trends. In the early development period of the channel system (S1 stage), high flow rate and abundant supply of sediments with strong erosion mostly contribute to form deep downcutting complex channels. They, as erosive channels, mainly transport sediments. In the middle development period of the channel system (S2 stage), the sea level begins to rise. The sediment supply is still rich, but the slowing flow rate leads to its slightly weakened downcutting capacity and strengthened aggradation, with mixed development of aggradational channels and erosion channels mainly under the effect of sedimentation. In the middle to late development period of the channel system (S3 stage), with the continual rise of the sea level, the sediment supply falls gradually (except in tsunamis, earthquakes and other unexpected events). The channel's downcutting capacity weakens, but the lateral accretion capacity becomes increasingly stronger, with aggrading highly sinuous channels mainly developed. In the late development period of the channel system (S4 stage), the sea level reached a high level and the sediment supply was the weakest, pointing to a sediment shrinkage stage where only a small number of highly sinuous aggraded channels and even isolated mudstone-filled single channels were developed.

6 Conclusions

1. The study is focused on the Tier 3–5 architectural unit of the reservoir in the study area. The channel sinuosity is controlled by continental shelf slope, a key factor to influence sinuosity, and there exists negative relations between single channel sinuosity and slope gradient. Affected by an autogenetic cycle of sedimentation, lithofacies inside the channel, from bottom up, form a configuration pattern of retention sediment ~ massive gravelly coarse sandstones (which may contain mud-sized grains) ~ massive middle-fine sandstone ~ interlaced bedded sandstone ~ fine-grained sediment, with sand bodies becoming thinner and finer from axis to the edge.

2. In plan view, two types of migration are found for single channels inside the complex channel—lateral migration and paleocurrent migration. From the profile, there are lateral, indented and swing migrations. Lateral migration is horizontal, with the thickness of sand bodies similar to that of single channel. While for indented and swing patterns, the thickness of sand bodies is basically larger than that of a single channel. There are weakly restricted and non-restricted channel systems in this area. From the provenance direction and vertical direction, the channel system tends to evolve from weakly restricted to non-restricted.

3. From bottom upward, the O73 channel system of the M oilfield can be subdivided into four phases of complex channel sedimentation S1–S4. The downcutting capacity of the channel weakens while the vertical and lateral migration strengthened gradually. Provenance and paleotopography slope are main factors to control sedimentary evolution.

Acknowledgements This paper is supported by the National Major Scientific and Technological Special Project during the Thirteenth Five-year Plan Period (2016ZX05033-003-002) and the Project of Sinopec Science and Technology Development Department (G5800-15-ZS-KJB016).

References

Abreu V, Sullivan M, Pirmez C, et al. Lateral accretion packages (LAPs): an important reservoir element in deep water sinuous channels. Mar Pet Geol. 2003;20(6–8):631–48. doi:10.1016/j.marpetgeo.2003.08.003.

Anka Z, Séranne M, Lopez M, et al. The long-term evolution of the Congo deep-sea fan: a basin-wide view of the interaction between a giant submarine fan and a mature passive margin (ZaiAngo Project). Tectonophysics. 2009;470(1–2):42–56. doi:10.1016/j.tecto.2008.04.009.

Bouma AH. Introduction to submarine fans and related turbidite systems. In: Bouma AH, Normark WR, Barnes NE, editors. submarine fans and related turbidite systems. New York: Springer; 1985.

Booth JR, Dean MC, DuVernay AE, et al. Paleo-bathymetric controls on the stratigraphic architecture and reservoir development of confined fans in the Auger Basin: central gulf of Mexico. Mar Pet Geol. 2003;20(6/8):563–86. doi:10.1016/j.marpetgeo.2003.03.008.

Beydoun W, Kerdraon Y, Lefeuvre F, Lancelin JP. Benefits of a 3D HR survey for Girassol field appraisal and development, Angola. Lead Edge. 2002;21:1152–5. doi:10.1190/1.1523744.

Broucke O, Temple F, Rouby D, Robin C, Calassou S, Nalpas T, Guillocheau F. The role of deformation processes on mud-dominated turbiditic systems, Oligocene and Lower–Middle Miocene of the Congo basin (West African margin). Mar Pet Geol. 2004;21(3):327–48. doi:10.1016/j.marpetgeo.2003.11.013.

Chen H, Xie XN, Mao KN. Deep-water contourite depositional system in vicinity of Yi'tong Shoal on northern margin of the South China Sea. Earth Sci. 2015;40(4):733–43 (**in Chinese**).

Clark JD, Pickering KT. Submarine channels: processes and architecture. London: Vallis Press; 1996.

Deptuck ME, Steffens GS, Barton M, et al. Architecture and evolution of upper fan channel-belts on the Niger Delta Slope and in the Arabian Sea. Mar Pet Geol. 2003;20(6–8):649–76. doi:10.1016/j.marpetgeo.2003.01.004.

Deng RJ, Deng YH, Yu S, et al. Hydrocarbon geology and reservoir formation characteristics of Niger Delta Basin. Pet Explor Dev. 2008;35(6):755–62 (**in Chinese**).

Habgood EL, Kenyon NH, Masson DG, et al. Deep-water sediment wave fields, bottom current sand channels and gravity flow channel-lobe systems: Gulf of Cadiz NE Atlantic. Sedimentology. 2003;50(3):483–510. doi:10.1046/j.1365-3091.2003.00561.x.

He YL, Xie XN, Lu YC, et al. Architecture and characteristics of mass transport deposits (Mtds) in Qiongdongnan Basin in Northern South China Sea. Earth Sci. 2011;5:905–13 (**in Chinese**).

Heiniö P, Davies RJ. Knickpoint migration in submarine channels in response to fold growth, Western Niger Delta. Mar Pet Geol. 2007;24(6–9):434–49. doi:10.1016/j.marpetgeo.2006.09.002.

Hubbard SM, de Ruig MJD, Graham SA. Confined channel-levee complex development in an elongate depo-center: deep-water Tertiary strata of the Austrian Molasse Basin. Mar Pet Geol. 2009;26(1):85–112. doi:10.1016/j.marpetgeo.2007.11.006.

Kane IA, Kneller BC, Dykstra M, et al. Anatomy of a submarine channel-levee: an example from Upper Cretaceous slope sediments, Rosario Formation, Baja California Mexico. Mar Pet Geol. 2007;24(6–9):540–63. doi:10.1016/j.marpetgeo.2007.01.003.

Kolla V, Bourges P, Urruty JM, et al. Evolution of deep-water Tertiary sinuous channels offshore Angola (West Africa) and implications for reservoir architecture. AAPG Bull. 2001;85(8):1373–405. doi:10.1306/8626cac3-173b-11d7-8645000102c1865d.

Kolla V. A review of sinuous channel avulsion patterns in some major deep-sea fans and factors controlling them. Mar Pet Geol. 2007;24(6–9):450–69. doi:10.1016/j.marpetgeo.2007.01.004.

Labourdette R. Integrated three-dimensional modeling approach of stacked turbidite channels. AAPG Bull. 2007;91(11):1603–18. doi:10.1306/06210706143.

Lamb MA. Stratigraphic architecture of a sand-rich, deep-sea depositional system: The Stevens sandstone, San Joaquin Basin, California. AAPG Special Publication; 2003.

Liu L, Zhang T, Zhao X, et al. Sedimentary architecture models of deepwater turbidite channel systems in the Niger Delta continental slope West Africa. Pet Sci. 2013;10(2):139–48. doi:10.1007/s12182-013-0261-x.

Li H, He YB, Wang ZQ. Morphology and characteristics of deep-water high sinuous channel-levee system. J Palaeogeogr. 2011;13(2):139–49 (in Chinese).

Li L, Wang YM, Huang ZC, et al. Study on sequence stratigraphy and seismic facies in deep-water Niger Delta. Acta Sedimentol Sin. 2008;26(3):407–16 (in Chinese).

Liu JP, Pan XH, Ma J, et al. Petroleum geology and resources in West Africa: an overview. Pet Explor Dev. 2008;35(3):378–84 (in Chinese).

Liu ZD, Li JH. Tectonic evolution and petroleum geology characteristics of petroliferous salt basins area along passive continental margin West Africa. Mar Orig Pet Geol. 2009;14(3):46–52 (in Chinese).

Lin Y, Wu SH, Wang X, et al. Research on architecture model of deepwater turbidity channel system: a case study of a deepwater research area in Niger Delta Basin West Africa. Geol Rev. 2013;59(3):510–20 (in Chinese).

Lü M, Wang Y, Chen Y. A discussion on origins of submarine fan deposition model and its exploration significance in Nigeria deep-water area. China Offshore Oil Gas. 2008;20(4):275–82 (in Chinese).

Menard HW Jr. Deep-sea channels, topography, and sedimentation. AAPG Bull. 1955;39(2):236–55. doi:10.1306/5ceae136-16bb-11d7-8645000102c1865d.

Mutti E, Normark WR. Comparing examples of modern and ancient turbidite systems: problems and concepts. Mar Clastic Sedimentol. 1987. doi:10.1007/978-94-009-3241-8_1.

Peakall J, McCaffrey B, Kneller B. A process model for the evolution, morphology, and architecture of sinuous submarine channels. J Sediment Res. 2000;70(3):434–48. doi:10.1306/2DC4091C-0E47-11D7-8643000102C1865D.

Posamentier HW, Kolla V. Seismic geomorphology and stratigraphy of depositional elements in deep-water settings. J Sediment Res. 2003;73(3):367–88. doi:10.1306/111302730367.

Pirmez C, Imran J. Reconstruction of turbidity currents in Amazon Channel. Mar Pet Geol. 2003;20(6–8):823–49. doi:10.1016/j.marpetgeo.2003.03.005.

Prather BF. Controls on reservoir distribution, architecture and stratigraphic trapping in slope settings. Mar Pet Geol. 2003;20(6–8):529–45. doi:10.1016/j.marpetgeo.2003.03.009.

Shanmugam G. The constructive functions of tropical cyclones and tsunamis on deep-water sand deposition during sea level highstand: implications for petroleum exploration. AAPG Bull. 2008;92(4):443–71. doi:10.1306/12270707101.

Slatt RM. Stratigraphic reservoir characterization for petroleum geologists, geophysicists, and engineers. Handb Pet Explor Prod. 2006. doi:10.1016/s1567-8032(06)x8035-7.

Stow DAV, Reading HG, Collinson JD. Deep seas. In: Reading HG, editor. Sedimentary environments: processes, facies and stratigraphy. Oxford: Blackwell Science; 1996.

Sun H, Jiang T, Li CF, et al. Characteristics of gravity flow deposits in slope basin of Nankai Trough and their responses to subduction tectonics. Earth Sci. 2014;39(10):1383–94 (in Chinese).

Violet J, Sheets B, Pratson L, et al. Experiment on turbidity currents and their deposits in a model 3D subsiding minibasin. J Sediment Res. 2005;75(5):820–43. doi:10.2110/jsr.2005.065.

Wang HR, Wang YM, Qiu Y, et al. The control of the multiple geomorphologic breaks on evolution of gravity flow dynamics in deep-water environment. Acta Geol Sin. 2009;83(6):812–9 (in Chinese).

Wood LJ, Mize-Spansky KL. Quantitative seismic geomorphology of a Quaternary leveed-channel system, offshore eastern Trinidad and TobagoNortheastern South America. AAPG Bull. 2009;93(1):101–25. doi:10.1306/08140807094.

Wynn RB, Cronin BT, Peakall J. Sinuous deep-water channels: genesis, geometry and architecture. Mar Pet Geol. 2007;24(6–9):341–87. doi:10.1016/j.marpetgeo.2007.06.001.

Xie XN, Chen ZH, Sun ZP, et al. Depositional architecture characteristics of deepwater depositional systems on the continental margins of northwestern South China Sea. Earth Sci. 2012;37(4):627–34 (in Chinese).

Xiong LP, Wang J, Yin JY, et al. Tectonic evolution and its control on hydrocarbon accumulation in West Africa. Oil Gas Geol. 2005;26(5):641–3 (in Chinese).

Yu S, Cheng T, Chen Y. Depositional characteristics of deepwater systems in the Niger Delta Basin. Earth Sci. 2012;4:763–70 (in Chinese).

Zhang WB, Liu ZQ, Chen ZH, et al. Establishment and application of geological data base on deep-water channels in Angola Block. Acta Sedimentol Sin. 2015;33(1):142–52 (in Chinese).

Zhao XM, Wu SH, Liu L. Sedimentary architecture model of deepwater channel complexes in slope area of West Africa. J China Univ Pet. 2012a;36(6):1–5 (in Chinese).

Zhuo HT, Wang YM, Xu Q, et al. Sedimentary characteristics and genesis of lateral accretion packages in the Pliocene of Dongfang area of Yinggehai Basin in northern South China Sea. J Palaeogeogr. 2013;15(6):787–94 (in Chinese).

Zhao XM, Wu SH, Yue DL, et al. Research on lithofacies types and identification method of deep-water submarine fan—taking one oilfield of West Africa as a case. Well Logging Technol. 2010;34(5):505–10 (in Chinese).

Zhao XM, Wu SH, Liu L, et al. Characterization of reservoir architecture for Neocene deepwater turbidity channels of Akpo oilfield Niger Delta basin. Acta Pet Sin. 2012b;33(6):1049–58 (in Chinese).

17

Performance improvement of ionic surfactant flooding in carbonate rock samples by use of nanoparticles

Mohammad Ali Ahmadi[1] · James Sheng[2]

Abstract Various surfactants have been used in upstream petroleum processes like chemical flooding. Ultimately, the performance of these surfactants depends on their ability to reduce the interfacial tension between oil and water. The surfactant concentration in the aqueous solution decreases owing to the loss of the surfactant on the rock surface in the injection process. The main objective of this paper is to inhibit the surfactant loss by means of adding nanoparticles. Sodium dodecyl sulfate and silica nanoparticles were used as ionic surfactant and nanoparticles in our experiments, respectively. AEROSIL® 816 and AEROSIL® 200 are hydrophobic and hydrophilic nanoparticles. To determine the adsorption loss of the surfactant onto rock samples, a conductivity approach was used. Real carbonate rock samples were used as the solid phase in adsorption experiments. It should be noted that the rock samples were water wet. This paper describes how equilibrium adsorption was investigated by examining adsorption behavior in a system of carbonate sample (solid phase) and surfactant solution (aqueous phase). The initial surfactant and nanoparticle concentrations were 500–5000 and 500–2000 ppm, respectively. The rate of surfactant losses was extremely dependent on the concentration of the surfactant in the system, and the adsorption of the surfactant decreased with an increase in the nanoparticle concentration. Also, the hydrophilic nanoparticles are more effective than the hydrophobic nanoparticles.

Keywords Adsorption · Hydrophobic silica nanoparticles · Hydrophilic silica nanoparticles · Ionic surfactant · Carbonate rock

1 Introduction

Owing to declining oil production rates around the world, it is important to improve the oil recovery factor (Zang et al. 2008). To obtain more oil from depleted oil fields, various methods called "enhanced oil recovery (EOR)" techniques should be utilized. Enhanced oil recovery approaches have different subsets, including thermal oil recovery, chemical oil recovery, and miscible and immiscible flooding. Chemical flooding has been attracted more attention in recent years, because it has various challenges such as wettability alteration, adsorption loss, interfacial tension reduction, and oil and water phase behavior (Kong and Ohadi 2010; Ahmadi and Shadizadeh 2013a, b, 2015; Ahmadi et al. 2014).

To improve the robustness and effectiveness of water flooding or chemical flooding, nanotechnology approaches have been implemented widely, such as mobility ratio improvement (Shah 2009; Suleimanov et al. 2011), interfacial tension reduction (Le et al. 2011), emulsion stability, wettability alteration (Al-Anssari et al. 2016), and resistance to adsorption onto reservoir rocks (Ahmadi and Shadizadeh 2012, 2013c). Le et al. investigated synergistic mixtures of surfactants and silica nanoparticles for enhanced oil recovery (EOR) in challenging reservoirs such as high-temperature reservoirs. To meet this goal, they carried out various tests including different mixtures

✉ Mohammad Ali Ahmadi
ahmadi6776@yahoo.com

[1]   Department of Petroleum Engineering, Ahwaz Faculty of Petroleum Engineering, Petroleum University of Technology, Ahwaz, Iran

[2]   Petroleum Department, Texas Tech University, P.O. Box 43111, Lubbock, TX 79409, USA

Edited by Yan-Hua Sun

of silica nanoparticles and surfactants. Their experiments divided into two types: (1) interfacial tension measurement and (2) contact angle measurement. They used a spinning drop tension meter (Temco 500) to investigate the effects of silica nanoparticles on IFT values. Moreover, they investigated the effect of silica nanoparticles on the oil displacement efficiency by contact angle measurements. Owing to their reported outcomes, some of the mixtures revealed appropriate agents for EOR purposes due to their thermal stability at 91 °C and infinitesimal loss on the rock surface by adsorption (Le et al. 2011). Suleimanov et al. (2011) conducted some experiments into the modification of interfacial properties in aqueous solutions by dispersing nanoparticles in the addressed solutions. They used different nonferrous nanoparticles in their experiments and draw a conclusion that the nanosuspension could increase the efficiency of oil displacement in porous media.

Onyekonwu and Ogolo (2010) investigated the effects of different polysilica nanoparticles (PSNP), on the wettability of reservoir rocks. They utilized water wet core samples and illustrated that silane-treated neutral and hydrophilic polysilica nanoparticles increased the recovery factor by 50 % over primary and secondary recoveries. Al-Anssari et al. (2016) studied the ability of silica nanoparticles to change the wettability of calcite rocks, including both oil-wet and mixed-wet calcite samples. They concluded that silica nanoparticles are able to change the wettability of such rocks from oil-wet to mixed/water wet, and this means that this type of nanoparticles is useful for EOR. Moreover, they pointed out that the concentration of nanoparticles and salinity of the solution were the most important factors in changing the wettability of the calcite rock samples. Furthermore, Ju and Fan (2009) demonstrated that untreated polysilica nanoparticles could change the wettability of sandstones from oil wet to water wet by an adsorption phenomenon. In addition, adding untreated polysilica nanoparticles could improve the effective water permeability, while decreasing the absolute permeability of the addressed sandstone samples.

Another characteristic of nanoparticles is the stabilization of droplets of emulsions that are small enough to move through the porous media without much retention (Zhang et al. 2010). The most-implemented fumed silica nanoparticles were spherically shaped, with a diameter of twenty to thirty nanometers. Also, the wettability of fumed silica nanoparticles is changed by coating materials, such as silanol. If silanol groups of the surface coating groups of the silica nanoparticle are greater than 90 %, the silica nanoparticle is considered a hydrophilic particle. Owing to this hydrophilic characteristic, silica nanoparticles could form a highly stabilized oil-in-water emulsion. On the other hand, if the coating groups on the surface of the silica nanoparticle are only 10 % silanol groups, the silica nanoparticle is considered hydrophobic and will form a stable water-in-oil emulsion (Zhang et al. 2010). Another property of nanoparticles is their high ability to stabilize oil-in-water emulsions, and the nanoemulsions can travel for a long distance in reservoirs without much retention (Kong and Ohadi 2010). In addition, nanoparticles can stabilize emulsions of supercritical CO_2 in water and emulsions of water in supercritical CO_2 (Dickson et al. 2004; Adkins et al. 2007).

Kanj et al. (2009) investigated the transport of nanoparticles in porous media and estimated the optimum size of nanoparticles effectively used in reservoir rocks. In addition, Skauge et al. (2010) studied the flow behavior of silica nanoparticles in porous media and found silica nanoparticles could move easily in porous media. Owing to their inherent conditions in reservoirs, they pose no environmental impacts. Due to their very small sizes, they also could not create tension or block pores, which make them an excellent advantage for EOR goals.

The huge potential of nanoparticles in upstream oil and gas is shown by various applications of nanoparticles in different oil and gas processes. Owing to the inherent characteristics of silica nanoparticles, they have been studied in recent years to improve the sweep efficiency of water flooding. Ogolo et al. (2012) studied the effect and potential of a combination of three nanoparticles, including Al_2O_3, MgO, Fe_2O_3, and SiO_2 nanoparticles. Some combinations of these nanoparticles were better than silica nanoparticles alone. Hendraningrat et al. (2012) found nanoparticles could decrease the interfacial tension between oil and brine/nanofluid. Also, the nanofluid could increase oil recovery by 13 % for both secondary and tertiary recoveries (Hendraningrat et al. 2012, 2013; Li et al. 2013).

In recent years, Ahmadi and Shadizadeh (2012, 2013c) studied the effect of nanoparticles on the adsorption loss of a surfactant derived from plant leaves on sandstone, shale sandstone, and sandstone mineral samples. Increasing the hydrophobicity of silica nanoparticles resulted in a reduction in adsorption loss of the surfactant (Ahmadi and Shadizadeh 2012, 2013c). This is due to the fact that the increasing hydrophobicity of nanoparticles may enhance hydrophobic bonds between the surfactant head and the hydrophobic part of silica nanoparticles. Consequently, fewer surfactant molecules are available to adsorb onto the rock surface.

The adsorption mechanisms of a combination of silica nanoparticles and sodium dodecyl sulfate on carbonate minerals have not been studied previously. The aim of this paper is to study the adsorption behavior of the mentioned ionic surfactant in the presence of silica nanoparticles in aqueous solutions. Moreover, the effects of silica nanoparticles on the oil sweep efficiency of the ionic surfactant in porous media were investigated with a core displacement apparatus. Two different types of silica nanoparticles were utilized in both adsorption and core displacement experiments. The experimental results were explained and discussed in detail.

2 Materials and methods

2.1 Chemicals

The ionic surfactant used was sodium dodecyl sulfate (SDS), which was identified as a good foaming agent. The ionic surfactant SDS was purchased from the Merck Company with a high degree of purification (99 %). It should be noted that the utilized chemicals were used as received without any further purification.

2.2 Nanoparticles

The nanoparticles were made from SiO_2 and an additive (Ahmadi and Shadizadeh 2012, 2013c). A transmission electron microscope (TEM) was used to measure the spherical shape and size of the silica nanoparticles as shown in Fig. 1. In order to investigate the effect of nanoparticle wettability on the inhibition of surfactant adsorption loss, two types of silica nanoparticles,

hydrophilic and hydrophobic silica nanoparticles, were used. AEROSIL® 816 and AEROSIL® 200 were used as hydrophobic and hydrophilic nanoparticles which were purchased from Degussa.

AEROSIL® 816 is a fumed silica after treated with hexadecylsilane based on AEROSIL® 200. It is used in water-based coating systems. AEROSIL® 816 can be applied in coating systems as an antisettling agent, for stabilization of pigments, and to enhance the effect of corrosion protection. It is also effective in controlling the rheology of complex liquid systems.

2.3 Core and crushed core samples

A core sample used in core displacement tests was cut from an Iranian carbonate reservoir rock, and its characteristics are illustrated in Table 1. To evaluate adsorption of the ionic surfactant in the presence of different silica nanoparticles, two core samples were crushed using a jaw crusher and then passed through specific sieves (50–70 mesh size) for repeatability of the experiments and to double check our adsorption experiments (Salari et al. 2011; Ahmadi and Shadizadeh 2012, 2013a, b, c, 2015; Ahmadi et al. 2014; Zendehboudi et al. 2013). X-ray diffraction (XRD) was conducted to analyze the phase composition of the core samples, and the results are shown in Fig. 2. As illustrated in Fig. 2, the rock samples were

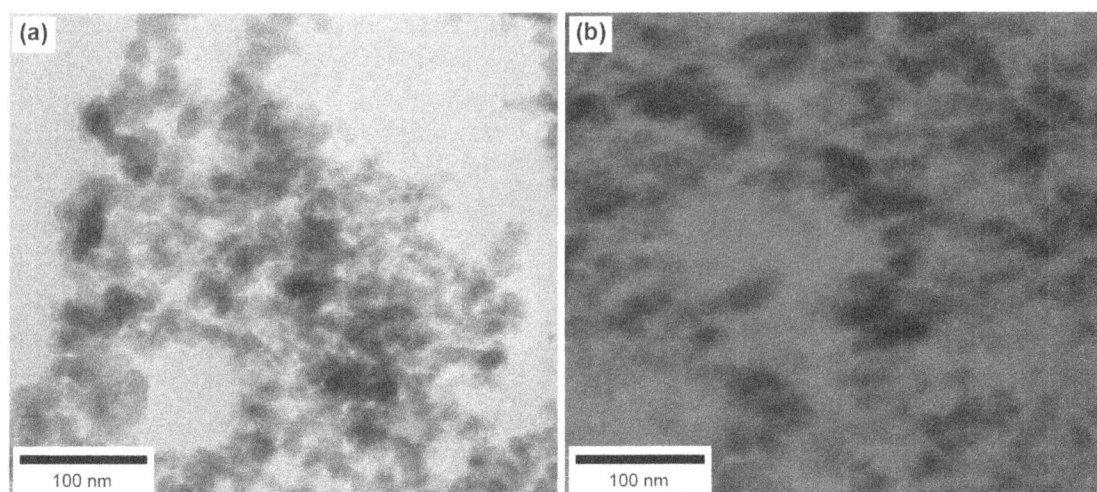

Fig. 1 Images of hydrophilic (**a**) and hydrophobic (**b**) silica nanoparticles observed with a TEM

Table 1 Characteristics of the utilized core sample

Core name	Length, cm	Average diameter, cm	Area, cm²	Bulk volume, cm³	Pore volume ($S_w = 1$), cm³	Porosity, %	Absolute permeability, mD
AB	8.51	3.8	11.40	97.1	13.7	14.1	1.98

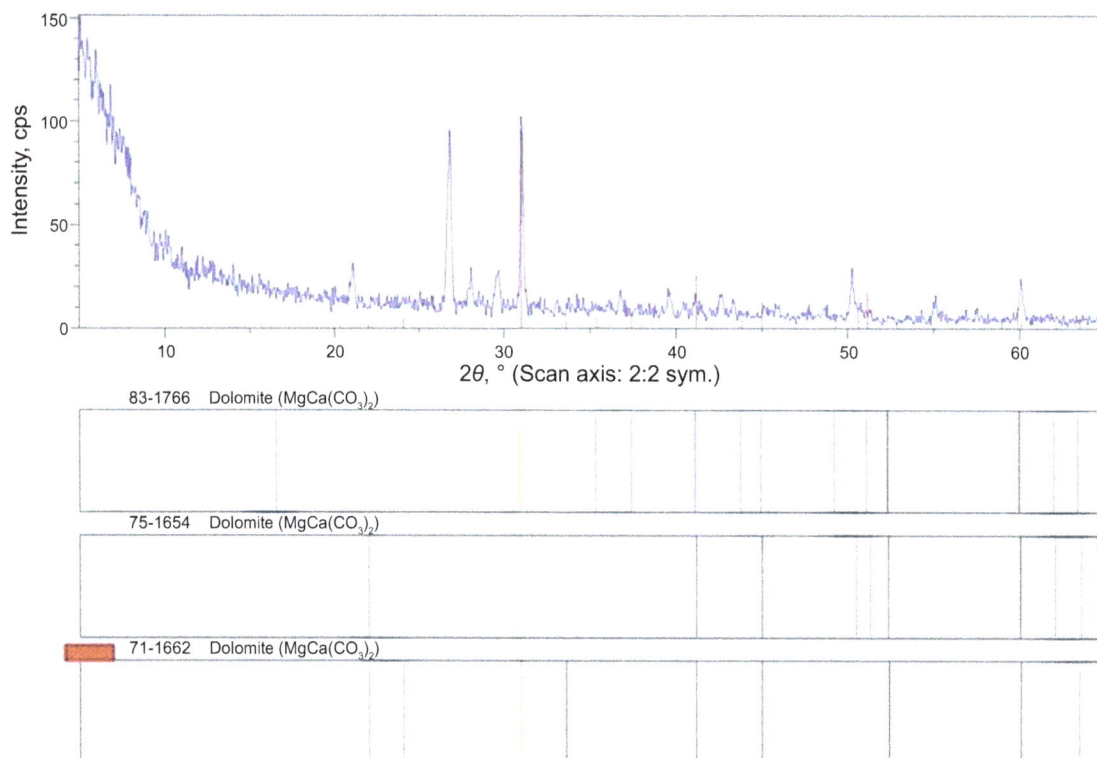

Fig. 2 X-ray diffraction (XRD) of the crushed rock samples

predominately dolomite. (As noted in the text, the major phase in the core sample is dolomite; however, it may contain some quartz. This is because the core is a real core sample, and it does not have a pure lithology.)

2.4 Oil sample

Crude oil used was taken from a light oil field located in the northern Persian Gulf. The properties of the crude oil sample are presented in Table 2 (Ahmadi and Shadizadeh 2013b).

2.5 Determination of critical micelle concentration (CMC)

Various methods were used to estimate the CMC of the surfactant in the aqueous solution based on different intrinsic characteristics of surface active agents, such as surface tension, interfacial tension, thermal conductivity, and electrical conductivity. Based on the high electrical conductance of the introduced surfactant in aqueous solutions, the electrical conductivity measurement was selected as a robust and precise method to determine the micellization behavior of the introduced surfactant with and without nanoparticles in aqueous solutions. To achieve the goals of this research, various concentrations of the

Table 2 Properties of crude oil used (after Ahmadi and Shadizadeh 2013b)

Property	Value
Component, mol%	
H_2S	0
CO_2	0
N_2	0
C_1	0
C_2	0.13
C_3	0.25
i-C_4	0.73
n-C_4	1.23
i-C_5	1.38
n-C_5	3.56
C_{6+}	92.72
Molecular weight of C_{6+}	203.24
Specific gravity of C_{6+}	0.8325

introduced surfactant were considered ranging from 500 to 5000 ppm, and a plot of electrical conductance versus surfactant concentration for each nanoparticle concentration was generated. It should be noted that a conductivity detector from Crison Company (EC-GLP 31[+]) was used in experiments (Salari et al. 2011; Ahmadi and Shadizadeh

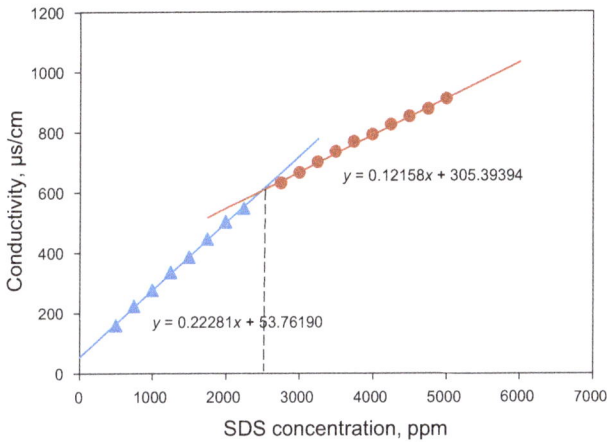

Fig. 3 Conductivity of SDS solutions versus corresponding surfactant concentration

2012, 2013c, 2015; Zendehboudi et al. 2013). The electrical conductivity trend of surfactant solutions at various concentrations without nanoparticles is illustrated in Fig. 3.

2.6 Core displacement experiments

To assess the performance of the chemical agents (surfactant, nanoparticle/surfactant) in enhanced oil recovery under reservoir conditions, a comprehensive series of core displacement experiments were carried out under high pressure and high temperature (HPHT). As shown in Fig. 4, the implemented setup consisted of two transfer vessels—including one nanofluid and one oil—and a core

holder mounted in a temperature-controlled air bath, which also enclosed an HPLC constant rate pump for high pressure injection of nanofluid or water. Also, to maintain the pressure of the system at the output of the core, a back-pressure regulator (BPR) was installed. A differential pressure transducer (DPT) was used to measure the pressure drop across the core. Before each displacement experiment, the core sample was initially saturated with brine (15,000 ppm NaCl) and then flooded with oil at a low flow rate (0.5 mL/h) until connate water saturation was reached under reservoir conditions. All core displacement tests were launched with the samples saturated with oil and connate water saturation, followed by the enhanced oil recovery (EOR) process. In each test, effluent fluids were collected for analysis. The displacement experiments were performed on several carbonate rocks which were water wet, 8.5 cm in length, and 3.8 cm in diameter. The scaling method proposed by Rapoport and Leas (1953) was carried out to cancel the dependency of oil recovery on the fluid injection rate and the core length. The mentioned scaling criterion is expressed by the following equation (Rapoport and Leas 1953; Kulkarni and Rao 2004; Mcelfresh et al. 2012; Ahmadi and Shadizadeh 2013b):

$$LV\mu \geq 1, \tag{1}$$

where L represents the core length, cm; μ stands for the viscosity of the displacing phase, cP; and V denotes the fluid velocity, cm/min (Rapoport and Leas 1953; Kulkarni and Rao 2004; Mcelfresh et al. 2012; Ahmadi and Shadizadeh 2013b).

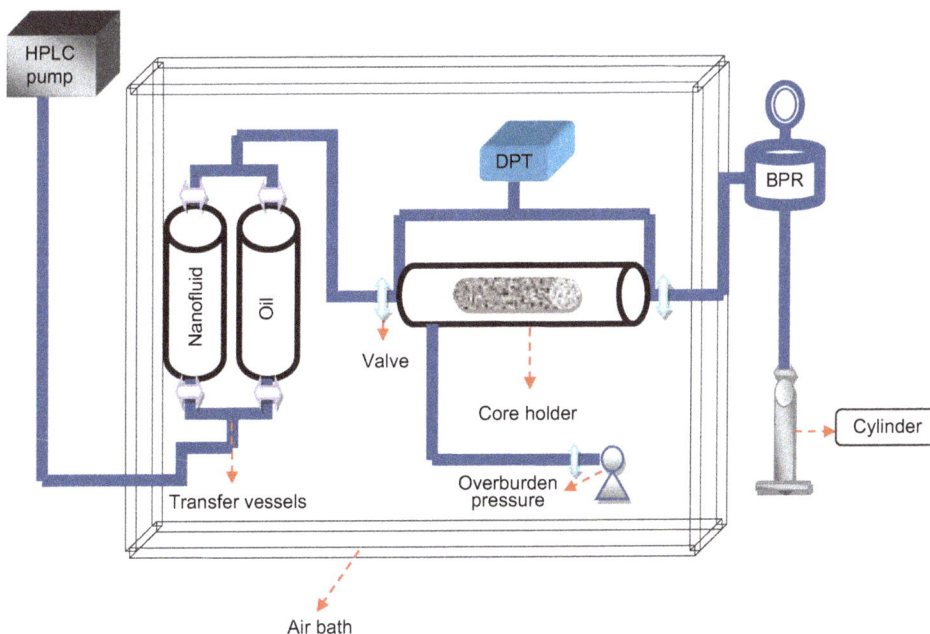

Fig. 4 Schematic picture of the core displacement apparatus

2.7 Adsorption experiment

According to the procedure of adsorption experiments reported by Ahmadi and his colleagues (Ahmadi and Shadizadeh 2012, 2013c, 2015; Zendehboudi et al. 2013) and Salari et al. (2011), the depletion solution or batch tests were conducted to indicate adsorption behavior of the ionic surfactant on the carbonate surface in the presence of silica nanoparticles. Two crucial points of the proposed adsorption experiments are the adsorption equilibrium time and the ratio of the solid and aqueous phases. Before explaining the details of the adsorption experiments, it is worth mentioning that the experiments were conducted at 25 °C and atmosphere pressure. It should be mentioned here that to determine the amount of adsorption loss of the addressed ionic surfactant onto the reservoir rock samples, a batch test was used. Due to this fact, the weight of the crushed rock samples and the volume of the aqueous solution with different surfactant concentrations should be consistent for all the adsorption experiments as illustrated in Ahmadi and Shadizadeh (2012, 2013a, c, 2015). Two more crucial points should be mentioned: First, the equilibrium time of adsorption was about 24 h, so adsorption experiments were conducted for 24 h. Second, the mass ratio of the surfactant solution to the crushed rock was 5:1. In addition, a wide range of surfactant concentrations, from 500 to 5000 ppm of surfactant, were used. As noted previously, the conductivity of the aqueous solution was utilized to indicate the surfactant concentration before and after adsorption loss onto the rock surface. For more details about the procedure of conductivity measurement for adsorption experiments, the authors referred to Ahmadi and Shadizadeh (2012, 2013a, c, 2015). Finally, the magnitude of the surfactant loss onto the rock surface (in terms of mg surfactant/g of rock) was calculated from the following formulation (Salari et al. 2011; Ahmadi and Shadizadeh 2012, 2013c, 2015; Zendehboudi et al. 2013):

$$\Gamma = ((C_i - C_e) \times M_s / M_c) / 1000, \qquad (2)$$

where Γ stands for the adsorption density, mg/g; C_i and C_e represent the initial and equilibrium SDS concentrations in the aqueous solution, respectively, ppm; M_s denotes the mass of the solution, g; and M_c represents the mass of the carbonate rock sample, g.

3 Results and discussion

3.1 Adsorption experiments

The CMC of SDS was determined by measuring the electrical conductivity of the solution, which was also used by Ahmadi and his coworkers (Salari et al. 2011; Ahmadi and Shadizadeh 2012, 2013c, 2015; Zendehboudi et al.

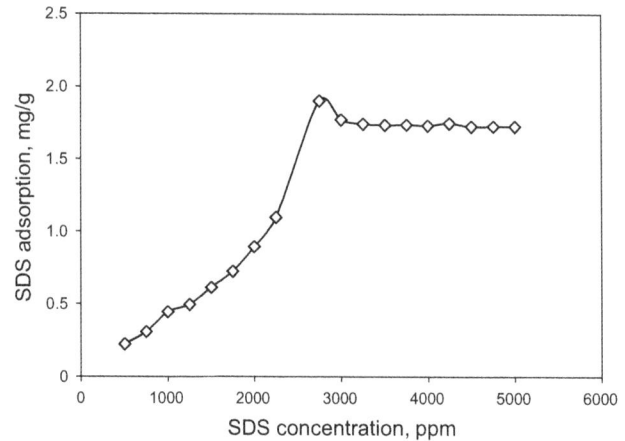

Fig. 5 Adsorption isotherm of SDS onto the crushed carbonate sample

2013). The CMC of SDS was 2485 ppm. The fate or loss of the surfactant in terms of mg/g was measured in the aqueous and solid phases of an initial surfactant concentration from 500 to 5000 ppm and a ratio of aqueous solution to solid of 5:1. Due to the achieved results of adsorption experiments under the mentioned conditions, the adsorption of SDS on the carbonate rock surface was different for the concentrations below and above the CMC.

As demonstrated in Fig. 5, by approaching the adsorption density of 1.90 mg/g at 2750 ppm of the SDS concentration, adsorption reached equilibrium (constant value). In other words, increasing the SDS concentration that is lower than or equal to the CMC of SDS caused the amount of adsorption loss to follow linear behavior; however, the adsorption loss above a concentration of 2750 ppm of SDS did not change significantly. The main reason for this phenomenon which can be explained by the number of SDS monomers does not change and remains constant after reaching the CMC value. On the other hand, when the SDS concentration is lower than the CMC value, the number of monomers is not a constant and increases with the SDS concentration. In this regard, increasing the SDS concentration, when it is lower than the CMC value, increases the adsorption density. So the maximum adsorption loss of SDS on the surface of crushed carbonate sample was about 1.90 mg/g. It was found that the adsorption loss below or near the CMC value in the system is a function of the SDS concentration. Figure 6 shows the effect of hydrophilic silica nanoparticles on the adsorption of SDS on the crushed carbonate sample at different SDS concentrations. As shown in Fig. 6, more reduction was observed in the adsorption loss of SDS on the crushed carbonate sample when increasing the concentration of silica nanoparticles in aqueous solutions. This reduction is caused by the increasing concentration of hydroxyl groups, which exist in aqueous solutions and can create hydrogen bonds with the tail of the surfactant and an electrostatic bond with the positively

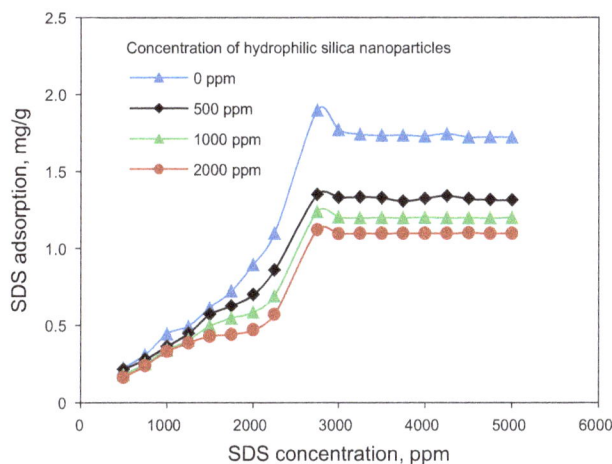

Fig. 6 Comparison of adsorption isotherms at different concentrations of hydrophilic silica nanoparticles

charged rock surface. Also another reason for the reduction is the adsorption of some silica nanoparticles on the crushed carbonate sample from the aqueous solution, but that is unfavorable for us because of the loss of silica nanoparticles from the solution. Figure 6 shows a comparison of the SDS adsorption efficiency at different concentrations of hydrophilic silica nanoparticles. As shown in Fig. 6, the effective concentration of hydrophilic silica nanoparticles was 2000 ppm, which could reduce the maximum value of the adsorption density from 1.90 to 1.12 mg/g. Figure 7 depicts the effect of hydrophobic silica nanoparticles on the adsorption loss of SDS on the crushed carbonate samples at different SDS concentrations. As illustrated in Fig. 7, more reduction was observed in adsorption of SDS on the crushed carbonate sample when increasing the concentration of hydrophobic silica nanoparticles in the aqueous solution, but the magnitude of reduction was lower than the hydrophilic silica nanoparticles, owing to a smaller number of hydroxyl groups in hydrophobic silica nanoparticles. Due to this fact, the hydrophobic silica nanoparticles were observed to be less

effective in inhibiting adsorption loss of SDS onto the surface of crushed carbonate samples. As depicted in Fig. 7, increasing the silica nanoparticle concentration would reduce the losses of SDS onto reservoir rock samples. 2000 ppm of silica nanoparticles could reduce the maximum value of adsorption loss from 1.90 to 1.24 mg/g.

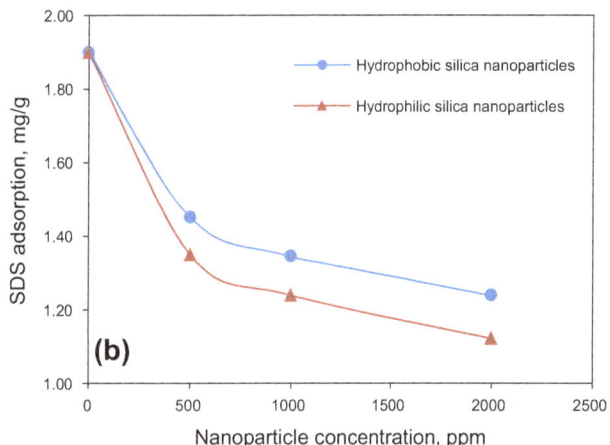

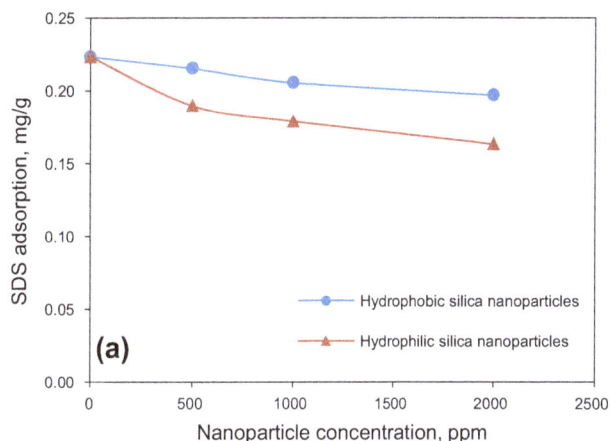

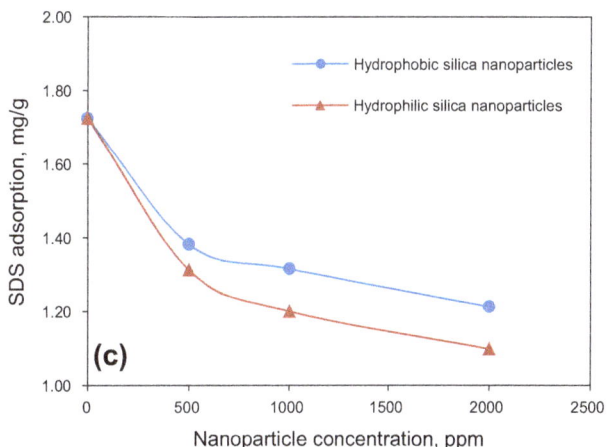

Fig. 8 Effects of hydrophilic and hydrophobic silica nanoparticles on the adsorption of SDS onto carbonate rock samples. **a** SDS concentration $C_i = 500$ ppm (lower than the CMC). **b** SDS concentration $C_i = 2750$ ppm (near the CMC). **c** SDS concentration $C_i = 5000$ ppm (above the CMC)

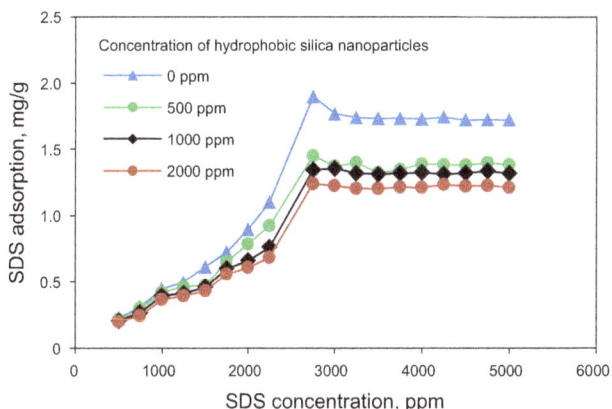

Fig. 7 Comparison of adsorption isotherms at different concentrations of hydrophobic silica nanoparticles

(a)

Hydrophobic tail of surfactant

Head of surfactant

(b)

Hydrophilic silica nanoparticle

Hydrophobic tail of surfactant

Head of surfactant

(c)

Hydrophobic silica nanoparticle

Hydrophobic tail of surfactant

Head of surfactant

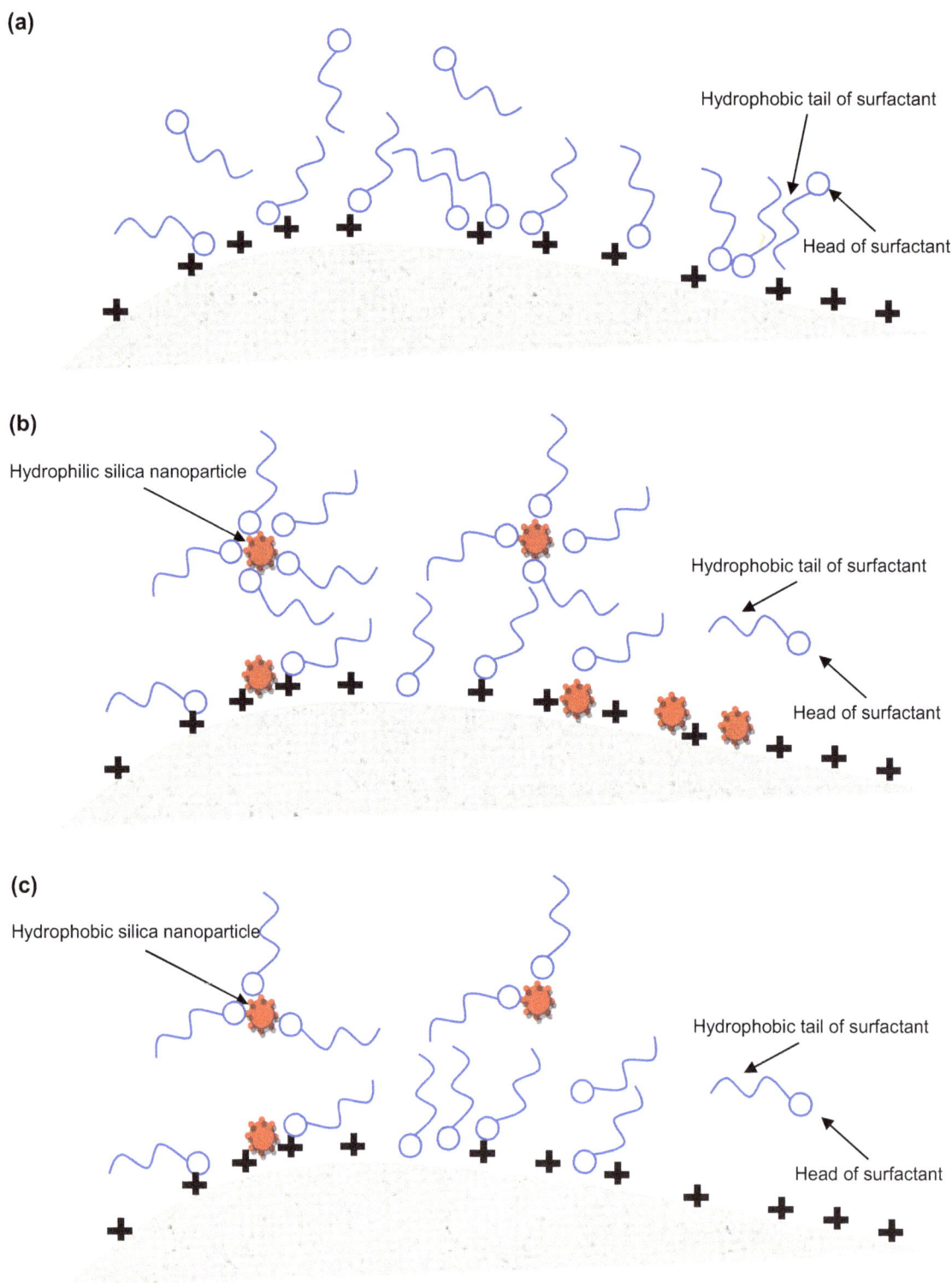

Fig. 9 Schematic of adsorption process of ionic surfactant. **a** Ionic surfactant only. **b** In the presence of hydrophilic silica nanoparticles. **c** In the presence of hydrophobic silica nanoparticles

Figure 8 compares the effects of hydrophilic and hydrophobic silica nanoparticles on the adsorption density of SDS at three levels of SDS concentration in the aqueous solution. When the SDS concentration (500 ppm) was lower than the CMC of SDS, the adsorption density decreases slightly with increasing concentration of silica nanoparticles (Fig. 8a). However, when the SDS concentration was near and above the CMC of SDS, the adsorption density decreases significantly with the increasing concentration of silica nanoparticles (Fig. 8b, c).

To better understand the adsorption mechanism of the ionic surfactant onto the carbonate rock surface, schematics of the adsorption of the surfactant alone and the surfactant in the presence of hydrophilic and hydrophobic silica nanoparticles are depicted in Fig. 9. As shown in Fig. 9a, the main mechanism of the surfactant adsorption onto the positively charged surface is electrostatic bonding between the negatively charged head of the surfactant and the positively charged surface. As mentioned previously, two main mechanisms exist to inhibit the surfactant loss onto positively charged carbonate rock samples. The first one is hydrogen bonding between hydroxyl groups of the silica nanoparticles and the tail of the ionic surfactant; the second is adsorption of the silica nanoparticles onto the carbonate rock surface due to a high magnitude of negative charges. In other words, silica nanoparticles were sacrificed to avoid adsorption loss of the surfactant at very low concentrations. These are depicted graphically in Fig. 9b, c.

3.2 Effect of silica nanoparticles on CMC

The changes in CMC that occur with increasing concentrations of hydrophobic and hydrophilic silica nanoparticles are depicted in Fig. 10. As mentioned earlier in the CMC determination section, a turning point in the plot of electrical conductivity against surfactant concentration represents the CMC of the surfactant. However, it seems that the hydrophobic and hydrophilic silica nanoparticles influenced the surfactant micellization properties, particularly its CMC. As can be seen in Fig. 10, the coexistence of SDS and hydrophilic silica nanoparticles (AEROSIL® 200) in a solution led to a CMC value lower than the one for just the ionic surfactant system. Figure 10 demonstrates the CMCs of different systems considered in this study, and the presence of both nanoparticles resulted in surfactant molecules aggregating into micelles at lower concentrations. This phenomenon is more severe for higher nanoparticle concentrations.

The observed phenomenon may be related to the surfactant–nanoparticle interactions. Ignoring the small amount of surfactant adsorption on the surface of nanoparticles, the similar negative electrical charge on the surfactant head groups and the surface of nanoparticles results in an electrostatic repulsion between surfactant molecules toward each other, prompting the micellization process. Moreover, the hydrophilic nanoparticles make the bulk solution unfavorable for hydrophobic surfactant tails and increase their affinity to form micelles. Obviously, in such a situation, micelle aggregates form at lower concentrations, and the CMC is reduced. When the concentration of nanoparticles increases, the repulsion forces

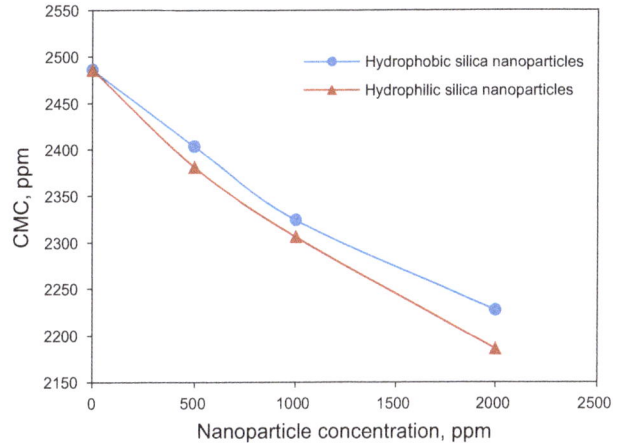

Fig. 10 Effect of silica nanoparticles on the CMC of SDS

become stronger (due to the larger number of nanoparticles). Also, the bulk solution becomes more hydrophilic. As a result, micellization occurs even at lower concentrations. Another important point that may be inferred from Fig. 10 is that the reduction in the CMC is more dramatic for hydrophilic silica nanoparticles. As mentioned earlier, the presence of these nanoparticles intensifies the hydrophilic characteristics of the solvent. In aqueous solutions, the greater dissimilarity between the hydrophobic chain of the surfactant and silica nanoparticles leads to higher aggregation. Consequently, a sharper decrease in the CMC value is observed than with the slightly hydrophobic nanoparticles (AEROSIL® 816). The previous discussions are illustrated in Fig. 11.

3.3 Core displacement results

The induced effects of the nanoparticles on the ultimate oil recovery and performance of the ionic surfactant in porous media were examined. Figure 12 demonstrates the ultimate oil recovery in terms of % original oil in place (% OOIP) versus volume of the fluid injected into the porous media for four different water–oil systems. As depicted in Fig. 12, the oil recovery was 51.1 % OOIP with water injection. Also, as demonstrated in Fig. 12, adding 5000 ppm of SDS to the aqueous phase resulted in more oil production and the recovery factor was about 78.8 % OOIP. This may be explained by the reduction in the interfacial tension between two immiscible fluids (water and oil). Moreover, the addition of hydrophobic and hydrophilic silica nanoparticles could improve the sweep efficiency of SDS, and more oil was recovered from the porous media due to inhibition of adsorption loss of the surfactant, but the magnitude of the ultimate oil recovery highly depended on the hydrophobicity of silica nanoparticles. For better understanding, it is worth mentioning that the hydrophobic

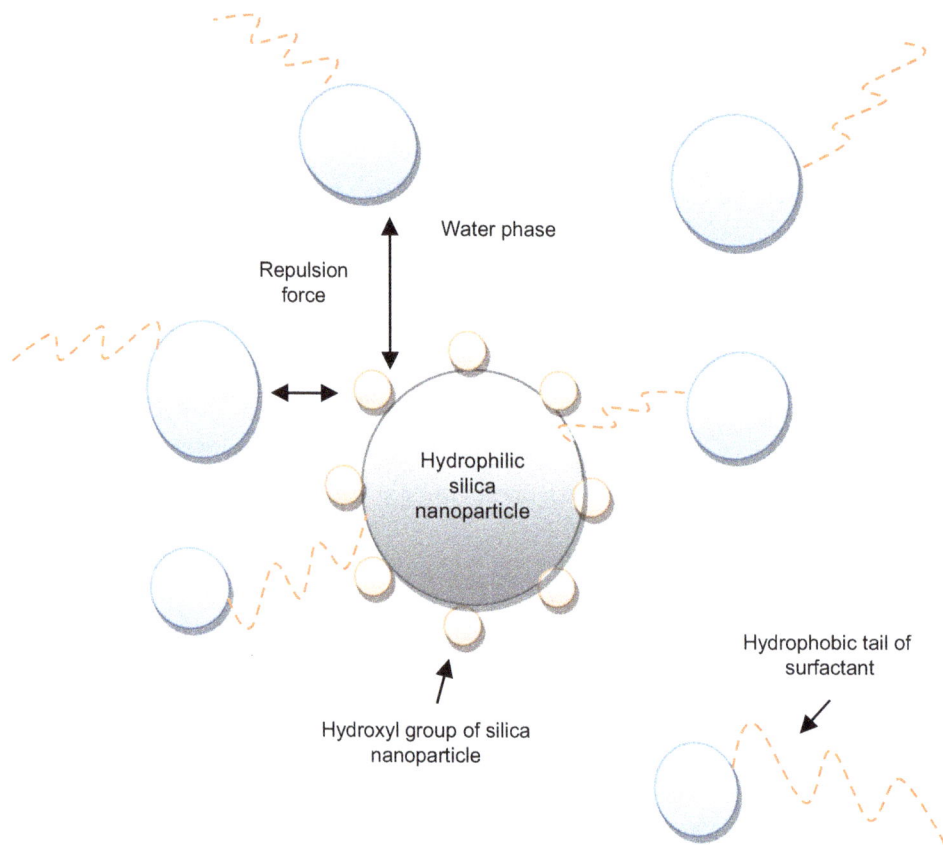

Fig. 11 Schematic of aggregation of silica nanoparticles and SDS

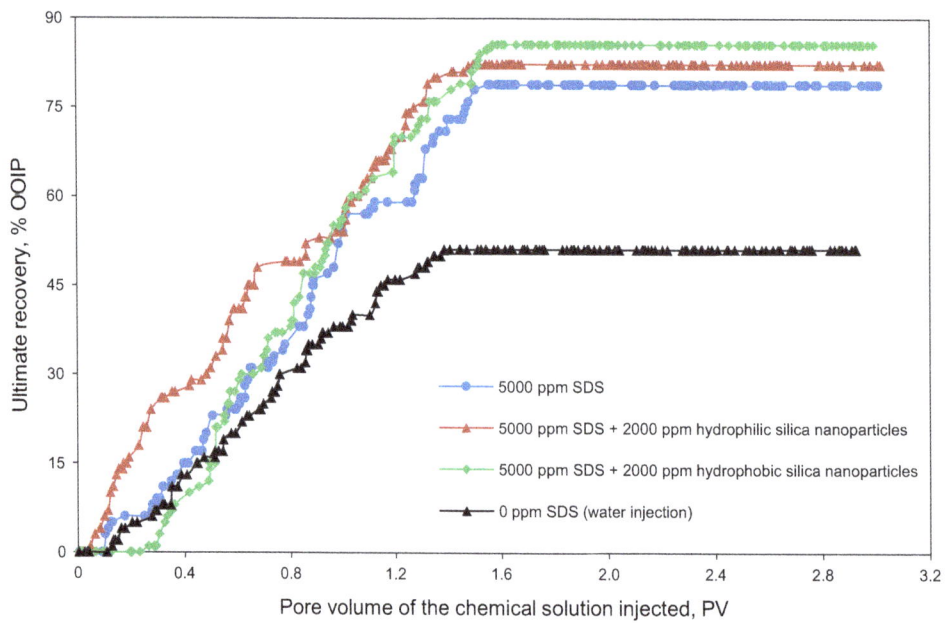

Fig. 12 Comparison of ultimate oil recovery for four systems including water, SDS solution only, and combinations of the SDS solution and silica nanoparticles

silica nanoparticles may reduce the interfacial tension between oil and water phases, but the effect of hydrophilic silica nanoparticles is the reverse. According to the noted facts, the ultimate oil recoveries were 82.3 and 85.6 % OOIP for the hydrophilic and hydrophobic silica nanoparticles, respectively.

4 Conclusions

1. The addition of both hydrophilic and hydrophobic silica nanoparticles could reduce adsorption loss of the ionic surfactant.
2. The hydrophilic silica nanoparticles could be more effective than the hydrophobic ones in reducing adsorption onto carbonate samples, because there are more hydroxyl groups in hydrophilic silica nanoparticles than in hydrophobic silica particles. Owing to this fact, more hydrogen bonds exist between hydroxyl groups and the tail of the ionic surfactant, while more electrostatic bonds are formed between hydroxyl groups of silica nanoparticles and positively charged rock surfaces are also observed.
3. When the SDS concentration was lower than the CMC of SDS, the adsorption density decreases slightly with the increasing concentration of silica nanoparticles. However, when the SDS concentration was near and above the CMC of SDS, the adsorption density decreases significantly with the increasing concentration of silica nanoparticles.
4. The addition of silica nanoparticles to the surfactant solution resulted in decreasing the CMC of the ionic surfactant. However, the magnitude of the CMC was noticeably dependent on the hydroxyl group and the magnitude of the negative charges on the silica nanoparticle surface. In other words, the hydrophilic silica nanoparticles could reduce the CMC value more than the hydrophobic silica nanoparticles in the aqueous solution.
5. The addition of silica nanoparticles could improve the sweep efficiency of the ionic surfactant. However, the magnitude of the additional oil recovery was highly dependent on the wettability of the silica nanoparticles. For hydrophobic silica nanoparticles, the reduction in interfacial tension and the inhibition of adsorption loss were two factors in favor, but these phenomena were different for hydrophilic silica nanoparticles. Hydrophilic silica nanoparticles could inhibit adsorption loss of the surfactant; however, it increased the interfacial tension between oil and water and hence did not improve the oil recovery as much as the hydrophobic nanoparticles.

References

Adkins SS, Gohil D, Dickson JL, Webber SE, Johnston KP. Water-in-carbon dioxide emulsion stabilized with hydrophobic silica particles. Phys Chem Chem Phys. 2007;9(48):6333–43. doi:10.1039/b711195a.

Ahmadi MA, Arabsahebi Y, Shadizadeh SR, Shokrollahzadeh Behbahani S. Preliminary evaluation of mulberry leaf-derived surfactant on interfacial tension in an oil-aqueous system: EOR application. Fuel. 2014;117(Part A):749–55. doi:10.1016/j.fuel.2013.08.081.

Ahmadi MA, Shadizadeh SR. Adsorption of novel nonionic surfactant and particles mixture in carbonates: enhanced oil recovery implication. Energy Fuels. 2012;26(8):4655–63. doi:10.1021/ef300154h.

Ahmadi MA, Shadizadeh SR. Experimental investigation of adsorption of a new nonionic surfactant on carbonate minerals. Fuel. 2013a;104:462–7. doi:10.1016/j.fuel.2012.07.039.

Ahmadi MA, Shadizadeh SR. Implementation of high performance surfactant for enhanced oil recovery from carbonate reservoir. J Pet Sci Eng. 2013b;110:66–73. doi:10.1016/j.petrol.2013.07.007.

Ahmadi MA, Shadizadeh SR. Induced effect of adding nano silica on adsorption of new surfactant onto sandstone rock: experimental and theoretical study. J Pet Sci Eng. 2013c;112:239–47. doi:10.1016/j.petrol.2013.11.010.

Ahmadi MA, Shadizadeh SR. Experimental and theoretical study of a new plant derived surfactant adsorption on quartz surface: kinetic and isotherm. J Dispers Sci Technol. 2015;36(3):441–52. doi:10.1080/01932691.2013.860035.

Al-Anssari S, Lebedev M, Wang S, Barifcani A, Iglauer S. Wettability alteration of oil-wet carbonate by silica nanofluid. J Colloid Interface Sci. 2016;461:435–42. doi:10.1016/j.jcis.2015.09.051.

Dickson J, Johnston K, Binks B. Stabilization of carbon dioxide-in-water emulsions with silica nanoparticles. Langmuir. 2004;20(19):7976–83. doi:10.1021/la0488102.

Hendraningrat L, Engeset B, Suwarno S, Torsæter O. Improved oil recovery by nanofluids flooding: an experimental study. In: SPE Kuwait international petroleum conference and exhibition, 10–12 December, Kuwait City, Kuwait, 2012. doi:10.2118/163335-MS.

Hendraningrat L, Li S, Torsæter O. A coreflood investigation of nanofluid enhanced oil recovery. J Pet Sci Eng. 2013;111:128–38. doi:10.1016/j.petrol.2013.07.003.

Ju B, Fan T. Experimental study and mathematical model of nanoparticle transport in porous media. Powder Technol. 2009;195(2):192–202. doi:10.1016/j.powtec.

Kanj MY, Al-Yousif Z, Funk J. Nano fluid core flood experiment in the ARAB-D. In: SPE Saudi Arabia section technical symposium and exhibition, 9–11 May, Aikhobar, Saudi Arabia, 2009. doi:10.2118/126161-MS.

Kong X, Ohadi MM. Application of micro and nano technologies in the oil and gas industry: an overview of the recent progress. In: Abu Dhabi international petroleum exhibition & conference, 1–4 November, Abu Dhabi, UAE, 2010. doi:10.2118/138241-MS.

Kulkarni MM, Rao DN. Experimental investigation of various methods of tertiary gas injection. In: SPE annual technical conference and exhibition, 26–29 September, Houston, TX, 2004. doi:10.2118/90589-MS.

Le N, Pham DK, Le KH, Nguyen PT. Design and screening of synergistic blends of SiO_2 nanoparticles and surfactants for enhanced oil recovery in high-temperature reservoirs. Adv Nat Sci. 2011;2(3):035013. doi:10.1088/2043-6262/2/3/035013.

Li S, Hendraningrat L, Torsæter O. Improved oil recovery by hydrophilic silica nanoparticles suspension: 2- phase flow experimental studies. In: International petroleum technology conference, 26–28 March, Beijing, China, 2013. doi:10.2523/IPTC-16707-MS.

Mcelfresh P, Holcomb D, Ector D. Application of nanofluid technology to improve recovery in oil and gas wells. In: SPE international oilfield technology conference, 12–14 June, Noordwijk, The Netherlands, 2012. doi:10.2118/154827-MS.

Ogolo N, Olafuyi O, Onyekonwu M. Enhanced oil recovery using nanoparticles. In: SPE Saudi Arabia section technical symposium and exhibition, 8–11 April, Al-Khobar, Saudi Arabia, 2012. doi:10.2118/160847-MS.

Onyekonwu MO, Ogolo NA. Investigating the use of nanoparticles in enhancing oil recovery. In: Nigeria Annual international conference and exhibition, 31 July–7 August, Tinapa-Calabar, Nigeria, 2010.

Rapoport LA, Leas WJ. Properties of linear waterfloods. J Pet Technol. 1953;5(5):139–48. doi:10.2118/213-G.

Salari Z, Ahmadi MA, Kharrat R, Abbaszadeh Shahri A. Experimental studies of cationic surfactant adsorption onto carbonate rocks. Aust J Basic Appl Sci. 2011;5(12):808–13.

Shah RD. Application of nanoparticle saturated injectant gases for EOR of heavy oil. In: SPE annual technical conference and exhibition, 4–7 October, New Orleans, Louisiana, USA, 2009. doi:10.2118/129539-STU.

Skauge T, Spildo K, Skauge A. Nano-sized particles for EOR. In: SPE improved oil recovery symposium, 24–28 April, Tulsa, Oklahoma, USA, 2010. doi:10.2118/129933-MS.

Suleimanov B, Ismalov F, Veliyev E. Nano fluid for enhanced oil recovery. J Pet Sci Eng. 2011;78(2):431–7. doi:10.1016/j.petrol.2011.06.014.

Zang L, Yuan J, Liang H, Le K. Energy from abandoned oil and gas reserves. In: SPE Asia Pacific oil and gas conference and exhibition, 20–22 October, Perth, Australia, 2008. doi:10.2118/115055-MS.

Zendehboudi S, Ahmadi MA, Rajabzadeh AR, Mahinpey N, Chatzis I. Experimental study on adsorption of a new surfactant onto carbonate reservoir samples-application to EOR. Can J Chem Eng. 2013;91(8):1439–49. doi:10.1002/cjce.21806.

Zhang T, Davidson A, Bryant SL, Huh C. Nanoparticle-stabilized emulsions for application in enhanced oil recovery. In: SPE improved oil recovery symposium, 24–28 April, Tulsa, Oklahoma, USA, 2010. doi:10.2118/129885-MS.

18

Fluidization characteristics of different sizes of quartz particles in the fluidized bed

Zi-Jian Wang[1] · Jun Tang[1] · Chun-Xi Lu[1]

Abstract Fluidization characteristics of quartz particles with different sizes are experimentally investigated in a fluidized bed with an inner diameter of 300 mm and height of 8250 mm. Results show that the average solid holdup increases with the increase in superficial gas velocity and the decrease in initial solid holdup in the dense zone of the fluidized bed. The average cross-sectional solid holdup decreases with increasing bed height and superficial gas velocity. The bed expansion coefficient increases with the increase in superficial gas velocity and the decrease in solid holdup. Correlations of average solid holdup, average cross-sectional solid holdup and bed expansion coefficient are also established and discussed. These correlations can provide guidelines for better understanding of the fluidization characteristics.

Keywords Fluidization characteristic · Solid holdup · Axial average section solid holdup · Bed expansion coefficient

1 Introduction

Oil sands are an alternative fossil fuel which is composed of 10 %–12 % (mass fraction) bitumen, 80 %–85 % sand and clay and 3 %–5 % water (Painter et al. 2010; Xu et al. 2008). In China, the total oil sands reserves are approximately 5.97 billion tons, but only 2.58 billion tons can be

extracted and utilized with current technology, meaning great development and utilization potentials. Conventionally, there exist two methods for separation of bitumen from oil sands, the hot water separation method (Fan and Bai 2015; Ren 2011) that can only be used for water-wet oil sands (Zhao et al. 2014) and the solvent extraction method that can be used to process oil-wet oil sands, but it requires high treatment costs and can result in environmental pollution.

The pyrolysis method has also been reported to improve the bitumen recovery from oil-wet oil sands with better operation flexibility than the two methods mentioned above. Recently, a lot of research has been focused on the pyrolysis of oil sands in fixed beds (Zhang et al. 2014; Wang 2015). Meng et al. (2007) studied the pyrolysis behaviors of Tumuji oil sands (from Inner Mongolia, China) in fixed beds by thermogravimetry (TG), which is used to investigate the effects of heating rate on pyrolysis and reaction kinetics. Lu et al. (2008) made an investigation on extraction of bitumen from oil sands by a direct fluidized-bed coking method, as shown in Fig. 1. The pyrolysis of oil sands is carried out in the fluidized bed. Then, the coked oil sands particles are conveyed to the burner to burn out the coke in the particles. After that, the burned oil sands particles are quickly returned to the reactor, and the heat produced in the burner is also taken to the reactor by the burned oil sand particles for heating the raw oil sand feedstock and for the pyrolysis. This process can improve the bitumen recovery with heat balance and good operation feasibility. Research has indicated that there are significant differences between the pyrolysis and solvent extraction methods in terms of qualities of product. Gao et al. (2013) compared the products of Inner Mongolia oil sands processed, respectively, by organic solvent extraction and fluidized-bed thermal reaction (pyrolysis)

✉ Chun-Xi Lu
  lcx725@sina.com

[1] State Key Laboratory of Heavy Oil Processing, China
  University of Petroleum (Beijing), Beijing 102249, China

Edited by Xiu-Qin Zhu

1. Semi-regenerative riser

2. Valve

3. Feed pipe

4. Striper

5. Gas distributor

6. Coker

7. Feeder

8. Valve

9. Regeneration sloped pipe

10. Burning reactor

11. Valve

12. Cyclone

13. Cyclone leg

14. Valve

15. Gas distributor

16. External heat exchanger

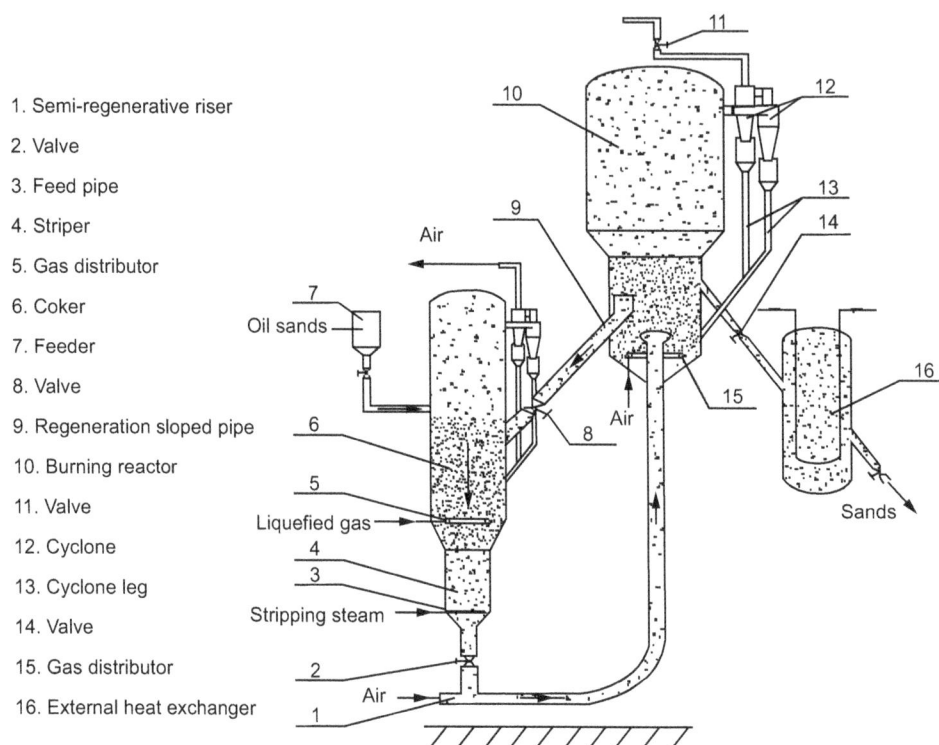

Fig. 1 Fluidized-bed coking process for oil sands

and found that the liquid product from the fluidized-bed thermal reaction had much lower density, viscosity and Conradson carbon residue than that from organic solvent extraction.

The unique features of direct fluidized-bed coking of oil sands enable it to be effectively used in the separation of bitumen from oil sands, where the fluidization characteristics of burned oil sands particles are of main concern. However, oil sands particles from different places and buried depths have wide and different size distributions. Therefore, the fluidization characteristics of different sizes of oil sand particles are critical for proper industrial design of fluidized beds.

The fluidization characteristics that have been studied mainly include the average solid holdup, the axial average solid holdup and the bed expansion coefficient (Ahuja and Patwardhan 2008; Sun et al. 2009; Zhang et al. 2015). The average solid holdup in the dense region is the key parameter for designing industrial fluidized beds. Avidan and Yerushalmi (1982) reviewed earlier studies on the effect of superficial gas velocity on the void ratio at high velocity. Lu et al. (1996a, b) have studied the average solid holdup in the dense zone in a turbulent bed and obtained a correlation of it. The axial average solid holdup distribution is crucial for investigation of the momentum transfer, mass transfer and heat transfer between gas and solid. Cai et al. (2008) found that the average dense zone solid holdup

decreased with increasing height of the fluidized bed. Recently, Cui et al. (2014) studied the axial distribution and evolution of solid holdup in a fluidized bed-Riser coupled reactor and the effect of superficial gas velocity on the axial distribution of solid holdup. Zhu et al. (2014) studied the axial distribution of solid holdup in a pre-lifting structure with two strands of catalyst inlets. The bed expansion coefficient is widely used to determine the height of the dense bed. Lu et al. (1996a, b) systematically studied the bed expansion coefficient in a turbulent fluidized bed and proposed the empirical equation for prediction of the expansion height in the turbulent fluidized bed. Tang et al. (2012) studied the expansion characteristics of particle mixtures in the dense region of fluidized beds using the bed height-to-dense bed ratio. However, most of these experiments are concentrated on the fluidization characteristics of single-component particles, and the fluidization characteristics of multi-component particles are rarely reported.

The purpose of this work is to contribute to a better understanding and modeling of the fluidization characteristics of multi-component particles. For this objective, four kinds of particles with different sizes were used in Plexiglas experimental equipment for the study of multi-sized mixed particles. The models for the average solid holdup, the axial average section solid holdup and the bed expansion coefficient were developed.

2 Experimental method

2.1 Experimental apparatus and method

Experiments were carried out in Plexiglas equipment with an inner diameter of 300 mm and a height of 8250 mm, as shown in Fig. 2. A plate distributor with 100 holes of diameter 3 mm was fixed in the bottom of the fluidized bed. The opening area ratio is 1.1 %.

The pressures at different positions along the bed height were measured by using a FXC-II/32 pressure transducer (Beijing Sensing Star Control Technology Co., Ltd. China), and the air superficial velocity was measured by a rotameter. The initial and dense bed height was measured by using a ruler adhered on the wall of the bed. As shown in Fig. 3, there were 16 measuring points on the wall along the bed height. More measuring points were installed in the dense bed. The average solid holdup ε_s can be calculated by the following two equations,

$$\Delta P = \Delta H \times g \times \left((1 - \varepsilon_s)\rho_g + \varepsilon_s\rho_p\right) \approx \Delta H \times g \times \varepsilon_s\rho_p \tag{1}$$

$$\varepsilon_s = \frac{\Delta P}{\Delta H \times g \times \rho_p} \tag{2}$$

where ΔP means the pressure drop, kPa; ΔH is the distance between two measure points, m; ρ_p is the density of particles, kg/m^3.

No.	Space, mm
1—2	100
2—3	200
3—4	200
4—5	200
5—6	200
6—7	200
7—8	400
8—9	400
9—10	600
10—11	600
11—12	600
12—13	1000
13—14	1000
14—15	1000
15—16	700

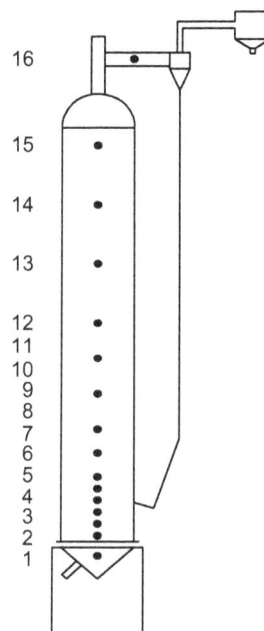

Fig. 3 Schematic diagram of axial measuring points

2.2 Experimental materials

In this experiment, the solid particles were Geldart A, B, C, and D quartz sand particles. The particle size distributions are shown in Fig. 4a–d, and the physical properties of the particles are given in Table 1, and Geldart has shown the difference between different types of Geldart particles (Geldart 1973). Ambient air was used as the fluidizing gas.

3 Results and discussion

3.1 Average solid holdup

Figures 5, 6 and 7 show the effect of different factors on the average solid holdup of A, B, C and D quartz sand particles in the dense phase. As shown in Fig. 5, the average solid holdup increased with increasing particle diameter. The slope of the curves decreased with the increase in particle diameter. This is reasonable because initial solid holdup increases with increasing particle diameter. When the particle diameter was small, initial solid holdup increased rapidly with increasing particle diameter. Thus, the average solid holdup increased with increasing initial solid holdup.

Figure 6 shows the effect of superficial gas velocity on the average solid content. It was clear that the average solid content decreased with the increasing superficial gas velocity because the solid holdup decreased with more gas passing through the dense phase. It was found that the average solid holdup of particles C and D decreased more greatly than that of particles A and B because of their different expansibilities.

1. Roots blower
2. Surge tank
3. Distributor
4. Rotameter
5. Distributor
6. Fluidized bed
7. Riser
8. Cyclone
9. Dust collector
10. Dipleg
11. Air valve

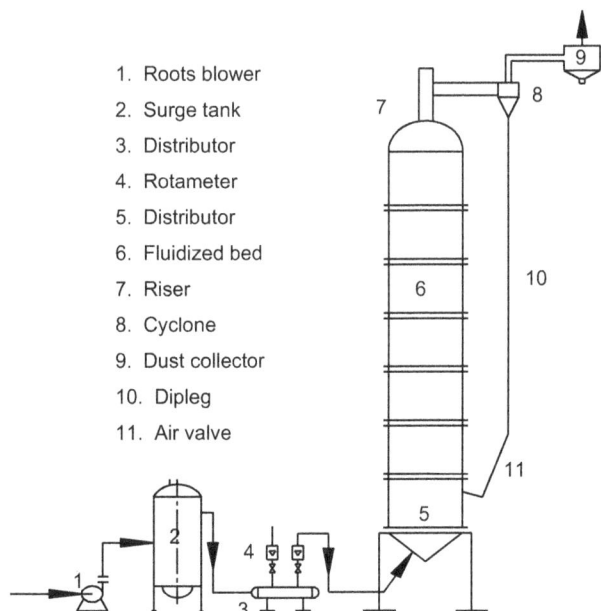

Fig. 2 Schematic diagram of the experimental setup

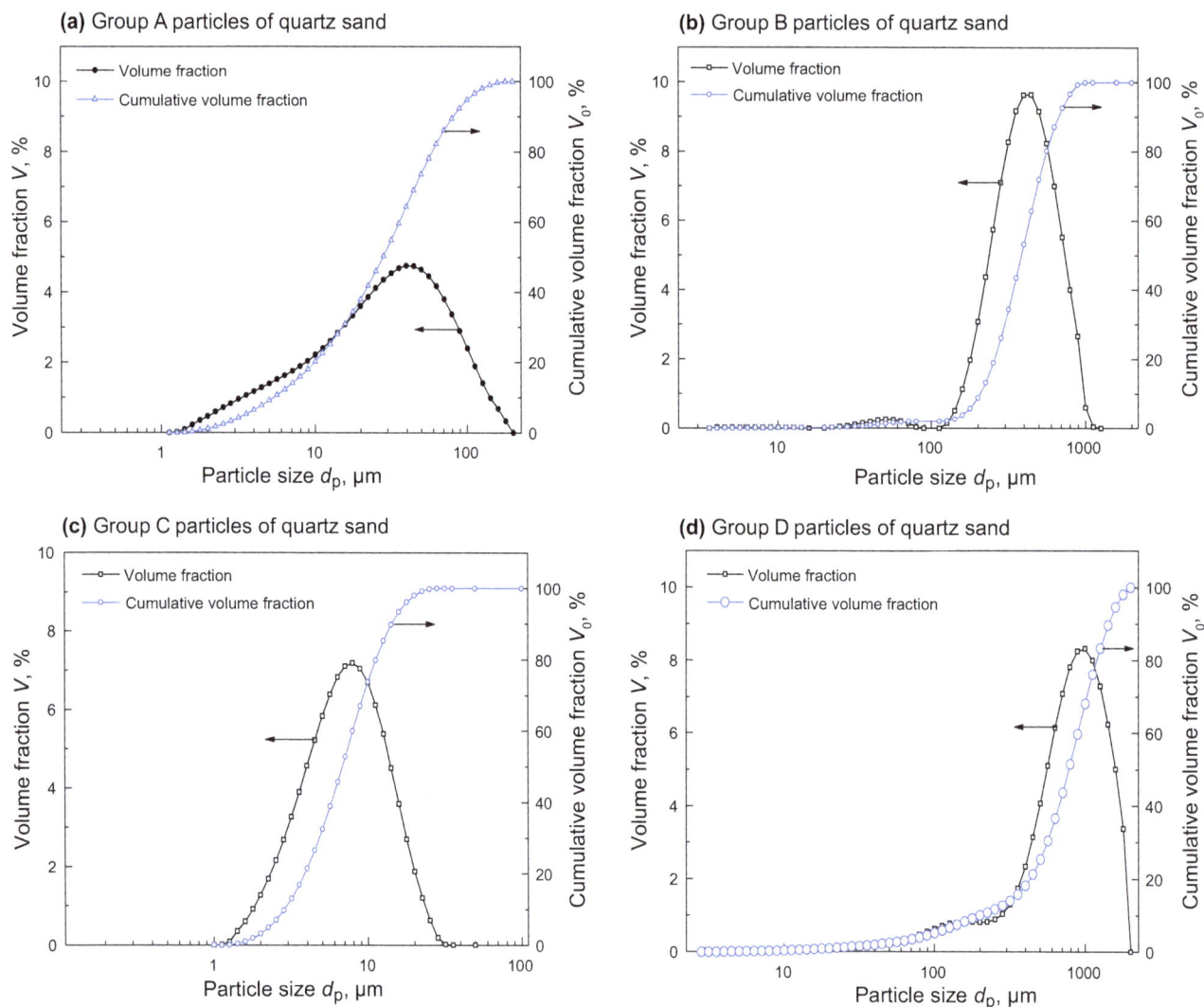

Fig. 4 Particle size distributions of the A, B, C and D quartz sand particles

Table 1 Physical properties of solid particles

Particle	Mean diameter, μm	Bulk density, kg m^{-3}	Particle density, kg m^{-3}
A quartz sand particle	36.80	885	2451
B quartz sand particle	411.70	1255	2451
C quartz sand particle	7.80	613	2451
D quartz sand particle	810.70	1413	2451

Figure 7 shows that the average solid content increased with increasing initial solid holdup. The initial solid holdup had a more significant effect on the average solid concentration than the particle diameter and superficial gas velocity shown in Figs. 5 and 6.

3.2 Axial average section solid holdup

Figure 8a–d shows the axial average solid holdup distribution of four kinds of different size quartz particles in the

dense phase (when the initial bed height is 450 and 650 mm). As shown in Fig. 8, the curves of the four different size ranges of particles were similar in shape. The average solid holdup decreased along the axial height and also decreased with an increase in superficial gas velocity. Since the density of the quartz particles is high, gravity has an appreciable impact on the axial average solid holdup distribution when particles travel against gravity. Gas began to accumulate into big bubbles along the axial height resulting in a higher void ratio along the axial height. When

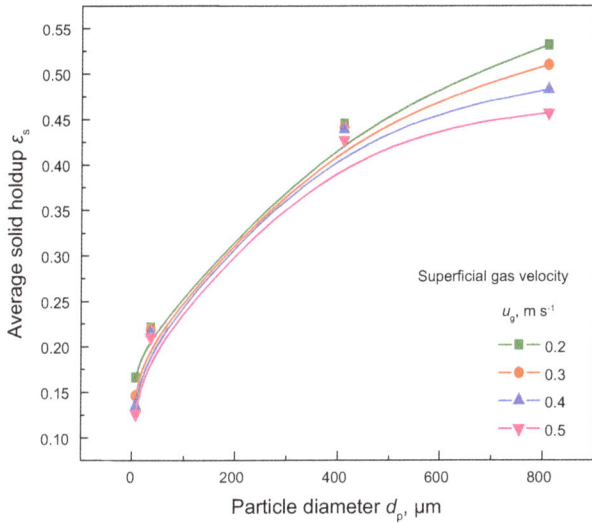

Fig. 5 Effect of particle diameter on average solid holdup in the dense phase

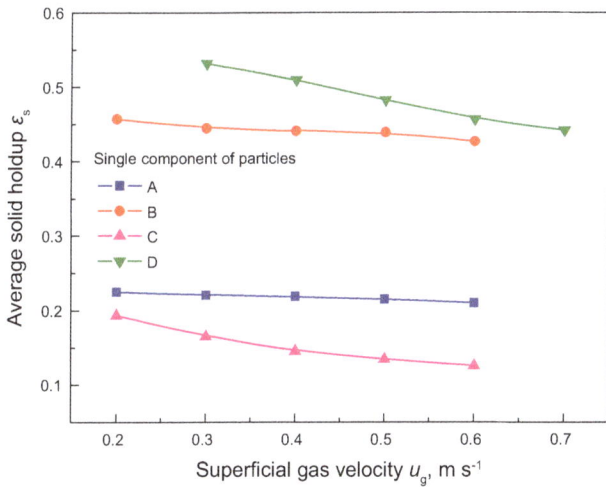

Fig. 6 Effect of superficial gas velocity on average solid holdup in the dense phase

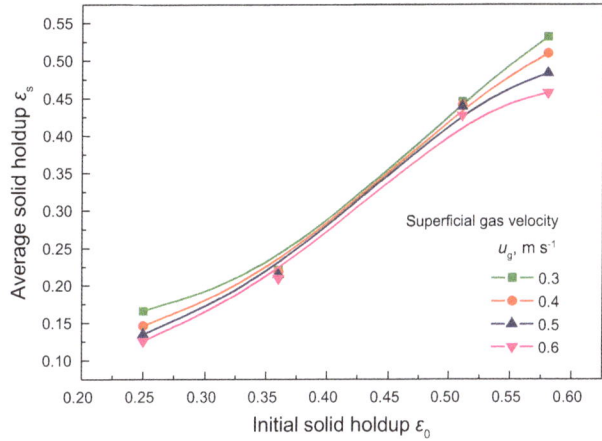

Fig. 7 Effect of initial solid holdup on average solid holdup in the dense phase

drastically with the increase in superficial gas velocity. This clearly verified that the smaller the diameter of particles was, the smaller the diameter of bubbles was.

3.3 Bed expansion coefficient

Two methods are generally used to calculate the bed expansion coefficient. One is based on the bed height ratio (R_h), which means the ratio of the dense bed height to the initial bed height which can be regarded as the bed expansion coefficient. The other is based on the solid content ratio (R_ε), which represents the ratio of initial solid holdup to average solid holdup in the dense phase. These two equations are shown as follows,

$$R_h = \frac{H}{H_0} \tag{3}$$

$$R_\varepsilon = \frac{\varepsilon_0}{\varepsilon_s} \tag{4}$$

where H means the dense bed height, m, and H_0 means the initial bed height, m. The method based on R_h can be used to calculate and measure bed expansion coefficient easily when the superficial gas velocity was low. As for high superficial gas velocity, which will cause more fine particles being carried into the dilute phase, the method above exposed shortcomings by getting the result that R_h decreased with increasing superficial gas velocity. It is contradictory to the actual fact that the bed expansion coefficient increases with the increasing superficial gas velocity. On the contrary, the method based on R_ε can be used under the condition of high superficial gas velocity. Thus, the bed expansion coefficient of the four sizes of quartz particles was calculated by using the method based on solid content ratio (R_ε).

the superficial gas velocity increased, two phenomena appeared. On the one hand, as a result of more and more bubbles appearing, the void ratio of the dense bed increased rapidly. On the other hand, the increasing diameters of bubbles followed by a rapid ascending motion led to the decrease in the void ratio. However, the first factor occupies the leading position. Because in this experiment, the flow regime was turbulent bed, on the impact of turbulent gas flow, bubbles were broken. Thus, when the superficial gas velocity increased, the diameter of bubbles decreased and the number of bubbles increased. As a result, the average solid holdup decreased with increasing superficial gas velocity. Compared with particles A, B and D, the average solid holdup of particles C decreased

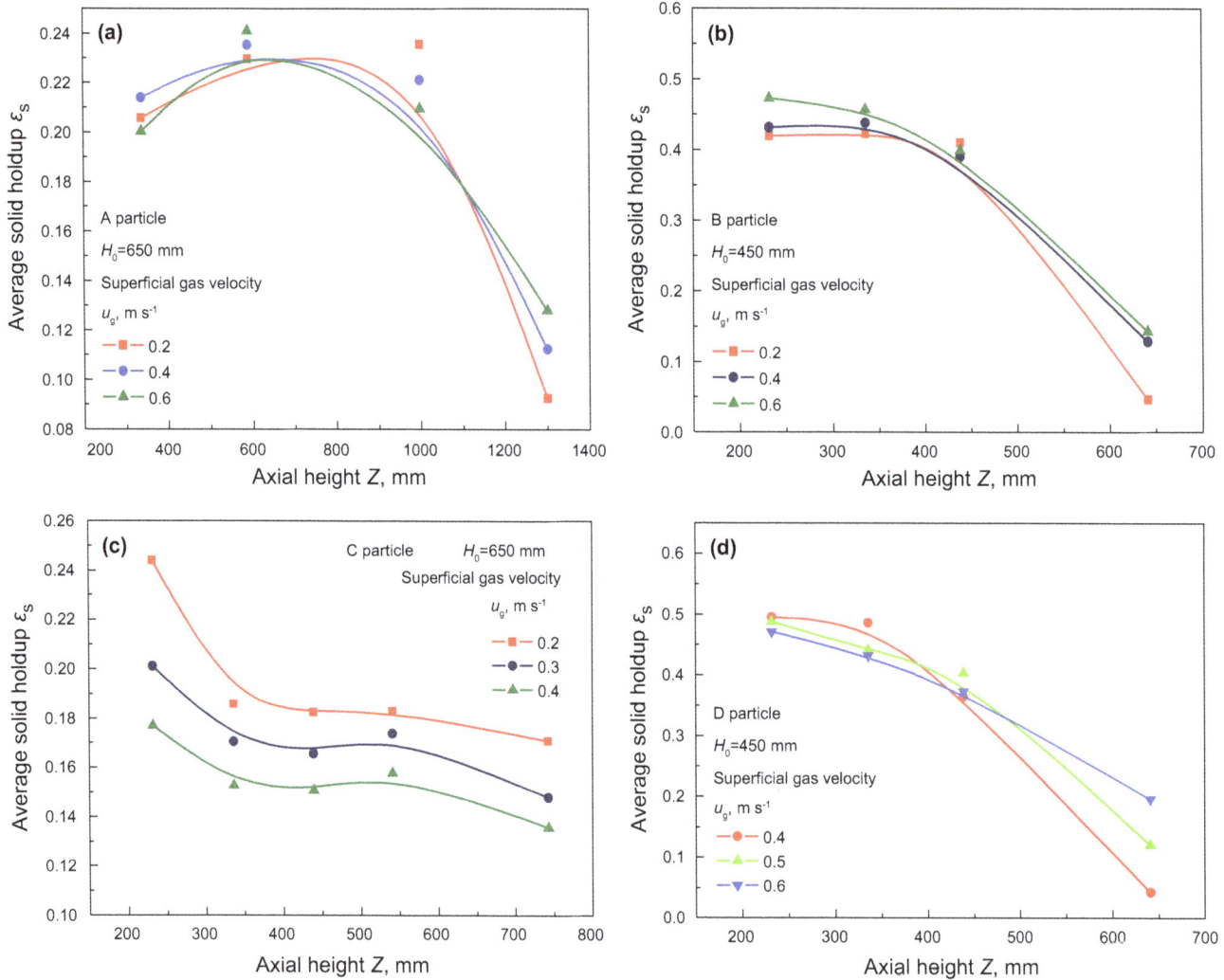

Fig. 8 Axial average solid holdup distribution in the dense phase

Figure 9 shows the effect of superficial gas velocity on the bed expansion coefficient based on R_ε. The bed expansion coefficient increased in proportion to the superficial gas velocity. That is why bed expansion coefficients of particles A and C were bigger than those of particles B and D. This phenomenon can be explained by the following aspects. On the one hand, the diameter of particles was in direct proportion to the weight of the particles, which indicated that heavier particles were more difficult to be expanded than fine particles. On the other hand, the increase in particles diameter further gave rise to the increase in the bubble diameter in the dense bed. As a result, big bubbles had higher rising velocity which weakened the expansion of the dense bed. This conclusion is similar to the study based on bed height ratio (R_h).

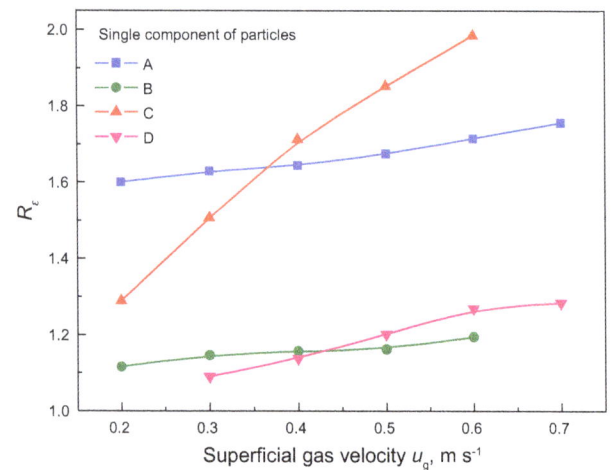

Fig. 9 Bed expansion coefficient based on solid content ratio

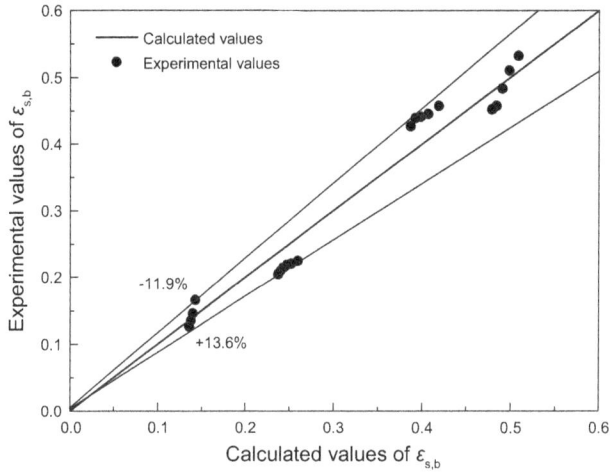

Fig. 10 Comparison between calculated and experimental values of average solid holdup in the dense phase

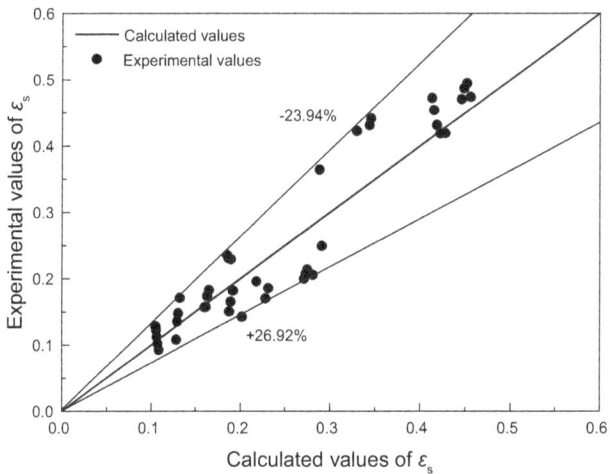

Fig. 11 Comparison between calculated and experimental values of axial average solid holdup distribution in the dense phase

4 Correlation development

Correlations of average solid holdup, axial average section solid holdup and bed expansion coefficient were proposed based on the analysis of experimental data and previous studies. Comparison between the calculated result and the experimental data was made to show the feasibility of the correlations.

4.1 The correlation of average solid holdup

Analysis of the experimental results clearly highlights the significant combined influence of superficial gas velocity, particle diameters and initial solid holdup on the average solid holdup of the particles. The Reynolds number ($Re = (d_p u_g \rho_g)/\mu$) was used to show the effect of operation

conditions and properties. The correlation built by the least squares method is represented as follows:

$$\varepsilon_{s,b} = 1.7715 Re^{-0.0714} \varepsilon_{s,0}^{1.8669} \tag{5}$$

where $\varepsilon_{s,b}$ means the solid holdup of the fluidized bed. As shown in Eq. (5), the average solid holdup of particles increased with increasing diameter and initial solid holdup and decreased with the increase in superficial gas velocity. Figure 10 shows how this was in good agreement with the experimental result.

Figure 10 shows that the calculated values were in good agreement with the experimental data. The deviations were within $-11.9\ \% \sim 13.6\ \%$, demonstrating the reliable fitting of this correlation to predict the average solid holdup of the particles.

4.2 The correlation of axial average section solid holdup

As shown above, superficial gas velocity (u_g), axial height of the dense phase (h), initial solid holdup ($\varepsilon_{s,0}$) and the properties of particles together affected the distribution of the axial average solid holdup of different size particles. The Reynolds number was used to illustrate the effect of superficial gas velocity and particle properties. The ratio of the height to the diameter of fluidized bed was used to show the effect of the axial height of the dense phase. The correlation is shown as follows:

$$\varepsilon_{s,b} = 0.6310 Re^{-0.0319} \varepsilon_{s,0}^{0.6732} \left(\frac{h}{D}\right)^{-0.7047} \tag{6}$$

where h means the height of the fluidized bed, m, and D means the diameter of the fluidized bed, m. As shown in Eq. (6), the average solid holdup of different component particles increased with increasing initial solid holdup and decreased with increasing superficial gas velocity. Meanwhile, the average solid holdup decreased along the axial height.

Figure 11 shows the comparison between the calculated average solid holdup and the experimental data. The average relative error was 15.4 %, according to which the correlation of the axial average solid holdup was feasible.

4.3 The correlation of bed expansion coefficient

The solid content ratio (R_ε) is used to calculate the bed expansion coefficient in the situation of high superficial gas velocity. The correlation is shown as follows:

$$R_\varepsilon = 0.1067 Re^{0.1861} \left(\frac{d_p}{D}\right)^{-0.2793} \tag{7}$$

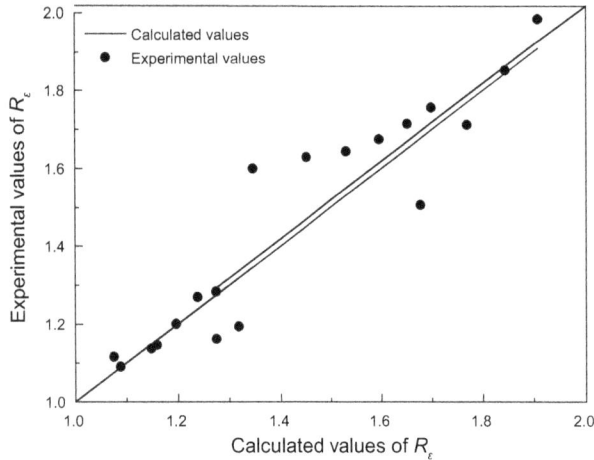

Fig. 12 Comparison between calculated and experimental values of R_ε

where D means the diameter of the fluidized bed, m, and d_p is the diameter of particles, m. The correlation shown in Eq. (7) indicated that the solid content ratio increased with increasing superficial gas velocity and decreased with the increase in particle diameter. This showed that Eq. (7) is in good agreement with the analysis above. Figure 12 shows the comparison between the calculated bed expansion efficient and the experimental data. The average relative error was only 4.35 %, which means that the bed expansion efficient correlation based on R_ε was reliable.

5 Conclusions

In this work, the fluidization characteristics of different sized particles were investigated at various superficial gas velocities in the dense phase. Predictive correlations between average solid holdup in the dense phase, axial average solid holdup and bed expansion coefficient were also established and discussed. The following conclusions are obtained:

(1) The average solid holdup in the dense zone decreases with increasing superficial gas velocity and decreases with a decrease in initial solid holdup.
(2) The axial average section solid holdup decreases with increasing bed height and increasing superficial gas velocity.
(3) The bed expansion coefficient increases with the increase in superficial gas velocity and increases with a decrease in initial solid holdup.

References

Ahuja GN, Patwardhan AW. CFD and experimental studies of solids hold-up distribution and circulation patterns in gas–solid fluidized beds. Chem Eng J. 2008;143(1–3):147–60.

Avidan AA, Yerushalmi J. Bed expansion in high velocity fluidization. Powder Technol. 1982;32(2):223–32.

Cai J, Li T, Sun QW, et al. Solid concentration distribution in a gas–solid fluidized bed. Process Eng. 2008;8(5):839–44.

Cui G, Liu MX, Lu CX. Axial distribution and development of solids hold-up in a fluidized bed-riser coupled reactor. Chin J Process Eng. 2014;14(4):556–61 (**in Chinese**).

Fan Q, Bai G. The evaluation of oil sand bitumen produced from inner Mongolia. Pet Sci Technol. 2015;33(4):437–42.

Gao JS, Xu T, Wang G, et al. Reaction behavior of oil sand in fluidized-bed pyrolysis. Pet Sci. 2013;10:562–70.

Geldart D. Types of gas fluidization. Powder Technol. 1973;7(5):285–92.

Lu CX, Xu CM, Li SY, et al. A process and apparatus of direct fluid coking for oil sands. China Patent CN 101358134A; 2008 (**in Chinese**).

Lu CX, Xu YF, Shi MX, et al. Bed expansion in the dense region of FCC turbulent fluidized bed. J Chem Ind Eng. 1996a;47(1):110–3 (**in Chinese**).

Lu CX, Xu YF, Shi MX, et al. Study of the radial distribution of voidage in gas–solid turbulent fluidized beds. J Chem Ind Eng. 1996b;47(1):110–3 (**in Chinese**).

Meng M, Hu HQ, Zhang QM, et al. Pyrolysis behaviors of Tumuji oil sand by thermogravimetry (TG) and in a fixed bed reactor. Energy Fuels. 2007;21(4):2245–9.

Painter P, Williams P, Lupinsky A. Recovery of bitumen from Utah tar sands using ionic liquids. Energy Fuels. 2010;24(9):5081–8.

Ren SL. Research progress in water-based bitumen extraction from oil sands. Chin J Chem Eng. 2011;9(62):2406–12 (**in Chinese**).

Sun D, Wang SY, Lu HL, et al. A second-order moment method of dense gas–solid flow for bubbling fluidization. Chem Eng Sci. 2009;64(23):5013–27.

Tang J, Chen XN, Zhang YM, et al. Bed expansion in the dense region of a FCC turbulent fluidized bed. J Chem Ind Eng. 2012;47(1):110–3 (**in Chinese**).

Wang ZC. Structure changes of asphaltene oil sands during pyrolysis. Doctoral dissertation. Northeast Dianli University; 2015.

Xu XQ, Wang HY, Zeng DW, et al. Research progress in application of the oil sands. Liaoning Chem Ind. 2008;37(4):268–71 (**in Chinese**).

Zhang C, Wang DM, Zhang BB, et al. Catalytic pyrolysis behaviors of Xinjiang Tuoli oil sand. Chin J Process Eng. 2014;14(3):456–61 (**in Chinese**).

Zhang YL, Ye M, Zhao YF, et al. Emulsion phase expansion of Geldart A particles in bubbling fluidized bed methanation reactors: a CFD-DEM study. Powder Technol. 2015;275(8):199–210.

Zhao RY, Wang T, Zhang C, et al. Research progress of oil sands wettability. Oilfield Chem. 2014;31(4):620–5 (**in Chinese**).

Zhu LY, Fan YP, Lu CX. Flow characteristics of catalytic particles in the pre-lifting structure with two strands of catalyst inlets. Chin J Process Eng. 2014;14(1):9–15 (**in Chinese**).

19

Fractional thermoelasticity applications for porous asphaltic materials

Magdy Ezzat[1,2] · **Shereen Ezzat**[1]

Abstract A new mathematical model of poro-thermoelasticity has been constructed in the context of a new consideration of heat conduction with fractional order. One-dimensional application for a poroelastic half-space saturated with fluid is considered. The surface of the half-space is assumed to be traction-free, permeable, and subjected to heating. The Laplace transform technique is used to solve the problem. The inversion of the Laplace transform will be obtained numerically and the numerical values of the temperature, stresses, strains, and displacements will be illustrated graphically for the solid and the liquid.

Keywords Generalized thermo-poroelasticity · Asphaltic material · Thermal shock · Fractional calculus · Numerical results

List of symbols

c_s, c_f	Specific heat of the solid and the fluid phases
e_s	Strain component of the solid phase
e_f	Strain component of the fluid phase
F_i	Components of the body forces per unit mass
k_s	Thermal conductivity of the solid phase
k_f	Thermal conductivity of the fluid phase
q_{is}^*, q_{if}^*	Intensities of heat fluxes of the solid and fluid phases
q_{is}	$q_{is} = (1 - \beta)q_{is}^*$, heat flux of the solid phase per unit area
q_{if}	$q_{if} = \beta q_{if}^*$, heat flux of the fluid phase per unit area
Q	Intensity of the heat sources per unit mass
R_{11}, R_{12}, R_{21}, R_{22}	Mixed and thermal coefficients
T_o	Reference temperature
u_{is}	ith component of the solid-phase displacement
u_{if}	ith component of the fluid-phase displacement
α_s, α_f	Coefficients of the thermal expansion of the phases
β	Porosity of the aggregate
σ_{ij}	Stress components in the solid
σ	Normal stress in the fluid
θ_s	$\theta_s = T_s - T_o$, temperature increment of the solid phase
θ_f	$\theta_f = T_f - T_o$, temperature increment of the fluid phase
ρ_s, ρ_f	Density of the solid and the liquid phases
ρ_1	$\rho_1 = (1 - \beta)\rho_s$, density of the solid phase per unit volume of bulk
ρ_2	$\rho_2 = \beta\rho_f$, density of the fluid phase per unit volume of bulk
ρ_{11}	$\rho_{11} = \rho_1 - \rho_{12}$, mass coefficient of the solid phase

✉ Magdy Ezzat
maezzat2000@yahoo.com

1 Department of Mathematics, Faculty of Science and Letters in Al Bukayriyyah, Al-Qassim University, Al-Qassim, Saudi Arabia

2 Faculty of Education, Alexandria University, Alexandria, Egypt

Edited by Yan-Hua Sun

ρ_{22}	$\rho_{22} = \rho_2 - \rho_{12}$, mass coefficient of the fluid phase
ρ_{12}	Dynamics coupling coefficient
ρ	Displacements of the skeleton and fluid phases
ρ	$\rho = \rho_1 + \rho_2$
κ	Interface coefficients of the interphase heat conduction
τ_s, τ_f	Relaxation times of the solid and the fluid phases
λ, μ, R, G	Poroelastic coefficients
η_s	$\eta_s = \frac{\rho_s c_s}{k_s}$, thermal viscosity of the solid
η_f	$\eta_f = \frac{\rho_f c_f}{k_f}$, thermal viscosity of the fluid
η	$\eta = \frac{\rho_{12} c_{sf}}{k}$, thermal viscosity couplings between the phases
P	$P = 3\lambda + 2\mu$
F_{11}	$F_{11} = \rho_s C_s$
F_{22}	$F_{22} = \rho_f C_f$
R_{11}	$R_{11} = \alpha_s P + \alpha_{fs} G$
R_{22}	$R_{22} = \alpha_f R + 3\alpha_{sf} G$
R_{12}	$R_{12} = \alpha_f G + \alpha_{sf} P$

1 Introduction

The concept of a poroelastic material was introduced by Biot (1955) in order to describe the mechanical behavior of water-saturated soil. Biot's material consists of a combination of a deformable solid and a fluid, the solid constituting a porous skeleton whose innumerable tiny cavities are interconnected and filled with the fluid. Apart from water-saturated soil, the class of poroelastic materials includes other materials such as polyurethane foam and sound-absorbing materials, osseous tissue, and so on.

Due to many applications in the fields of geophysics, plasma physics and related topics, increasing attention is being devoted to the interaction between fluid such as water and thermo elastic solids, which is the domain of the theory of poro-thermoelasticity. The field of poro-thermoelasticity has a wide range of applications especially in studying the effect of using waste materials on disintegration of asphalt concrete mixture (ACM).

Asphalt concrete pavements are made of asphalt concrete mixtures (ACMs) consisting of graded aggregates, asphalt binders, and air voids. Due to the complexity of internal composition, mechanical characteristics of ACM exhibit extremely non-linear constitutive behavior with respect to loading time, loading rate, and temperature. The experimental observations suggest that the total deformation behavior has recoverable and irrecoverable parts,

which could occur simultaneously and are time-dependent functions of stress levels, strain rates, and temperature (Krishnan and Rajagopal 2003). Time-dependent properties of ACM are not only one of the main sources of pavement rutting (e.g., permanent deformation), but also plays a critical role in fatigue and low-temperature cracking, roughness, and corrugation.

Coupled thermal and poromechanical processes play an important role in a number of problems of interest in geomechanics such as stability of boreholes and hydraulic fracturing in geothermal reservoirs or high temperature petroleum-bearing formations. This is due to that fact that when rocks are heated/cooled, the bulk solid as well as the pore fluid tends to undergo expansion/contraction. A volumetric expansion can result in significant pressurization of the pore fluid depending on the degree of containment and the thermal and hydraulic properties of the fluid as well as the solid. The net effect is a coupling of thermal and poromechanical processes. A few analytical procedures have been developed and used to solve geomechanics problems of interest involving coupled thermal and poromechanical problems (Delaney 1982; Wang and Papamichos 1994; Li et al. 1998; Ghassemi and Diek 2002; Ghassemi and Zhang 2004). However, many problems formulated within the framework of poro-thermoelasticity are not amenable to analytical treatment and need to be solved numerically.

The coupled rock deformation and liquid flow and various factors that affect it need to be studied both theoretically and experimentally. Implementation of theoretical rheologies in numerical models enables scientific interpretation of field measurements and provides a powerful means for understanding and predicting phenomena related to the internal earth processes. This is of paramount significance to the development of renewable geothermal energy resources as well as the prediction of natural hazards. However, the traditional isotropic elastic, plastic, and viscous rheological models are inadequate to study Earth and its tectonics. This is especially true at transform plate boundaries, where deformation-enhanced weakening is thought to be responsible for localization of strain (Simakin and Ghassemi 2005). Therefore, there is strong scientific and industrial interest in developing new mechanical models of liquid saturated rocks such as poro-viscoelasticity with damage mechanics that can include the effects of fluid phase pressurization and transport.

Porous materials make their appearance in a wide variety of settings, natural, and artificial, and in diverse technological applications. As a consequence a number of problems arise dealing with, among other issues, statics and strength, fluid flow and heat conduction, and dynamics. In connection with the latter, we note that problems of this kind are encountered in the prediction of the behavior of sound-absorbing materials and in the area of exploration geophysics, the steadily growing literature bearing witness to the importance of the subject. The existence and uniqueness of the generalized solutions for the boundary value problems in elasticity of initially stressed bodies with voids (porous materials) are proved (Marin 2008).

The field of poro-thermoelasticity has a wide range of applications especially in studying the effect of using waste materials on disintegration of asphalt concrete mixture (Pecker and Deresiewicz 1973). The problem of a fluid-saturated porous material has been studied for many years. A short list of papers pertinent to the present study includes Biot (1956a, b), Biot and Willis (1957), Deresiewicz and Skalak (1963), Nur and Byerlee (1971), Sherief and Hussein (2012), Youssef (2007).

Mathematical modeling is the process of constructing mathematical objects whose behavior or properties correspond in some way to a particular real-world system. The term real-world system could refer to a physical system, a financial system, a social system, an ecological system, or essentially any other system whose behavior can be observed. In this description, a mathematical object could be a system of equations, a stochastic process, a geometric or algebraic structure, an algorithm or any other mathematical apparatus like a fractional derivative, integral or fractional system of equations. Fractional calculus and fractional differential equations serve as mathematical objects describing many real-world systems.

Fractional calculus has been used successfully to modify many existing models of physical processes. One can state that the whole theory of fractional derivatives and integrals was established in the second half of the 19th century. The first application of fractional derivatives was given by Abel, who applied fractional calculus in the solution of an integral equation that arises in the formulation of the tautochrone problem. The generalization of the concept of derivatives and integrals to a non-integer order has been subjected to several approaches and some various alternative definitions of fractional derivatives appeared elsewhere (Gorenflo and Mainardi 1997; Miller and Ross 1993; Samko et al. 1993; Oldham and Spanier 1974). In the last few years, fractional calculus has been applied successfully in various areas to modify many existing models of physical processes, e.g., chemistry, biology, modeling and identification, electronics, wave propagation, and viscoelasticity (Rossikhin and Shitikova 1967; Bagley and Torvik 1986). Fractional order models often work well, particularly for dielectrics and viscoelastic materials over extended ranges of time and frequency (Lakes 1999; Grimnes and Martinsen 2000). In heat transfer and electrochemistry, for example, the half-order fractional integral is the natural integral operator connecting the applied gradients (thermal or material) with the diffusion of ions

with heat (Gorenflo et al. 2002). One can refer to Podlubny (1999) for a survey of applications of fractional calculus.

Ezzat (2010, 2011a, b, c) was the first writer who established a new formula of heat conduction law by using the new Taylor-Riemann series expansion of time-fractional order α developed by Jumarie (2010) as follows:

$$q(x, t) + \frac{\tau^{\upsilon}}{\upsilon!} \frac{\partial^{\upsilon} q}{\partial t^{\upsilon}} = -k \nabla T, \quad 0 < \upsilon \leq 1 \tag{1}$$

where q the heat flux vector and τ is the relaxation time.

Sherief et al. (2010) introduced a fractional formula of heat conduction and proved a uniqueness theorem and derived a reciprocity relation and a variational principle. El-Karamany and Ezzat (2011) introduced two general models of fractional heat conduction law for a non-homogeneous anisotropic elastic solid. Uniqueness and reciprocal theorems are proved and the convolutional variational principle is established and used to prove a uniqueness theorem with no restriction on the elasticity or thermal conductivity tensors except symmetry conditions. Ezzat and El-Karamany (2011a, b, c) introduced a new mathematical model for electro-thermoelasticity equations using the methodology of fractional calculus. The model is applied to one-dimensional problems to investigate the thermal behavior in thermoelectric solids. The verification process was done by comparison of model predictions with the previous work and shows good agreement is achieved. Abbas (2014, 2015) solved some problems on fractional order theory of thermoelasticity for a functional graded material. Some applications of fractional calculus to various problems in continuum mechanics are reviewed in the literature (Ezzat et al. 2012a, b; 2013a, b; 2014a, b, c, d; 2015).

In the current work, a modified law of heat conduction including fractional order of the time derivative is constructed and replaces the conventional Fourier's law in poro-thermoelasticity. The resulting non-dimensional coupled equations for poroelastic half-space saturated with fluid are considered. The general solution in the Laplace transform domain is obtained and applied in a certain asphalt material which is thermally shocked on its bounding plane. The inversion of the Laplace transform will be obtained numerically and the numerical values of the temperature, displacement, and stress will be illustrated graphically.

2 Derivation of fractional heat conduction equation in poro-thermoelasticity

Much effort has been devoted recently for determining conditions which guarantee that the assumption of local thermal equilibrium (LTE) is accurate when modeling heat transfer in porous media. When it is accurate, then the thermal field is well approximated by a single thermal energy equation. In other circumstances, the local thermal non-equilibrium (LTNE) prevails, and it is necessary to employ two energy equations, one for each phase (Nouri-Borujerdi et al. 2007). Fourie and Du Plessis (2003) showed that the intrinsic volume-averaged equilibrium temperature of the solid phase is equal to that of the fluid phase everywhere though that their values may differ locally.

The conventional poro-thermoelasticity is based on the principles of the classical theory of heat conductivity, specifically on classical Fourier's law, in which relates the heat flux vector q to the temperature gradient (Nowinski 1978)

$$q_{is}(x_i, t) = -k_s \theta_{s,i} \tag{2}$$

$$q_{if}(x_i, t) = -k_f \theta_{f,i} \tag{3}$$

The energy equation in terms of the heat conduction vectors q_{is} and q_{if} are

$$\frac{\partial}{\partial t}(F_{11}\,\theta_s + T_o R_{11}\,e_s + T_o R_{21}\,e_f) = -\nabla q_{is} + \rho_1\,Q_s - \kappa(\theta_s - \theta_f) \tag{4}$$

$$\frac{\partial}{\partial t}(F_{22}\,\theta_f + T_o R_{12}\,e_s + T_o R_{22}\,e_f) = -\nabla q_{if} + \rho_2\,Q_f + \kappa(\theta_s - \theta_f) \tag{5}$$

Although Fourier's law of heat conduction is well tested for most practical problems, it fails to describe the transient temperature field in situations involving short times, high frequencies, and small wavelengths (Lebon et al. 2008). To eliminate these anomalies, Cattaneo (1948) and Vernotte (1958) proposed a damped version of Fourier's law by introducing a heat flux relaxation term, by taking Taylor's series to expand $q_{is}(x_i,\ t+\tau_s)$, $q_{if}(x_i,\ t+\tau_f)$ and retaining terms up to the first order in τ_s and τ_f. The first well-known generalization of such a type (Sherief and Hussein 2012)

$$q_{is} + \tau_s \dot{q}_{is} = -k_s \theta_{s,i} \tag{6}$$

$$q_{if} + \tau_f \dot{q}_{if} = -k_f \theta_{f,i} \tag{7}$$

leads to the hyperbolic-type heat transport equation in the theory of poro-thermoelasticity

$$k_s\theta_{s,ii} + \left(1 + \tau_s\frac{\partial}{\partial t}\right)[\rho_1\,Q_s - \kappa(\theta_s - \theta_f)]$$
$$= \frac{\partial}{\partial t}\left(1 + \tau_s\frac{\partial}{\partial t}\right)(F_{11}\,\theta_s + T_o R_{11}\,e_s + T_o R_{21}\,e_f) \tag{8}$$

$$k_f\theta_{f,ii} + \left(1 + \tau_f\frac{\partial}{\partial t}\right)[\rho_2\,Q_f + \kappa(\theta_s - \theta_f)]$$
$$= \frac{\partial}{\partial t}\left(1 + \tau_f\frac{\partial}{\partial t}\right)(F_{22}\,\theta_f + T_o R_{12}\,e_s + T_o R_{22}\,e_f). \tag{9}$$

In the present work, the new fractional Taylor-Riemann series of time-fractional order v is adopted to expand $q_{is}(x_i,\ t+\tau_s)$, $q_{if}(x_i,\ t+\tau_f)$ and retaining terms up to order v in the thermal relaxation times τ_s and τ_f, we get

$$q_{is}(x_i, t + \tau_s) = q_{is}(x_i, t) + \frac{\tau_s^v}{v!}\frac{\partial^v q_{is}}{\partial t^v}, \quad 0 < v \leq 1 \tag{10}$$

$$q_{if}(x_i, t + \tau_f) = q_{if}(x_i, t) + \frac{\tau_f^v}{v!}\frac{\partial^v q_{if}}{\partial t^v}, \quad 0 < v \leq 1 \tag{11}$$

From a mathematical viewpoint, Fourier's laws (2) and (3) in poro-thermoelasticity theory of generalized fractional heat conduction are given by

$$q_{is}(x_i, t) + \frac{\tau_s^v}{v!}\frac{\partial^v q_{is}}{\partial t^v} = -k_s\theta_{s,i}, \quad 0 < v \leq 1 \tag{12}$$

$$q_{if}(x_i, t) + \frac{\tau_f^v}{v!}\frac{\partial^v q_{if}}{\partial t^v} = -k_f\theta_{f,i}, \quad 0 < v \leq 1. \tag{13}$$

Taking the partial time derivative of fraction order v of Eqs. (4) and (5), we get (Tarasov 2008)

$$\frac{\partial^{v+1}}{\partial t^{v+1}}(F_{11}\,\theta_s + T_o R_{11}\,e_s + T_o R_{21}\,e_f)$$
$$= -\nabla\left(\frac{\partial^v q_{is}}{\partial t^v}\right) + \frac{\partial^v}{\partial t^v}[\rho_1\,Q_s - \kappa(\theta_s - \theta_f)] \tag{14}$$

$$\frac{\partial^{v+1}}{\partial t^{v+1}}(F_{22}\,\theta_f + T_o R_{12}\,e_s + T_o R_{22}\,e_f)$$
$$= -\nabla\left(\frac{\partial^v q_{if}}{\partial t^v}\right) + \frac{\partial^v}{\partial t^v}[\rho_2\,Q_f + \kappa(\theta_s - \theta_f)] \tag{15}$$

Multiplying Eqs. (14) and (15) by $\frac{\tau_s^v}{v!}$ and $\frac{\tau_f^v}{v!}$ and adding to Eqs. (4) and (5), respectively, we have

$$k_s\theta_{s,ii} + \left(1 + \frac{\tau_s^v}{v!}\frac{\partial^v}{\partial t^v}\right)[\rho_1\,Q_s - \kappa(\theta_s - \theta_f)]$$
$$= \frac{\partial}{\partial t}\left(1 + \frac{\tau_s^v}{v!}\frac{\partial^v}{\partial t^v}\right)(F_{11}\,\theta_s + T_o R_{11}\,e_s + T_o R_{21}\,e_f) \tag{16}$$

$$k_f\theta_{f,ii} + \left(1 + \frac{\tau_s^v}{v!}\frac{\partial^v}{\partial t^v}\right)[\rho_2\,Q_f + \kappa(\theta_s - \theta_f)]$$
$$= \frac{\partial}{\partial t}\left(1 + \frac{\tau_s^v}{v!}\frac{\partial^v}{\partial t^v}\right)(F_{22}\,\theta_f + T_o R_{12}\,e_s + T_o R_{22}\,e_f) \tag{17}$$

Equations (16) and (17) are the generalized energy equations of poro-thermoelasticity with fractional derivatives and taking into account the relaxation times τ_s and τ_f. Some theories of heat conduction law follow as limit cases for different values of the parameters τ_s, τ_f, and v.

Taking into consideration

$$\frac{\partial^v}{\partial t^v}f(y,t) = \begin{cases} f(y,t) - f(y,0) & v \to 0 \\ I^{v-1}\dfrac{\partial f(y,t)}{\partial t} & 0 < v < 1 \\ \dfrac{\partial f(y,t)}{\partial t} & v = 1, \end{cases}$$

where the notion I^v is the Riemann–Liouville fractional integral introduced as a natural generalization of the well-known n-fold repeated integral $I^n f(t)$ written in a convolution-type form (Mainardi and Gorenflo 2000)

$$I^\upsilon f(t) = \int_0^t \frac{(t-\varsigma)^{\upsilon-1}}{\Gamma(\upsilon-1)} f(\varsigma) d\varsigma \left.\begin{array}{c}\\ \\ \\ \\ \end{array}\right\} \upsilon > 0$$

$$I^0 f(t) = f(t)$$

In the limit as υ tends to 1, Eqs. (16) and (17) reduce to the well-known Cattaneo-Vernotte (1948) law used by Lord and Shulman (1967) to derive the equation of the generalized theory of poro-thermoelasticity with one relaxation time. It is known that the classical entropy derived using this law instead of monotonically increasing behaves in an oscillatory way (Lebon et al. 2008; Jou et al. 1988). Strictly speaking, this result is not incompatible with Clausius' formulation of the second law, which states that the entropy of the final equilibrium state must be higher than the entropy of the initial equilibrium state. However, the non-monotonic behavior of the entropy is in contradiction with the local equilibrium formulation of the second law, which requires that the entropy production must be positive everywhere at any time (Lebon et al. 2008). During the last two decades, this became the subject of many research papers and resulted in the introduction of what is known now as extended irreversible thermodynamics. A review can be found in Jou et al. (1988).

2.1 Limiting cases

1. The heat Eqs. (16) and (17) for the two-phase system in the limiting case $\upsilon = 0$ transforms to the work of Biot (1984) in the context of coupled thermoelasticity (Biot 1956b).
2. The heat Eqs. (16) and (17) for the two-phase system in the limiting case $\upsilon = 1$ transforms to the work of Sherief and Hussein (2012) in the context generalized thermoelasticity with one relaxation time (Lord and Shulman 1967).

3 Mathematical models

The linear governing equations of isotropic, generalized poro-thermoelasticity in the absence of body forces and heat sources are as follows:

(i) Equations of motion

$$\mu u_{is,jj} + [(\lambda + \mu) e_s + G e_f - R_{11}\theta_s - R_{12}\theta_f]_i = \rho_{11}\ddot{u}_{si} + \rho_{12}\ddot{u}_{fi}$$

$$(18)$$

$$[G e_s + R e_f - R_{21}\theta_s - R_{22}\theta_f]_i = \rho_{12}\ddot{u}_{is} + \rho_{22}\ddot{u}_{if}$$

$$(19)$$

(ii) Fractional heat equation

$$k_s\theta_{s,ii} - \kappa\left(1 + \frac{\tau_s^\upsilon}{\upsilon!}\frac{\partial^\upsilon}{\partial t^\upsilon}\right)(\theta_s - \theta_f)$$
$$= \frac{\partial}{\partial t}\left(1 + \frac{\tau_s^\upsilon}{\upsilon!}\frac{\partial^\upsilon}{\partial t^\upsilon}\right)(F_{11}\theta_s + T_oR_{11}e_s + T_oR_{21}e_f)$$

$$(20)$$

$$k_f\theta_{f,ii} + \kappa\left(1 + \frac{\tau_s^\upsilon}{\upsilon!}\frac{\partial^\upsilon}{\partial t^\upsilon}\right)(\theta_s - \theta_f)$$
$$= \frac{\partial}{\partial t}\left(1 + \frac{\tau_s^\upsilon}{\upsilon!}\frac{\partial^\upsilon}{\partial t^\upsilon}\right)(F_{22}\theta_f + T_oR_{12}e_s + T_oR_{22}e_f)$$

$$(21)$$

(iii) Constitutive equations

$$\sigma_{ij} = 2\mu e_{ij} + \lambda e_s\delta_{ij} + (Ge_f - R_{11}\theta_s - R_{12}\theta_f)\delta_{ij}$$

$$(22)$$

$$\sigma = Re_f + Ge_s - R_{21}\theta_s - R_{22}\theta_f$$

$$(23)$$

$$e_{ij} = \frac{1}{2}(u_{is,j} + u_{js,i}), \quad e_s = e_{ii} = u_{is,i}$$

$$(24)$$

$$e_f = u_{if,i}$$

$$(25)$$

4 Formulation of the problem

We shall consider a homogeneous isotropic thermo-poroelastic medium occupying the region $x \geq 0$, where the x-axis is taken perpendicular to the bounding plane of half-space pointing inwards, subjected to traction-free time-dependent heating. The initial conditions of the problem are taken to be quiescent.

For the one-dimensional problems, all the considered functions will depend only on the space variables x and t. The displacement components take the form:

$$u_{xs} = u_s(x,t), \quad u_{ys} = u_{zs} = 0, \quad u_{xf} = u_f(x,t),$$
$$u_{yf} = u_{zf} = 0$$

Let us introduce the following non-dimensional variables

$$(x', u_s', u_f') = c_1\eta(x, u_s, u_f),$$
$$(t', \tau_s', \tau_s') = c_1^2\eta(t, \tau_s, \tau_s),$$
$$c_1 = \sqrt{M/\rho_{11}},$$
$$(\sigma_{ij}', \sigma') = (\sigma_{ij}, \sigma)/M,$$
$$(\theta_s', \theta_f') = R_{11}(\theta_s, \theta_f)/M,$$
$$\eta = F_{11}/k_s,$$
$$(q_s', q_f') = R_{11}M(k_sq_s', k_fq_f')/c_1\eta,$$
$$M = \lambda + 2\mu$$

In terms of these non-dimensional variables, Eqs. (18)–(25) take the following form (dropping primes for convenience):

$$\frac{\partial^2 u_s}{\partial x^2} + a_1 \frac{\partial^2 u_f}{\partial x^2} - \frac{\partial \theta_s}{\partial x} - A_1 \frac{\partial \theta_f}{\partial x} = \frac{\partial^2 u_s}{\partial t^2} + b_1 \frac{\partial^2 u_f}{\partial t^2} \quad (26)$$

$$a_1 \frac{\partial^2 u_s}{\partial x^2} + a_2 \frac{\partial^2 u_f}{\partial x^2} - A_2 \frac{\partial \theta_s}{\partial x} - A_3 \frac{\partial \theta_f}{\partial x} = b_1 \frac{\partial^2 u_s}{\partial t^2} + b_2 \frac{\partial^2 u_f}{\partial t^2} \quad (27)$$

$$\frac{\partial^2 \theta_s}{\partial x^2} = \left(\frac{\partial}{\partial t} + \frac{\tau_s^v}{v!} \frac{\partial^{v+1}}{\partial t^{v+1}} \right) \left(\theta_s + \varepsilon_2 \frac{\partial}{\partial x} [u_s + A_2 u_f] \right)$$
$$+ \varepsilon_1 \left(1 + \frac{\tau_s^v}{v!} \frac{\partial^v}{\partial t^v} \right) (\theta_s - \theta_f) \quad (28)$$

$$\frac{\partial^2 \theta_f}{\partial x^2} = \left(\frac{\partial}{\partial t} + \frac{\tau_s^v}{v!} \frac{\partial^{v+1}}{\partial t^{v+1}} \right) \left(F\theta_f + \omega \varepsilon_2 \frac{\partial}{\partial x} [A_1 u_s + A_3 u_f] \right)$$
$$- \omega \varepsilon_1 \left(1 + \frac{\tau_s^v}{v!} \frac{\partial^v}{\partial t^v} \right) (\theta_s - \theta_f) \quad (29)$$

$$\sigma_{xx} = \frac{\partial u_s}{\partial x} + a_1 \frac{\partial u_f}{\partial x} - \theta_s - A_1 \theta_f \quad (30a)$$

$$\sigma_{yy} = \sigma_{zz} = a_3 \frac{\partial u_s}{\partial x} + a_1 \frac{\partial u_f}{\partial x} - \theta_s - A_1 \theta_f \quad (30b)$$

$$\sigma_{xy} = \sigma_{xz} = \sigma_{zy} = 0 \quad (30c)$$

$$\sigma = a_2 \frac{\partial u_f}{\partial x} + a_1 \frac{\partial u_s}{\partial x} - A_3 \theta_f - A_2 \theta_s \quad (30d)$$

$$q_s + \frac{\tau_s^v}{v!} \frac{\partial^v q_s}{\partial t^v} = -\frac{\partial \theta_s}{\partial x}, \quad 0 < v \le 1 \quad (31a)$$

$$q_f + \frac{\tau_f^v}{v!} \frac{\partial^v q_f}{\partial t^v} = -\frac{\partial \theta_f}{\partial x}, \quad 0 < v \le 1 \quad (31b)$$

The boundary conditions can be expressed as

$$\sigma_{xx}(0, t) = \sigma(0, t) = 0 \quad (32a)$$

$$q_s(0, t) + B_s \theta_s(0, t) = f(t) \quad (32b)$$

$$q_f(0, t) + B_f \theta_f(0, t) = f(t), \quad (32c)$$

where B_s and B_f are the Biot's numbers and $f(t)$ represents the magnitude of surface heating.

5 The analytical solutions in the Laplace transform domain

Performing the Laplace transform defined by the relation

$$\bar{g}(s) = \int_0^\infty e^{-st} g(t) \, dt$$

of both sides Eqs. (26)–(32), with the homogeneous initial conditions

$$D^2 \bar{u}_s + a_1 D^2 \bar{u}_f - D\bar{\theta}_s - A_1 D\bar{\theta}_f = s^2 (\bar{u}_s + b_1 u_f) \quad (33)$$

$$a_1 D^2 \bar{u}_s + a_2 D^2 \bar{u}_f - A_2 D\bar{\theta}_s - A_3 D\bar{\theta}_f = s^2 (b_1 \bar{u}_s + b_2 u_f) \quad (34)$$

$$D^2 \bar{\theta}_s = s \left(1 + \frac{\tau_s^v}{v!} s^v \right) (\bar{\theta}_s + \varepsilon_2 D[\bar{u}_s + A_2 \bar{u}_f])$$
$$+ \varepsilon_1 \left(1 + \frac{\tau_s^v}{v!} s^v \right) (\bar{\theta}_s - \bar{\theta}_f) \quad (35)$$

$$D^2 \bar{\theta}_f = s \left(1 + \frac{\tau_s^v}{v!} s^v \right) (F\bar{\theta}_f + \omega \varepsilon_2 D[A_1 \bar{u}_s + A_3 \bar{u}_f])$$
$$+ \omega \varepsilon_1 \left(1 + \frac{\tau_s^v}{v!} s^v \right) (\bar{\theta}_s - \bar{\theta}_f) \quad (36)$$

$$\bar{\sigma}_{xx} = D\bar{u}_s + a_1 D\bar{u}_f - \bar{\theta}_s - A_1 \bar{\theta}_f \quad (37a)$$

$$\bar{\sigma}_{yy} = \bar{\sigma}_{zz} = a_3 D\bar{u}_s + a_1 D\bar{u}_f - \bar{\theta}_s - A_1 \bar{\theta}_f \quad (37b)$$

$$\bar{\sigma}_{xy} = \bar{\sigma}_{xz} = \bar{\sigma}_{zy} = 0 \quad (37c)$$

$$\bar{\sigma} = a_2 D\bar{u}_f + a_1 D\bar{u}_s - A_3 \bar{\theta}_f - A_2 \bar{\theta}_s \quad (37d)$$

$$\left(1 + \frac{\tau_s^v}{v!} s^v \right) \bar{q}_s = -D\bar{\theta}_s, \quad 0 < v \le 1 \quad (38a)$$

$$\left(1 + \frac{\tau_s^v}{v!} s^v \right) \bar{q}_f = -D\bar{\theta}_f, \quad 0 < v \le 1 \quad (38b)$$

$$\bar{\sigma}_{xx}(0, t) = \bar{\sigma}(0, t) = 0 \quad (39a)$$

$$\left(1 + \frac{\tau_s^v}{v!} s^v \right) [f(t) - B_s \theta_s(0, t)] = -D\bar{\theta}_s, \quad 0 < v \le 1 \quad (39b)$$

$$\left(1 + \frac{\tau_f^v}{v!} s^v \right) [f(t) - B_f \theta_f(0, t)] = -D\bar{\theta}_f, \quad 0 < v \le 1 \quad (39c)$$

Eliminating $\bar{u}_s$, $\bar{u}_f$, and $\bar{\theta}_f$ between Eqs. (33) and (36), we obtain the following equation satisfied by $\bar{\theta}_s$:

$$\left(D^8 + \ell_1 D^6 + \ell_2 D^4 + \ell_3 D^2 + \ell_4 \right) \bar{\theta}_s = 0. \quad (40a)$$

In a similar manner, we can get the following Eqs. for $\bar{u}_s$, $\bar{u}_f$, and $\bar{\theta}_f$ which satisfy Eq. (40) as follows:

$$\left(D^8 + \ell_1 D^6 + \ell_2 D^4 + \ell_3 D^2 + \ell_4 \right) \bar{u}_s = 0 \quad (40b)$$

$$\left(D^8 + \ell_1 D^6 + \ell_2 D^4 + \ell_3 D^2 + \ell_4 \right) \bar{u}_f = 0 \quad (40c)$$

$$\left(D^8 + \ell_1 D^6 + \ell_2 D^4 + \ell_3 D^2 + \ell_4 \right) \bar{\theta}_f = 0, \quad (40d)$$

where ℓ_i, $i = 1, 2, 3, 4$ are the parameters depending on s.

Equation (40a) can be factorized as

$$\left(D^2 - k_1^2 \right) \left(D^2 - k_2^2 \right) \left(D^2 - k_3^2 \right) \left(D^2 - k_4^2 \right) \bar{\theta}_s = 0, \quad (41)$$

where k_m, $m = 1, 2, 3, 4$ are the roots of the characteristic equation of the system (40) which takes the form

$$k^8 + \ell_1 k^6 + \ell_2 k^4 + \ell_3 k^2 + \ell_4 = 0. \tag{42}$$

We can consider the general solutions of Eqs. (40), according to the bounded state of functions at infinity, in the following forms

$$\left(\bar{u}_s, \ \bar{u}_f, \ \bar{\theta}_s, \ \bar{\theta}_s\right) = \sum_{m=1}^{4} \left(L_m, \ M_m, \ N_m, \ H_m\right) e^{-k_m x}, \tag{43}$$

where L_m, M_m, N_m, and H_m are the parameters that depend on s.

Hence, the velocity for the fluid is given by

$$\bar{v}_f(x, s) = \sum_{m=1}^{4} s M_m(s) \, e^{-k_m x}. \tag{44}$$

Thus, the skin friction function τ_w (shear stress on the bounding plane surface of half-space) takes the expression (Ezzat 1994; Ezzat and Abd-Elaal 1997)

$$\tau_w(s) = -\left.\frac{\partial \bar{v}_f(x, s)}{\partial x}\right|_{x=0} = \sum_{m=1}^{4} s \, k_m(s) \, M_m(s). \tag{45}$$

Putting

$$M_m = g_{m1} \, L_m, N_m = g_{m2} \, L_m, \ H_m = g_{m3} \, L_m \tag{46}$$

Hence, from Eqs. (43) and (33)–(35), one can obtain the following relations:

$$g_{m1} = [k_m r_{m3}(A_1 A_2 - A_3) - A_1 r_{m2} Z_3 - A_2 r_{m1} Z_4$$
$$+ A_3 r_{m1} Z_3 + r_{m2} Z_4]/\Lambda,$$

$$g_{m2} = [k_m(A_1(r_{m2} Z_2 - r_{m3} Z_1) - A_3(r_{m1} Z_2 + r_{m2} r_{m3}))$$
$$+ Z_4(r_{m1} Z_1 + r_{m2}^2)]/k_m \Lambda,$$

$$g_{m3} = [k_m(A_2(r_{m1} Z_2 + r_{m2} r_{m3}) - r_{m2} Z_2 + r_{m3} Z_1)$$
$$- Z_3(r_{m1} Z_1 + r_{m2}^2)]/k_m \Lambda.$$

where all constants are given in the "Appendix."

From conditions (37) and Eqs. (37a, 37d), (43) and (44), we obtain a linear system of equations whose solution (numerically) gives the parameters L_m, $m = 1, 2, 3, 4$.

This completes the solution in the Laplace transform domain.

6 Numerical results and discussion

The method, which is based on a Fourier series expansion proposed by Honig and Hirdes (1984) and is developed in detail in some literature (Ezzat et al. 1999; Sherief et al. 2010), is adopted to invert the Laplace transform in Eqs. (43). Numerical code has been prepared using the Fortran 77 programming language.

The asphaltic material saturated with water was chosen for the purpose of numerical evaluations. The basic data for

this two-phase system are shown in Table 1 (Sherief and Hussein 2012).

The computations were carried out for the functions $f(t)$ where

$$f(t) = \begin{cases} \sin\left(\dfrac{\pi t}{a_o}\right) & 0 \le t \le a_o \\ 0 & \text{otherwise} \end{cases} \quad \text{or}$$

$$\bar{f}(s) = \frac{\pi a_o(s + e^{-a_o s})}{a_o^2 s^2 + \pi^2}$$

The temperature, stress, and displacement as well as the velocity of the fluid and the skin friction values for the two phases were calculated by using the numerical method of the inversion of the Laplace transform. The FORTRAN programming language was used on a personal computer. The precision maintained was five digits for the numerical program. The roots of the characteristic Eq. (42) were obtained analytically using the Mathematika 6.0 software. The system of linear equations resulting from applying the boundary conditions was solved numerically using the LU decomposition method. The results are displayed graphically at different positions of x as shown in Figs. 1, 2, 3, 4, 5, 6. In these figures, the solid line represents the solution obtained in the frame of the dynamic coupled theory ($\upsilon = 0$) (Biot 1955), and the dotted lines represent the solution obtained in the frame of the generalized thermoelasticity with thermal relaxation time ($\upsilon = 1.0, \ \tau_o = 0.02$) (Sherief and Hussein 2012), while the dashed lines represent the solution obtained in the frame of generalized thermoelasticity with fractional heat transfer ($0 < \upsilon < 1$, $\tau_o = 0.02$) (Ezzat 2012).

The important phenomenon observed in all figures is that the solution of any of the considered functions in the fractional theory is restricted in a bounded region. Beyond this region, the variations of these distributions do not take place. This means that the solutions according the generalized fractional thermoelasticity theory exhibit the behavior of finite speeds of wave propagation.

In the frame of the fractional theory, Figs. 1 and 2 indicate the variation in temperature in the elastic body and fluid for different values of fractional order $\upsilon = 0.0, \ 1.0, \ 0.5$. We notice that the temperature fields have been affected by the fractional order υ and the thermal waves cut the x-axis more rapidly when υ decreases. In the fractional theory of generalized poro-thermoelasticity, we observed that the thermal waves are continuous functions, smooth and reach steady state depending on the value of fractional υ, which means that the particles transport the heat to the other particles easily and this makes the decreasing rate of the temperature greater than the other ones.

Figures 3, 4, 5, 6 display the stress and displacement distributions in the elastic body and fluid, with distance

Table 1 Value of the constants

T_o, °C	27	ρ_s, mg cm^{-3}	2.35	τ_f, s	0.001
G, dyne cm^{-2}	0.4853×10^{11}	ρ_f, mg cm^{-3}	0.82	κ, W m^{-2} °C^{-1}	1.84×10^3
R, dyne cm^{-2}	0.0362×10^{11}	c_s, kJ kg^{-1} °C^{-1}	1.17152	ρ_{12}	-0.001ρ
α_s, °C^{-1}	0.0001	c_f, kJ mg^{-1} °C^{-1}	2.092	β	0.25
α_f, °C^{-1}	2.16×10^{-5}	k_s, W m^{-1} °C^{-1}	1.83	B_s	1
λ, dyne cm^{-2}	0.2160×10^{11}	k_f, W m^{-1} °C^{-1}	0.148	B_f	1
μ, dyne cm^{-2}	0.0926×10^{11}	τ_s, s	0.02		

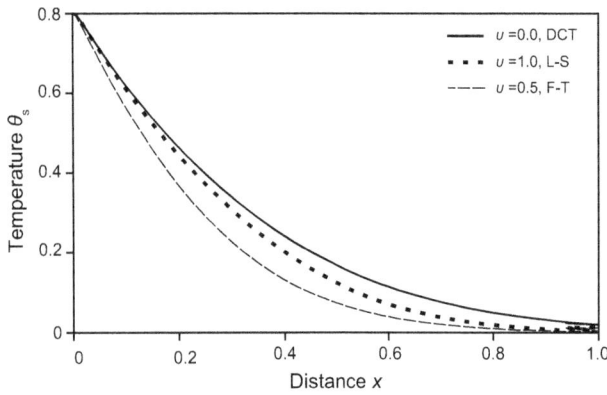

Fig. 1 Temperature distribution for the solid phase for different theories. (DCT: Dynamic coupled theory, L-S: Lord-Shulman theory, F-T: Fractional heat transfer model)

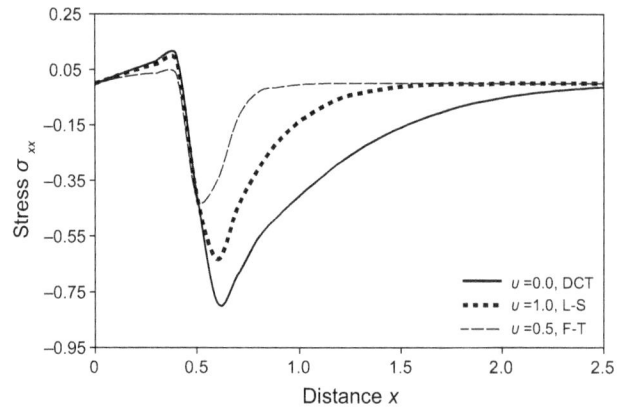

Fig. 3 Normal stress distribution for the solid phase for different theories

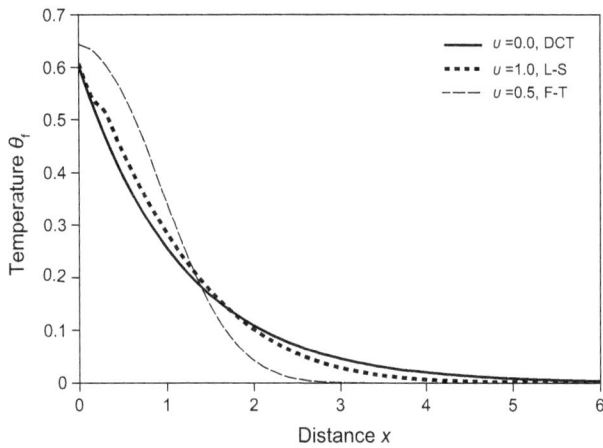

Fig. 2 Temperature distribution for the fluid phase for different theories

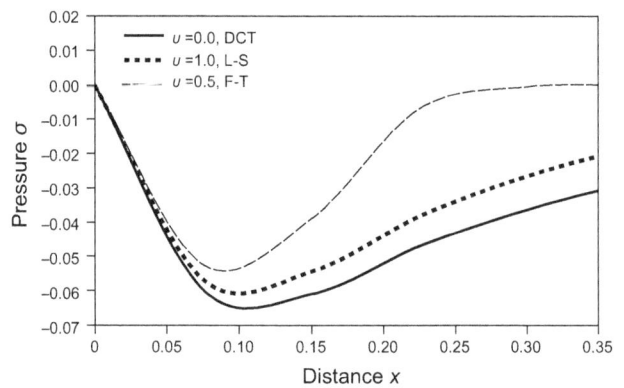

Fig. 4 Pressure for the fluid phase for different theories

x for different values of fractional order $\upsilon = 0.0, \ 1.0, \ 0.5$. We observe that the stress and displacement fields have the same behavior as the temperature and the absolute value of the maximum stress decreases. We note here that some of

these figures show broken lines indicating discontinuities of the solutions. These discontinuities indicate the locations of the wave fronts of the shock waves resulting from the sudden heating of the surface.

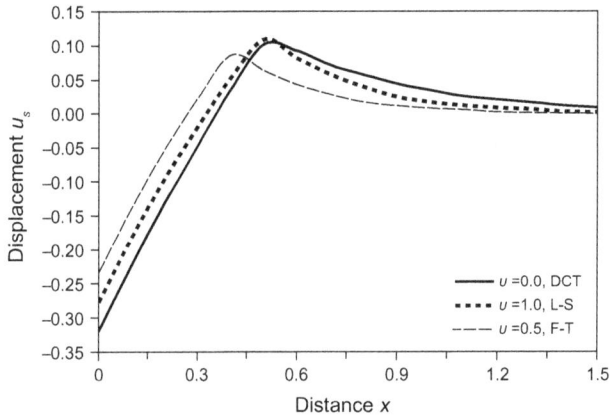

Fig. 5 Displacement distribution for the solid phase for different theories

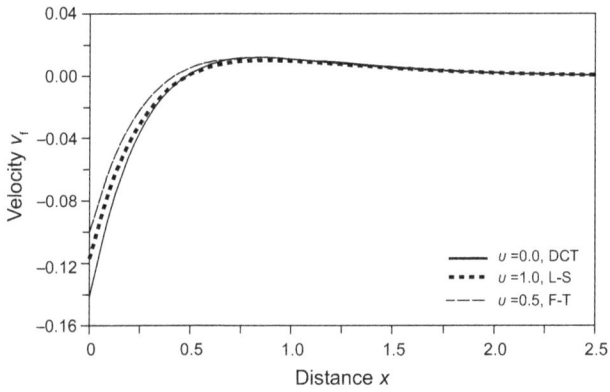

Fig. 6 Velocity distribution for the fluid phase

7 Conclusions

- The main goal of this work is to introduce a new mathematical model for Fourier law of heat conduction with time-derivative fractional order α (Fractional model). According to this new theory, we have to construct a new classification for materials according to their fractional parameter where this parameter becomes a new indicator of its ability to conduct heat.
- For the fractional model of thermoelasticity $0 < \upsilon < 1$, the solution seems to behave like the generalized theory of generalized thermoelasticity (Lord-Shulman theory). This result is very important that the new theory may preserve the advantage of the generalized theory that the velocity of waves is finite.
- As was mentioned in (Povstenko 2011) "From numerical calculations, it is difficult to say whether the solution for υ approaching 1 has a jump at the wave

front or it is continuous with very fast changes. This aspect invites further investigation."

- The present model avoids the negative temperature defect of the Lord-Shulman model.
- The result provides a motivation to investigate thermoelastic materials as a new class of applicable materials.

Acknowledgments The first author is grateful for the support of Al-Qassim University (No. 2367), Al-Qassim, Saudi Arabia.

Appendix

$$F = F_{22}\omega/F_{11},$$

$$\omega = k_s/k_f,$$

$$b_1 = \rho_{12}/\rho_{11}, \quad b_2 = \rho_{22}/\rho_{11},$$

$$a_1 = G/M, \quad a_2 = R/M, \quad a_3 = \lambda/M,$$

$$A_1 = R_{12}/R_{11}, \quad A_2 = R_{21}/R_{11} A_3 = R_{22}/R_{11},$$

$$\varepsilon_1 = \kappa k_s \rho_{11}/MF_{11}^2, \quad \varepsilon_2 = T_0 R_{11}^2/MF_{11},$$

$$\ell = a_1^2 - a_2,$$

$$\ell_1 = -\left(s^2(2b_1 a_1 - b_2 - a_2) + s\varepsilon_2(2a_1(A_1 A_3 \omega \zeta_1 + A_2 \zeta_0) \right.$$
$$\left. -a_2(A_1^2 \omega \zeta_1 + \zeta_0) - A_2^2 \zeta_0 - A_3^2 \omega \zeta_1) + a(\zeta_2 + \zeta_3)\right)/\ell,$$

$$\ell_2 = \left(s^4(b_1^2 - b_2) + \varepsilon_2 s^3 \zeta_4 + s^2 \zeta_5 + \varepsilon_2 s \zeta_7 \right.$$
$$\left. + \ell(\zeta_2 \zeta_3 - \omega \zeta_0 \zeta_1 \varepsilon_1^2)\right)/a,$$

$$\ell_3 = -s^2\left(s^2(b_1^2 - b_2)(\zeta_2 + \zeta_3) + s\varepsilon_2 \zeta_8 \right.$$
$$\left. + (2a_1 b_1 - b_2 - a_2)(\zeta_2 \zeta_3 - \omega \zeta_0 \zeta_1 \varepsilon_1^2)\right)/\ell,$$

$$\ell_4 = s^4(b_1^2 - b_2)(\zeta_2 \zeta_3 - \omega \zeta_0 \zeta_1 \varepsilon_1^2)/\ell,$$

$$\zeta_0 = 1 + \frac{\tau_s^\upsilon}{\upsilon!}s^\upsilon, \quad \zeta_1 = 1 + \frac{\tau_f^\upsilon}{\upsilon!}s^\upsilon,$$

$$\zeta_2 = \zeta_0(\varepsilon_1 + s), \quad \zeta_3 = \zeta_1(\varepsilon_1 \omega + sA_3),$$

$$\zeta_4 = 2b_1(A_1 A_3 \omega \zeta_1 + A_2 \zeta_0) - b_2(A_1^2 \omega \zeta_1 + \zeta_0) - A_2^2 \zeta_0$$
$$- A_3^2 \omega \zeta_1,$$

$$\zeta_5 = (\zeta_2 + \zeta_3)(2a_1 b_1 - b_2 - a_2) - \omega \zeta_0 \zeta_1 \varepsilon_2^2(A_1 A_2 - A_3)^2,$$

$$\zeta_6 = A_1 \omega \zeta_1(A_2 \zeta_0 \varepsilon_1 + A_3 \zeta_2) + \zeta_0(A_2 \zeta_3 + A_3 \omega \zeta_1 \varepsilon_1),$$

$$\zeta_7 = 2a_1\zeta_6 - a_2\left(A_1^2\omega\zeta_1\zeta_2 + 2A_1\omega\zeta_0\zeta_1 + \zeta_0\zeta_3\right) - A_2^2\zeta_0\zeta_3 \\ - A_3\omega\zeta_1(2A_2\zeta_0\varepsilon_1 + A_3\zeta_2),$$

$$\zeta_8 = 2b_1\zeta_6 - b_2(A_1^2\omega\zeta_1\zeta_2 + 2A_1\omega\zeta_0\zeta_1\varepsilon_1 + \zeta_0\zeta_3) - A_2^2\zeta_0\zeta_3 \\ - A_3\omega\zeta_1(2A_2\zeta_0\varepsilon_1) + A_3\zeta_2,$$

$$Z_1 = a_2k_m^2 - b_2s^2, \quad Z_2 = s\zeta_0A_2k_m\varepsilon_2,$$

$$Z_3 = k_m^2 - \zeta_0(\varepsilon_1 + s), \quad Z_4 = \varepsilon_1\zeta_0,$$

$$r_{m1} = s^2 - k_m^2, \quad r_{m2} = b_1s^2 - a_1k_m^2, \quad r_{m3} = -sk_m\varepsilon_2\zeta_0,$$

$$\Lambda = k_mZ_2(A_1A_2 - A_3) - A_1Z_1Z_3 + A_2r_{m2}Z_4 - A_3r_{m2}Z_3 \\ + Z_1Z_4.$$

References

Abbas IA. A problem on functional graded material under fractional order theory of thermoelasticity. Theor Appl Frac Mech. 2014;74(8):18–22. doi:10.1016/j.tafmec.2014.05.005.

Abbas IA. Generalized thermoelastic interaction in functional graded material with fractional order three-phase lag heat transfer. J Cen South Univ. 2015;22(5):1606–13. doi:10.1007/s11771-015-2677-5.

Bagley RL, Torvik PJ. On the fractional calculus model of viscoelastic behavior. J Rheol. 1986;30(1):133–55. doi:10.1122/1.549887.

Biot MA. Theory of elasticity and consolidation for a porous anisotropic solid. J Appl Phys. 1955;26(2):182–98. doi:10.1063/1.1721956.

Biot MA. Theory of propagation of elastic waves in a fluid-saturated porous solid. I: Low-frequency range. J Acoust Soc Am. 1956a;28(2):168–78. doi:10.1121/1.1908239.

Biot MA. Thermoelasticity and irreversible thermodynamics. J Appl Phys. 1956b;27(3):240–53. doi:10.1063/1.1722351.

Biot MA. New variational-Lagrangian irreversible thermodynamics with application to viscous-flow, reaction diffusion, and solid mechanics. Adv Appl Mech. 1984;24:1–91. doi:10.1016/S0065-2156(08)70042-5.

Biot MA, Willis DG. The elastic coefficients of the theory of consolidation. J Appl Mech. 1957;24(11):594–601.

Cattaneo C. Sulla conduzione del calore. Atti Sem Mat Fis Univ Modena. 1948;3(3):83–101.

Delaney PT. Rapid intrusion of magma into wet rock: groundwater flow due to pore pressure increases. J Geophys Res. 1982;87(B9):7739–56. doi:10.1002/(ISSN)2156-2202.

Deresiewicz H, Skalak R. On uniqueness in dynamic poroelasticity. Bull Seismol Soc Am. 1963;53(4):783–8.

El-Karamany AS, Ezzat MA. On fractional thermoelasticity. Math Mech Solids. 2011;16(3):334–46. doi:10.1177/1081286510397228.

Ezzat MA. State space approach to unsteady two-dimensional free convection flow through a porous medium. Can J Phys. 1994;72(5–6):311–7.

Ezzat MA. Thermoelectric MHD non-Newtonian fluid with fractional derivative heat transfer. Phys B. 2010;405(7):4188–94. doi:10.1016/j.physb.2010.07.009.

Ezzat MA. Thermoelectric MHD with modified Fourier's law. Int J Therm Sci. 2011a;50(4):449–55. doi:10.1016/j.ijthermalsci.2010.11.005.

Ezzat MA. Theory of fractional order in generalized thermoelectric MHD. Appl Math Model. 2011b;35(10):4965–78. doi:10.1016/j.apm.2011.04.004.

Ezzat MA. Magneto-thermoelasticity with thermoelectric properties and fractional derivative heat transfer. Phys B. 2011c;406(1):30–5. doi:10.1016/j.physb.2010.10.005.

Ezzat MA. State space approach to thermoelectric fluid with fractional order heat transfer. Heat Mass Trans. 2012;48(1):71–82. doi:10.1007/s00231-011-0830-8.

Ezzat MA, Abd-Elaal M. Free convection effects on a viscoelastic boundary layer flow with one relaxation time through a porous medium. J Frank Inst. 1997;334B(4):685–6.

Ezzat MA, El-Karamany AS. Fractional order theory of a perfect conducting thermoelastic medium. Can J Phys. 2011a;89(3):311–8. doi:10.1139/P11-022.

Ezzat MA, El-Karamany AS. Fractional order heat conduction law inmagneto-thermoelasticity involving two temperatures. ZAMP. 2011b;62(3):937–52. doi:10.1007/s00033-011-0126-3.

Ezzat MA, El-Karamany AS. Theory of fractional order in electro-thermoelasticity. Euro J Mech A/Solid. 2011c;30(4):491–500. doi:10.1016/j.euromechsol.2011.02.004.

Ezzat MA, El-Karamny AS, Ezzat SM, et al. Two-temperature theory in magneto-thermoelasticity with fractional order dual-phase-lag heat transfer. Nucl Eng Des. 2012a;252(11):267–77. doi:10.1016/j.nucengdes.2012.06.012.

Ezzat MA, El-Karamny AS, Fayik M, et al. Fractional ultrafast laser-induced thermo-elastic behavior in metal films. J Therm Stress. 2012b;35(7):637–51. doi:10.1080/01495739.2012.688662.

Ezzat MA, El-Bary AA, Fayik MA, et al. Fractional Fourier law with three-phase lag of thermoelasticity. Mech Adv Mater Struct. 2013a;20(8):593–602. doi:10.1080/15376494.2011.643280.

Ezzat MA, El-Karamny AS, El-Bary AA, Fayik M, et al. Fractional calculus in one-dimensional isotropic thermo-viscoelasticity. CR Mec. 2013b;341(7):553–66. doi:10.1016/j.crme.2013.04.001.

Ezzat MA, Abbas IM, El-Bary AA, Ezzat SM, et al. Numerical study of the Stokes' first problem for thermoelectric micropolar fluid with fractional derivative heat transfer. MHD. 2014a;50(3):263–77.

Ezzat MA, Alsowayan NS, Al-Mohiameed ZI, Ezzat SM, et al. Fractional modelling of Pennes' bioheat transfer equation. Heat Mass Trans. 2014b;50(7):907–14. doi:10.1007/s00231-014-1300-x.

Ezzat MA, El-Karamny AS, El-Bary AA, Fayik M, et al. Fractional ultrafast laser- induced magneto-thermoelastic behavior in perfect conducting metal films. J Electromagn Waves Appl. 2014c;28(1–2):64–82. doi:10.1080/09205071.2013.855616.

Ezzat MA, Sabbah AS, El-Bary AA, Ezzat SM, et al. Stokes' first problem for a thermoelectric fluid with fractional-order heat transfer. Rep Math Phys. 2014d;74(2):145–58. doi:10.1016/S0034-4877(15)60013-1.

Ezzat MA, El-Karamny AS, El-Bary AA, et al. On thermo-viscoelasticity with variable thermal conductivity and fractional-order heat transfer. Int J Thermophys. 2015;36(7):1684–97. doi:10.1007/s10765-015-1873-8.

Ezzat MA, Othman MI, Helmy K, et al. A problem of a micropolar magneto hydrodynamic boundary-layer flow Canad. J Phys. 1999;77(10):813–27.

Fourie J, Du Plessis J. A two-equation model for heat conduction in porous media. Transp Porous Med. 2003;53(2):145–61. doi:10.1023/A:1024071928123.

Ghassemi A, Diek A. Poro-thermoelasticity for swelling shales. J Pet Sci Eng. 2002;34(4–5):123–35. doi:10.1016/S0920-4105(02)00159-6.

Ghassemi A, Zhang Q. A transient fictitious stress boundary element method for porothermoelastic media. Eng Anal Boun Elem. 2004;28(11):1363–73. doi:10.1016/j.enganabound.2004.05.003.

Gorenflo R, Mainardi F. Fractional calculus: integral and differential equations of fractional orders, fractals and fractional calculus in continuum mechanics, vol. 378. Wien: Springer; 1997. p. 223–76.

Gorenflo R, Mainardi F, Moretti D, Paradisi P, et al. Time fractional diffusion: a discrete random walk approach. Nonlinear Dyn. 2002;29(1):129–43. doi:10.1023/A:1016547232119.

Grimnes S, Martinsen OG. Bioimpedance and bioelectricity basics. San Diego: Academic Press; 2000.

Honig G, Hirdes U. A method for the numerical inversion of the Laplace transform. J Comput Appl Math. 1984;10(1):113–32. doi:10.1016/0377-0427(84)90075-X.

Jou D, Casas-Vazquez J, Lebon G. Extended irreversible thermodynamics. Rep Prog Phys. 1988;51(9):1105–79. doi:10.1088/0034-4885/51/8/002.

Jumarie G. Derivation and solutions of some fractional Black-Scholes equations in coarse-grained space and time. Application to Merton's optimal portfolio. Comput Math Appl. 2010;59(3):1142–64. doi:10.1016/j.camwa.2009.05.015.

Krishnan JM, Rajagopal KR. Review of the uses and modeling of bitumen from ancient to modern times. Appl Mech Rev. 2003;56(2):149–214. doi:10.1115/1.1529658.

Lakes RS. Viscoelastic solids. Boca Raton: CRC Press; 1999.

Lebon G, Jou D, Casas-Vázquez J. Understanding non-equilibrium thermodynamics: foundations, applications, frontiers. Berlin: Springer-Verlag; 2008.

Li X, Cui L, Roegiers J-C. Thermoporoelastic modeling of wellbore stability in non-hydrostatic stress field. Int J Rock Mech Min Sci. 1998;35(4–5):584–8. doi:10.1016/s0148-9062(98)00079-5.

Lord H, Shulman Y. A generalized dynamical theory of thermoelasticity. J Mech Phys Solids. 1967;15(5):299–309.

Mainardi F, Gorenflo R. On Mittag-Leffler-type function in fractional evolution processes. J Comput Appl Math. 2000;118(1–2):283–99. doi:10.1016/S0377-0427(00)00294-6.

Marin M. Weak solutions in elasticity of dipolar porous materials. Math Prob Eng. 2008;. doi:10.1155/2008/158908.

Miller KS, Ross B. An introduction to the fractional integrals and derivatives—theory and applications. New York: John Wiley & Sons Inc; 1993.

Nouri-Borujerdi A, Noghrehabadi A, Rees A, et al. The effect of local thermal non-equilibrium on conduction in porous channels with a uniform heat source. Transp Porous Med. 2007;69(2):281–8. doi:10.1007/s11242-006-9064-5.

Nur A, Byerlee JD. An exact effective stress law for elastic deformation of rock with fluids. J Geophys Res. 1971;76(26):6414–9. doi:10.1029/JB076i026p06414.

Nowinski JL. Theory of thermoelasticity with applications. Alphen aan den Rijn: Sijthoff & Noordhoff International Publishers; 1978.

Oldham SG, Spanier J. The fractional calculus. New York: Academic Press; 1974.

Pecker C, Deresiewiez H. Thermal effects on wave in liquid-filled porous media. J Acta Mech. 1973;16(1):45–64. doi:10.1007/BF01177125.

Podlubny I. Fractional differential equations. New York: Academic Press; 1999.

Povstenko YZ. Fractional Cattaneo-type equations and generalized thermoelasticity. J Therm Stress. 2011;34(2):97–114. doi:10.1080/01495739.2010.511931.

Rossikhin YA, Shitikova MV. Applications of fractional calculus to dynamic problems of linear and nonlinear heredity mechanics of solids. Appl Mech Rev. 1967;50(1):15–67.

Samko SG, Kilbas AA, Marichev OI, et al. Fractional integrals and derivatives—theory and applications. Longhorne: Gordon & Breach; 1993.

Sherief HH, El-Said A, Abd El-Latief A, et al. Fractional order theory of thermoelasticity. Int J Solid Struct. 2010;47(2):269–75. doi:10.1016/j.ijsolstr.2009.09.034.

Sherief HH, Hussein EM. A mathematical model for short-time filtration in poroelastic media with thermal relaxation and two temperatures. Trans. Porous med. 2012;91(1):199–223. doi:10.1007/s11242-011-9840-8.

Simakin A, Ghassemi A. Modelling deformation of partially melted rock using a poroviscoelastic rheology with dynamic power law viscosity. Tectonophys. 2005;397(3–4):195–9. doi:10.1016/j.tecto.2004.12.004.

Tarasov VE. Fractional vector calculus and fractional Maxwell's equations. Ann Phys. 2008;323(11):2756–78. doi:10.1016/j.aop.2008.04.005.

Vernotte MP. Les paradoxes de la théorie continue de l'équation de la chaleur. CR Acad Sci. 1958;246(22):3154–5.

Wang Y, Papamichos E. Conductive heat flow and thermally induced fluid flow around a well bore in a poroelastic medium. Water Resour Res. 1994;30(12):3375–84. doi:10.1029/94WR01774.

Youssef HM. Theory of generalized porothermoelasticity. Int J Rock Mech Min Sci. 2007;44(2):222–7. doi:10.1016/j.ijrmms.2006.07.001.

PERMISSIONS

LIST OF CONTRIBUTORS

Jian Hou
State Key Laboratory of Heavy Oil Processing, China University of Petroleum, Qingdao 266580, Shandong, China

Nu Lu, Chuan-Jin Yao and Guang-Lun Lei
School of Petroleum Engineering, China University of Petroleum, Qingdao 266580, Shandong, China

Yan-Hui Zhang
CNOOC LTD., Tianjin Bohai Oilfield Institute, Tianjin 300452, China

Song Han, Bao-Sheng Zhang, Xu Tang and Ke-Qiang Guo
School of Business Administration, China University of Petroleum, Beijing 102249, China

Wei-Ji Wang, Zheng-Song Qiu, Han-Yi Zhong, Wei-An Huang and Wen-Hao Dai
School of Petroleum Engineering, China University of Petroleum, Qingdao 266580, Shandong, China

Jian Xiong, Xiang-Jun Liu and Li-Xi Liang
State Key Laboratory of Oil and Gas Reservoir Geology and Exploitation, Southwest Petroleum University, Chengdu 610500, Sichuan, China

Qun Zeng
Institute of Chemical Materials, Engineering Physical Academy of China, Mianyang 621999, Sichuan, China

S. Rushd and R. S. Sanders
Department of Chemical and Materials Engineering, University of Alberta, Edmonton, AB, Canada

Xi-Xiang Liu and Shao-Nan Zhang
State Key Laboratory of Oil and Gas Reservoir Geology and Exploitation, Southwest Petroleum University, Chengdu 610500, Sichuan, China
School of Geoscience and Technology, Southwest Petroleum University, Chengdu 610500, Sichuan, China

Xiao-Qi Ding
College of Energy, Chengdu University of Technology, Chengdu 610059, Sichuan, China

Hao He
Second Production Plant, PetroChina Changqing Oilfield Company, Qinyang 745000, Gansu, China

Deepak M. Kirpalani and Dipti Prakash Mohapatra
Energy Mining and Environment Portfolio, National Research Council of Canada, 1200 Montreal Road, Ottawa, ON K1A 0R6, Canada

Zhi-Gang Wang, Jia-Ning Pei, Sheng-Li Chen, Zheng Zhou and Gui-Mei Yuan
State Key Laboratory of Heavy Oil Processing, College of Chemical Engineering, China University of Petroleum, Beijing 102249, People's Republic of China

Zhi-Qing Wang, Guo-Qiang Ren and Hong-Jun Jiang
Shanghai Petrochemical Company of Sinopec, Jinshan, Shanghai 200540, People's Republic of China

Shao-Hua Dong, He-Wei Zhang and Lai-Bin Zhang
The Pipeline Technology Research Center, China University of Petroleum (Beijing), Beijing 102249, China

Li-Jian Zhou and Lei Guo
PetroChina R&D Center, Langfang 065000, Hebei, China

Ren-Jin Sun and Xiu-Cheng Dong
School of Business Administration, China University of Petroleum-Beijing, Beijing 102249, China

Hui Li
School of Business Administration, China University of Petroleum-Beijing, Beijing 102249, China
Energy Systems Research Center, University of Texas at Arlington, Arlington, TX 76019, USA

Kang-Yin Dong
School of Business Administration, China University of Petroleum-Beijing, Beijing 102249, China
Department of Agricultural, Food and Resource Economics, Rutgers, State University of New Jersey, New Brunswick, NJ 08901, USA

Zhong-Bin Zhou
School of Management, Yangtze University, Hubei 434023, China

Xia Leng
Sinopec Offshore Oilfield Services Company, Shanghai 200000, China

Maliheh Dargahi-Zaboli and Eghbal Sahraei
Department of Chemical Engineering, Petroleum Research Center, Sahand University of Technology, Tabriz, Iran

Behzad Pourabbas
Department of Polymer Engineering, Nanostructured Materials Research Center, Sahand University of Technology, Tabriz, Iran

Sheng-Zhu Zhang, Ru-Jun Wang, Ying-Quan Duo and Si-Ning Chen
China Academy of Safety Science and Technology, Beijing 100012, China

Song-Yang Li
College of Mechanical and Transportation Engineering, China University of Petroleum, Beijing 102249, China

Zong-Zhi Wu
State Administration of Work Safety, Beijing 100713, China

Ren-Jin Sun and Hong-Dian Jiang
School of Business Administration, China University of Petroleum-Beijing, 102249 Beijing, China

Kang-Yin Dong
School of Business Administration, China University of Petroleum-Beijing, 102249 Beijing, China
Department of Agricultural, Food and Resource Economics, Rutgers, State University of New Jersey, New Brunswick, NJ 08901, USA

Hui Li
School of Business Administration, China University of Petroleum-Beijing, 102249 Beijing, China
Energy Systems Research Center, University of Texas at Arlington, Arlington, TX 76019, USA

Yu Chen, Chun-Mao Chen, Shao-Hui Guo, Ping Wang, Shi-Jie Dong and Qing-Hong Wang
State Key Laboratory of Petroleum Pollution Control, Beijing Key Laboratory of Oil and Gas Pollution Control, China University of Petroleum, Beijing 102249, China

Brandon A. Yoza
Hawaii Natural Energy Institute, University of Hawaii at Manoa, Honolulu, HI 96822, USA

Qing X. Li
Department of Molecular Biosciences and Bioengineering, University of Hawaii at Manoa, Honolulu, HI 96822, USA

Elinda McKenna
Energy and Geoscience Institute, University of Utah, Salt Lake City, UT 84108, USA

Shu Jiang
Energy and Geoscience Institute, University of Utah, Salt Lake City, UT 84108, USA

Research Institute of Unconventional Oil and Gas and Renewable Energy, China University of Petroleum (East China), Qingdao 266580, Shandong, China

You-Liang Feng
Research Institute of Petroleum Exploration and Development, PetroChina, Beijing 100083, China

Lei Chen
School of Geoscience and Technology, Southwest Petroleum University, Chengdu 610500, Sichuan, China

Yue Wu
Sinopec Petroleum Exploration and Production Research Institute, Beijing 100083, China

Zheng-Yu Xu
PetroChina Hangzhou Institute of Geology, Hangzhou 310023, Zhejiang, China

Zheng-Long Jiang and Dong-Sheng Zhou
School of Ocean Sciences, China University of Geosciences, Beijing 100083, China

Dong-Sheng Cai
CNOOC, Beijing 100027, China

Wen-Biao Zhang, Tai-Zhong Duan, Zhi-Qiang Liu, Yan-Feng Liu, Lei Zhao and Rui Xu
Petroleum Exploration and Production Research Institute, SINOPEC, Beijing 100083, China

Mohammad Ali Ahmadi
Department of Petroleum Engineering, Ahwaz Faculty of Petroleum Engineering, Petroleum University of Technology, Ahwaz, Iran

James Sheng
Petroleum Department, Texas Tech University, P.O. Box 43111, Lubbock, TX 79409, USA

Zi-Jian Wang, Jun Tang and Chun-Xi Lu
State Key Laboratory of Heavy Oil Processing, China University of Petroleum (Beijing), Beijing 102249, China

Shereen Ezzat
Department of Mathematics, Faculty of Science and Letters in Al Bukayriyyah, Al-Qassim University, Al-Qassim, Saudi Arabia

Magdy Ezzat
Department of Mathematics, Faculty of Science and Letters in Al Bukayriyyah, Al-Qassim University, Al-Qassim, Saudi Arabia
Faculty of Education, Alexandria University, Alexandria, Egypt

Index